U0216095

SILKWORM

家蚕功能基因组研究

下册

夏庆友　主编

西南师范大学出版社
国家一级出版社　全国百佳图书出版单位

第六章

家蚕生物反应器研究

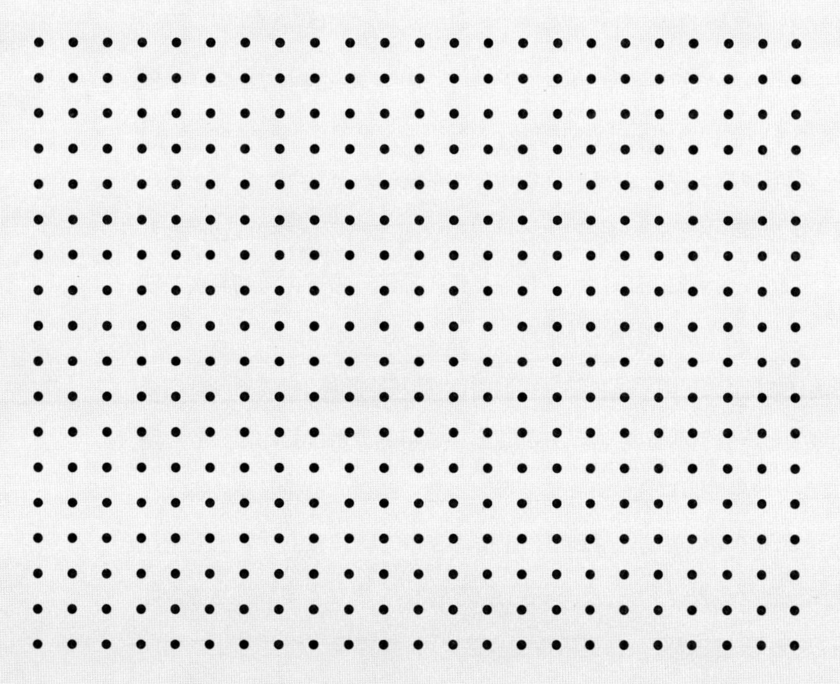

Chapter 6

家蚕(*Bombyx mori*)属于鳞翅目蚕蛾科，是被人类完全驯化的经济昆虫。家蚕具有生长周期短，遗传背景清楚，以及饲养成本低和可规模化等优点，在昆虫工厂和生物医药研究领域备受青睐，有望成为21世纪规模化生产蛋白药物的新模式。

昆虫杆状病毒表达系统是最早被开发用于在家蚕体内高效生产外源重组蛋白的系统，其原理是以携带有外源基因的重组杆状病毒感染家蚕的幼虫或蛹体得以实现。研究人员不断优化和改进家蚕杆状病毒表达系统，进一步提高了重组蛋白的生产效率以及稳定性，且缩短了重组病毒的制备周期。此外，共表达分子伴侣的策略也使重组蛋白的得率和生物活性进一步得到提高。目前，基于杆状病毒表达系统的家蚕生物反应器已经成功生产了数百种重组蛋白，其中部分重组蛋白已经进入产业化生产阶段，除此以外，该系统在病毒样颗粒、基因治疗、亚单位疫苗、组织工程以及药物筛选等领域也得到高度关注与发展。

随着家蚕转基因技术的突破，利用家蚕的组织器官(丝腺和脂肪体)合成外源蛋白成为拓展家蚕应用领域和经济价值的重要方向，并展现出了独特的技术优势和巨大的应用潜力。其中，尤以家蚕丝腺生物反应器的开发与利用为特色。转基因家蚕丝腺生物反应器是一类丝蛋白基因启动子实现驱动外源基因在家蚕丝腺中的特异表达并分泌到蚕丝中的系统。科研人员围绕丝腺表达系统的效率进行系统性优化，建立了高效的丝腺表达系统，随后对数十种高附加值的外源蛋白进行重组表达研究，为探索家蚕丝腺生物反应器用于生物制药以及功能性蚕丝的开发积累了宝贵经验。此外，研究人员还建立了脂肪体特异的表达系统，并成功在脂肪体中特异表达植酸酶，有望将蚕蛹开发为新型的饲料添加源，为提高蚕蛹的综合利用提供技术参考。

王峰

Development of a rapid and efficient *Bm*NPV baculovirus expression system for application in mulberry silkworm, *Bombyx mori*

Cao CP[1] Wu XF[1*] Zhao N[1]
Yao HP[1] Lu XM[1] Tan YP[2]

Abstract: Silkworm-baculovirus gene expression system is one of the most powerful eukaryotic expression systems. However, due to very low recombination frequency, the traditional method to construct and obtain pure recombinant baculovirus requires plaque assay that is time-consuming and also utilizes skillful techniques. In order to overcome this disadvantage, a rapid *Bm*NPV expression system applicable to silkworm was constructed based on the working principle of *Ac*MNPV Bac-to-Bac system. A large 8.6 kb fragment containing the low-copy-number mini-F replicon, a kanamycin resistance marker and a segment of DNA encoding the *lac*Zα peptide from the *Ac*MNPV bacmid, was cloned into polyhedrin locus of *Bm*NPV genome to replace the polyhedrin gene. This recombinant, designated as *BmBacmid*, was transformed into *Escherichia coli* DH10β strain, in which a helper plasmid encoding the transposase was already transformed. We designated the DH10β strain containing *BmBacmid* and helper as DH10BmBac. With this bacterium, the recombinant baculovirus can be rapidly and easily generated through gene transposition. This system has an advantage of high recombination frequency, simple manipulation, high-efficiency and is time-saving. This approach permits large potential value in recombinant protein production using silkworm as a "biofactory" in the future biotechnological industry and can become a powerful tool for structural and functional analysis of protein in post-genomic era.

Published On: Current Science, 2006, 91(12), 1692-1697.

1 Lab of Bioresource and Biotechnology, College of Animal Sciences, Zhejiang University, Hangzhou 310029, China.
2 Department of Entomology, University of California, Riverside, CA 92521, USA.
*Corresponding author E-mail: wuxiaofeng@zju.edu.cn.

应用于桑叶蚕的快速有效的家蚕 NPV 杆状病毒表达系统的开发

曹翠萍[1]　吴小锋[1*]　赵　娜[1]
姚慧鹏[1]　鲁兴萌[1]　谭叶萍[2]

摘要： 家蚕杆状病毒基因表达系统是一种最强大的真核生物表达系统。但是由于其极低的重组频率，采用传统的方法来构建和获得纯的重组杆状病毒需要噬菌斑测定，需要娴熟的技术并且耗费大量的时间。为了克服这个缺点，基于 *Ac*MNPV（苜蓿银纹夜蛾多核衣壳核型多角体病毒）Bac-to-Bac 表达系统的工作原理构建了适用于家蚕 *Bm*NPV 快速表达的系统。将包含一个 8.6 kb 片段低拷贝数 mini-F 复制子、卡那霉素抗性标记和编码 *LacZ*α 多肽的部分 DNA 重组入家蚕 *Bm*NPV 基因组的多角体蛋白位点，替换多角体蛋白基因。该重组基因命名为 *BmBacmid*，转化入大肠杆菌 DH10β 菌株，其中 helper 质粒编码的转座酶也一同被转化。我们把包含 *BmBacmid* 和 helper 的 DH10β 菌株命名为 DH10BmBac。通过这种细菌，进行基因替换，重组的杆状病毒能够快速、容易地生成。这个系统具有高重组频率，操作简单，高效率和节约时间的特点。这个方法在未来生物工程产业中，用家蚕作为生物反应器生产重组蛋白上有潜在的价值，在基因时代成为分析蛋白结构和功能的有力工具。

1　生物资源与生物技术实验室，动物科学学院，浙江大学，杭州
2　昆虫系，加利福尼亚大学，加利福尼亚，美国

Expression, purification and characterization of human GM-CSF using silkworm pupae (*Bombyx mori*) as a bioreactor

Chen J[1,2] Wu XF[3] Zhang YZ[2*]

Abstract: To date, many recombinant proteins have been expressed in *Bombyx mori* cells or silkworm larvae, apart from in pupae. Silkworm pupae may be more suitable for the expression of heterologous proteins as a bioreactor. If maintained at an appropriate temperature, silkworm pupae could be inoculated with recombinant baculovirus for the expression of a protein of interest. In this study, human granulocyte-macrophage colony-stimulating factor was successfully expressed in silkworm pupae using *B. mori* nucleopolyhedrovirus, purified and characterized with respect to its physico-chemical properties. The target protein expressed had an apparent molecular mass of 29 kDa and an isoelectric point of 5.1. The protein was purified using three chromatographic steps with a final recovery of 10.3%. Finally, approximately 3.5 mg of the protein was obtained with a biological activity of up to 8.4×10^6 cfu/mg. The results of this study suggest that silkworm pupae represent a convenient and low-cost bioreactor for the expression of heterologous proteins.

Published On: Journal of Biotechnology, 2006,123(2), 236-247.

1 College of Life Sciences, Zhejiang University, Hangzhou 310029, China.

2 Institute of Biochemistry, Zhejiang Sci-Tech University, Hangzhou 310018, China.

3 Shanghai Institute of Biochemistry and Cell Biology, Chinese Academy of Sciences, Shanghai 200031, China.

*Corresponding author E-mail: yaozhou@chinagene.com.

以家蚕蛹作为生物反应器表达、纯化和鉴定人粒细胞巨噬细胞集落刺激因子(hGM-CSF)

陈　健[1,2]　吴祥甫[3]　张耀洲[2*]

摘要：迄今为止，除蛹外，许多重组蛋白已在家蚕细胞或幼虫中得到表达。家蚕蛹可能更适合作为生物反应器表达外源蛋白。如果在适当温度下，家蚕蛹能接种重组杆状病毒以表达目的蛋白。本研究利用家蚕核型多角体病毒在家蚕蛹中成功表达了人粒细胞巨噬细胞集落刺激因子(hGM-CSF)，并对其理化性质进行了纯化和鉴定。表达的人粒细胞巨噬细胞集落刺激因子约29 kDa，等电点为5.1。该蛋白质经三步柱层析而纯化，最终得率10.3%。获得生物活性达8.4×10^6 cfu/mg，质量共约3.5 mg的蛋白质。研究结果表明，家蚕蛹可作为一种用于外源蛋白表达的方便而低成本的生物反应器。

1　生命科学学院，浙江大学，杭州
2　生物化学研究所，浙江理工大学，杭州
3　生物化学与细胞生物学研究所，中国科学院，上海

Incorporation of partial polyhedrin homology sequences (PPHS) enhances the production of cloned foreign genes in a baculovirus expression system

Gong ZH[1] Jin YF[1*] Zhang YZ [2*]

Abstract: Baculovirus expression vector systems (BEVSs) have been used extensively for high-level expression of cloned foreign genes. In many instances, the levels of recombinant protein(s) produced in insect cells and larvae are insufficient for experimental purposes. Thus new techniques and methods are needed to increase significantly the protein expression levels in BEVS. In the present paper, we describe the incorporation of a 15 bp element derived from the 5'-end partial sequence of the polyhedrin gene, which contains the non-coding sequence ATAAAT and the coding sequence ATGCCGAAT, into the 5'-end of the CTB (cholera toxin B subunit)-INS (insulin) fusion gene. With the addition of the PPHS (partial polyhedrin homology sequences), two extra amino acids (Pro-Asn) were added to the N-terminus of the mCTB-INS (modified CTB-INS) fusion protein. This new fusion protein was expressed in both insect cells and larvae using BEVSs. We found that the addition of PPHS enhanced 4-fold the expression of CTB-INS in both insect cells and larvae. Further analysis revealed that the additional two amino acids in mCTB-INS did not significantly affect binding affinity for G_{M1} ganglioside. Therefore, the PPHS can be used as a constitutive element immediately downstream of the polyhedrin promoter to induce significant increases in the expression levels of cloned foreign genes.

Published On: Biotechnology and Applied Biochemistry, 2006, 43, Pt3, 165-170.

1 College of Life Sciences, Zhejiang University, Hangzhou 310029, China.

2 College of Life Sciences, Zhejiang Sci-Tech University, Hangzhou 310018, China.

*Corresponding author E-mail: jinyf@zju.edu.cn; yaozhou@chinagene.com.

部分多角体同源序列(PPHS)的整合增强杆状病毒载体系统克隆外源基因的表达

龚朝辉[1]　金勇丰[1*]　张耀洲[2*]

摘要：杆状病毒表达载体系统(BEVSs)已广泛用于外源基因的高效表达。在许多情况下，昆虫细胞和幼虫产生的重组蛋白表达水平是不足以用于实验的。因此，需要新技术和新方法显著提高BEVS系统重组蛋白的表达水平。在本文中，我们描述了将一个来源于多角体蛋白基因5′端部分序列(包含非编码序列ATAAAT和编码序列ATGCCGAAT)的15 bp(PPHS)元件整合到霍乱毒素B亚基-胰岛素(CTB-INS)融合基因5′-端的影响。随着PPHS的添加，两个额外的氨基酸(Pro-Asn)被添加到了融合蛋白(CTB-INS)的N-末端。这种新的融合蛋白可以用BEVSs在昆虫细胞和幼虫中表达。我们发现，PPHS的添加使CTB-INS在昆虫细胞和幼虫的表达增强了4倍。进一步分析表明，mCTB-INS额外的两个氨基酸没有显著影响蛋白与神经节苷脂G_{M1}的结合亲和力。因此，PPHS可以作为一个紧接多角体启动子下游以促进外源基因表达水平显著提高的构成元件。

1　生命科学学院，浙江大学，杭州
2　生命科学学院，浙江理工大学，杭州

Recombinant functional human lactoferrin expressed in baculovirus system

Liu T[1,2] Zhang YZ [2*] Wu XF[3]

Abstract: Human lactoferrin (hLf) is a multifunctional iron-binding glycoprotein. In this study, we amplified *hLf* cDNA by reverse transcription-polymerase chain reaction from normal human mammary gland. The nucleotide sequence of the *hLf* was identical to the known *hLf*. We constructed a recombinant virus, vBm-hLf, harboring the *hLf* gene and exploited the *BmN* cells as host to produce recombinant human lactoferrin (rhLf). It was found that a recombinant protein with a molecular mass of approximately 78 kDa was expressed. Approximately 13.5 μg rhLf was purified from 1×10^5-2×10^5 *BmN* cells infected by vBm-hLf and the rhLf proved to be biologically active. This method established in our study will pave the way for efficient production of rhLf for further application of this protein in the future.

Published On: Acta Biochimica et Biophysica Sinica, 2006, 38(3), 201-206.

1 Institute of Biochemistry, College of Life Sciences, Zhejiang University, Hanghzou 310029, China.
2 Institute of Biochemistry, Zhejiang Sci-Tech University, Hangzhou 310018, China.
3 ShanghaiInstitute of Biochemistry and Cell Biology, Chinese Academy of Sciences, Shanghai 200031, China.
*Corresponding author E-mail: yaozhou@chinagene.com.

在杆状病毒表达系统中表达功能性重组人乳铁蛋白

刘 涛[1,2] 张耀洲[2*] 吴祥甫[3]

摘要： 人乳铁蛋白(Human lactoferrin, hLf)是一种多功能的铁结合糖蛋白。本研究通过逆转录PCR技术从正常人乳腺中扩增得到*hLf*基因的cDNA。获得的*hLf*基因核酸序列和已知的*hLf*一致。我们构建了包含*hLf*基因的重组病毒vBm-hLf，以*BmN*细胞为宿主细胞表达重组人乳铁蛋白(rhLf)，表达的重组蛋白分子质量约78 kDa。从大约被1×10^5—2×10^5个vBm-hLf病毒感染的*BmN*细胞中纯化出13.5 μg具有生物活性的rhLf蛋白。本研究建立了一种高效生产rhLf蛋白的方法，为该蛋白的进一步应用打下基础。

1 生物化学研究所，生命科学学院，浙江大学，杭州

2 生物化学研究所，浙江理工大学，杭州

3 生物化学与细胞生物学研究所，中国科学院，上海

Expression of polyhedrin–hEGF fusion protein in cultured cells and larvae of *Bombyx mori*

Yu W Chen J Zhao XL
Lü ZB Nie ZM Zhang YZ *

Abstract: For mass production of human epidermal growth factor (hEGF), silkworm baculovirus expression vector system (BEVS) was adopted in this study, *hEGF* gene was in-frame fused with *polyhedrin* (*Ph*) gene under the control of *Ph* promoter and was used to co-transfect *BmN* cell with the modified *Bombyx mori* baculovirus DNA to obtained recombinant virus. The ELISA showed a maximum expression on day 4 in larvae and pupae. Both cellular extracts and haemolymph of silkworm larvae infected with rBacPh-EGF could all support the proliferation of *Balb/c 3T3* cell. The corresponding materials from *BmN* cell and silkworm larvae infected with wild virus also indicated a weak effect of upregulation on *3T3* cells proliferation. The animal study showed that both pupae infected with rBacPh-EGF virus and wild virus protected the gastric mucosa against ethanol-induced damage in rats, although the protection from pupae infected with wild virus was slightly weak. The mechanism is under investigating.

Published On: African Journal of Biotechnology, 2006, 5(11), 1034-1040.

Institute of Biochemistry, Zhejiang Sci-Tech University, Hangzhou 310018, China.
*Corresponding author E-mail: yaozhou@zstu.edu.cn.

多角体-人表皮生长因子融合蛋白在家蚕细胞及幼虫中的表达

于 威 陈 健 赵秀玲
吕正兵 聂作明 张耀洲*

摘要：本实验采用家蚕杆状病毒表达系统（baculovirus expression vector system，BEVS）来进行人表皮生长因子（human epidermal growth factor，hEGF）大规模生产的研究。将hEGF基因与多角体基因（*polyhedrin*，*Ph*）构建于受*Ph*启动子控制的同一阅读框内，随后与经过修饰的家蚕杆状病毒DNA共转染*BmN*细胞，得到重组病毒。ELISA结果显示，家蚕幼虫和蚕蛹中的表达量在第4天达到最大。感染rBacPh-EGF的家蚕幼虫细胞和血淋巴提取物皆可促进*Balb/c 3T3*细胞的增殖，感染野生病毒的*BmN*细胞和家蚕幼虫的相应材料也显示出了对*3T3*细胞增殖的微弱上调作用。动物实验结果显示，感染rBacPh-EGF病毒和感染野生病毒的蚕蛹皆可在一定程度上保护大鼠胃黏膜免受乙醇诱导的损伤，但是感染野生病毒的蚕蛹的保护作用较弱。该作用机制仍在研究中。

生物化学研究所，浙江理工大学，杭州

Can 29 kDa rhGM-CSF expressed by silkworm pupae bioreactor bring into effect as active cytokine through orally administration?

Zhang YZ [1*] Chen J[1] Lü ZB[1]
Nie ZM[1] Zhang XY[1*] Wu XF[1,2]

Abstract: In order to study the effect of human granulocyte-macrophage colony-stimulating factor (hGM-CSF) as active cytokine through orally administration, we expressed hGM-CSF within silkworm pupae bioreactor. The purified rhGM-CSF named as *Bm*rhGM-CSF is characterized as 29 kDa glycoprotein, and its biological activity was measured both *in vitro* and *in vivo*. We found out *Bm*rhGM-CSF could stimulate the colony formation of human bone marrow cells in a dose-dependent manner whether which were treated with or without γ-ray 24 h before. The ability of colony formation induced by *Bm*rhGM-CSF is negatively correlated with γ-ray intensity. As soon as 15 min post oral administration with *Bm*rhGM-CSF labeled with ^{125}I, an approximately 20 kDa protein fragment was detected within mice blood by SDS-PAGE followed by autoradiography. In blood sample of test mice, a protein was also recognized by anti-hGM-CSF antibody using ELISA. The immunohistochemical analysis showed that *Bm*rhGM-CSF was detected within intestinal histiocyte. This indicated it might be absorbed into blood via intestinal microvillus.

Pharmacokinetics analysis after orally administered *Bm*rhGM-CSF in animal model of leucopenia including mice, Beagle dogs and macaques showed that: (1) *Bm*rhGM-CSF pro- moted the CFU-S formation in mice spleen and the synthesis of DNA in bone marrow cells of mice; (2) *Bm*rhGM-CSF induced bone marrow karyocyta granulocyte growth significantly in both macaques and Beagle dogs compared to the negative control group. On the 9th day of orally administration, the animal WBC significantly increased in a dose-dependant manner, in which neutrophilic granulocyte was predominant. The WBC level of dogs in high dose group was about 1.5×10^9 cells/L more than that in the negative control. And the bone marrow smear revealed that the percents of both myloblast and progranulocyte in WBC in the hGM-CSF group were obviously higher than those in the negative control. These results proved that *Bm*rhGM-CSF, a 29 kDa glycoprotein expressed by Silkworm pupae bioreactor, could bring into the effect as active cytokine through oral administration.

Published On: European Journal of Pharmaceutical Sciences, 2006, 28(3), 212-223.

1 Institute of Biochemistry, Zhejiang Sci-Tech University, Hangzhou 310018, China.
2 Shanghai Institute of Biochemistry and Cell Biology, Chinese Academy of Sciences, Shanghai 200031, China.
*Corresponding author E-mail: yaozhou@chinagene.com; zhangxy@chinaids.org.cn.

蚕蛹生物反应器表达的29 kDa rhGM-CSF作为活性细胞因子能否通过口服发挥作用?

张耀洲[1*] 陈 健[1] 吕正兵[1]
聂作明[1] 张晓燕[1*] 吴祥甫[1,2]

摘要: 为了研究人类粒细胞–巨噬细胞集落刺激因子(hGM-CSF)通过口服作为细胞活性因子的作用,我们在蚕蛹生物反应器内表达hGM-CSF。纯化后的rhGM-CSF称为*Bm*rhGM-CSF,其为29 kDa的糖蛋白。我们测定了该蛋白质在生物体内和体外的活性。发现*Bm*rhGM-CSF能够以剂量依赖的方式刺激人类骨髓细胞集落的生成,并且与是否有24 h的γ–射线照射有关。*Bm*rhGM-CSF诱导形成集落的能力与γ–射线强度呈现负相关。SDS-PAGE放射自显影结果显示,在口服15 min之后,^{125}I标记的*Bm*rhGM-CSF就能在小鼠血液中检测到,检测到的蛋白质片段大约20 kDa。在受试小鼠的血液样本中,ELISA也能够检测到*Bm*rhGM-CSF的存在。免疫组化分析表明,在肠组织细胞中也检测到*Bm*rhGM-CSF,这表明它可能是通过肠道微绒毛被吸收到血液中的。药物动力学分析采用口服*Bm*rhGM-CSF方式检测,受试动物为白细胞减少症模式动物,包括小鼠、比格犬和猕猴,结果表明:(1)*Bm*rhGM-CSF可促进小鼠脾CFU-S的形成和小鼠骨髓细胞内DNA的合成;(2)与阴性对照相比,比格犬和猕猴试验组的*Bm*rhGM-CSF诱导骨髓核型粒细胞生长的效果明显。口服给药后第9天,以中性粒细胞为主,动物白细胞呈剂量依赖性增加。在高剂量组,犬的白细胞水平约为$1.5×10^9$个/L,高于阴性对照。骨髓涂片实验显示,与阴性对照组相比,试验组的原始粒细胞和早幼粒细胞都明显要高。这些结果表明,蚕蛹生物反应器生产的29 kDa大小糖蛋白*Bm*rhGM-CSF可作为活性细胞因子,通过口服给药发挥作用。

1 生物化学研究所,浙江理工大学,杭州
2 生物化学与细胞生物学研究所,中国科学院,上海

Large-scale purification of human granulocyte-macrophage colony-stimulating factor expressed in *Bombyx mori* pupae

Chen J[1] Nie ZM[1] Lü ZB[1] Zhu CG[2]
Xu CZ[1] Jin YF[2] Wu XF[3] Zhang YZ[1*]

Abstract: Human granulocyte-macrophage colony-stimulating factor (hGM-CSF) acts on many different kinds of cells, including monocytes, macrophages, granulocytes, eosinophils, and multipotential stem cells. To explore further explore pharmaceutical action, we expressed hGM-CSF by the *Bombyx mori* nucleopolyhedrovirus expression system in silkworm pupae. However, purifying recombinant proteins from silkworm pupae on a large scale has been a big challenge. To establish purification methods suitable for mass production, we tried two crude preparation methods: $(NH_4)_2SO_4$ fractional precipitation and isoelectric precipitation with a combination of gel filtration and ion-exchange chromatography. The isoelectric precipitation method was found to be more efficient. With this method, we eventually obtained approx 11.7 mg of 95% pure product from 1 000 g of infected silkworm pupae. The recovery of purified protein was greatly increased, by approx 40%, compared with the other method. The biologic activity of this protein was determined up to 9.0×10^6 colony-forming units/mg in the final purified product.

Published On: Applied Biochemistry and Biotechnology, 2007, 141(1), 149-159.

1 Institute of Biochemistry, Zhejiang Sci-Tech University, Hangzhou 310018, China.
2 College of Life Sciences, Zhejiang University, Hangzhou 310029, China.
3 Shanghai Institute of Biochemistry and Cell Biology, Chinese Academy of Sciences, Shanghai 200031, China.
*Corresponding author E-mail: yaozhou@chinagene.com.

人粒细胞-巨噬细胞集落刺激因子(hGM-CSF)在蚕蛹中的大规模表达纯化

陈　健[1]　聂作明[1]　吕正兵[1]　朱成刚[2]
徐承志[1]　金勇丰[2]　吴祥甫[3]　张耀洲[1*]

摘要：人粒细胞-巨噬细胞集落刺激因子(hGM-CSF)作用于许多不同种类的细胞,包括单核巨噬细胞、粒细胞、嗜酸性粒细胞和多能干细胞。为进一步探索该蛋白质的药物作用,我们利用家蚕核型多角体病毒在蚕蛹中表达hGM-CSF。然而,大规模纯化蚕蛹蛋白是一大挑战。为了建立适合大规模生产的纯化方法,我们尝试了两种粗提物制备方法:结合凝胶过滤和离子交换层析的$(NH_4)_2SO_4$分步沉淀法和等电点沉淀法。其中,等电点沉淀法较$(NH_4)_2SO_4$分步沉淀法更有效。用这种组合纯化方法,我们从1 000 g感染蚕蛹中最终纯化获得约11.7 mg纯度为95%的蛋白。与其他方法相比,纯化的蛋白质回收率大大增加,达到约40%。在最终纯化产物中,该蛋白的生物活性高达9×10^6 cfu/mg。

1　生命化学研究所,浙江理工大学,杭州
2　生命科学学院,浙江大学,杭州
3　生物化学与细胞生物学研究所,中国科学院,上海

Suppression of diabetes in non-obese diabetic (NOD) mice by oral administration of a cholera toxin B subunit-insulin B chain fusion protein vaccine produced in silkworm

Gong ZH[1,2] Jin YF[2*] Zhang YZ [3*]

Abstract: Oral tolerance has been applied successfully as a potential therapeutic strategy for preventing and treating autoimmune diseases including type Ⅰ diabetes. In this paper we constructed an edible vaccine consisting of a fusion protein composed of cholera toxin B subunit (CTB) and insulin B chain (InsB) that was produced in silkworm larvae. The silkworm larvae produced this fusion protein at levels of up to 0.97 mg/mL of hemolymph as the pentameric CTB-InsB form, which retained the GM1-ganglioside binding affinity and the native antigenicity of CTB. Non-obese diabetic mice fed hemolymph containing microgram quantities of the CTB-InsB fusion protein showed a prominent reduction in pancreatic islet inflammation and a delay in the development of diabetic symptoms. This study demonstrates that silkworm-produced CTB-InsB fusion protein can be used as an ideal oral protein vaccine for induction of immunological tolerance against autoimmune diabetes.

Published On: Vaccine, 2007, 25(8), 1444-1451.

1 School of Medicine, Ningbo University, Ningbo 315211, China.

2 Institute of Biochemistry, College of Life Sciences, Zhejiang University, Hangzhou 310029, China.

3 Institute of Biochemistry, College of Life Sciences, Zhejiang Sci-Tech University, Hangzhou 310018, China.

*Corresponding author E-mail: jinyf@zju.edu.cn; yaozhou@chinagene.com.

家蚕表达的霍乱毒素B亚基-胰岛素B链融合蛋白疫苗对非肥胖糖尿病(NOD)小鼠糖尿病的口服抑制作用

龚朝辉[1,2]　金勇丰[2*]　张耀洲[3*]

摘要：口服耐受已成为应用于预防和治疗自身免疫性疾病的一种潜在治疗策略，包括Ⅰ型糖尿病。本文中，我们构造了一种可口服疫苗，该疫苗由包含霍乱毒素B亚基（CTB）和胰岛素B链（InsB）的融合蛋白（CTB-InsB）组成，通过家蚕幼虫产生。家蚕幼虫产生的这种融合蛋白以五聚体的形式存在，在血淋巴中表达水平可达0.97 mg/mL。它保留了GM1-神经节苷脂结合亲和力和CTB的天然抗原性。饲喂含有微量CTB-InsB融合蛋白蚕血淋巴的非肥胖糖尿病小鼠显示胰岛炎症明显减轻，糖尿病症状的发展延缓。这项研究表明，家蚕表达的CTB-InsB融合蛋白可用于诱导自身免疫性糖尿病的免疫耐受。

1　医学院，宁波大学，宁波
2　生物化学研究所，生命科学学院，浙江大学，杭州
3　生物化学研究所，生命科学学院，浙江理工大学，杭州

Oral administration activity determination of recombinant osteoprotegerin from silkworm larvae

Xiao HL[1] Zhang YZ[2*] Wu XF[3]

Abstract: Osteoprotegerin (OPG) regulates the formation of osteoclasts and is involved in the regulation of bone resorption and remodeling. To investigate the feasibility of using silkworm (*Bombyx mori*) larvae to produce recombinant osteoprotegerin as an oral administration drug, the rh-OPG was expressed in the larvae of silkworm through the silkworm baculovirus expression system, and was orally administered to mice. Compared with the control, oral administration of rh-OPG was effective to decrease serum calcium concentration in normal mice, and block the bone loss induced by the loss of estrogen in ovariectomized mice. These results indicated that oral administration of rh-OPG expressed in silkworm larvae had the proper bioactivity.

Published On: Molecular Biotechnology, 2007, 35(2), 179-184.

1 Hangzhou Institute of Calibration and Testing for Quality and Technical Supervision, Hangzhou 310004, China.

2 Institute of Biochemistry, College of Life Sciences, Zhejiang Sci-Tec University, Hangzhou 310018, China.

3 Shanghai Institute of Biochemistry and Cell Biology, Shanghai Institute for Biological Sciences, Chinese Academy of Sciences, Shanghai 200025, China.

*Corresponding author E-mail: yaozhou@chinagene.com.

家蚕幼虫表达重组骨保护素的口服活性测定

肖海龙[1]　张耀洲[2*]　吴祥甫[3]

摘要：骨保护素（OPG）调节破骨细胞的形成，并参与骨吸收和重塑的调节。为了研究家蚕（*Bombyx mori*）幼虫生产的重组骨保护素作为口服药物的可行性，将rh-OPG通过家蚕杆状病毒表达系统在蚕幼虫中表达，并对小鼠进行口服给药实验。与对照组相比，口服rh-OPG可有效降低正常小鼠的血清钙浓度，同时可阻断卵巢切除小鼠固雌激素缺失引起的骨丢失。这些结果表明，口服的在家蚕幼虫中表达的rh-OPG具有适当的生物活性。

1　杭州市质量技术监督检测院，杭州
2　生物化学研究所，生命科学学院，浙江理工大学，杭州
3　生物化学与细胞生物研究所，上海生命科学研究院，中国科学院，上海

Anti-oxidation and immune responses in mice upon exposure to manganese superoxide dismutase expressed in silkworm larvae, *Bombyx mori* L.

Yue WF[1] Li GL[1] Liu JM[1] Sun JT[1]
Sun HX[2] Li XH[1] Wu XF[1] Miao YG[1*]

Abstract: With manganese superoxide dismutase expressed in silkworm larvae, *Bombyx mori* L., we investigate the effects of silkworm larvae powder containing SOD on the antioxidation and the immune system of mouse. The contents of MDA both in mice plasma or liver organ treated with silkworm larvae powder containing manganese superoxide dismutase were reduced compare to control. The superoxide dismutase (SOD) and glutathione peroxidase (GSH-Px) activities both in plasma or liver organ of the treated mice were significantly higher than that of both control and bromobenzene treated mice (group-BM), suggesting the silkworm larvae powder containing SOD play a positive role in anti-oxidation in mice. This experiment was also designed to investigate the effects of silkworm larvae powder containing SOD on the immune system of mouse, focused on hemolysin response, hemagglutination against SRBC and the activity of natural killer (NK) cells. All treated mice showed significant increase in hemolysin response to SRBC and demonstrated an activation of NK cell function by the SOD-contained silkworm larvae powder, which suggest a promotion in humoral immunity. The results suggested the SOD expressed in silkworm maybe have potential application in medicine.

Published On: Cell Biology International, 2007, 31(9), 974-978.

1 Department of Special Economic Animals, College of Animal Sciences, Zhejiang University, Hangzhou 310029, China.
2 Department of Animal Medical, College of Animal Sciences, Zhejiang University, Hangzhou 310029, China.
*Corresponding author E-mail: miaoyg@zju.edu.cn.

家蚕表达的SOD对小鼠的抗氧化和免疫反应的影响

岳万福[1]　李广立[1]　刘剑梅[1]　孙建通[1]
孙红祥[2]　李兴华[1]　吴小锋[1]　缪云根[1*]

摘要： 我们在家蚕(*Bombyx mori* L.)幼虫中表达了锰超氧化物歧化酶(SOD)，研究了含有SOD的幼虫全蚕粉对小鼠抗氧化和免疫系统的影响。在含锰SOD的蚕幼虫粉处理的小鼠血浆或肝脏器官中MDA的含量均降低。在蚕幼虫粉处理的小鼠血浆或肝脏器官中的SOD和谷胱甘肽过氧化物酶(GSH-Px)活性均显著高于对照组和溴苯处理组(BM组)，表明含有SOD的蚕幼虫粉在小鼠抗氧化中发挥积极作用。本实验旨在研究含有SOD的蚕幼虫粉对小鼠免疫系统的影响，重点在于溶血素反应，抗SRBC的血细胞凝集和自然杀伤(NK)细胞的活性。所有处理的小鼠显示出对SRBC的溶血素反应显著增加，激活了NK细胞的功能，这表明在体液免疫中含有SOD的蚕幼虫粉有促进作用。结果表明，在家蚕中表达的SOD可能在医学上具有潜在应用价值。

1　特种经济动物系，动物科学学院，浙江大学，杭州
2　动物医学系，动物科学学院，浙江大学，杭州

Improvement of recombinant baculovirus infection efficiency to express manganese superoxide dismutase in silkworm larvae through dual promoters of *Pph* and *Pp10*

Yue WF[1] Li XH[1] Wu WC[2] Bhaskar Roy[1] Li GL[1]
Liu JM[1] Wu XF[1] Zhou JY[3] Zhang CX[4]
Wan Chi Cheong David[5] Miao YG[1,6*]

Abstract: The silkworm, *Bombyx mori*, has been used as an important bioreactor for the production of recombinant proteins through baculovirus expression system (BES). There are several problems which will probably be the bottleneck for practical and industrial utilization of silkworm bioreactor. Traditionally, the recombinant virus should infect the larvae through individual dorsal injection by a syringe. This is a time- and labor-consuming procedure. This drawback has become a bottleneck for practical and industrial utilization of baculovirus expression system in the silkworm bioreactor. In this paper, we constructed a dual expression baculovirus to express the renovated polyhedron and target manganese superoxide dismutase (SOD) gene under *P10* and polyhedron promoters, respectively, through oral infection. The results showed that the direct injection of recombinant rBacmid/*Bm*NPV/SOD DNA with cellfectin reagent infected the silkworm larvae partially. When next batches of larvae were fed orally with hemolymph, which was collected from first batch of injected and infected larvae, the obvious symptom of infection was found and high target SOD was expressed. These results imply it is feasible to express target genes through combination of recombinant bacmid DNA injection and oral feeding by a dual expression bacmid baculovirus.

Published On: Applied Microbiology and Biotechnology, 2008, 78(4), 651-657.

1 Institute of Sericulture and Apiculture, Zhejiang University, Hangzhou 310029, China.
2 Zhejiang Academy of Agricultural Sciences, Hangzhou 310021, China.
3 Institute of Preventive Veterinary Medicine, Zhejiang University, Hangzhou 310029, China.
4 Institute of Insect Sciences, Zhejiang University, Hangzhou 310029, China.
5 Department of Biochemistry, The Chinese University of Hong Kong, Hong Kong 999077, China.
6 College of Animal Sciences, Zhejiang University, Hangzhou 310029, China.
*Corresponding author E-mail: miaoyg@zju.edu.cn.

利用双启动子*Pph*与*Pp10*改善重组杆状病毒感染效率在家蚕幼虫中表达锰超氧化物歧化酶

岳万福[1] 李兴华[1] 吴蔚成[2] Bhaskar Roy[1] 李广立[1]
刘剑梅[1] 吴小锋[1] 周继勇[3] 张传溪[4]
Wan Chi Cheong David[5] 缪云根[1,6*]

摘要：家蚕是利用杆状病毒表达系统(BES)生产重组蛋白的重要的生物反应器。但是,将家蚕生物反应器投入到生产和工业应用仍然存在许多瓶颈。传统上,重组病毒需要通过个体单独注射的方法才能感染家蚕,这是一个既耗时又耗力的过程。这已经成了制约家蚕生物反应器在生产和工业应用的一个瓶颈。本文通过构建双启动子杆状病毒表达载体,在*P10*启动子的驱动下表达改造过的多角体基因,在*Ph*启动子驱动下表达超氧化物歧化酶(SOD)基因,从而使得该病毒可以通过食下感染。结果显示,直接将重组rBacmid/*Bm*NPV/SOD的DNA与转染试剂混合注射家蚕,可以使部分幼虫感染重组病毒。收集感染后家蚕血淋巴,喂食正常家蚕,正常家蚕表现出明显的感染特征,并且高量表达了SOD蛋白。这些结果显示,通过双启动子表达系统改造的杆状病毒可以采用注射和食下两种感染方式,从而高量表达外源蛋白。

1 蚕蜂研究所,浙江大学,杭州
2 浙江省农业科学院,杭州
3 动物预防医学研究所,浙江大学,杭州
4 昆虫科学研究所,浙江大学,杭州
5 生物化学系,香港中文大学,香港
6 动物科学学院,浙江大学,杭州

The transgenic *BmN* cells with polyhedrin gene: a potential way to improve the recombinant baculovirus infection *per os* to insect larvae

Chen L[1] Shen WD[2] Wu Y[1] Li B[2]
Gong CL[2] Wang WB[1*]

Abstract: The principle of baculovirus expression system is that substitute exogenous gene for polyhedrin (*polh*) gene, and the recombinant baculovirus lacks the ability to infect insect larvae by oral inoculation. In this study, we cloned the *polh* gene with immediate early gene 1 (*ie1*) promoter of *Bombyx mori* nucleopolyhedrovirus (*Bm*NPV) into transposon pigA3GFP vector, transported it into *BmN* cells by lipofectamine and obtained the transgenic *BmN* cell line. The mRNA transcription of the polyhedrin gene was demonstrated by reverse transcription-polymerase chain reaction. Then the *polh* gene negative viruses (*Bm*PAK6 and *Bm*GFP), infected the transgenic *BmN* cells and Polyhedrin-like structures were observed in the infected cells. Subsequently, the viruses (v*Bm*PAK6 and v*Bm*GFP) from infected cells were used to orally inoculate the fifth instar larvae of *B. mori*, respectively. The results showed that *B. mori* larvae could be infected *per os* with the recombinant baculoviruses v*Bm*PAK6 and v*Bm*GFP, respectively. These results suggest that the products of *polh* gene expressed in the transgenic *BmN* cells could package the recombinant baculoviruses when the viruses infected the cells and raise the pathogenicity of the recombinant virus in orally infected *B. mori* larvae.

Published On: Applied Biochemistry and Biotechnology, 2009, 158(2), 277-284.

1 Institute of Life Sciences, Jiangsu University, Zhenjiang 212013, China.
2 School of Life Sciences, Soochow University, Suzhou 215123, China.
*Corresponding author E-mail: wenbingwang@ujs.edu.cn.

携带多角体基因的转基因*BmN*细胞：一种提高重组杆状病毒经口感染昆虫幼虫的潜在途径

陈　璐[1]　沈卫德[2]　吴　岩[1]
李　兵[2]　贡成良[2]　王文兵[1*]

摘要： 通常杆状病毒表达系统是利用外源基因取代多角体基因，这样的重组病毒不能经口感染昆虫幼虫。在本研究中，我们将家蚕杆状病毒的极早期基因*ie1*启动子启动的多角体基因克隆到转座质粒pigA3GFP中，利用脂质体转染进*BmN*细胞，建立了表达多角体蛋白的*BmN*细胞系。通过逆转录聚合酶链式反应证实了多角体蛋白基因的转录。然后以不含多角体基因的重组家蚕杆状病毒*Bm*PAK6及*Bm*GFP感染转基因的细胞系，在感染的细胞中观察到多角体的结构。将获得的病毒颗粒（v*Bm*PAK6和v*Bm*GFP）经口感染家蚕幼虫，发现家蚕均出现感染症状。结果表明通过这种方式，可以使重组杆状病毒获得多角体，提高重组病毒对经口感染的家蚕幼虫的致病率。

1　生命科学研究院，江苏大学，镇江
2　生命科学学院，苏州大学，苏州

Safety and immunogenicity of H5N1 influenza vaccine based on baculovirus surface display system of *Bombyx mori*

Jin RZ[1,2#] Lü ZB[2#] Chen Q[2#] Quan YP[2] Zhang HH[2]
Li S[2] Chen GG[2] Zheng QL[2] Jin LR[2] Wu XF[2]
Chen JG[3*] Zhang YZ[1,2*]

Abstract: Avian influenza virus (H5N1) has caused serious infections in human beings. This virus has the potential to emerge as a pandemic threat in humans. Effective vaccines against H5N1 virus are needed. A recombinant *Bombyx mori* baculovirus, *Bm*g64HA, was constructed for the expression of HA protein of H5N1 influenza virus displaying on the viral envelope surface. The HA protein accounted for approximately 3% of the total viral proteins in silkworm pupae infected with the recombinant virus. Using a series of separation and purification methods, pure *Bm*gp64HA virus was isolated from these silkworm pupae bioreactors. Aluminum hydroxide adjuvant was used for an H5N1 influenza vaccine. Immunization with this vaccine at doses of 2 mg/kg and 0.67 mg/kg was carried out to induce the production of neutralizing antibodies, which protected monkeys against influenza virus infection. At these doses, the vaccine induced 1∶40 antibody titers in 50% and 67% of the monkeys, respectively. The results of safety evaluation indicated that the vaccine did not cause any toxicity at the dosage as large as 3.2 mg/kg in cynomolgus monkeys and 1.6 mg/kg in mice. The results of dose safety evaluation of vaccine indicated that the safe dose of the vaccine were higher than 0.375 mg/kg in rats and 3.2 mg/kg in cynomolgus monkeys. Our work showed the vaccine may be a candidate for a highly effective, cheap, and safe influenza vaccine for use in humans.

Published On: PLoS ONE. 2008, 3(12).

1 College of Life Sciences, Zhejiang University, Hangzhou 310029, China.
2 Institute of Biochemistry, College of Life Sciences, Zhejiang Sci-Tech University, Hangzhou 310018, China.
3 College of Life Sciences, Peking University, Beijing 100871, China.
#These authors contributed equally.
*Corresponding author E-mail: yaozhou@chinagene.com; chenjg@pku.edu.cn.

基于家蚕杆状病毒表面展示系统的 H5N1 流感疫苗的安全性和免疫原性

金荣仲[1,2#] 吕正兵[2#] 陈 琴[2#] 全滟平[2] 张海花[2]
李 司[2] 陈国刚[2] 郑青亮[2] 金来荣[2] 吴祥甫[2]
陈建国[3*] 张耀洲[1,2*]

摘要： 禽流感病毒(H5N1)已在人类中造成严重感染。这种病毒具有在人类之间广泛传播的潜在威胁，因此制备针对H5N1病毒有效的疫苗迫在眉睫。我们构建了重组家蚕杆状病毒*Bmg*64HA，用于表达H5N1流感病毒囊膜表面上的HA蛋白。在重组病毒感染的蚕蛹中，HA蛋白占总病毒蛋白质约3%。通过使用一系列分离纯化方法，从这些蚕蛹生物反应器中分离纯化了*Bmg*p64HA病毒。使用氢氧化铝作为H5N1流感疫苗的佐剂，以2 mg/kg和0.67 mg/kg的剂量注射，诱导产生中和抗体，可使猴子能够抵抗流感病毒感染。在这两种剂量下，疫苗分别在50%和67%的猴中能诱导产生抗体，抗体滴度为1∶40。安全性评价结果表明，当食蟹猴的接种剂量不超过3.2 mg/kg和小鼠的使用剂量不超过1.6 mg/kg时，两种动物均没有发生毒性反应。疫苗剂量安全性评价结果表明，疫苗安全剂量在大鼠体内高于0.375 mg/kg，在食蟹猴体内高于3.2 mg/kg。结果表明，这种疫苗可作用于人类，并且具有高效、廉价和安全的特点。

1 生命科学研究院，浙江大学，杭州
2 生物化学研究所，生命科学学院，浙江理工大学，杭州
3 生命科学学院，北京大学，北京

High-level expression of orange fluorescent protein in the silkworm larvae by the Bac-to-Bac system

Liu JM[1] Wan Chi Cheong David[2] Denis Tsz-Ming Ip[2] Li XH[1]
Li GL[1] Wu XF[1] Yue WF[1] Zhang CX[3] Miao YG[1*]

Abstract: This novel orange fluorescent protein (OFP) emits brilliant orange fluorescent light. OFP has high fluorescence quantum yield, fast maturation rate, and stability, which imply this protein should be the most favorable biotechnological tools used to investigate the function of target gene by visualizing, monitoring, and quantifying in living cells. *Bombyx mori*, silkworm has been used as an important bioreactor for the production of recombinant proteins through baculovirus expression system (BES). In this paper, we used infection technique which introduced the baculovirus DNA into silkworms using a cationic lipofectin reagent instead of directly injecting the virus, and demonstrated a high-level expression of the orange fluorescent protein (OFP) gene in the *Bombyx mori*, silkworm larvae. When recombinant rBacmid/*Bm*NPV/OFP DNA ranging from 50-100 ng / larval was injected, a sufficient OFP expression in hemolymph was harvested. The recombinant viruses could be obtained from the hemolymph of infected larvae and stored as seed which could be used for the large-scale expression. This procedure omitted the costly and labor-consumed insect cell culture. Further investigation of OFP should provide us with more insight in unlocking the mystery of the mechanisms of autocatalytic bioluminescence and its utilization in biotechnology.

Published On: Molecular Biology Reports, 2009, 36(2), 329-335.

1 Institute of Sericulture and Apiculture, College of Animal Sciences, Zhejiang University, Hangzhou 310029, China.
2 Department of Biochemistry, The Chinese University of Hong Kong, Hong Kong 999077, China.
3 Institute of Insect Sciences, Zhejiang University, Hangzhou 310029, China.
*Corresponding author E-mail: miaoyg@zju.edu.cn.

利用杆状病毒表达系统在家蚕幼虫体内高效表达橙色荧光蛋白

刘剑梅[1] Wan Chi Cheong David[2] Denis Tsz-Ming Ip[2] 李兴华[1]
李广立[1] 吴小锋[1] 岳万福[1] 张传溪[3] 缪云根[1*]

摘要：新型橙色荧光蛋白(OFP)可以发射出明亮的橙色荧光。OFP具有荧光量子产率高、成熟速度快和稳定性高的特性，这些特性使其成为研究活细胞中目的基因功能可视化、监控和定量的最佳生物技术工具。利用杆状病毒表达系统，家蚕常被作为一种生产重组蛋白的重要生物反应器。在本文中，我们利用阳离子脂质体试剂将杆状病毒DNA导入家蚕而不是直接注射病毒，并证明了在家蚕幼虫中橙色荧光蛋白(OFP)基因可以高量表达。当每头幼虫注射50—100 ng的重组rBacmid/*Bm*NPV/OFP DNA时，从家蚕血淋巴中可以收获足够的OFP。重组病毒可以从感染幼虫的血淋巴中获得，并作为种子储存，可用于大规模表达。该方法省略了昂贵且耗费劳力的昆虫细胞培养过程。对OFP的进一步研究将会为我们提供更多的知识，以揭开荧光蛋白自催化生物发光机制之谜及拓展其在生物技术中的应用。

1 蚕蜂研究所，动物科学学院，浙江大学，杭州
2 生物化学系，香港中文大学，香港
3 昆虫科学研究所，浙江大学，杭州

A novel way to purify recombinant baculoviruses by using bacmid

Su WJ[1] Shen WD[2] Li B[2] Wu Y[1]
Gao G[1] Wang WB[1*]

Abstract: In the present study, we studied the feasibility of deleting essential genes in insect cells by using bacmid and purifying recombinant bacmid in *Escherichia coli* DH10B cells. To disrupt the *orf4* (open reading frame 4) gene of *Bm*NPV [*Bm* (*Bombyx mori*) nuclear polyhedrosis virus], a transfer vector was constructed and co-transfected with *Bm*NPV bacmid into *Bm* cells. Three passages of viruses were carried out in *Bm* cells, followed by one round of purification. Subsequently, bacmid DNA was extracted and transformed into competent DH10B cells. A colony harbouring only *orf4*-disrupted bacmid DNA was identified by PCR. A mixture of recombinant (white colonies) and non-recombinant (blue colonies) bacmids were also transformed into DH10B cells. PCR with M13 primers showed that the recombinant and non-recombinant bacmids were separated after transformation. The result confirmed that purification of recombinant viruses could be carried out simply by transformation and indicated that this method could be used to delete essential genes. *Orf4*-disrupted bacmid DNA was extracted and transfected into *Bm* cells. Viable viruses were produced, showing that *orf4* was not an essential gene.

Published On: Bioscience Reports, 2009, 29(2), 71-75.

1 Institute of Life Sciences, Jiangsu University, Zhenjiang 212013, China.
2 School of Life Sciences, Soochow University, Suzhou 215123, China.
*Corresponding author E-mail: wenbingwang@ujs.edu.cn.

利用Bacmid纯化重组杆状病毒的新方法

苏武杰[1]　沈卫德[2]　李　兵[2]
吴　岩[1]　高　广[1]　王文兵[1*]

摘要：在本研究中，我们研究了使用在大肠杆菌DH10B细胞中纯化的重组杆粒在昆虫细胞中删除必需基因的可行性。为了破坏家蚕核型多角体病毒*Bm*NPV的*orf4*（开放阅读框4）基因，我们构建了转移载体，并与*Bm*NPV Bacmid共转染*Bm*细胞。在*Bm*细胞中进行3次病毒传代，然后进行一轮纯化。随后，提取bacmid DNA并转化DH10B感受态细胞。通过PCR鉴定仅含有*orf4*被破坏的bacmid DNA的菌落。此外，还将重组杆菌（白色菌落）和非重组杆菌（蓝色菌落）的混合物转化DH10B细胞。用M13引物进行PCR，结果显示重组和非重组的bacmid在转化后分离。结果证实，重组病毒的纯化可以简单地通过转化进行，并表明该方法可用于删除必需基因。提取*Orf4*被破坏的bacmid DNA并转染*Bm*细胞，在细胞中产生了活性病毒，表明*orf4*不是必需基因。

1　生命科学研究院，江苏大学，镇江
2　生命科学学院，苏州大学，苏州

Expression of EGFP driven by *Bm*NPV *orf4* promoter, a novel immediate early promoter

Su WJ[1] Li B[2] Shen WD[2] Wu Y[1]

Zhu SY[1] Sun Y[1] Wang WB[1*]

Abstract: *Bombyx mori* nucleopolyhedrovirus (*Bm*NPV) *orf*4 has been shown to be expressed at very early stage of *Bm*NPV infection cycle. In this study, using transient expression experiment, we demonstrated for the first time that *orf*4 promoter is an immediate early promoter, indicating that *orf*4 may play a role in the immediate-early stage of *Bm*NPV infection. Moreover, with the recently developed Bac-to-Bac / *Bm*NPV baculovirus expression system and a modified pFast-Bac1 whose polyhedrin promoter was replaced with *orf*4 promoter, a recombinant bacmid baculovirus expressing enhanced green fluorescent protein (EGFP) under the control of *orf*4 promoter in *Bombyx mori* (*Bm*) cells was successfully constructed. The result not only showed that the polyhedrin promoter can be replaced easily with other promoters to direct the expression of foreign genes by using this novel system but also laid the foundation for the rescue experiment of *orf4* deletion mutant.

Published On: Biologia, 2009, 64(2), 383-387.

1 Institute of Life Sciences, Jiangsu University, Zhenjiang 212013, China.

2 School of Life Sciences, Soochow University, Suzhou 215123, China.

*Corresponding author E-mail: wenbingwang@ujs.edu.cn.

利用家蚕核型多角体病毒一个新的极早期基因*orf4*启动子驱动EGFP的表达

苏武杰[1] 李 兵[2] 沈卫德[2] 吴 岩[1]
朱姗颖[1] 孙 莹[1] 王文兵[1*]

摘要：*orf4*是家蚕核型多角体病毒感染极早期表达的基因。在本文中，利用瞬时表达实验，我们首次证明了*orf4*启动子是一种即刻早期启动子，可能在病毒感染的即刻早期阶段发挥作用。同时，我们构建了一个修饰的pFast-Bac1，将其中的多角体蛋白启动子替换为*orf4*启动子，利用最近开发的Bac-to-Bac/*Bm*NPV杆状病毒表达系统成功构建了表达增强型绿色荧光蛋白（EGFP）的重组杆状病毒。该病毒可以在*orf4*启动子的控制下在家蚕（*Bombyx mori*）细胞中表达EGFP。结果表明，多角体蛋白启动子可以很容易地被其他启动子取代，通过这种新系统介导外源基因的表达，同时也为*orf4*缺失突变体的拯救实验奠定了基础。

1 生命科学研究院，江苏大学，镇江
2 生命科学学院，苏州大学，苏州

Bioavailability of orally administered rhGM-CSF: a single-dose, randomized, open-label, two-period crossover trial

Zhang WP[1,2#] Lü ZB[2#] Nie ZM[2] Chen GG[3] Chen J[2]
Sheng Q[2] Yu W[2] Jin YF[1] Wu XF[2,4] Zhang YZ [2*]

Abstract:

Background: Recombinant human granulocyte-macrophage colony-stimulating factor (rhGM-CSF) is usually administered by injection, and its oral administration in a clinical setting has been not yet reported. Here we demonstrate the bioavailability of orally administered rhGM-CSF in healthy volunteers. The rhGM-CSF was expressed in *Bombyx mori* expression system (*Bm*rhGM-CSF).

Methods and Findings: Using a single-dose, randomized, open-label, two-period crossover clinical trial design, 19 healthy volunteers were orally administered with *Bm*rhGM-CSF (8 μg/kg) and subcutaneously injected with rhGM-CSF (3.75 μg/kg) respectively. Serum samples were drawn at 0.0 h, 0.5 h, 0.75 h, 1.0 h, 1.5 h, 2.0 h, 3.0 h, 4.0 h, 5.0 h, 6.0 h, 8.0 h, 10.0 h and 12.0 h after administrations. The hGM-CSF serum concentrations were determined by ELISA. The AUC was calculated using the trapezoid method. The relative bioavailability of *Bm*rhGM-CSF was determined according to the AUC ratio of both orally administered and subcutaneously injected rhGM-CSF. Three volunteers were randomly selected from 15 orally administrated subjects with ELISA detectable values. Their serum samples at the 0.0 h, 1.0 h, 2.0 h, 3.0 h and 4.0 h after the administrations were analyzed by Q-Trap MS/MS TOF. The different peaks were revealed by the spectrogram profile comparison of the 1.0 h, 2.0 h, 3.0 h and 4.0 h samples with that of the 0.0 h sample, and further analyzed using both Enhanced Product Ion (EPI) scanning and Peptide Mass Fingerprinting Analysis. The rhGM-CSF was detected in the serum samples from 15 of 19 volunteers administrated with BmrhGM-CSF. Its bioavailability was observed at an average of 1.0%, with the highest of 3.1%. The rhGM-CSF peptide sequences in the serum samples were detected by MS analysis, and their sizes ranging from 2 039 to 7 336 Da.

Conclusions: The results demonstrated that the oral administered *Bm*rhGM-CSF was absorbed into the blood. This study provides an approach for an oral administration of rhGM-CSF protein in clinical settings.

Published On: PLoS ONE, 2009, 4(5).

1 Institute of Biochemistry, College of Life Sciences, Zhejiang University, Hangzhou, China.

2 Key Laboratory of Bioreactor and Biopharmacy of Zhejiang Province, Institute of Biochemistry, Zhejiang Sci-Tech University, Hangzhou 310018, China.

3 Zhejiang Chinagene Biopharmaceutical Co., Ltd., Haining 314400, China.

4 Institute of Biochemistry, Chinese Academy of Sciences, Shanghai 200025, China.

[#]These authors contributed equally.

[*]Corresponding author E-mail: yaozhou@chinagene.com.

口服rhGM-CSF生物利用度：单剂量、随机、开放、两期交叉临床试验

张文平[1,2#]　吕正兵[2#]　聂作明[2]　陈国刚[3]　陈　健[2]
盛　清[2]　于　威[2]　金勇丰[1]　吴祥甫[2,4]　张耀洲[2*]

摘要： 背景——重组人粒细胞-巨噬细胞集落刺激因子(rhGM-CSF)通常是通过注射给药，口服制剂的方法尚未在临床上应用。本文报道了健康志愿者口服rhGM-CSF的生物利用度，这种rhGM-CSF是通过家蚕生物反应器制备的口服制剂(*Bm*rhGM-CSF)。方法和结果——设计单剂量、随机、开放、两期交叉临床试验，19名健康志愿者分别口服*Bm*rhGM-CSF(8 μg/kg)和皮下注射rhGM-CSF(3.75 μg/kg)。在给药后的0 h、0.5 h、0.75 h、1 h、1.5 h、2 h、3 h、4 h、5 h、6 h、8 h、10 h、12 h抽取血样，并通过ELISA方法测定血清中hGM-CSF的浓度，利用梯形法计算曲线下面积(the area under a curve，AUC)，根据口服和皮下注射rhGM-CSF的AUC的比值计算*Bm*rhGM-CSF的相对生物利用度。从具有ELISA可检测值的15名口服给药受试者中随机选择3名志愿者，通过Q-Trap MS/MS TOF质谱分析给药后0 h、1.0 h、2.0 h、3.0 h和4.0 h的血清样品。与0 h的样本相比，1.0 h、2.0 h、3.0 h和4.0 h的样本存在不同的质谱峰，进一步用增强产物离子(EPI)扫描和肽质量指纹对这些不同的质谱峰进行鉴定。结果表明，在19名志愿者中的15名的血清样本里能够检测到rhGM-CSF，其平均生物利用度为1%，最高为3.1%。通过质谱分析检测血清样品中的rhGM-CSF肽序列，其分子质量大小为2 039—7 336 Da。结论——结果表明*Bm*rhGM-CSF可以通过口服吸收进入血液。本研究为临床口服rhGM-CSF提供了一种方法。

1　生物化学研究所，生命科学学院，浙江大学，杭州
2　浙江省家蚕生物反应器和生物医药重点实验室，生物化学研究所，浙江理工大学，杭州
3　浙江中奇生物医药股份有限公司，海宁
4　生物化学研究所，中国科学院，上海

A transgenic *Bm* cell line of *piggy*Bac transposon-derived targeting expression of humanized glycoproteins through *N*-glycosylation

Hu JB Zhang P Wang MX Zhou F Niu YS Miao YG*

Abstract: Glycoproteins have been implicated in a wide variety of important biochemical and biological functions, including protein stability, immune function, enzymatic function, cellular adhesion and others. Unfortunately, there is no therapeutic protein produced in insect system to date, due to the expressed glycoproteins are paucimannosidic *N*-glycans, rather than the complex, terminally sialylated *N*-glycans in mammalian cells. In this paper, we cloned the necessary genes in glycosylation of mammalian cells, such as *N-acetylglucosaminyltransferase II (Gn-TII), galactosyltransferases (Gal-Ts)*, *2, 6-Sial-T (ST6 GalII)* and *2, 3-Sial-T (ST3GalIII)*, and transformed them to silkworm genome of *BmN* cell line through transgenesis to establish a transgenic *Bm* cell line of *piggy*Bac transposon-derived targeting expression of humanized glycoproteins. The study supplied a new insect cell line which is practically to produce "bisected" complex *N*-glycans like in mammalian cells.

Published On: Molecular Biology Reports, 2012, 39(8), 8405-8413.

Key Laboratory of Animal Virology of Ministry of Agriculture, College of Animal Sciences, Zhejiang University, Hangzhou 310058, China.

*Corresponding author E-mail: miaoyg@zju.edu.cn.

利用*piggy*Bac转座子获得表达*N*-糖基化的人源化糖蛋白的转基因*Bm*细胞系

胡嘉彪　张　芃　王梅仙　周　芳　牛艳山　缪云根*

摘要：糖蛋白具有多种重要的生物化学和生物学功能，包括蛋白质稳定性、免疫功能、酶功能、细胞黏附等。但是，迄今为止，在昆虫表达系统中还没有合成治疗性蛋白质，这是因为昆虫表达系统表达的低聚*N*-糖蛋白，而不是哺乳动物细胞中复杂的末端唾液酸化的糖蛋白。本研究中，我们克隆了哺乳动物细胞糖基化中必需的基因，如*N*-乙酰葡糖胺基转移酶Ⅱ(*Gn-TII*)、半乳糖基转移酶(*Gal-Ts*)、2,6-*Sial-T*(*ST6 GalII*)和2,3-*Sial-T*(*ST3GalIII*)，并通过转基因方法将其转入*BmN*细胞系的蚕基因组中，建立了*piggy*Bac转座子靶向表达人源化糖蛋白的转基因*Bm*细胞系。该研究提供了一种新的昆虫细胞系，可应用于产生哺乳动物细胞中那样均一复杂的*N*-聚糖蛋白。

农业部动物病毒学重点实验室，动物科学学院，浙江大学，杭州

Construction of the ie1-Bacmid expression system and its use to express EGFP and *Bm*AGO2 in *BmN* cells

Zhou F Gao Z Lü ZB Chen J Hong YT Yu W
Wang D Jiang CY Wu XF Zhang YZ Nie ZM*

Abstract: The presently available expression tools and vectors (e.g., eukaryotic expression vectors and the adenovirus expression system) for studying the functional genes in *Bombyx mori* are insufficient. The baculovirus expression system is only used as a protein production tool; therefore, recombinant proteins expressed by *B. mori* using the baculovirus expression system equipped with a *polyhedrin* promoter cannot be used for *in vivo* research applications. In this work, we constructed and screened a eukaryotic expression vector for silkworm cells. The EGFP and *B. mori* Argonaute2 proteins were found to be efficiently expressed using the screened pIEx-1 vector with the FuGENE 6 transfection reagent. Additionally, we constructed a novel nucleopolyhedrovirus *ie1*-Bacmid expression system for the production of recombinant protein; we then used the system to highly express the EGFP and *B. mori* Argonaute2 proteins. In this system, the protein of interest can be efficiently expressed 13 h after infection by controlling the *B. mori* nucleopolyhedrovirus immediate early *ie1* promoter. The *ie1*-Bacmid system provides a powerful "adenovirus-like" expression tool; not only can the tool be used to study baculovirus molecular biology for the silkworm but it is also useful in other research applications as well, such as the study of gene functions involved in cellular physiological processes.

Published On: Applied Biochemistry and Biotechnology, 2013,169(8), 2237-2247.

Institute of Biochemistry, College of Life Sciences, Zhejiang Sci-Tech University, Hangzhou 310018, China.
*Corresponding author E-mail: wuxinzm@126.com.

ie1-Bacmid 表达系统的构建及其在*BmN*细胞中表达EGFP和*Bm*AGO2的应用

周 芳 高 珍 吕正兵 陈 健 洪叶挺 于 威
王 丹 蒋彩英 吴祥甫 张耀洲 聂作明*

摘要:目前家蚕中用于研究基因功能的表达工具和载体(如真核表达载体和腺病毒表达系统)很少。常规的杆状病毒表达系统仅仅只能作为蛋白表达的工具,因此,利用多角体蛋白启动子的杆状病毒系统表达的重组蛋白不能用于体内研究应用。在本文中,我们构建并筛选了一种家蚕细胞的真核表达载体。发现 pIEx-1 载体结合 FuGENE 6 转染试剂可以高效表达 EGFP 和家蚕 Argonaute2 (*Bm*AGO2)蛋白。此外,我们构建了一种新型的核型多角体杆状病毒 *ie1*-Bacmid 表达系统,用于生产重组蛋白,利用该系统我们高效表达了 EGFP 和家蚕 *Bm*AGO2 蛋白。由于该系统采用了家蚕杆状病毒极早期启动子 *ie1*,目的蛋白在感染病毒 13 h 后即可获得高效表达。*ie1*-Bacmid 系统为家蚕基因功能研究提供了一种强大的类似于腺病毒的表达工具,该工具不仅可以用于家蚕杆状病毒分子生物学研究,还可用于家蚕基因涉及的细胞内生理功能研究。

生物化学研究所,生命科学学院,浙江理工大学,杭州

A highly efficient and simple construction strategy for producing recombinant baculovirus *Bombyx mori* nucleopolyhedrovirus

Liu XJ[1] Wei YL[2] Li YN[1] Li HY[1]
Yang X[1] Yi YZ[3] Zhang ZF[1*]

Abstract: The silkworm baculovirus expression system is widely used to produce recombinant proteins. Several strategies for constructing recombinant viruses that contain foreign genes have been reported. Here, we developed a novel defective-rescue *Bm*NPV Bacmid (re*Bm*Bac) expression system. A *CopyControl* origin of replication was introduced into the viral genome to facilitate its genetic manipulation in *Escherichia coli* and to ensure the preparation of large amounts of high quality re*Bm*Bac DNA as well as high quality recombinant baculoviruses. The *ORF1629*, *cathepsin* and *chitinase* genes were partially deleted or rendered defective to improve the efficiency of recombinant baculovirus generation and the expression of foreign genes. The system was validated by the successful expression of *luciferase* reporter gene and porcine *interferon γ*. This system can be used to produce batches of recombinant baculoviruses and target proteins rapidly and efficiently in silkworms.

Published On: PLoS ONE, 2016, 11(3).

1 Biotechnology Research Institute, Chinese Academy of Agricultural Sciences, Beijing 100081, China.
2 State Key Laboratory of Biomembrane and Membrane Biotechnology, Institute of Zoology, Chinese Academy of Sciences, Beijing 100101, China.
3 Sericultural Research Institute, Chinese Academy of Agricultural Sciences, Zhenjiang 212018, China.
*Corresponding author E-mail: zhifangzhang@yahoo.com.

一种高效、简单的家蚕核型多角体病毒重组策略

刘兴健[1] 韦永龙[2] 李轶女[1] 李皓洋[1]
杨 鑫[1] 易咏竹[3] 张志芳[1*]

摘要：蚕杆状病毒表达系统广泛用于生产重组蛋白。已经报道了构建含有外来基因的重组病毒的策略。在这里，我们开发了一种新颖的缺陷拯救 *Bm*NPV Bacmid（re*Bm*Bac）表达系统。将 *CopyControl* 复制起点引入病毒基因组，以方便其在大肠杆菌中的遗传操作，确保制备大量高质量的 re*Bm*Bac DNA 和高质量的重组杆状病毒。*ORF1629*、组织蛋白酶和几丁质酶基因部分缺失或缺陷，可以提高重组杆状病毒产生和外源基因表达效率。荧光素酶报告基因和猪干扰素γ的成功表达验证了该系统。该系统可用于在家蚕中快速有效地批量生产重组杆状病毒和靶蛋白。

1 生物技术研究所，中国农业科学院，北京
2 生物膜与膜生物工程国家重点实验室，动物研究所，中国科学院，北京
3 蚕业研究所，中国农业科学院，镇江

Human insulin gene expressing with *Bombyx mori* multiple nucleopolyhedrovirus (*Bm*MNPV) expression system

Yue WF[1,2] Zhou F[1] Hu JB[1] Enoch Y. Park[2]
Joe Hull[3] Miao YG[1,3*]

Abstract: Using human genomic DNA as a template, the human insulin gene was cloned and used to construct various re*Bm*MNPV bacmids. Cysteine protease gene deletion (CPD-*Bm*MNPV bacmid) and cysteine protease- and chitinase-deficient (CPPD-*Bm*MNPV bacmid) baculoviruses were used to express both native and FLAG-tagged human insulin. Silkworm larvae were infected with the above recombinant bacmid DNAs, and the expressed insulin was purified and identified from infected silkworm haemolymph. The highest expression was shown with the CPPD *Bm*MNPV bacmid, which was about two times that of the wild type of re*Bm*MNPV bacmid, reaching 15.827 ng/mL haemolymph.

Published On: World Journal of Microbiology and Biotechnology, 2011, 27(2), 393-399.

1 Key Laboratory of Animal Epidemic Etiology & Immunological Prevention of Ministry of Agriculture, College of Animal Sciences, Zhejiang University, Hangzhou 310029, China.

2 Laboratory of Biotechnology, Department of Applied Biological Chemistry, Faculty of Agriculture, Shizuoka University, 836 Ohya Suruga-ku, Shizuoka 422-8529, Japan.

3 Department of Molecular Biology, University of Wyoming, Laramie, WY 82071-3944, USA.

*Corresponding author E-mail: miaoyg@zju.edu.cn.

利用家蚕多核型多角体病毒(*Bm*MNPV)表达系统表达人胰岛素基因

岳万福[1,2]　周　芳[1]　胡嘉彪[1]　Enoch Y.Park[2]
Joe Hull[3]　缪云根[1,3*]

摘要:以人基因组DNA作为模板,克隆人胰岛素基因并用于构建各种re*Bm*MNPV杆粒。半胱氨酸蛋白酶基因缺失(CPD-*Bm*MNPV bacmids)、半胱氨酸蛋白酶和几丁质酶缺陷(CPPD-*Bm*MNPV bacmids)的杆状病毒用于表达天然和FLAG标记的人胰岛素。用上述重组bacmids DNAs感染家蚕幼虫,并从感染的家蚕血淋巴中纯化并鉴定表达的胰岛素。CPPD-*Bm*MNPV bacmids中胰岛素的表达量最高,约为野生型re*Bm*MNPV bacmids中的2倍,血淋巴中浓度达到15.827 ng/mL。

1　农业部动物疫病病原学与免疫控制重点实验室,动物科学学院,浙江大学,杭州
2　生物技术实验室,应用生物化学系,农学院,静冈大学,静冈,日本
3　分子生物系,怀俄明大学,拉勒米,美国

Cloning and expression of a cellulase gene in the silkworm, *Bombyx mori* by improved Bac-to-Bac/*Bm*NPV baculovirus expression system

Li XH Wang D Zhou F Yang HJ Roy Bhaskar
Hu JB Sun CG Miao YG*

Abstract: Cellulases catalyze the hydrolysis of cellulose which are mainly three types: endoglucanases, cellobiohydrolases and β-glucosidases. It can be used in converting cellulosic biomass to glucose that can be used in different applications such as production of fuel ethanol, animal feed, waste water treatment and in brewing industry. In this paper, we cloned a 1 380 bp endoglucanase I (*EG I*) gene from mycelium of filamentous fungus *Trichoderma viride* strain AS 3.3711 using PCR-based exon splicing methods, and expressed the recombinant EG I mature peptide protein in both silkworm *BmN* cell line and silkworm larvae with a newly established Bac-to-Bac/*Bm*NPV mutant baculovirus expression system, which lacks the virus-encoded chitinase (*chi*A) and cathepsin (*v-cath*) genes of *Bombyx mori* nucleopolyhedrovirus(*Bm*NPV). An around 49 kDa protein was visualized after mBacmid/*Bm*NPV/EG I infection, and the maximum expression in silkworm larvae was at 84 h post-infection. The ANOVA showed that the enzymes from recombinant baculoviruses infected silkworms exhibited significant maximum enzyme activity at the environmental condition of pH 7.0 and temperature 50 ℃. It was stable at pH range from 5.0 to 10.0 and at temperature range from 50 to 60 ℃, and increased 24.71 and 22.84% compared with that from wild baculoviruses infected silkworms and normal silkworms, respectively. The availability of large quantities of EG I that the silkworm provides maybe greatly facilitate the future research and the potential application in industries.

Published On: Molecular Biology Reports, 2010, 37(8), 3721-3728.

Key Laboratory of Animal Epidemic Etiology & Immunological Prevention of Ministry of Agriculture, College of Animal Sciences, Zhejiang University, Hangzhou 310029, China.
*Corresponding author E-mail: miaoyg@zju.edu.cn.

克隆和在家蚕中利用改进的Bac-to-Bac/*Bm*NPV杆状病毒表达系统表达纤维素酶基因

李兴华　王　丹　周　芳　杨华军　Roy Bhaskar
胡嘉彪　孙春光　缪云根*

摘要： 纤维素酶催化纤维素的水解，其主要包括3种类型：内切葡聚糖酶、纤维二糖水解酶和β-葡糖苷酶。它可以将纤维素生物质转化为葡萄糖，其可用于不同的应用中，例如生产燃料乙醇、动物饲料、废水处理和酿造工业。在本文中，我们使用基于PCR的外显子剪接方法从丝状真菌绿色木霉菌株AS 3.3711的菌丝体中克隆了1 380 bp的内切葡聚糖酶I（*EG I*）基因，并在家蚕*BmN*细胞系中表达了重组EG I成熟肽蛋白。此外，我们还利用新建立的Bac-to-Bac/*Bm*NPV突变体杆状病毒表达系统在家蚕幼虫中表达了该蛋白。该突变体系统缺乏家蚕核型多角体病毒（*Bm*NPV）编码的几丁质酶（*chi*A）和组织蛋白酶（*v-cath*）基因。经mBacmid/*Bm*NPV/EG I感染后，观察到大约49 kDa的蛋白质，在感染后84 h，该蛋白表达量达到最高。方差分析表明，来自于重组杆状病毒感染家蚕的酶在pH 7.0和50 ℃下表现出了显著的最大酶活性。该酶在pH 5.0—10.0和50—60 ℃的范围内稳定，分别比在野生杆状病毒感染的蚕和正常蚕中表达量增加24.71%和22.84%。家蚕表达的大量EG I将可能极大地促进EG I在未来的研究和各行业中的潜在应用。

农业部动物疫病病原学与免疫控治重点实验室，动物科学学院，浙江大学，杭州

Expression of UreB and HspA of *Helicobacter pylori* in silkworm pupae and identification of its immunogenicity

Zhang XL Shen WD Lu Y Zheng XJ

Xue RY Cao GL Pan ZH Gong CL*

Abstract: For mass production of urease B subunit (UreB) and heat shock protein A subunit (HspA) of *Helicobacter pylori* with *Bombyx mori* nuclear polyhedrosis virus (*Bm*NPV) baculovirus expression system (BES) and to determine whether they could be used as an oral vaccine against *H. pylori*, besides, to determine the time course of expressed recombinant protein and the optimum acquisition time directly through green fluorescence, HspA and enhanced green fluorescence protein (*EGFP*) genes were cloned into vector pFastBacDual to form donor vector pFastBacDual - (EGFP) (HspA), *UreB* gene was cloned into vector pFastBacDual to form donor vector pFastBacDual-*UreB,* then they were transformed into *E. coli Bm*DH10Bac to obtain the recombinant Bacmid - (EGFP) (HspA) and Bacmid-UreB respectively. They were used to transfect *BmN* cells and generated the recombinant baculovirus *Bm*NPV-(EGFP) (HspA) and *Bm*NPV-UreB. Using these recombinant baculovirus *Bm*NPV-(EGFP) (HspA) and *Bm*NPV-UreB inoculated the silkworm pupae, a recombinant HspA and UreB protein were expressed in silkworm pupae, which were around 13 and 62 kDa in sodium dodecyl sulfate-polyacrylamide gel electrophoresis (SDS-PAGE) and western blot analysis. After oral immunization of mice, serum specific IgG antibodies against HspA and UreB in vaccine group were much higher than that in mock and native silkworm powder control groups. The results indicated that the expressed recombinant HspA and UreB in silkworm pupae would possess good immunogenicity. In addition, when EGFP and HspA proteins were expressed, a direct correlation between the increase in intensity of fluorescence and HspA concentration.

Published On: Molecular Biology Reports, 2011, 38(5), 3173-3180.

School of Pre-clinical Medicine and Biological Science, Medical College, Soochow University, Suzhou 215123, China.

*Corresponding author E-mail: gongcl@suda.edu.cn.

幽门螺杆菌尿素酶B亚基(UreB)和热休克蛋白A亚基(HspA)在蚕蛹中的表达以及其免疫原性鉴定

张孝林　沈卫德　陆　叶　郑小坚
薛仁宇　曹广力　潘中华　贡成良*

摘要：本研究利用家蚕核型多角体杆状病毒（*Bm*NPV）表达系统（BES）大规模生产幽门螺杆菌尿素酶B亚基（UreB）和热击蛋白A亚基（HspA），并同时评估了它们作为幽门螺杆菌口服疫苗的可行性。为直接通过绿色荧光强度确定重组蛋白在蚕蛹中表达的时期和蚕蛹的最佳收获时间，我们将*HspA*基因和增强型绿色荧光蛋白（EGFP）基因克隆至pFastBacDual载体，构建了pFastBacDual-（EGFP）（HspA）供体载体，*ureB*基因克隆至pFastBacDual载体构建了pFastBacDual-*UreB*，然后将上述载体分别转化*E. coli Bm*DH10Bac细胞，获得了重组的Bacmid-（EGFP）（HspA）和Bacmid-UreB，将它们转染*BmN*细胞获得重组杆状病毒*Bm*NPV-（EGFP）（HspA）和*Bm*NPV-UreB。SDS-PAGE和Western blot分析结果显示，在感染重组杆状病毒的蚕蛹中可检测到13 kDa和62 kDa的特异性条带，表明幽门螺杆菌的HspA和UreB在蚕蛹中成功表达。用感染重组杆状病毒的蚕蛹冻干粉口服免疫小鼠，免疫组小鼠血清中HspA与UreB特异性抗体IgG的滴度明显高于模拟和天然蚕蛹粉对照组小鼠，表明蚕蛹中表达的UreB和HspA具有良好的免疫原性。此外，我们发现接种重组病毒的蚕蛹的荧光强度与重组HspA的表达水平有直接的相关性。

基础医学与生物科学学院，医学部，苏州大学，苏州

In vivo bioassay of recombinant human growth hormone synthesized in *Bombyx mori* pupae

Lan HL[1] Nie ZM[2] Liu Y[3] Lü ZB[2] Liu YS[2]
Quan YP[2] Chen JQ[2] Zhen QL[2] Chen Q[2] Wang D[2]
Sheng Q[2] Yu W[2] Chen J[2] Wu XF[2] Zhang YZ[2*]

Abstract: The human growth hormone (hGH) has been expressed in prokaryotic expression system with low bioactivity previously. Then the effective *B. mori* baculovirus system was employed to express hGH identical to mature hGH successfully in larvae, but the expression level was still limited. In this work, the hGH was expressed in *B. mori* pupae by baculovirus system. Quantification of recombinant hGH protein (*Bm*rhGH) showed that the expression of *Bm*rhGH reached the level of approximately 890 μg/mL pupae supernatant solution, which was five times more than the level using larvae. Furthermore, Animals were gavaged with *Bm*rhGH at the dose of 4.5 mg/(rat·day), and the body weight gain (BWG) of treated group had a significant difference ($P < 0.01$) compared with the control group. The other two parameters of liver weight and epiphyseal width were also found to be different between the two groups ($P < 0.05$). The results suggested that *Bm*rhGH might be used as a protein drug by oral administration.

Published On: Journal of Biomedicine and Biotechnology, 2010.

1 College of Life Sciences, Zhejiang University, Hangzhou 310058, China.
2 Institute of Biochemistry, Zhejiang Sci-Tech University, Hangzhou 310018, China.
3 School of Applied Technology, Zhejiang Economic & Trade Polytechnic, Hangzhou 310018, China.
*Corresponding author E-mail: yaozhou@chinagene.com.

家蚕蛹生产的重组人生长激素体内生物活性测定

蓝航莲[1] 聂作明[2] 刘悦[3] 吕正兵[2] 刘颖硕[2]
全滟平[2] 陈剑清[2] 郑青亮[2] 陈琴[2] 王丹[2]
盛清[2] 于威[2] 陈健[2] 吴祥甫[2] 张耀洲[2*]

摘要：在原核表达系统中表达的人生长激素(hGH)具有较低的生物活性。随后利用高效的杆状病毒系统在家蚕幼虫中成功表达与成熟hGH相同的hGH，但其表达水平仍然有限。本研究通过杆状病毒系统在家蚕蛹中表达hGH。重组hGH蛋白(*Bm*rhGH)的定量实验显示，*Bm*rhGH在蛹上清液的表达量达到约890 μg/mL，是使用幼虫的表达水平的5倍。此外，用4.5 mgH/(头·d)的剂量饲养大鼠，与对照组相比，治疗组的体重增加(BWG)显著($P<0.01$)。两组之间肝脏重量和骨骺宽度两个参数也有不同($P<0.05$)。结果表明，*Bm*rhGH可作为通过口服给药的蛋白质药物。

1 生命科学学院，浙江大学，杭州
2 生物化学研究所，浙江理工大学，杭州
3 应用工程系，浙江经贸职业技术学院，杭州

Immobilization of foreign protein into polyhedra of *Bombyx mori* nucleopolyhedrovirus (*Bm*NPV)

Xiang XW Yang R Chen L Hu XL

Yu SF Cao CP Wu XF*

Abstract: In the late phase of *Bombyx mori* nucleopolyhedrovirus (*Bm*NPV) infection, a large amount of polyhedra appear in the infected cell nucleolus, these polyhedra being dense protein crystals protecting the incorporated virions from the harsh environment. To investigate whether the foreign protein could be immobilized into the polyhedra of *Bm*NPV, two recombinant baculoviruses were generated by a novel *Bm*NPV polyhedrin-plus (polh$^+$) Bac-to-Bac system, designated as v*Bm*Bac(polh$^+$)-enhanced green fluorescent protein (EGFP) and v*Bm*Bac(polh$^+$) - *Lac*Z, which can express the polyhedrin and foreign protein simultaneously. Light microscopy analysis showed that all viruses produced polyhedra of normal appearance. Green fluorescence can be apparently detected on the surface of the v*Bm*Bac(polh$^+$)-EGFP polyhedra, but not the *Bm*NPV polyhedra. Fluorescence analysis and anti-desiccation testing confirmed that EGFP was embedded in the polyhedra. As expected, the v*Bm*Bac(polh$^+$)-*Lac*Z polyhedra contained an amount of LacZ and had a higher β-galactosidase activity. Sodium Dodecyl Sulfate Polyacrylamide Gel Electrophoresis (SDS-PAGE) and Western blot were also performed to verify if the foreign proteins were immobilized into polyhedra. This study provides a new inspiration for efficient preservation of useful proteins and development of new pesticides with toxic proteins.

Published On: Journal of Zhejiang University Science B, 2012, 13(2), 111-117.

College of Animal Sciences, Zhejiang University, Hangzhou 310058, China.

*Corresponding author E-mail: wuxiaofeng@zju.edu.cn.

利用家蚕核型多角体病毒(*Bm*NPV)固定外源蛋白的研究

相兴伟　杨　锐　陈　琳　胡小龙
于少芳　曹翠萍　吴小锋*

摘要： 家蚕核型多角体病毒(*Bm*NPV)感染晚期，在感染的细胞核中产生大量多角体蛋白，并形成致密的蛋白质晶格，保护包埋的病毒颗粒免受外界恶劣环境的影响。为了研究外源蛋白是否可以被固定入 *Bm*NPV 多角体内，利用新型 *Bm*NPV Polh$^+$ Bac-to-Bac 系统构建了 v*Bm*Bac(polh$^+$)-EGFP 和 v*Bm*Bac(polh$^+$)-*LacZ* 两种重组杆状病毒，可同时表达多角体蛋白和外源蛋白。光学显微镜分析显示，所有病毒均可产生正常形态的多角体。在 v*Bm*Bac(polh$^+$)-EGFP 的多角体表面可以检测到明显的绿色荧光，而 *Bm*NPV 与之相反。荧光分析和抗干燥实验证实，EGFP 嵌入多角体中。与预期相符，v*Bm*Bac(polh+)-*LacZ* 多角体含有一定量的 LacZ 并具有更高的β-半乳糖苷酶活性。同时进行 SDS-PAGE 和 Western blot 实验以验证外源蛋白是否被固定在多角体中。这项研究为有效保存有用蛋白质和开发含有毒素蛋白的新型生物杀虫剂提供了新的启示。

动物科学学院，浙江大学，杭州

2A self-cleaving peptide-based multi-gene expression system in the silkworm *Bombyx mori*

Wang YC[1,2#] Wang F[1,2#] Wang RY[1]

Zhao P[1,2] Xia QY[1,2*]

Abstract: Fundamental and applied studies of silkworms have entered the functional genomics era. Here, we report a multi-gene expression system (MGES) based on 2A self-cleaving peptide (2A), which regulates the simultaneous expression and cleavage of multiple gene targets in the silk gland of transgenic silkworms. First, a glycine-serine-glycine spacer (GSG) was found to significantly improve the cleavage efficiency of 2A. Then, the cleavage efficiency of six types of 2As with GSG was analyzed. The shortest porcine teschovirus-1 2A (P2A-GSG) exhibited the highest cleavage efficiency in all insect cell lines that we tested. Next, P2A-GSG successfully cleaved the artificial human serum albumin (66 kDa) linked with human acidic fibroblast growth factor (20.2 kDa) fusion genes and vitellogenin receptor fragment (196 kDa) of silkworm linked with *EGFP* fusion genes, importantly, vitellogenin receptor protein was secreted to the outside of cells. Furthermore, P2A-GSG successfully mediated the simultaneous expression and cleavage of a *DsRed* and *EGFP* fusion gene in silk glands and caused secretion into the cocoon of transgenic silkworms using our sericin1 expression system. We predicted that the MGES would be an efficient tool for gene function research and innovative research on various functional silk materials in medicine, cosmetics, and other biomedical areas.

Published On: Scientific Reports, 2015, 5.

1 State Key Laboratory of Silkworm Genome Biology, Southwest University, Chongqing 400716, China.

2 College of Biology and Technology, Southwest University, Chongqing 400716, China.

[#]These authors contributed equally.

[*]Corresponding author E-mail: xiaqy@swu.edu.cn.

基于2A自剪切肽的家蚕多基因表达系统

王元成[1,2#]　王　峰[1,2#]　王日远[1]
赵　萍[1,2]　夏庆友[1,2*]

摘要：家蚕基础研究和应用研究已进入功能基因组学时代。因此，我们以2A自剪切肽（2A）为基础在转基因家蚕丝腺中建立了多基因表达系统，该系统可以同时调节多个基因靶标的表达和切割。首先，我们发现甘氨酸-丝氨酸-甘氨酸间隔（GSG）显著提高了2A的切割效率。其次，我们分析了带GSG的6种不同类型2As的切割效率。在我们检测的所有昆虫细胞系中，最短的猪捷申病毒-1 2A（P2A-GSG）的切割效率最高。再次，P2A-GSG成功切割人造人血清白蛋白（66 kDa）与人酸性成纤维细胞生长因子（20.2 kDa）、家蚕卵黄原蛋白受体片段（196 kDa）与*EGFP*形成的融合基因。更重要的是，卵黄原蛋白受体蛋白被分泌到细胞外。此外，P2A-GSG成功介导了*DsRed*和*EGF*P融合基因在丝腺中的表达和切割，并通过我们的丝胶1表达系统分泌到转基因家蚕茧中。我们预测MGES将会是一种高效的基因功能研究手段，同时也可应用于医学、化妆品和其他生物医学领域中各种功能丝材料的创新性研究。

1　家蚕基因组生物学国家重点实验室，西南大学，重庆
2　生物技术学院，西南大学，重庆

The advances and perspectives of recombinant protein production in the silk gland of silkworm *Bombyx mori*

Xu HF*

Abstract: The silk gland of silkworm *Bombyx mori*, is one of the most important organs that has been fully studied and utilized so far. It contributes finest silk fibers to humankind. The silk gland has excellent ability of synthesizing silk proteins and is a kind tool to produce some useful recombinant proteins, which can be widely used in the biological, biotechnical and pharmaceutical application fields. It's a very active area to express recombinant proteins using the silk gland as a bioreactor, and great progress has been achieved recently. This review recapitulates the progress of producing recombinant proteins and silk-based biomaterials in the silk gland of silkworm in addition to the construction of expression systems. Current challenges and future trends in the production of valuable recombinant proteins using transgenic silkworms are also discussed.

Published On: Transgenic Research, 2014, 23(5), 697-706.

State Key Laboratory of Silkworm Genome Biology, Southwest University, Chongqing 400716, China
*Corresponding author E-mail: xuhf@swu.edu.cn.

利用家蚕丝腺生产重组蛋白的研究进展及展望

徐汉福*

摘要：家蚕丝腺是迄今被充分研究和利用的一个重要器官，它为人类贡献了最好的丝纤维。丝腺具有出色的高效合成丝蛋白的能力，是生产有用重组蛋白的理想工具，利用其生产的重组蛋白可广泛应用于生物学、生物技术以及制药应用等领域。利用家蚕丝腺作为生物反应器生产重组蛋白是一个非常热门的研究领域，并在近年取得了很大的进步。本文概述了丝腺表达系统以及利用丝腺生产重组蛋白和蚕丝蛋白生物材料的研究进展，并对当前研发中存在的一些挑战以及利用转基因蚕生产高价值重组蛋白的发展趋势进行了讨论。

家蚕基因组生物学国家重点实验室，西南大学，重庆

Expression of the hGM-CSF in the silk glands of germline of gene-targeted silkworm

Li YM[1#] Cao GL[1,2#] Chen HM[1]
Jia HF[1] Xue RY[1,2] Gong CL[1,2*]

Abstract: To express human granulocyte-macrophage colony-stimulating factor (*hGM-CSF*) gene in the silk glands of transformation silkworm (*Bombyx mori*) based on gene-targeting, two fragments from fibroin heavy chain gene (*fib*-H) of silkworm were cloned and sequenced. One fragment contains the 1st exon and its downstream 1st intron's partial sequence; and the other fragment contains the 1st intron's partial sequence and the 2nd exon's partial sequence. Then the two fragments, as homologous arm, were inserted into *p*SK to generate a gene-targeted vector, pSK-HL-A3GFP-FLP-GM-CSF-FLPA-HR in which a *gfp* gene driven by *A3* promoter and an *hGM–CSF* gene under the control of fibroin light chain (*fib*-L) promoter were included. The vector was transferred into the silkworm eggs using sperm-mediated gene transfer. After being screened for green fluorescent, the transformation silkworm was obtained, whose genome was verified by PCR and dot hybridization to confirm whether the target genes had been integrated into the silkworm genome. Furthermore, in the posterior silk glands of the G_4 generation transformation silkworms, a specific band with the molecular weight of 22 kDa could be detected by Western blot with an antibody against hGM-CSF, and the expression level of the hGM-CSF estimated by ELISA was approximately 1.26 ng per gram fresh posterior silk gland.

Published On: Biochemical and Biophysical Research Communications, 2010, 391(3), 1427-1431.

1 Pre-clinical Medical and Biological Science College, Soochow University, Suzhou 215123, China.
2 National Engineering Laboratory for Modern Silk, Soochow University, Suzhou 215123, China.
#These authors contributed equally.
*Corresponding author E-mail: gongcl@suda.edu.cn.

家蚕丝腺靶向基因hGM-CSF的表达

李艳梅[1#]　曹广力[1,2#]　陈慧梅[1]
贾海芳[1]　薛仁宇[1,2]　贡成良[1,2*]

摘要： 为了通过基因靶向实现人粒细胞—巨噬细胞集落刺激因子（hGM-CSF）基因在家蚕丝腺组织中的特异表达，我们克隆并测定了来自家蚕丝素重链基因（*fib*-H）的两个片段，其中一个片段包括第一外显子及其下游第一内含子的部分序列，另一片段包括第一内含子的部分序列和第二外显子部分序列，以这两个片段作为基因靶向的同源臂插入*p*SK载体中，并将*A3*启动子控制的*gfp*报告基因以及由丝素轻链（*fib*-L）启动子控制的*hGM-CSF*基因插入二同源臂之间构建基因靶向载体pSK-HL-A3GFP-FLP-GM-CSF-FLPA-HR，利用精子介导的基因转移将载体导入家蚕卵中，通过绿色荧光筛选鉴定转基因家蚕，通过PCR和点杂交验证其基因组以确认靶基因是否已整合到家蚕基因组中。用hGM-CSF抗体进行Western blot检测，发现G_4代转基因家蚕的后部丝腺中可以检测到分子质量为22 kDa的特异性条带。ELISA检测结果显示，每克新鲜的丝腺组织中hGM-CSF的表达量为1.26 ng。

1　基础医学与生物科学学院，苏州大学，苏州
2　现代丝绸国家工程实验室，苏州大学，苏州

Lowering the blood glucose of diabetes mellitus mice by oral administration with transgenic human insulin-like growth factor I silkworms

Xue RY[1,2] Wang Y[1] Cao GL[1,2] Pan ZH[1,2]

Zheng XJ[1] Zhou WL[3] Gong CL[1,2*]

Abstract: To evaluate the biological activity of the posterior silk glands of transgenic silkworms expressing human insulin-like growth factor I (hIGF-I), we bred hIGF-I-transgenic silkworms through eight generations by continuously selecting with green fluorescence and G418. The G_8 transgenic silkworms were confirmed by polymerase chain reaction and dot blotting, and their posterior silk glands were removed from the fifth instar larvae to make freeze-dried powders. Enzyme-Linked Immunosorbent Assay results showed that the expression level of hIGF-I in the posterior silk glands of G_8 transgenic silkworm is approximately 493 ng/g of freeze-dried powder. When the freeze-dried powder was administrated by gavage to diabetes mellitus (DM) mice, the blood glucose in DM mice significantly decreased ($P < 0.05$) in a time- and dose-dependent manner compared with that of DM mice orally administrated with distilled water and normal freeze-dried powders made of untreated silk glands. These results demonstrated that hIGF-I expressed in posterior silk glands of transgenic silkworms could reduce blood glucose by oral administration.

Published On: Journal of Agricultural and Food Chemistry, 2012, 60(26), 6559-6564.

1 School of Biology and Basic Medical Sciences, Soochow University, Suzhou 215123, China.

2 National Engineering Laboratory for Modern Silk, Soochow University, Suzhou 215123, China.

3 Sericulture Research Institute, Zhejiang Academy of Agricultural Sciences, Hangzhou 310021, China.

*Corresponding author E-mail: gongcl@suda.edu.cn.

口服转基因家蚕表达的人胰岛素生长因子I降低糖尿病小鼠的血糖

薛仁宇[1,2] 王 洋[1] 曹广力[1,2] 潘中华[1,2]
郑小坚[1] 周文林[3] 贡成良[1,2*]

摘要： 为检测表达人胰岛素生长因子I(hIGF-I)的转基因家蚕后部丝腺的生物活性，我们通过连续选择绿色荧光和G418培育了8代hIGF-I转基因蚕。利用聚合酶链式反应及斑点杂交实验证实第8代(G_8)蚕为转hIGF-I基因家蚕后，取该五龄蚕的后部丝腺制备成冻干粉。酶联免疫吸附实验结果显示，第8代转基因家蚕后部丝腺中hIGF-I的表达量为每克冻干粉中约含493 ng。通过给糖尿病小鼠灌胃冻干粉的方法证明，相较于口服双蒸水和非转基因家蚕的正常丝腺冻干粉的糖尿病小鼠，服用转基因丝腺冻干粉的糖尿病小鼠的血糖明显下调($P<0.05$)，同时呈现剂量依赖性。这些结果表明，在转基因家蚕的丝腺中表达的hIGF-I可以通过口服给药的方式来降低血糖。

1 基础医学与生物科学学院，苏州大学，苏州
2 现代丝绸国家工程实验室，苏州大学，苏州
3 蚕桑研究所，浙江省农业科学院，杭州

Expression of human granulocyte-macrophage colony-stimulating factor in stably-transformed *BmN* and *Sf-9* cells and silkworms by a non-transposon vector

Zhang HK[1] Cao GL[1,2] Li YM Xue RY[1,2] Gong CL[1,2*]

Abstract: This study aimed to explore the possibility of non-transposon vector mediated foreign gene expression in cultured insect cells and transgenic silk worms. To this end, the human Granulocyte-Macrophage Colony-Stimulating Factor (hGM−CSF) gene was inserted into the insect cell expression vector pIZT-V5-His to generate the recombinant vector pIZT-hGM-CSF. After transfection of *BmN* and Sf-9 cells with the *p*IZT-hGM-CSF vector, stably-transformed cells expressing the *hGM−CSF* gene were selected using the antibiotic zeocin at a final concentration of 300-400 μg/mL. Expression of a 22 kDa protein band representing hGM-CSF was detected in the transformed cells by Sodium Dodecyl Sulfate-Polyacrylamide Gel Electrophoresis and Western blot. The expression levels of hGM-CSF in *BmN* and *Sf-9* cells were determined by Enzyme-Linked Immunosorbent Assay (ELISA) to be about 0.7 ng and 0.3 ng/10^6 cells, respectively. The transgenic vector pIZT-hGM-CSF was transferred into silkworm eggs using sperm-mediated gene transfer. Transgenic silkworms were obtained after screening for the *gfp* gene and were verified by polymerase chain reaction, dot hybridization and Western blot. The expression level of hGM-CSF determined by ELISA was about 4.7 ng /g of freeze-dried silk glands in the G_5 generation. These results suggest that heterologous genes can be integrated into cultured *BmN* and *Sf-9* cells and into the silkworm genome using a non-transposon vector and can be expressed successfully.

Published On: Journal of Animal and Veterinary Advances, 2012, 11(16), 2890-2897.

1 School of Biology and Basic Medical Sciences ,Soochow University, Suzhou 215123, China.

2 National Engineering Laboratory for Modern Silk, Soochow University, Suzhou 215123, China.

*Corresponding author E-mail: gongcl@suda.edu.cn.

通过非转座子载体在稳定转化的*BmN*、*Sf9*细胞及家蚕中表达人粒细胞-巨噬细胞集落刺激因子

张昊堃[1]　曹广力[1,2]　李艳梅[1]　薛仁宇[1,2]　贡成良[1,2*]

摘要：本研究旨在探讨非转座子载体介导的外源基因在昆虫细胞及转基因家蚕中表达的可能性。为此，将人粒细胞-巨噬细胞集落刺激因子（hGM-CFS）基因插入到昆虫表达载体pIZT-V5-His上，获得重组质粒pIZT-hGM-CSF。将该重组质粒转染*BmN*和*Sf-9*细胞，使用终浓度为300-400 μg/mL的抗生素zeocin选择表达*hGM-CSF*基因的稳定转化的细胞。通过SDS-PAGE和Westem blot方法在转化的细胞中检测到代表hGM-CSF的22 kDa蛋白质条带的表达。通过ELISA测定每10^6个*BmN*和*Sf-9*细胞中*hGM-CSF*的表达量分别为0.7 ng和0.3 ng。使用精子介导的基因转移法将转基因载体pIZT-hGM-CSF转移到家蚕卵中。筛选*gfp*基因后获得转基因蚕，并通过聚合酶链式反应、点杂交和蛋白质印迹验证。通过ELISA测定在G_5代中冻干丝腺的hGM-CSF表达量约为4.7 ng/g。这些结果表明外源基因可以使用非转座子载体整合到培养的*BmN*、*Sf-9*细胞和家蚕基因组中并且可以成功表达。

1　基础医学与生物科学学院，苏州大学，苏州

2　现代丝绸国家工程实验室，苏州大学，苏州

An optimized sericin-1 expression system for mass-producing recombinant proteins in the middle silk glands of transgenic silkworms

Wang F[#] Xu HF[#] Yuan L Ma SY Wang YC

Duan XL Duan JP Xiang ZH Xia QY[*]

Abstract: The middle silk gland (MSG) of silkworm is thought to be a potential host for mass-producing valuable recombinant proteins. Transgenic MSG expression systems based on the usage of promoter of *sericin1* gene (sericin-1 expression system) have been established to produce various recombinant proteins in MSG. However, further modifying the activity of the sericin-1 expression system to yield higher amounts of recombinant proteins is still necessary. In this study, we provide an alternative modification strategy to construct an efficient sericin-1 expression system by using the hr3 enhancer (*hr3 CQ*) from a Chongqing strain of the *Bombyx mori* nuclear polyhedrosis virus(*Bm*NPV) and the 3' UTRs of the fibroin heavy chain(*Fib-HPA*), the fibroin light chain (*Fib-LPA*), and *Sericin1* (*Ser1PA*) genes. We first analyzed the effects of these DNA elements on expression of luciferase, and found that the combination of *hr3 CQ* and *Ser1PA* was most effective to increase the activity of luciferase. Then, *hr3 CQ* and *Ser1PA* were used to modify the sericin1 expression system. Transgenic silkworms bearing these modified sericin1 expression vectors were generated by a *piggy*Bac transposon mediated genetic transformation method. Our results showed that mRNA level of *DsRed* reporter gene in transgenic silkworms containing *hr3 CQ* and *Ser1PA* significantly increased by 9 fold to approximately 83% of that of endogenous sericin1. As the results of that, the production of recombinant RFP increased by 16 fold to 9.5% (w/w) of cocoon shell weight. We conclude that this modified sericin-1 expression system is efficient and will contribute to the MSG as host to mass produce valuable recombinant proteins.

Published On: Transgenic Research, 2013, 22(5), 925-938.

State Key Laboratory of Silkworm Genome Biology, Southwest University, Chongqing 400716, China.

[#]These authors contributed equally.

[*]Corresponding author E-mail: xiaqy@swu.edu.cn.

转基因家蚕中部丝腺中丝胶1表达系统高量表达重组蛋白的优化

王　峰[#]　徐汉福[#]　袁　林　马三垣
王元成　段小利　段建平　向仲怀　夏庆友[*]

摘要：家蚕中部丝腺(MSG)是大量合成高价值重组蛋白的潜在生物反应器。我们已经建立了基于丝胶蛋白1基因启动子的转基因中部丝腺(MSG)表达系统(丝胶蛋白-1表达系统),并在MSG中表达了多种重组蛋白。但是,非常有必要提高中部丝腺表达系统的活性从而产生更高表达量的重组蛋白。在本研究中,我们提供了另一种修饰策略,通过使用来自家蚕核型多角体病毒(*Bm*NPV)重庆品系的hr3增强子(*hr3 CQ*)和丝素重链(*Fib-HPA*)、丝素轻链(*Fib-LPA*)和丝胶蛋白1(*Ser1PA*)基因的3′UTRs来构建高效的丝胶1表达系统。我们首先分析了这些DNA元件对荧光素酶表达的影响,发现*hr3 CQ*和*Ser1PA*的组合对增加荧光素酶的活性最有效。然后,我们利用*hr3 CQ*和*Ser1PA*修饰丝胶1表达系统。我们通过*piggy*Bac转座子介导的遗传转化方法构建了携带优化丝胶1表达系统的转基因家蚕。结果显示,含有*hr3 CQ*和*Ser1PA*的转基因蚕中*DsRed*报告基因的mRNA水平显著提高了9倍,为内源性丝胶蛋白1表达量的83%左右,重组RFP的产量提高了16倍,为茧壳重量的9.5%(w/w)。我们认为,这种优化的丝胶1表达系统非常高效,将有助于中部丝腺作为大量生产高价值的重组蛋白的生物反应器。

家蚕基因组生物学国家重点实验室,西南大学,重庆

Immobilization of foreign protein in *Bm*NPV polyhedra by fusion expression with partial polyhedrin fragments

Chen L　Xiang XW　Yang R　Hu XL
Cao CP　Firdose Ahmad Malik　Wu XF*

Abstract: *Bombyx mori* nucleopolyhedrovirus (*Bm*NPV) produces large, proteinaceous crystal matrix named polyhedra, which occlude progeny virions which are produced during infection and protect virions from hostile environmental conditions. In this study, five overlapping N-terminal fragments of the *Bm*NPV polyhedrin ORF were cloned and ligated with the foreign gene *egfp*, and five recombinant baculoviruses were constructed by *Bm*NPV (Polh$^+$) Bac-to-Bac baculovirus expression system was used to co-express the polyhedrin and fused protein. The results showed that the fusion proteins were highly expressed, and the foreign proteins fused with the 100 aa fragment of polyhedrin could be embedded into polyhedra at a higher ratio. This study provides a new method for efficient preservation of useful proteins for the development of new biopesticide with toxin protein and delivery vector system of vaccines.

Published On: Journal of Virological Methods, 2013, 194(1-2), 185-189.

Lab of Silkworm Biotechnology, College of Animal Sciences, Zhejiang University, Hangzhou 310058, China.
*Corresponding author E-mail: wuxiaofeng@zju.edu.cn.

通过与部分多角体片段的融合表达将外源蛋白固定在*Bm*NPV多角体中

陈　琳　相兴伟　杨　锐　胡小龙
曹翠萍　Firdose Ahmad Malik　吴小锋*

摘要： 家蚕核型多角体病毒(*Bm*NPV)在感染期间产生大分子蛋白晶体——多角体，包被子代病毒粒子保护其免受恶劣环境的影响。在本研究中，我们克隆了*Bm*NPV多角体蛋白ORF的5个重叠N末端片段，并将其与外源基因*egfp*连接，利用*Bm*NPV（$Polh^{+}$）Bac-to-Bac杆状病毒表达系统构建了5个重组杆状病毒，使多角体蛋白与融合蛋白共表达。结果表明，融合蛋白被大量表达，融合了多角体蛋白100个氨基酸片段的外源蛋白可以以更高的比例嵌入多角体中。本研究为有效保存有用蛋白质提供了一种新方法，为开发新型生物农药与毒素蛋白及疫苗的递送载体系统提供了保障。

家蚕生物技术实验室，动物科学学院，浙江大学，杭州

Advanced silk material spun by a transgenic silkworm promotes cell proliferation for biomedical application

Wang F[1,2] Xu HF[1,2] Wang YC[1,2] Wang RY[1,2] Yuan L[1,2] Ding H[1,2]
Song CN[1,2] Ma SY[1,2] Peng ZX[1,2] Peng ZC[1,2] Zhao P[1,2] Xia QY[1,2*]

Abstract: Natural silk fiber spun by the silkworm *Bombyx mori* is widely used not only for textile materials, but also for biofunctional materials. In the present study, we genetically engineered an advanced silk material, named hSFSV, using a transgenic silkworm, in which the recombinant human acidic fibroblast growth factor (hFGF1) protein was specifically synthesized in the middle silk gland and secreted into the sericin layer to surround the silk fiber using our previously optimized sericin 1 expression system. The content of the recombinant hFGF1 in the hSFSV silk was estimated to be approximate 0.07% of the cocoon shell weight. The mechanical properties of hSFSV raw silk fiber were enhanced slightly compared to those of the wild-type raw silk fiber, probably due to the presence of the recombinant of hFGF1 in the sericin layer. Remarkably, the hSFSV raw silk significantly stimulated the cell growth and proliferation of *NIH/3T3* mouse embryonic fibroblast cells, suggesting that the mitogenic activity of recombinant hFGF1 was well maintained and functioned in the sericin layer of hSFSV raw silk. These results show that the genetically engineered raw silk hSFSV could be used directly as a fine biomedical material for mass application. In addition, the strategy whereby functional recombinant proteins are expressed in the sericin layer of silk might be used to create more genetically engineered silks with various biofunctions and applications.

Published On: Acta Biomaterialia, 2014, 10(12), 4947-4955.

1 State Key Laboratory of Silkworm Genome Biology, Southwest University, Chongqing 400716, China.
2 College of Biotechnology, Southwest University, Chongqing 400716, China.
*Corresponding author E-mail: xiaqy@swu.edu.cn.

通过转基因家蚕获得的先进蚕丝材料可以促进细胞增殖从而应用于生物医学领域

王　峰[1,2]　徐汉福[1,2]　王元成[1,2]　王日远[1,2]　袁　林[1,2]　丁　欢[1,2]
宋春暖[1,2]　马三垣[1,2]　彭芷昕[1,2]　彭章川[1,2]　赵　萍[1,2]　夏庆友[1,2*]

摘要：由家蚕分泌的天然丝纤维不仅广泛应用于纺织材料，而且还应用于生物功能材料。在本研究中，我们使用先前优化的家蚕 sericin 1 表达系统在家蚕中部丝腺中特异性合成并分泌人酸性成纤维细胞生长因子（hFGF1）到丝胶层中，形成了一种名为 hSFSV 的先进材料。hSFSV 中重组 hFGF1 的含量估计约为茧壳重量的 0.07%。与野生型生丝相比，hSFSV 生丝纤维的机械性能略有增加，这可能是由于丝胶层中存在重组的 hFGF1。值得注意的是，hSFSV 生丝显著刺激了 *NIH/3T3* 小鼠胚胎成纤维细胞的生长和增殖，暗示在 hSFSV 蚕丝中，丝胶层中表达的重组 hFGF1 保持着促进有丝分裂的活性。结果表明，遗传改良的生丝 hSFSV 作为优秀的生物材料可以直接大规模应用于精细生物医学中。此外，功能性重组蛋白在丝胶层中表达的策略可能被用于创建更多的包含各种生物功能和应用的基因工程蚕丝。

1　家蚕基因组生物学国家重点实验室，西南大学，重庆
2　生物技术学院，西南大学，重庆

Reducing blood glucose levels in TIDM mice with an orally administered extract of sericin from *hIGF-I*-transgenic silkworm cocoons

Song ZW[1#] Zhang MY[1#] Xue RY[1,2#] Cao GL[1,2] Gong CL[1,2*]

Abstract: In previous studies, we reported that the blood glucose levels of mice with type I diabetes mellitus (TIDM) was reduced with orally administered silk gland powder from silkworms transgenic for human insulin-like growth factor-I (*hIGF-I*). However, potential safety hazards could not be eliminated because the transgenic silk gland powder contained heterologous DNA, including the green fluorescent protein (*gfp*) and neomycin resistance (*neo*) genes. These shortcomings might be overcome if the recombinant hIGF-I were secreted into the sericin layer of the cocoon. In this study, silkworm eggs were transfected with a novel *piggy*Bac transposon vector, *pig*A3GFP-serHS-hIGF-I-neo, containing the *neo*, *gfp*, and *hIGF-I* genes controlled by the sericin 1 (*ser-1*) promoter with the signal peptide DNA sequence of the fibrin heavy chain (Fib-H) and a helper plasmid containing the *piggy*Bac transposase sequence under the control of the *Bombyx mori actin 3* (*A3*) promoter, using sperm-mediated gene transfer to generate the transformed silkworms. The hIGF-I content estimated by enzyme-linked immunosorbent assay was approximately 162.7 ng/g. To estimate the biological activity of the expressed hIGF-I, streptozotocin-induced TIDM mice were orally administered sericin from the transgenic silkworm. The blood glucose levels of the mice were significantly reduced, suggesting that the extract from the transgenic hIGF-I silkworm cocoons can be used as an orally administered drug.

Published On: Food and Chemical Toxicology, 2014, 67, 249-254.

1 School of Biology and Basic Medical Science, Soochow University, Suzhou 215123, China.
2 National Engineering Laboratory for Modern Silk, Soochow University, Suzhou 215123, China.
[#]These authors contributed equally.
[*]Corresponding author E-mail: gongcl@suda.edu.cn.

口服*hIGF-I*转基因家蚕蚕茧丝胶蛋白的提取物降低TIDM小鼠血糖水平

宋作伟[1#]　张梦窈[1#]　薛仁宇[1,2#]　曹广力[1,2]　贡成良[1,2*]

摘要： 在先前的研究中，我们发现I型糖尿病小鼠口服人体类胰岛素生长因子I(*hIGF-1*)转基因家蚕的丝腺粉后血糖水平降低。然而，潜在的安全隐患并没有消除，因为转基因蚕丝腺粉中含有外源DNA，包括绿色荧光蛋白(*gfp*)和新霉素抗性(*neo*)基因。如果重组hIGF-I被分泌到茧层丝胶中，这些缺点可能被克服。在本研究中，我们通过精子介导法向家蚕蚕卵导入新型*piggy*Bac转座子质粒*pig*A3GFP-serHS-hIGF-I-neo，该质粒包含*neo*、*gfp*以及家蚕丝胶(*ser-1*)启动子控制丝素重链Fib-H信号肽DNA序列控制的*hIGF-I*融合基因和一个在家蚕肌动蛋白3(*A3*)启动子控制下的*piggy*Bac转座酶序列的辅助质粒。酶联免疫吸附实验结果显示，hIGF-I的表达量约为162.7 ng/g。为评估表达的hIGF-1的生物活性，给链脲霉素诱导的TIDM小鼠口服转基因蚕的丝胶蛋白，小鼠的血糖水平显著降低，表明来自转基因hIGF–I家蚕茧的提取物可用作口服药物。

1　基础医学与生物科学学院，苏州大学，苏州
2　现代丝绸国家工程实验室，苏州大学，苏州

Nonvirus encoded proteins could be embedded into *Bombyx mori* cypovirus polyhedra

Zhang YL[1#] Xue RY[1,2#] Cao GL[1,2] Meng XK[1]
Zhu YX[1] Pan ZH[1,2] Gong CL[1,2*]

Abstract: To explore whether the nonvirus encoded protein could be embedded into *Bombyx mori* cypovirus (*Bm*CPV) polyhedra. The stable transformants of *BmN* cells expressing a polyhedrin (*Polh*) gene of *Bm*CPV were constructed by transfection with a non-transposon derived vector containing a *polh* gene. The polyhedra were purified from the midguts of *Bm*CPV-infected silkworms and the transformed *BmN* cells, respectively. The proteins embedded into polyhedra were determined by mass spectrometry analysis. Host derived proteins were detected in the purified polyhedra. Analysis of structure and hydrophilicity of embedded proteins indicated that the hydrophilic proteins, in structure, were similar to the left-handed structure of polyhedrin or the N-terminal domain of *Bm*CPV structural protein VP3, which were easily embedded into the *Bm*CPV polyhedra. The lysate of polyhedra purified from the infected transformation of *BmN* cells with modified *B. mori* baculovirus *Bm*PAK6 could infect *BmN* cells, indicating that *B. mori* baculovirus could be embedded into *Bm*CPV polyhedra. Both the purified polyhedra and its lysate could be coloured by X-gal, indicating that the β-galactosidase expressed by *Bm*PAK6 could be incorporated into *Bm*CPV polyhedra. These results suggested that some heterologous proteins and baculovirus could be embedded into polyhedra in an unknown manner.

Published On: Molecular Biology Reports, 2014, 41(4), 2657-2666.

1 School of Biology and Basic Medical Sciences, Soochow University, Suzhou 215123, China.
2 National Engineering Laboratory for Modern Silk, Soochow University, Suzhou 215123, China.
#These authors contributed equally.
*Corresponding author E-mail: gongcl@suda.edu.cn.

非病毒编码的蛋白质能被包被进家蚕质型多角体

张铁岭[1#]　薛仁宇[1,2#]　曹广力[1,2]　孟祥坤[1]
朱越雄[1]　潘中华[1,2]　贡成良[1,2*]

摘要： 为了解非病毒蛋白质是否能被家蚕质型多角体包被，我们将包含 *Bm*CPV 多角体蛋白 *polh* 基因的非转座子载体转染以获得稳定表达 *polh* 基因的 *BmN* 细胞。分别从 *Bm*CPV 感染的家蚕中肠和转化的 *BmN* 细胞中纯化该多角体。被包被进病毒多角体的蛋白质通过质谱分析鉴定，发现在纯化的多角体中有来源于宿主的蛋白质。对包被的蛋白质的结构和亲水性分析表明亲水性蛋白在结构上类似多角体蛋白的左手结构，或者是 *Bm*CPV 的结构蛋白 VP3 的 N 端结构域，因此容易被包被进 *Bm*CPV 多角体中。改造过的家蚕杆状病毒 *Bm*PAK6 感染 *BmN* 细胞后，纯化获得的多角体的裂解产物能感染 *BmN* 细胞，表明改造过的家蚕杆状病毒能被包被进 *Bm*CPV 多角体中。纯化的多角体及多角体裂解液均能被 X-gal 染色，表明 *Bm*PAK6 表达的 β-半乳糖苷酶也能被 *Bm*CPV 的多角体蛋白包被。这些结果表明一些异源蛋白和杆状病毒能以未知的方式被多角体包被。

1　基础医学与生物科学学院，苏州大学，苏州
2　现代丝绸国家工程实验室，苏州大学，苏州

Overexpression of recombinant infectious bursal disease virus (IBDV) capsid protein VP2 in the middle silk gland of transgenic silkworm

Xu HF[#*] Yuan L[#] Wang F Wang YC
Wang RY Song CN Xia QY Zhao P

Abstract: Infectious bursal disease virus (IBDV) is the causative agent of a highly contagious disease affecting young chickens and causes serious economic losses to the poultry industry worldwide. Development of subunit vaccine using its major caspid protein, VP2, is one of the promising strategies to protect against IBDV. This study aims to test the feasibility of using silkworm to produce recombinant VP2 protein (rVP2) derived from a very virulent strain of IBDV (vvIBDV). A total of 16 transgenic silkworm lines harboring a codon-optimized *VP2* gene driven by the sericin 1 promoter were generated and analyzed. The results showed that the rVP2 was synthesized in the middle silk gland of all lines and secreted into their cocoons. The content of rVP2 in the cocoon of each line was ranged from 0.07% to 16.10% of the total soluble proteins. The rVP2 was purified from 30 g cocoon powders with a yield of 3.33 mg and a purity >90%. Further analysis indicated that the rVP2 was able to tolerate high temperatures up to 80 ℃, and exhibited specific immunogenic activity in mice. To our knowledge, this is the first report of overexpressing rVP2 in the middle silk gland of transgenic silkworm, which demonstrates the capability of silkworm as an efficient tool to produce recombinant immunogens for use in new vaccines against animal diseases.

Published On: Transgenic Research, 2014, 23(5), 809-816.

State Key Laboratory of Silkworm Genome Biology, Southwest University, Chongqing 400716, China.
[#]These authors contributed equally.
[*]Corresponding author E-mail: xuhf@swu.edu.cn.

利用转基因家蚕中部丝腺过表达重组传染性法氏囊病毒衣壳蛋白VP2

徐汉福[#*]　袁　林[#]　王　峰　王元成
王日远　宋春暖　夏庆友　赵　萍

摘要：传染性法氏囊病病毒(IBDV)是侵染雏鸡的高传染性病原体，给全世界的家禽业造成了严重的经济损失。利用IBDV的主要衣壳蛋白VP2制作亚单位疫苗是防控IBDV的一个希望策略。本研究的目的是探索利用家蚕生产来源IBDV高致病性毒株(vvIBDV)的重组VP2蛋白(rVP2)的可行性。共制作获得了16个由丝胶1启动子控制密码子优化的*VP2*基因的转基因系。分析表明，rVP2在所有转基因系的中部丝腺中合成并分泌到茧壳中；每个转基因系的茧壳中rVP2的含量占总的可溶性蛋白的0.07%到16.10%；从30 g茧壳粉中能够纯化出纯度超过90%的rVP2约3.33 mg。进一步分析发现，rVP2能够耐受80 ℃的高温，并且在小鼠中具有免疫活性。据我们所知，这是首次报道利用转基因家蚕中部丝腺过表达rVP2蛋白，本研究证实了利用家蚕作为有效工具生产用于动物疾病新疫苗制备的重组免疫原的可行性。

家蚕基因组生物学国家重点实验室，西南大学，重庆

Large-scale production of bioactive recombinant human acidic fibroblast growth factor in transgenic silkworm cocoons

Wang F[1,2#] Wang RY[1#] Wang YC[1,2] Zhao P[1,2] Xia QY[1,2*]

Abstract: With an increasing clinical demand for functional therapeutic proteins every year, there is an increasing requirement for the massive production of bioactive recombinant human acidic fibroblast growth factor (r-haFGF). In this present study, we delicately explore a strategy for the mass production of r-haFGF protein with biological activity in the transgenic silkworm cocoons. The sequence-optimized *haFGF* was inserted into an enhanced sericin 1 expression system to generate the original transgenic silkworm strain, which was then further crossed with a PIG jumpstarter strain to achieve the remobilization of the expression cassette to a "safe harbor" locus in the genome for the efficient expression of r-haFGF. In consequence, the expression of r-haFGF protein in the mutant line achieved a 5.6-fold increase compared to the original strain. The high content of r-haFGF facilitated its purification and large-scald yields. Furthermore, the r-haFGF protein bioactively promoted the growth, proliferation and migration of *NIH/3T3* cells, suggesting the r-haFGF protein possessed native mitogenic activity and the potential for wound healing. These results show that the silk gland of silkworm could be an efficient bioreactor strategy for recombinant production of bioactive haFGF in silkworm cocoons.

Published On: Scientific Reports, 2015, 5.16323.

1 State Key Laboratory of Silkworm Genome Biology, Southwest University, Chongqing 400716, China.

2 College of Biotechnology, Southwest University, Chongqing 400716, China.

#These authors contributed equally.

*Corresponding author E-mail: xiaqy@swu.edu.cn.

具有生物活性的重组人酸性成纤维细胞生长因子在转基因家蚕茧中的大规模生产

王　峰[1,2#]　王日远[1#]　王元成[1,2]　赵　萍[1,2]　夏庆友[1,2*]

摘要：随着功能性医疗蛋白临床需求的逐年增加，大规模生产具有生物活性的重组人酸性成纤维细胞生长因子（r-haFGF）日益迫切。在本研究中，我们在转基因蚕茧中大量生产具有生物活性的r-haFGF。将序列优化的*haFGF*插入到sericin 1增强表达系统形成原始的家蚕转基因系，该品系的蚕再与PIG jumpstarter品系蚕杂交使得表达框活化并转座到基因组的"安全"位置，从而实现r-haFGF的高效表达。因此，与原始品系相比，突变品系中的r-haFGF表达量提高了5.6倍。含量的r-haFGF更容易纯化，且更适合大规模生产。此外，具有生物活性的r-haFGF可促进*NIH/3T3*细胞的生长、繁殖和迁移，表明r-haFGF蛋白具有天然的促有丝分裂活性，在伤口愈合方面具有应用前景。这些结果表明家蚕丝腺是一个高效的生物反应器，可应用于具有生物活性的haFGF的重组生产中。

1　家蚕基因组生物学国家重点实验室，西南大学，重庆

2　生物技术学院，西南大学，重庆

Development of a VLP-based vaccine in silkworm pupae against rabbit hemorrhagic disease virus

Zheng XX[1,2] Wang SK[2] Zhang WJ[2,3] Liu XJ[2] Yi YZ[2]
Yang ST[2,3] Xia XZ[2,3] Li YN[2*] Zhang ZF[2*]

Abstract: Rabbit hemorrhagic disease virus (RHDV) is the etiological agent behind rabbit hemorrhagic disease (RHD), which is lethal and contagious in rabbits. The virus does not replicate in cell culture and the only commercial inactivated vaccine available is derived from infected rabbit livers. RHDV capsid protein, VP60, is the main antigen comprising the virion. We used a baculovirus-silkworm pupae system to express VP60, which self-assembled into virus-like particles (VLPs) with a similar size and morphology to RHDV. Hemagglutination assays (HAs) showed that VP60 expression levels of VP60 reached as high as 10^7 HA units (HAU) per pupa. A single intramuscular injection with 10^4 HAU of VLPs completely protected rabbits for at least 180 days against RHDV challenge, and for at least 360 days when the VLPs were emulsified with Freund's complete adjuvant. These data suggest that silkworm pupae can be used to develop VLP-based vaccines which confer durable protection against RHD.

Published On: International Immunopharmacology, 2016, 40, 164-169.

1 School of Public Health, Shandong University, Jinan 250012, China.
2 Biotechnology Research Institute, Chinese Academy of Agricultural Sciences, Beijing 10081, China.
3 Institute of Military Veterinary, Academy of Military Medical Sciences, Changchun 130122, China.
*Corresponding author E-mail: liyinv@caas.cn; zhifangzhang@yahoo.com.

基于VLP进行家蚕蛹抗兔出血症病毒疫苗开发

郑学星[1,2] 王树坤[2] 张渭蛟[2,3] 刘兴健[2] 易咏竹[2]
杨松涛[2,3] 夏咸柱[2,3] 李轶女[2*] 张志芳[2*]

摘要：兔出血症病毒(RHDV)是兔出血性疾病(RHD)的病原体,其在兔子中是传染性的、致命的。病毒不会在细胞培养物中复制,唯一可用的商业灭活疫苗来自受感染的兔肝。RHDV衣壳蛋白VP60是包含病毒颗粒的主要抗原。我们使用杆状病毒——家蚕蛹系统表达的VP60能自组装成与RHDV大小和形态相似的病毒样颗粒(VLPs)。血凝试验(HAs)显示,VP60的表达水平在每头蛹中高达10^7个血凝单位(HAU)。单独肌内注射10^4个HAU的VLP能保护兔子免受至少180 d的RHDV攻击,而注射经弗氏完全佐剂乳化的VLP免疫时长至少能达到360 d。这些数据表明,家蚕蛹可用于持久防御RHD的VLP疫苗开发。

1 公共卫生学院,山东大学,济南
2 生物技术研究所,中国农业科学院,北京
3 军事兽医研究所,军事医学科学院,长春

Transgenic silkworms secrete the recombinant glycosylated MRJP1 protein of Chinese honeybee, *Apis cerana cerana*

You ZY[1#] Qian QJ[1,2#] Wang YR[3] Che JQ[1]

Ye LP[1] Shen LR[3*] Zhong BX[1*]

Abstract: Major royal jelly protein-1 (MRJP1) is the most abundant glycoprotein of royal jelly (RJ) and is considered a potential component of functional foods. In this study, we used silkworm transgenic technology to obtain five transgenic silkworm lineages expressing the exogenous recombinant Chinese honeybee, *Apis cerana cerana*, protein-1 (rAccMRJP1) under the control of a *fibroin light chain* (*Fib-L*) promoter in the posterior silk glands. The protein was successfully secreted into cocoons; specifically, the highest rAccMRJP1 protein content was 0.78% of the dried cocoons. Our results confirmed that the protein band of the exogenous rAccMRJP1 protein expressed in the transgenic silkworm lineages was a glycosylated protein. Therefore, this rAccMRJP1 protein could be used as an alternative standard protein sample to measure the freshness of RJ. Moreover, we also found that the overall trend between the expression of the endogenous and exogenous genes was that the expression level of the endogenous *Fib-L* gene declined as the expression of the exogenous *rAccMRJP1* gene increased in the transgenic silkworm lineages. Thus, by employing genome editing technology to reduce silk protein expression levels, a silkworm bioreactor expression system could be developed as a highly successful system for producing various valuable heterologous proteins, potentially broadening the applications of the silkworm.

Published On: Transgenic Research, 2017, 26(5), 653-663.

1 College of Animal Sciences, Zhejiang University, Hangzhou 310058, China

2 Haining Sericulture Technology Extension Station of Zhejiang Province, Haining 314400, China

3 Department of Food Science and Nutrition, Fuli Institute of Food Science, Zhejiang Key Laboratory for Agro-Food Processing, Zhejiang University, Hangzhou 310058, China

#These authors contributed equally.

*Corresponding author E-mail: bxzhong@zju.edu.cn; shenlirong@zju.edu.cn.

转基因家蚕分泌中华蜜蜂的重组糖基化MRJP1蛋白

尤征英[1#]　钱秋杰[1,2#]　王一然[3]　车家倩[1]
叶露鹏[1]　沈立荣[3*]　钟伯雄[1*]

摘要：蜂王浆主蛋白1（MRJP1）是最丰富的蜂王浆糖蛋白，被认为是功能性食品的可能组成部分。本研究通过转基因技术成功获得由*Fib-L*基因启动子驱动中华蜜蜂蜂王浆主蛋白1（rAccMRJP1）在家蚕后部丝腺表达的5个转基因系。该蛋白成功分泌到蚕茧中，最高可占蚕茧干重的0.78%。研究结果证实表达的外源rAccMRJP1是糖基化的蛋白。因此，这种rAccMRJP1蛋白可以作为测量RJ新鲜程度的标准品。而且我们发现在转基因家蚕品系中，随着外源的*rAccMRJP1*的表达增加，内源性的*Fib-L*基因的表达减少。因此，通过基因编辑的技术减少丝蛋白的表达，能够使家蚕成为高效的生物反应器表达系统以开发生产各种有价值的异源蛋白，这将拓宽家蚕的应用。

1　动物科学学院，浙江大学，杭州

2　浙江省海宁蚕桑技术服务站，海宁

3　食品与营养系，馥莉食品研究院，浙江省农产品加工技术研究重点实验室，浙江大学，杭州

Analysis of the *sericin1* promoter and assisted detection of exogenous gene expression efficiency in the silkworm, *Bombyx mori* L.

Ye LP Qian QJ Zhang YY You ZY Che JQ Song J Zhong BX*

Abstract: In genetics, the promoter is one of the most important regulatory elements controlling the spatiotemporal expression of a target gene. However, most studies have focused on core or proximal promoter regions, and information on regions that are more distant from the 5′-flanking region of the proximal promoter is often lacking. Here, approximately 4 kb of the *sericin1* (*Ser1*) promoter was predicted to contain many potential transcriptional factor binding sites (TFBSs). Transgenic experiments have revealed that more TFBSs included in the promoter improved gene transcription. However, multi-copy proximal *Ser1* promoter combinations did not improve gene expression at the transcriptional level. Instead, increasing the promoter copy number repressed transcription. Furthermore, a correlation analysis between two contiguous genes, *firefly Luciferase* (*FLuc*) and *EGFP*, was conducted at the transcriptional level; a significant correlation was obtained regardless of the insertion site. The ELISA results also revealed a significant correlation between the transcriptional and translational *EGFP* levels. Therefore, the exogenous gene expression level can be predicted by simply detecting an adjacent *EGFP*. In conclusion, our results provided important insights for further investigations into the molecular mechanisms underlying promoter function. Additionally, a new approach was developed to quickly screen transgenic strains that highly expressing exogenous genes.

Published On: Scientific Reports, 2015, 5.

College of Animal Sciences, Zhejiang University, Hangzhou 310058, China.

*Corresponding author E-mail: bxzhong@zju.edu.cn.

家蚕*sericin1*启动子的活性分析及外源基因表达的辅助检测

叶露鹏　钱秋杰　张玉玉　尤征英　车佳倩　宋　佳　钟伯雄*

摘要：启动子是调控目的基因时空表达的重要原件之一。然而，绝大多数的研究关注于核心启动子或近端启动子区域，近端启动子5′端的序列却常常被人们所忽视。我们的研究通过生物信息学方法从4 kb长度的家蚕丝胶1（*sericin 1*，*Ser1*）启动子序列中预测到种类众多的潜在的转录因子结合位点，采用转基因实验发现启动子序列包含的转录因子结合位点越多，启动子的转录活性就越高。但是多拷贝的近端*Ser1*启动子组合并不能提高下游基因的转录水平，而且重复的近端启动子拷贝数越多，下游基因的转录水平反而越低。此外，我们构建了萤火虫荧光素酶（Firefly Luciferase，*FLuc*）和绿色荧光蛋白（Green fluorescent protein，*EGFP*）两个相邻的基因，做了转录水平的相关分析，结果表明即使插入位点不同，这两个基因的转录水平总是显著相关的。ELISA实验进一步显示*EGFP*的转录水平和翻译水平同样也是显著相关的。所以，外源基因的表达水平可以通过检测标志基因（*EGFP*）的表达水平来简单地预测。总之，我们的研究为今后构建高效家蚕丝腺生物反应器时选择合适长度的启动子提供了参考，也为转基因品种的快速筛选提供了一个简便的方法。

动物科学学院，浙江大学，杭州

Expression of hIGF-I in the silk glands of transgenic silkworms and in transformed silkworm cells

Zhao Y Li X Cao GL Xue RY Gong CL*

Abstract: To express human insulin-like growth factor-I (hIGF-I) in transformed *Bombyx mori* cultured cells and silk glands, the transgenic vector pigA3GFP-hIGF-ie-neo was constructed with a neomycin resistance gene driven by the baculovirus *ie-1* promoter, and with the *hIGF-I* gene under the control of the silkworm *sericin* promoter *Ser-1*. The stably transformed *BmN* cells expressing hIGF-I were selected by using the antibiotic G418 at a final concentration of 700-800 μg/mL after the *BmN* cells were transfected with the *piggy*Bac vector and the helper plasmid. The specific band of hIGF-I was detected in the transformed cells by Western blot. The expression level of hIGF-I, determined by ELISA, was about 7 800 pg in 5×10^5 cells. Analysis of the chromosomal insertion sites by inverse PCR showed that exogenous DNA could be inserted into the cell genome randomly or at TTAA target sequence specifically for *piggy*Bac element transposition. The transgenic vector pigA3GFP-hIGF-ie-neo was transferred into the eggs using sperm-mediated gene transfer. Finally, two transgenic silkworms were obtained after screening for the *neo* and *gfp* genes and verified by PCR and dot hybridization. The expression level of hIGF-I determined by ELISA was about 2 440 pg/g of silk gland of the transgenic silkworms of the G_1 generation.

Published On: Science in China Series C—Life Sciences, 2009, 52(12), 1131-1139.

Medical College, Soochow University, Suzhou 215123, China.

*Corresponding author E-mail: gongcl@suda.edu.cn.

转基因家蚕丝腺组织和转化家蚕细胞表达 hIGF-I

赵 越 李 曦 曹广力 薛仁宇 贡成良*

摘要：为在家蚕细胞和家蚕丝腺组织中表达人胰岛素样生长因子(hIGF-I)，以家蚕丝胶基因启动子(*ser-1*)驱动 *hIGF-I* 基因，构建了带有家蚕杆状病毒 *ie-1* 启动子控制新霉素抗性基因(*neo*)表达盒的转基因载体 pigA3GFP-hIGF-ie-neo。在表达转座酶的辅助质粒存在下，分别转染 *BmN* 细胞，以 700–800 μg/mL 的 G418 筛选，获得了稳定转化细胞系。Western blot 检测到 hIGF-I 的特异性条带，ELISA 检测结果显示，hIGF-I 在 5×10^5 个细胞中的表达水平约 7 800 pg。通过反向 PCR 分析表明，在转化细胞中外源 DNA 通过随机整合或按照 *piggy*Bac 特定的靶位点序列 TTAA 插入细胞基因组。转基因载体 pigA3GFP-hIGF-ie-neo 通过精子介导法导入蚕卵，利用 *neo*、*gfp* 基因的双重筛选，经过 PCR 和点杂交鉴定，获得了 2 头转基因家蚕。ELISA 检测结果显示，在 G_1 代 hIGF-I 在每克中部丝腺组织中的表达水平约为 2 440 pg。

医学部，苏州大学，苏州

The promoter of *Bmlp3* gene can direct fat body-specific expression in the transgenic silkworm, *Bombyx mori*

Deng DJ[1,2#] Xu HF[1#] Wang F[1] Duan XL[1]
Ma SY[1] Xiang ZH[1] Xia QY[1*]

Abstract: The fat body plays multiple, crucial roles in the life of silkworms. Targeted expression of transgenes in the fat body of the silkworm, *Bombyx mori*, is important not only for clarifying the function of endogenous genes expressed in this tissue, but also for producing valuable recombinant proteins. However, fat body-specific gene expression remains difficult due to a lack of suitable tissue-specific promoters. Here we report the isolation of the fat body-specific promoter of *Bmlp3*, a member of the 30K protein family of silkworms. The 1.1 kb fragment from —374 to +738 of *Bmlp3* displayed strong promoter activity in the cell lines *BmE* and *Spli-221*. In transgenic silkworms, a *DsRed* reporter gene controlled by the 1.1 kb *Bmlp3* promoter fragment was expressed specifically in the fat body in a stage-specific pattern that was nearly identical to the endogenous *Bmlp3* gene. We conclude that the 1.1 kb *Bmlp3* promoter fragment is sufficient to direct tissue and stage specific expression of transgenes in the fat body of silkworms, highlighting the potential use of this promoter for both functional genomics research and biotechnology applications.

Published On: Transgenic Research, 2013, 22(5), 1055-1063.

1 State Key Laboratory of Silkworm Genome Biology, Southwest University, Chongqing 400716, China.
2 Institute of Forensic Science, Chongqing Pulic Security Bureau, Chongqing 401147, China.
#These authors contributed equally.
*Corresponding author E-mail: xiaqy@swu.edu.cn.

*Bmlp3*启动子可以启动基因在转基因家蚕的脂肪体中特异性表达

邓党军[1,2#] 徐汉福[1#] 王 峰[1] 段小利[1]
马三垣[1] 向仲怀[1] 夏庆友[1*]

摘要：脂肪体在家蚕的生命活动中起着多种重要作用。在转基因家蚕脂肪体中特异性表达目的基因，不仅有助于解释该组织中表达的内源基因的功能，而且可以用于合成高价值重组蛋白。然而，由于缺乏合适的组织特异性启动子，脂肪体特异性基因表达仍然很困难。在本研究中，作者鉴定了家蚕30K蛋白家族的成员*Bmlp3*基因的启动子可以在脂肪体特异性表达。在*BmE*和*Spli-221*细胞系中*Bmlp3*的−374至+738片段（约1.1 kb）显示出很强的启动子活性。在转基因家蚕中，由1.1 kb *Bmlp3*启动子片段启动的*DsRed*报告基因和内源基因*Bmlp3*几乎以相同的时期特异性模式在脂肪体中表达。文章显示，1.1 kb *Bmlp3*启动子片段足以启动外源基因在家蚕脂肪体中以组织和时期特异性模式表达，说明该启动子在功能基因组学研究和生物技术应用中具有潜在用途。

1 家蚕基因组生物学国家重点实验室，西南大学，重庆
2 物证鉴定所，重庆市公安局，重庆

Analysis of the activity of virus internal ribosome entry site in silkworm *Bombyx mori*

Ye LP[1] Zhuang LF[1] Li JS[1,2] You ZY[1]
Liang JS[3] Wei H[1] Lin JR[4] Zhong BX[1*]

Abstract: Internal ribosome entry site (IRES) has been widely used in genetic engineering; however, the application in silkworm (*Bombyx mori*) has hardly been reported. In this study, the biological activity of partial sequence of Encephalomyocarditis virus (EMCV) IRES, *Rhopalosiphum padi* virus (RhPV) IRES, and the hybrid of IRES of EMCV and RhPV were investigated in *Spodoptera frugiperda* (*Sf9*) cell line and silkworm tissues. The hybrid IRES of EMCV and RhPV showed more effective than EMCV IRES or RhPV IRES in promoting downstream gene expression in insect and silkworm. The activities of all IRESs in middle silk gland of silkworm were higher than those in the fat body and posterior silk gland. The hybrid IRES of EMCV and RhPV was integrated into silkworm genome by transgenic technology to test biological activity of IRES. Each of the positive transgenic individuals had significant expression of report gene *EGFP*. These results suggested that IRES has a potential to be used in the genetic engineering research of silkworm.

Published On: Acta Biochim et Biophysica Sinica, 2013, 45(7), 534-539.

1 College of Animal Sciences, Zhejiang University, Hangzhou 310029, China.
2 Institute of Sericulture, Chengde Medical College, Chengde 067000, China.
3 College of Environmental & Resource Sciences, Zhejiang University, Hangzhou 310029, China.
4 College of Animal Sciences, South China of Agricultural University, Guangzhou 510642, China.
*Corresponding author E-mail: bxzhong@zju.edu.cn.

病毒核糖体进入位点在家蚕中的生物学活性检测

叶露鹏[1] 庄兰芳[1] 李季生[1,2] 尤征英[1] 梁建设[3]
危 浩[1] 林健荣[4] 钟伯雄[1*]

摘要： 核糖体进入位点(Internal ribosome entry site，IRES)在基因工程中具有广泛的应用。然而，IRES在家蚕中的应用却鲜有报道。本实验研究了猪脑心肌炎病毒(Encephalomyocarditis virus，EMCV)IRES、禾谷缢管蚜虫病毒(*Rhopalosiphum padi* virus，RhPV)IRES的部分序列，两者IRES的组合序列在斜纹夜蛾细胞(*sf9*)以及家蚕组织中的生物学活性。结果表明EMCV和RhPV组合的IRES促进下游基因的表达活性更强。此外，相比脂肪体和后部丝腺，IRES在家蚕中部丝腺的活性最强。最后，我们用转基因技术将EMCV和RhPV组合的IRES插到家蚕基因组中并检测其下游基因的表达情况。在所筛选到的家蚕阳性个体中都发现*EGFP*显著表达。这些结果表明IRES可以广泛地应用于家蚕基因工程研究。

1 动物科学学院，浙江大学，杭州
2 蚕业研究所，承德医学院，承德
3 环境与资源学院，浙江大学，杭州
4 动物科学学院，华南农业大学，广州

Overexpression and functional characterization of an *Aspergillus niger* phytase in the fat body of transgenic silkworm, *Bombyx mori*

Xu HF[1#] Liu YW[1,2#] Wang F[1] Yuan L[1]
Wang YC[1] Ma SY[1] Helen Beneš[3*] Xia QY[1*]

Abstract: In a previous study, we isolated 1 119 bp of upstream promoter sequence from *Bmlp3*, a gene encoding a member of the silkworm 30K storage protein family, and demonstrated that it was sufficient to direct fat body-specific expression of a reporter gene in a transgenic silkworm, thus highlighting the potential use of this promoter for both functional genomics research and biotechnology applications. To test whether the *Bmlp3* promoter can be used to produce recombinant proteins in the fat body of silkworm pupae, we generated a transgenic line of *Bombyx mori* which harbors a codon-optimized *Aspergillus niger phytase* gene (*phyA*) under the control of the *Bmlp3* promoter. Here we show that the *Bmlp3* promoter drives high levels of *phyA* expression in the fat body, and that the recombinant phyA protein is highly active (99.05 U/g and 54.80 U/g in fat body extracts and fresh pupa, respectively). We also show that the recombinant phyA has two optimum pH ranges (1.5-2.0 and 5.5-6.0), and two optimum temperatures (55 and 37 ℃). The activity of recombinant phyA was lost after high-temperature drying, but treating with boiling water was less harmful, its residual activity was approximately 84% of the level observed in untreated samples. These results offer an opportunity not only for better utilization of large amounts of silkworm pupae generated during silk production, but also provide a novel method for mass production of low-cost recombinant phytase using transgenic silkworms.

Published On: Transgenic Research, 2014, 23(4), 669-677.

1 State Key Laboratory of Silkworm Genome Biology, Southwest University, Chongqing 400716, China.

2 State Key Laboratory of Genetic Resources and Evolution, Kunming Institute of Zoology, Chinese Academy of Sciences, Kunming 650223, China.

3 Department of Neurobiology and Developmental Sciences, College of Medicine, University of Arkansas for Medical Sciences, Arkansas 72205, USA.

#These authors contributed equally.

*Corresponding author E-mail: beneshelen@uams.edu; xiaqy@swu.edu.cn.

黑曲霉植酸酶基因在转基因家蚕脂肪体中的表达及功能特征

徐汉福[1#] 刘耀文[1,2#] 王 峰[1] 袁 林[1]
王元成[1] 马三垣[1] Helen Beneš[3*] 夏庆友[1*]

摘要： 在前期的研究中，我们从家蚕中分离获得了30K蛋白家族成员*Bmlp3*基因的启动子上游的1 119 bp序列，证实了该启动子能够在转基因家蚕脂肪体中特异调控报告基因的表达，预示*Bmlp3*启动子在功能分析及生物技术领域具有应用价值。为探究*Bmlp3*启动子在蚕蛹脂肪体中调控重组蛋白表达的可行性，我们制作了一个由*Bmlp3*启动子驱动，经密码子优化的来源于黑曲霉的植酸酶基因(*phyA*)的转基因家蚕品系。实验结果表明，*Bmlp3*启动子能够驱动*phyA*在转基因家蚕脂肪体中高量表达，并且表达的重组phyA具有较高的生物活性(在脂肪体抽提物和鲜蛹中的酶活分别为99.05 U/g和54.80 U/g)；重组phyA具有两个最适pH范围(1.5—2.0和5.5—6.0)和两个最适温度(55 ℃和37 ℃)；高温烘干处理后重组phyA的活性丧失，但采用沸水处理对重组phyA的活性影响较小，其酶活约为未经沸水处理样品的84%。这些研究不但为更好地利用缫丝过程产生大量的蛹，而且还为利用转基因蚕大规模生产低成本的重组植酸酶提供了一个新方法。

1 家蚕基因组生物学国家重点实验室，西南大学，重庆
2 遗传资源与进化国家重点实验室，昆明动物研究所，中国科学院，昆明
3 神经生物学与发育科学系，医学院，阿肯色大学医学院，阿肯色州，美国

Surface display of human serum albumin on *Bacillus subtilis* spores for oral administration

Mao LY Jiang ST Li G He YQ
Chen HQ Yao Q Chen KP*

Abstract: Human serum albumin (HSA) is the major protein component of human plasma. To date, HSA for clinical uses is mostly produced by fractionation of human whole blood, which is accompanied by a lot of limitations. To obtain long-term bioactive albumin, we used *hsa* as a foreign gene and constructed a recombinant plasmid pJS700-HSA which carries a recombinant gene *cotC-hsa* under the control of *cotC* promoter. Plasmid pJS700-HAS was transformed into *Bacillus subtilis* by double cross-over and an amylase inactivated mutant was produced. After induction of spore formation, Western blot and fluorescence immunoassay were used to monitor HSA surface expression on spores. We estimated that HSA displayed on the spore accounted for 0.135 % of the total spore proteins and about 0.023 fg HSA were exposed on the surface of each spore. Oral administration to mice with spores displaying HSA implied that the recombinant spores may have potential ability to increase the serum albumin level in vivo due to the resistant characters of spores.

Published On: Current Microbiology, 2012, 64(6), 545-551.

Institute of Life Sciences, Jiangsu University, Zhenjiang 212013, China.
*Corresponding author E-mail: kpchen@ujs.edu.cn.

用于口服的表面展示含有人血清白蛋白重组芽孢及其制备方法

毛浪勇　姜珊彤　李　刚　何远清
陈慧卿　姚　勤　陈克平*

摘要：人血清白蛋白(HSA)是人血浆中的主要蛋白组成。迄今为止，临床使用的HSA大多是通过人全血分离产生的，这就存在很大的局限性。为了长期获得具有生物活性的蛋白，我们使用*hsa*作为外源基因，并且构建了在*cotC*启动子控制下携带重组基因*cotC-hsa*的重组质粒pJS700-HSA。重组质粒通过双交换法转化到枯草芽孢杆菌中，产生了淀粉酶失活的突变体。诱导孢子形成后，使用蛋白免疫印迹实验和免疫荧光监测HSA在孢子表面的表达。我们估计，孢子上展示的HSA占总孢子蛋白的0.135%，每个孢子表面暴露约0.023 fg HSA。小鼠口服含有HSA的孢子，结果显示由于孢子的抗性特征，重组孢子可能具有在体内提高血清白蛋白水平的潜能。

生命科学研究院，江苏大学，镇江

Surface display of human growth hormone on *Bacillus subtilis* spores for oral administration

Lian CQ[1,2*] Zhou Y[3] Feng F[1] Chen L[3]
Tang Q[3] Yao Q[3] Chen KP[1]

Abstract: Human growth hormone (hGH) is the major and important hormone component of human being. At present, hGH for clinical uses is mostly produced in *Escherichia coli*, which requires costly denaturation and refolding to recover functionality. To obtain long-term bioactive hormone, we used *hGH* as a foreign gene and constructed a recombinant plasmid pJS700-hGH which carries a recombinant gene *COTC-HGH* with an enterokinase site under the control of *cotC* promoter. Plasmid pJS700-hGH was transformed into *Bacillus subtilis* by double crossover and an amylase-inactivated mutant was produced. After spore formation, Western blot and fluorescence immunoassay were used to monitor hGH surface expression on spores. Oral administration to silkworm with spores displaying hGH further showed that the recombinant spores may have potential ability to be digested and absorbed into the silkworm's hemolymph due to both the resistant characters of spores and the addition of enterokinase site.

Published On: Current Microbiology, 2014, 68(4), 463-471.

1 School of Food and Biological Engineering, Jiangsu University, Zhenjiang 212013, China.
2 Department of Clinical Laboratory, Bengbu Medical College, Bengbu 233030, China.
3 Institute of Life Sciences, Jiangsu University, Zhenjiang 212013, China.
*Corresponding author E-mail: c.q.Lian@163.com.

枯草芽孢杆菌表面展示口服用人生长激素

连超群[1,2*] 周 阳[3] 冯 凡[1] 陈 亮[3]
唐 琦[3] 姚 勤[3] 陈克平[1]

摘要：人类生长激素(hGH)是人类激素中主要并且重要的成分。目前，临床应用的生长激素主要由大肠杆菌生产，这需要昂贵的变性和复性过程来恢复其功能。为了获得具有长期生物活性的激素，我们使用*hGH*作为外源基因来构建重组质粒pJS700-hGH，该重组质粒携带*cotC*启动子控制的肠激酶位点*COTC-HGH*。通过双交叉和淀粉酶灭活突变转化质粒pJS700-hGH进入枯草芽孢杆菌。孢子形成后，用Western blot和荧光免疫测定技术监测孢子表面表达的hGH。用孢子喂食家蚕，由于孢子的抗性特征和肠激酶位点的添加，重组孢子可能具有潜在的被消化和吸收到蚕的血淋巴中的能力。

1 食品与生物工程学院，江苏大学，镇江
2 临床实验室，蚌埠医学院，蚌埠
3 生命科学研究院，江苏大学，镇江

第七章

表观遗传与miRNA

Chapter 7

表观遗传是环境因素和细胞内的遗传物质之间发生相互作用，在基因的DNA序列不发生变化的情况下，导致可遗传的表现型变化。目前表观遗传研究主要集中在DNA甲基化修饰、组蛋白修饰、染色质重塑和非编码RNA调控方面。家蚕是经过人类驯化并得以充分利用的一种重要经济昆虫，也是鳞翅目昆虫研究的模式生物。与野桑蚕相比，家蚕受到了环境和人工选择的重要影响，而表观遗传也将参与其基因表达调控过程。在家蚕基因组计划完成之后，项目组较早地开展了家蚕的表观遗传调控研究，尤其在组蛋白修饰、核苷酸修饰和非编码RNA调控方面取得了较大的进展。

家蚕组蛋白修饰研究。参与组蛋白甲基化修饰的酶是一类包含有SET结构域的蛋白家族组成。项目组通过基因组同源比对分析，在家蚕中系统性地鉴定到24个SET蛋白，其中17个蛋白属于6个已知的家族，包括SUV39、SET1、SET2、SUV4-20、EZ和SMYD，这些蛋白都含有保守的催化核心“NHSC”基序和底物结合位点。进一步研究发现具有SET结构域的EZ家族成员*Bm*E(Z)可以靶向家蚕组蛋白H3的K27位点，并催化其发生三甲基化修饰，这种修饰可以作为家蚕基因表达的失活标记。另外家蚕SET结构域基因BmSu(*var*)*3-9*的表达特征和表达模式不同于其他无脊椎动物。这些结果为SET结构域家族成员的功能和进化特征研究提供了新的视角。与组蛋白甲基化酶功能相反，生物体中还存在一类组蛋白去甲基化酶，它能够特异性地识别发生甲基化修饰的组蛋白，并去除其甲基基团。项目组在家蚕中鉴定并克隆了组蛋白H3赖氨酸去甲基化酶*BmLid*，功能分析显示*Bm*Lid具有相比于其他物种更广谱的酶作用特性，不仅对H3K4和H3K9位点的三甲基化或二甲基化状态具有去甲基化活性，而且对甲基化修饰的H3K27也有活性，这种广谱性的去甲基化活性很可能是由*Bm*Lid中PLU结构域缺失所造成的。从组蛋白修饰入手，对家蚕与野桑蚕进行比较组蛋白修饰研究，相信会为我们揭示更多人工驯化所引起的表观遗传变化。

家蚕核苷酸修饰研究。核苷酸是核糖核酸（RNA）和脱氧核糖核酸（DNA）的基本组成单位，是体内合成核酸的前体。目前研究的核苷酸修饰主要以甲基化修饰形式为主，是核苷酸序列上的碱基在甲基化转移酶的作用下，以S-腺苷甲硫氨酸作为甲基供体，通过共价键结合的方式获得一个甲基基团的化学修饰过程。项目组构建了首个单碱基分辨率的家蚕丝腺5mC甲基化图谱，研究发现大约0.11%的家蚕基因组胞嘧啶被甲基化修饰，比哺乳动物和植物低至少50倍。甲基化区域主要富集在基因区，并且与基因表达水平成正相关。同时，研究还鉴定了家蚕中主要参与5mC甲基化修饰的酶DNMT1和DNMT2，表明其表达具有发育与组织调控性特征。项目组在家蚕中还鉴定到了参与RNA甲基化的甲基转移酶*Bm*RNAMTase，免疫组化表明其主要定位于细胞质中，利用RNA干扰技术下调*BmRNAMTase*基因的表达，结果显示其具有抑制家蚕细胞凋亡活性。基于新一代测序技术，家蚕核苷酸修饰将有望在基因组和转录组水平上获得更多组学数据及功能解析，推动家蚕表观遗传研究。

家蚕miRNA研究。miRNA是一类由内源基因编码的长度约为22个核苷酸的非编码单链RNA分子，广泛存在于真核生物中，能够参与个体发育、细胞分化、转录后基因表达调控等过程。项目组首次在家蚕中设计并制作了miRNA寡核苷酸芯片，并利用该芯片确认了354个miRNA表达，绘制了不同发育阶段的miRNA表达特征图谱，结果表明家蚕的miRNA在胚胎形成和变态发育过程中发挥着重要的作用。项目组对家蚕幼虫5龄第3天不同组织，家蚕变态期间的体壁、丝腺、中肠和脂肪体进行了miRNA表达模式分析，发现总共有63个miRNA在从幼虫到蛹的转变期间至少1个组织中表现出显著变化，家蚕miRNA的时空表达分析表明其在不同组织器官发育过程中发挥了重要功能。利用Solexa测序技术对家蚕4龄第2天和5龄第3天幼虫后部丝腺进行了miRNA测序分析，发现有29个miRNA存在表达差异，推测这些差异miRNA对丝素蛋白基因表达具有调控作用。总之，通过多种分子生物学技术对家蚕miRNA的鉴定与表达模式分析，为后续miRNA的调控研究奠定了基础。

李志清

Single base-resolution methylome of the silkworm reveals a sparse epigenomic map

Xiang H[1,2#] Zhu JD[3,4#] Chen Q[2#] Dai FY[5#] Li X[1#] Li MW[6] Zhang HY[3] Zhang GJ[2] Li D[5] Dong Y[1] Zhao L[1] Lin Y[5] Cheng DJ[5] Yu J[3] Sun JF[3] Zhou XY[3] Ma KL[3] He YH[3] Zhao YX[3] Guo SC[3] Ye MZ[2] Guo GW[2] Li YR[2] Li RQ[2] Zhang XQ[2] Ma LJ[2] Karsten Kristiansen[7] Guo QH[8] Jiang JH[8] Stephan Beck[9] Xia QY[5*] Wang W[1*] Wang J[2,7*]

Abstract: Epigenetic regulation in insects may have effects on diverse biological processes. Here we survey the methylome of a model insect, the silkworm *Bombyx mori*, at single-base resolution using Illumina high-throughput bisulfite sequencing (MethylC-Seq). We conservatively estimate that 0.11% of genomic cytosines are methylcytosines, all of which probably occur in CG dinucleotides. CG methylation is substantially enriched in gene bodies and is positively correlated with gene expression levels, suggesting it has a positive role in gene transcription. We find that transposable elements, promoters and ribosomal DNAs are hypomethylated, but in contrast, genomic loci matching small RNAs in gene bodies are densely methylated. This work contributes to our understanding of epigenetics in insects, and in contrast to previous studies of the highly methylated genomes of *Arabidopsis* and human, demonstrates a strategy for sequencing the epigenomes of organisms such as insects that have low levels of methylation.

Published On: Nature Biotechnology, 2010, 28(7), 756-756.

1 CAS-Max Planck Junior Research Group, State Key Laboratory of Genetic Resources and Evolution, Kunming Institute of Zoology, The Chinese Academy of Sciences, Kunming 650223, China.
2 BGI-Shenzhen, Shenzhen 518083, China.
3 Cancer Epigenetics and Gene Therapy Program, The State-key Laboratory for Oncogenes and Related Genes, Shanghai Cancer Institute, Shanghai Jiaotong University, Shanghai 200240, China.
4 Cancer Epigenetics Laboratory, Obstetrics and Gynecology Hospital, Fudan University, Shanghai 200433, China.
5 Key Sericultural Laboratory of Agricultural Ministry, College of Biotechnology, Institute of Sericulture and Systems Biology, Southwest University, Chongqing 400716, China.
6 Sericultural Research Institute, Chinese Academy of Agricultural Sciences, Zhenjiang 212000, China.
7 Department of Biology, University of Copenhagen, Copenhagen DK-2100kbh O, Denmark.
8 Shanghai Institute of Plant Physiology and Ecology, Shanghai Institutes for Biological Sciences, The Chinese Academy of Sciences, Shanghai 200032, China.
9 UCL Cancer Institute, University College London, London WCIE 6BT, UK.
[#]These authors contributed equally.
[*]Corresponding author E-mail: xiaqy@swu.edu.cn; wwang@mail.kiz.ac; wangj@genomics.org.cn.

家蚕单碱基分辨率甲基化谱揭示了一个稀疏的表观基因组图谱

相　辉[1,2#]　朱景德[3,4#]　陈　泉[2#]　代方银[5#]　李　昕[1#]　李木旺[6]　张宏宇[3]
张国捷[2]　李　东[5]　董　扬[1]　赵　莉[1]　林　英[5]　程道军[5]　余　坚[3]　孙晋枫[3]
周小宇[3]　马克龙[3]　何英华[3]　赵阳星[3]　郭世成[3]　叶明智[2]　郭光武[2]　李英睿[2]
李瑞强[2]　张秀清[2]　马丽佳[2]　Karsten Kristiansen[7]　郭秋红[8]　蒋建豪[8]
Stephan Beck[9]　夏庆友[5*]　王　文[1*]　王　俊[2,7*]

摘要：表观遗传调控在昆虫中可能影响着多种生物学过程。本研究中我们使用Illumina高通量亚硫酸氢盐测序(MethylC-Seq)在单碱基分辨率水平调查了模式生物家蚕的DNA甲基化。我们保守估计大约0.11%的基因组胞嘧啶被甲基化修饰，且都可能发生在CG二核苷酸处。CG甲基化主要富集在基因区，并且与基因的转录水平呈正相关，表明其在基因转录中具有促进作用。同时，我们发现转座子区、启动子区、核糖体DNA区都是低甲基化的，相反，与基因体中的小RNA匹配的基因组位点则是高度甲基化的。这项工作有助于了解昆虫的表观遗传学，与之前对拟南芥和人类高度甲基化基因组的研究形成鲜明对比，提供了一种对具有低甲基化水平的生物(如昆虫等)表观基因组进行测序的策略。

1　遗传资源与进化重点实验室，昆明动物研究所，中国科学院，昆明
2　北京基因组研究所(深圳)，深圳
3　癌基因及相关基因国家重点实验室，上海市肿瘤研究所，上海交通大学，上海
4　癌症表观遗传实验室，复旦大学，上海
5　农业部蚕桑学重点实验室，生物技术学院，蚕学和系统生物学研究所，西南大学，重庆
6　蚕业研究所，中国农业科学院，镇江
7　生物学系，哥本哈根大学，哥本哈根，丹麦
8　植物生理生态研究所，上海生命科学研究院，中国科学院，上海
9　癌症研究所，伦敦大学学院，伦敦

Genome-wide identification of Polycomb target genes reveals a functional association of Pho with Scm in *Bombyx mori*

Li ZQ[1] Cheng DJ[2] Hiroaki Mon[1] Tsuneyuki Tatsuke[1] Zhu L[1]
Xu J[1] Jae Man Lee[1] Xia QY[2] Takahiro Kusakabe[1*]

Abstract: Polycomb group (PcG) proteins are evolutionarily conserved chromatin modifiers and act together in three multimeric complexes, Polycomb repressive complex 1 (PRC1), Polycomb repressive complex 2 (PRC2), and Pleiohomeotic repressive complex (PhoRC), to repress transcription of the target genes. Here, we identified Polycomb target genes in *Bombyx mori* with holocentric centromere using genome-wide expression screening based on the knockdown of *BmSCE*, *BmESC*, *BmPHO*, or *BmSCM* gene, which represent the distinct complexes. As a result, the expressions of 29 genes were up-regulated after knocking down 4 PcG genes. Particularly, there is a significant overlap between targets of *BmPho* (331 out of 524) and *BmScm* (331 out of 532), and among these, 190 genes function as regulator factors playing important roles in development. We also found that *Bm*Pho, as well as *Bm*Scm, can interact with other Polycomb components examined in this study. Further detailed analysis revealed that the C-terminus of *Bm*Pho containing zinc finger domain is involved in the interaction between *Bm*Pho and *Bm*Scm. Moreover, the zinc finger domain in *Bm*Pho contributes to its inhibitory function and ectopic overexpression of *Bm*Scm is able to promote transcriptional repression by Gal4-Pho fusions including *Bm*Scm-interacting domain. Loss of *Bm*Pho expression causes relocalization of *Bm*Scm into the cytoplasm. Collectively, we provide evidence of a functional link between *Bm*Pho and *Bm*Scm, and propose two Polycomb-related repression mechanisms requiring only *Bm*Pho associated with *Bm*Scm or a whole set of PcG complexes.

Publish On: PLoS ONE. 2012; 7(4): e34330.

1 Laboratory of Silkworm Science, Kyushu University Graduate School of Bioresource and Bioenvironmental Sciences, Fukuoka 8128581, Japan

2 State Key Laboratory of Silkworm Genome Biology, Southwest University, Chongqing 400716, China

*Corresponding author E-mail: kusakabe@agr.kyushu-u.ac.jp

家蚕Polycomb靶基因的全基因组鉴定揭示Pho与Scm功能的相关性

李志清[1] 程道军[2] Hiroaki Mon[1] Tsuneyuki Tatsuke[1] 祝力[1]
徐剑[1] Jae Man Lee[1] 夏庆友[2] Takahiro Kusakabe[1*]

摘要：Polycomb 蛋白复合物是在进化过程中非常保守的组蛋白修饰因子，通常由PRC1，PRC2和PhoRC三个亚型复合物共同抑制目标基因的转录。本研究中，我们通过基因沉默技术将*BmSCE*，*BmESC*，*BmPHO*和*BmSCM*基因下调表达，并通过芯片筛选在家蚕中鉴定了其调控的目标基因。结果发现，在以上4个基因下调后鉴定到了29个基因被上调表达。重要的是，*BmPHO*和*BmSCM*基因下调表达后，共有331个基因是其共同调控的靶基因，在这些基因中，有190个基因作为调控因子参与家蚕发育过程。研究发现*Bm*Pho和*Bm*Scm都可以与其他Polycomb蛋白发生相互作用，而且*Bm*Pho蛋白C端含有的锌指结构直接参与了与*Bm*Scm蛋白的相互作用。进一步研究发现，该锌指结构具有抑制基因表达功能。细胞定位研究表明，在*Bm*Pho不存在的情况下，*Bm*Scm的细胞核定位发生了改变，部分被定位于细胞质。综合以上研究，我们认为在家蚕中*Bm*Pho和*Bm*Scm是作为一个复合物而发挥功能，家蚕中存在两种类型的Polycomb蛋白复合物作用机制，一种是完整的Polycomb蛋白复合物，另一种是*Bm*Pho和*Bm*Scm形成的复合物。

1　家蚕科研室，生物资源环境科学研究院，九州大学，福冈，日本
2　家蚕基因组生物学国家重点实验室，西南大学，重庆

Cell Cycle-Dependent Recruitment of Polycomb Proteins to the ASNS Promoter Counteracts C/ebp-Mediated Transcriptional Activation in *Bombyx mori*

Li ZQ[1] Cheng DJ[2] Hiroaki Mon[1] Zhu L[1] Xu J[1] Tsuneyuki Tatsuke[1]

Jae Man Lee[1] Xia QY[2] Takahiro Kusakabe[1*]

Abstract: Epigenetic modifiers and transcription factors contribute to developmentally programmed gene expression. Here, we establish a functional link between epigenetic regulation by Polycomb group (PcG) proteins and transcriptional regulation by C/ebp that orchestrates the correct expression of *Bombyx mori asparagine synthetase* (*BmASNS*), a gene involved in the biosynthesis of asparagine. We show that the *cis*-regulatory elements of YY1-binding motifs and the CpG island present on the *BmASNS* promoter are required for the recruitment of PcG proteins and the subsequent deposition of the epigenetic repression mark H3K27me3. RNAi-mediated knockdown of *PcG* genes leads to derepression of the *BmASNS* gene via the recruitment of activators, including *BmC*/ebp, to the promoter. Intriguingly, we find that PcG proteins and *BmC*/ebp can dynamically modulate the transcriptional output of the *BmASNS* target in a cell cycle-dependent manner. It will be essential to suppress *BmASNS* expression by PcG proteins at the G_2/M phase of the cell cycle in the presence of *BmC*/ebp activator. Thus, our results provide a novel insight into the molecular mechanism underlying the recruitment and regulation of the PcG system at a discrete gene locus in *Bombyx mori*.

Publish On: PLoS ONE,2013,8(1).

1 Laboratory of Silkworm Science, Kyushu University Graduate School of Bioresource and Bioenvironmental Sciences, Fukuoka 8128581, Japan

2 State Key Laboratory of Silkworm Genome Biology, Southwest University, Chongqing 400716, China

*Corresponding author E-mail: kusakabe@agr.kyushu-u.ac.jp

家蚕Polycomb蛋白与C/ebp转录因子调控天冬酰胺合成酶表达的机制研究

李志清[1] 程道军[2] Hiroaki Mon[1] 祝力[1] 徐剑[1] Tsuneyuki Tatsuke[1]
Jae Man Lee[1] 夏庆友[2] Takahiro Kusakabe[1*]

摘要：表观修饰因子和转录调控因子在基因表达过程中发挥重要作用。本研究中，我们对表观调控因子Polycomb蛋白和转录因子C/ebp进行了研究，发现其能够共同调控天冬酰胺合成酶ASNS的表达，该酶参与天冬酰胺的合成。通过生物信息学分析发现，*BmASNS*基因启动子区域含有YY1结合基序和CpG岛，而且研究表明YY1结合基序和CpG岛对于Polycomb蛋白的募集以及其组蛋白的H3K27me3修饰至关重要。我们将Polycomb蛋白进行基因沉默后发现*BmASNS*基因上调表达，而这种上调表达是由于转录激活因子*Bm*C/ebp在启动子区域的富集所致。有趣的是，Polycomb蛋白与*Bm*C/ebp可以细胞周期性地动态调控BmASNS基因的表达，在*Bm*C/ebp存在的情况下，Polycomb蛋白在细胞分裂G_2/M期对*BmASNS*进行了抑制，使得*BmASNS*基因处于一个较低的转录水平。因此，该研究在家蚕中建立了一种新的Polycomb蛋白募集与调控靶基因的调控机制。

1 生物资源环境科学研究院家蚕科研室，九州大学，福冈，日本
2 家蚕基因组生物学国家重点实验室，西南大学，重庆

Subcellular localization and RNA interference of an RNA methyltransferase gene from silkworm, *Bombyx mori*

Nie ZM Zhou RB Chen J Wang D Lü ZB

He PA Wang XD Shen HD Wu XF Zhang YZ *

Abstract: RNA methylation, which is a form of posttranscriptional modification, is catalyzed by S-adenosyl -L-methionone-dependent RNA methyltransterases (RNA MTases). We have identified a novel silkworm gene, *BmRNAMTase*, containing a 369 bp open reading frame that encodes a putative protein containing 122 amino acid residues and having a molecular weight of 13.88 kDa. We expressed a recombinant His-tagged *Bm*RNAMTase in *E. coli* BL21 (DE3), purified the fusion protein by metal-chelation affinity chromatography, and injected a New Zealand rabbit with the purified protein to generate anti-*Bm*RNAMTase polyclonal antibodies. Immunohistochemistry revealed that *Bm*RNAMTase is abundant in the cytoplasm of *Bm5* cells. In addition, using RNA interference to reduce the intracellular activity and content of *Bm*RNAMTase, we determined that this cytoplasmic RNA methyltransferase may be involved in preventing cell death in the silkworm.

Published On: Comparative and Functional Genomics, 2008.

Institute of Biochemistry, College of Life Sciences, Zhejiang Sci-Tech University, Hangzhou 310018, China.

*Corresponding author E-mail: yaozhou@zstu.edu.cn.

家蚕RNA甲基转移酶基因的亚细胞定位和RNA干扰

聂作明　周若冰　陈　健　王　丹　吕正兵
贺平安　王雪冬　沈红丹　吴祥甫　张耀洲*

摘要：RNA甲基化是转录后修饰的一种，由S-腺苷-L-甲硫氨酸依赖的RNA甲基转移酶(RNA MTases)催化形成。我们鉴定出一种新型家蚕基因*BmRNAMTase*，该基因包含一个369 bp的开放阅读框，编码一个含有122个氨基酸残基，分子质量为13.88 kDa的蛋白质。我们在大肠杆菌BL21(DE3)中重组表达了带有His标签的*Bm*RNAMTase，并通过亲和层析的方法纯化融合蛋白，之后注射到新西兰兔体内产生多克隆抗体。免疫组织化学实验表明*Bm*RNAMTase在*Bm5*细胞中含量丰富。此外，利用RNA干扰技术下调*Bm*RNAMTase的胞内活性和含量，结果表明这种细胞质RNA甲基转移酶可能参与了防止家蚕细胞死亡过程。

生物化学研究所，生命科学学院，浙江理工大学，杭州

Comprehensive profiling of lysine acetylation suggests the widespread function is regulated by protein acetylation in the silkworm, *Bombyx mori*

Nie ZM[1,2] Zhu HL[1] Zhou Y[1] Wu CC[1] Liu Y[3] Sheng Q[1]
Lü ZB[1] Zhang WP[1] Yu W[1] Jiang CY[1] Xie LF[4] Zhang YZ[1] Yao JM[2*]

Abstract: Lysine acetylation in proteins is a dynamic and reversible PTM and plays an important role in diverse cellular processes. In this study, using lysine-acetylation (Kac) peptide enrichment coupled with nano HPLC-MS / MS, we initially identified the acetylome in the silkworms. Overall, a total of 342 acetylated proteins with 667 Kac sites were identified in silkworm. Sequence motifs analysis around Kac sites revealed an enrichment of Y, F, and H in the +1 position, and F was also enriched in the +2 and –2 positions, indicating the presences of preferred amino acids around Kac sites in the silkworm. Functional analysis showed the acetylated proteins were primarily involved in some specific biological processes. Furthermore, lots of nutrient-storage proteins, such as apolipophorin, vitellogenin, storage proteins, and 30 K proteins, were highly acetylated, indicating lysine acetylation may represent a common regulatory mechanism of nutrient utilization in the silkworm. Interestingly, Ser2 proteins, the coating proteins of larval silk, were found to contain many Kac sites, suggesting lysine acetylation may be involved in the regulation of larval silk synthesis. This study is the first to identify the acetylome in a lepidoptera insect, and expands greatly the catalog of lysine acetylation substrates and sites in insects.

Published On: Proteomics, 2015, 15(18), 3253-3266.

1 College of Life Sciences, Zhejiang Sci-Tech University, Hangzhou 310018, China.
2 College of Materials and Textile, Zhejiang Sci-Tech University, Hangzhou 310018, China.
3 Zhejiang Economic and Trade Polytechnic, Hangzhou 310018, China.
4 Jingjie PTM Biolabs, Hangzhou 310018, China.
*Corresponding author E-mail: wuxinzm@zstu.edu.cn.

蛋白乙酰化谱显示乙酰化修饰广泛参与调控家蚕蛋白功能

聂作明[1,2]　朱红琳[1]　周　雍[1]　吴程程[1]　刘　悦[3]　盛　清[1]
吕正兵[1]　张文平[1]　于　威[1]　蒋彩英[1]　谢龙飞[4]　张耀洲[1]　姚菊明[2*]

摘要：乙酰化修饰是一种动态和可逆的蛋白翻译后修饰，其在多种细胞生物过程中起很重要的调控作用。本研究首次使用乙酰化多肽富集与nano-HPLC-MS/MS结合，大规模鉴定了家蚕的乙酰化组。首次从家蚕中鉴定了342个乙酰化蛋白，包含667个赖氨酸乙酰化修饰（Lysine acetylation，Kac）位点。Kac位置偏好性分析发现存在5个显著富集的Kac位点基序，分别为+1位的KH、KY、KF和±2位的K*F和F*K，表明家蚕蛋白Kac位点周围的氨基酸分布存在一定的偏好性。功能分析表明，乙酰化蛋白参与多个特定的生物学过程。进一步分析发现，营养储藏蛋白存在高度的乙酰化修饰，如载脂蛋白（apolipophorin）、卵黄蛋白（vitellogenin）、储藏蛋白（SP）以及30 K蛋白，可能暗示了一种新的赖氨酸乙酰化修饰调控家蚕营养贮藏和水解利用的新机制。令人感兴趣的是，蚕丝表层覆盖的丝胶蛋白Ser2也存在大量的Kac位点，暗示乙酰化修饰可能参与了蚕丝蛋白的合成调控。本研究首次鉴定了鳞翅目昆虫的乙酰化组，极大地扩充了昆虫赖氨酸乙酰化修饰蛋白的类别。

1　生命科学学院，浙江理工大学，杭州
2　材料与纺织学院，浙江理工大学，杭州
3　浙江经贸职业技术学院，杭州
4　景杰生物技术公司，杭州

Silkworm (*Bombyx mori*) *Bm*Lid is a histone lysine demethylase with a broader specificity than its homolog in *Drosophila* and mammals

Zhou B[1,2] Yang XN[1*] Jiang JH[1,2] Wang YB[1]
Li MH[3] Li MW[4] Miao XX[1] Huang YP[1]

Abstract: Histone methylation is a dynamic process that plays important roles in gene transcription regulation, and a number of enzymes have been shown to catalyze the removal of methyl marks identified one of the amino oxidases, lysine-specific demethylase 1 (LSD1), as the first specific demethylase for both mono-(me) and dimethylation (me2) of H3K4 and H3K9 in humans. Subsequently, a total of 27 JmjC-domain-containing proteins have been discovered within the human genome, and 15 of them exhibit demethylation activities for specific lysines in the H3 tail. *Dm*UTX, *Dm*KDM2, *Dm*Lid, and *Dm*KDM4 are JmjC-domain-containing histone demethylases found in *Drosophila*. JmjC-domain-containing histone demethylases have also been reported in *Arabidopsis*. Here, we examined the lysine methylation patterns of histone H3 in *Bombyx mori*, a biological model for lepidopteran insects, and showed that a putative histone H3 lysine demethylase, *Bm*Lid, has a much broader *in vitro* substrate specificity than its homolog in other species, exhibiting demethylase activity toward both tri - (me3) and dimethylated (me2) K4 and K9, as well as dimethylated K27. Evidence is also presented suggesting that the absence of a PLU domain in *Bm*Lid may be partially responsible for this broader *in vitro* H3 lysine substrate specificity.

Published On: Cell Research, 2010, 20(9), 1079-1082.

1 Shanghai Institute of Plant Physiology and Ecology, Shanghai Institute for Biological Sciences, Chinese Academy of Sciences, Shanghai 200032, China.
2 Graduate School of the Chinese Academy of Sciences, Beijing 100049, China.
3 Shanghai Biochip Company, Shanghai 201203, China.
4 Sericultural Research Institute, Chinese Academy of Agriculture Sciences, Zhenjiang 212018, China
*Corresponding author E-mail: xnyang@sippe.ac.cn.

家蚕组蛋白赖氨酸去甲基化酶*Bm*Lid比其在果蝇和哺乳动物中的同源物具有更广泛的特异性

周　博[1,2]　杨晓楠[1*]　蒋剑豪[1,2]　王玉冰[1]
李明辉[3]　李木旺[4]　苗雪霞[1]　黄勇平[1]

摘要：组蛋白甲基化是一个不间断的过程，并且在基因转录调节中起到重要的作用，一定数量的酶被证明具有等同于一种氨基氧化酶催化移除甲基标记的功能。赖氨酸特异性去甲基化酶1(LSD1)是发现的第一个特异作用于人类H3K4和H3K9，具有单甲基化和多甲基化作用的酶。随后，总计27个JmjC端结构域调控蛋白在人类基因组中被发现，并且其中的15个针对H3末端特异的赖氨酸具有去甲基化活性。*Dm*UTX，*Dm*KDM2，*Dm*Lid和*Dm*KDM4是在果蝇中发现的JmjC端结构域调控组蛋白去甲基化酶。JmjC端结构域调控组蛋白去甲基化酶同样在拟南芥中被报道。本文我们在鳞翅目模式动物家蚕中，测试组蛋白H3赖氨酸去甲基化的模式，并且证明了预测的家蚕组蛋白H3赖氨酸去甲基化酶*Bm*Lid相比于其他物种在体外有更强的酶作用特异性，对K4和K9有三甲基化或二甲基化活性，对K27有双甲基化活性。实验数据同时也表明*Bm*lid中PLU结构域的缺失可能是造成这种体外H3赖氨酸酶特异性差异的原因。

1　上海生命科学研究院植物生理生态研究所，中国科学院，上海
2　中国科学院研究生院，北京
3　上海生物芯片公司，上海
4　蚕业研究所，中国农业科学院，镇江

Identification and analysis of the SET-domain family in silkworm, *Bombyx mori*

Zhao HL　Zheng CQ　Cui HJ*

Abstract: As an important economic insect, *Bombyx mori* is also a useful model organism for lepidopteran insect. SET-domain-containing proteins belong to a group of enzymes named after a common domain that utilizes the cofactor S-adenosyl-L-methionine (SAM) to achieve methylation of its substrates. Many SET-domain-containing proteins have been shown to display catalytic activity towards particular lysine residues on histones, but emerging evidence also indicates that various nonhistone proteins are specifically targeted by this clade of enzymes. To explore their diverse functions of SET-domain superfamily in insect, we identified, cloned, and analyzed the SET-domains proteins in silkworm, *Bombyx mori.* Firstly, 24 genes containing SET domain from silkworm genome were characterized and 17 of them belonged to six subfamilies of SUV39, SET1, SET2, SUV4-20, EZ, and SMYD. Secondly, SET domains of silkworm SET-domain family were intraspecifically and interspecifically conserved, especially for the catalytic core"NHSC" motif, substrate binding site, and catalytic site in the SET domain. Lastly, further analyses indicated that silkworm SET-domain gene *BmSu(var)3-9* owned different characterization and expression profiles compared to other invertebrates. Overall, our results provide a new insight into the functional and evolutionary features of SET-domain family.

Published On: BioMed Research International, 2015.

State Key Laboratory of Silkworm Genome Biology, Southwest University, Chongqing 400716, China.
*Corresponding author E-mail: hcui@swu.edu.cn.

分析鉴定家蚕的SET结构域家族

赵海龙　郑春琴　崔红娟*

摘要：家蚕不仅是一种重要的经济昆虫，同时也是一个有用的模式生物。包含SET结构域的蛋白质属于利用S-adenosyl-L-methionine（SAM）来实现其底物的甲基化的一类酶。许多包含SET结构域的蛋白质已被证实对组蛋白上特定的赖氨酸残基具有催化活性，但也有新证据表明，多种非组蛋白的蛋白质是这类酶进化枝的特定的靶标。为了探究昆虫SET结构域家族的不同的功能，我们鉴定并克隆分析了家蚕的SET结构域蛋白。首先，共鉴定到家蚕中包含SET结构域的24个基因，其中17个基因属于SUV39，SET1，SET2，SUV4-20，EZ，SMYD 6个亚家族。其次，家蚕SET结构域家族中的SET结构域具有种内和种间保守性，特别是催化核心"NHSC"的基序、底物结合位点，以及SET结构域中的催化位点。另外，进一步的分析表明，家蚕SET结构域基因*BmSu*（*var*）*3-9*的表达特征和表达模式不同于其他无脊椎动物。总的来说，我们的研究结果为SET结构域家族成员的功能和进化特征提供了一个新的视角。

家蚕基因组生物学国家重点实验室，西南大学，重庆

Involvement of microRNAs in infection of silkworm with *Bombyx mori* cytoplasmic polyhedrosis virus (*Bm*CPV)

Wu P [1,2] Han SH[1] Chen T [1,2] Qin GX[1] Li L[1,2] Guo XJ[1,2*]

Abstract: *Bombyx mori* cytoplasmic polyhedrosis virus (*Bm*CPV) is one of the most important pathogens of silkworm. MicroRNAs (miRNAs) have been demonstrated to play key roles in regulating host-pathogen interaction. However, there are limited reports on the miRNAs expression profiles during insect pathogen challenges. In this study, four small RNA libraries from *Bm*CPV-infected midgut of silkworm at 72 h post-inoculation and 96 h post-inoculation and their corresponding control midguts were constructed and deep sequenced. A total of 316 known miRNAs (including miRNA*) and 90 novel miRNAs were identified. Fifty-eight miRNAs displayed significant differential expression between the infected and normal midgut ($P \leqslant 0.01$ and fold change$\geqslant$2.0 or$\leqslant$0.5), among which ten differentially expressed miRNA were validated by qRT-PCR method. Further bioinformatics analysis of predicted target genes of differentially expressed miRNAs showed that the miRNA targets were involved in stimulus and immune system process in silkworm.

Published On: PLoS ONE, 2013, 8(7).

1 Sericultural Research Institute, Jiangsu University of Science and Technology, Zhenjiang 212018, China.

2 Quality Inspection Center for Sericulture Products, Ministry of Agriculture, Zhenjiang 212018, China.

*Corresponding author E-mail: guoxijie@126.com.

家蚕感染质型多角体病毒(*Bm*CPV)相关的microRNAs

吴 萍[1,2] 韩韶华[1] 陈 涛[1,2] 覃光星[1] 李 龙[1,2] 郭锡杰[1,2*]

摘要:家蚕质型多角体病毒(*Bm*CPV)是感染家蚕重要的病毒之一。MicroRNAs (miRNAs)已被证明在调控宿主—病毒的互作中发挥重要作用。然而,有关昆虫受病原物侵染后宿主miRNAs的表达谱变化的相关报道非常有限。本研究构建了4个小RNA文库(分别来自于感染家蚕质型多角体病毒72 h,96 h以及相应的对照家蚕的中肠)并对之进行了深度测序。共检测到316个已知miRNAs (包括miRNA*)及90个新的miRNAs。58个miRNAs在感染家蚕和对照家蚕的中肠中表现出了显著的表达差异(P≤0.01;表达差异倍数≥2.0或≤0.5),其中,qRT-PCR方法验证了10个miRNAs的表达差异。差异miRNAs的靶基因预测表明这些靶基因的功能涉及家蚕的应激和免疫过程。

1 蚕业研究所,江苏科技大学,镇江

2 农业部蚕桑产业产品质量监督检验测试中心,镇江

Characterization and expression patterns of *let*-7 microRNA in the silkworm (*Bombyx mori*)

Liu SP　Xia QY*　Zhao P

Cheng TC　Hong KL　Xiang ZH

Abstract:

Background: *lin*-4 and *let*-7, the two founding members of heterochronic microRNA genes, are firstly confirmed in *Caenorhabditis elegans* to control the proper timing of developmental programs in a heterochronic pathway. *let*-7 has been thought to trigger the onset of adulthood across animal phyla. Ecdysone and *Broad-Complex* are required for the temporal expression of *let*-7 in *Drosophila melanogaster*. For a better understanding of the conservation and functions of *let*-7, we seek to explore how it is expressed in the silkworm (*Bombyx mori*).

Results: One member of *let*-7 family has been identified in silkworm computationally and experimentally. All known members of this family share the same nucleotides at ten positions within the mature sequences. Sequence logo and phylogenetic tree show that they are not only conserved but diversify to some extent among some species. The *bmo-let*-7 was very lowly expressed in ova harvested from newborn unmated female adult and in individuals from the first molt to the early third instar, highly expressed after the third molt, and the most abundant expression was observed after mounting, particularly after pupation. The expression levels were higher at the end of each instar and at the beginning of each molt than at other periods, coinciding with the pulse of ecdysone and *BR-C* as a whole. Using cultured ovary cell line, *BmN-SWU1*, we examined the effect of altered ecdysone levels on *bmo-let*-7 expression. The expression was also detected in various tissues of day 3 of the fifth instar and of from day 7 of the fifth to pupa, suggesting a wide distributing pattern with various signal intensities.

Conclusion: *bmo-let*-7 is stage- and tissue-specifically expressed in the silkworm. Although no signals were detected during embryonic development and first larval instar stages, the expression of *bmo-let*-7 was observed from the first molt, suggesting that it might also function at early larval stage of the silkworm. The detailed expression profiles in the whole life cycle and cultured cell line of silkworm showed a clear association with ecdysone pulse and a variety of biological processes.

Published On: BMC Developmental Biology, 2007, 7.

Key Sericultural Laboratory of Agricultural Ministry, College of Biotechnology, Southwest University, Chongqing 400716, China.

*Corresponding author E-mail: xiaqy@swu.edu.cn.

家蚕*let*-7 microRNA 基因的特征和表达模式

刘仕平　夏庆友*　赵　萍
程廷才　洪开丽　向仲怀

摘要：背景——作为重要的异时性microRNA基因，*lin*-4和*let*-7首先在线虫中被证实是通过时序调控途径来控制生物发育进程的。*let*-7被认为是启动成虫发育的开关。在果蝇中，*let*-7的表达与蜕皮激素和*BR-C*的作用有关。研究家蚕*let*-7的时空表达，旨在更好地深入研究*let*-7的生物学功能。结果——我们在家蚕基因组上找到了*let*-7家族的一个成员，命名为*bmo-let*-7，并通过Northern杂交实验证实了它的存在。比较*let*-7家族成员的成熟体序列，我们发现其中10个位点的核苷酸相同，但序列标志和系统发育树表明：它们在某些物种中不仅具有一定的保守性，而且具有一定的多样性。*bmo-let*-7的表达在未交配雌蛾体内取出的卵中，以及第一次脱皮到三龄初期的个体中均处于很低的水平，然而，三眠后，其表达量迅速上升，进入了高水平表达阶段。上蔟后，特别是在化蛹后，*bmo-let*-7的表达量达到了新的高度。进一步研究发现，其表达量在每一龄的末期和脱皮开始时达到高峰，这与蜕皮激素和*BR-C*的波动基本上是相一致的。用蜕皮激素处理卵巢细胞系*BmN-SWU1*后，发现*let*-7的表达与蜕皮激素的浓度及培养时间有关。通过对五龄第3天和五龄第7天到蛹期的不同组织进行研究，发现了*bmo-let*-7广泛地分布在所有组织中，但是组织之间的表达水平却有很大的差异。结论——*bmo-let*-7在家蚕中表现出明显的时空特异性或差异性表达规律；虽然在胚胎发育过程和一龄阶段检测不到*bmo-let*-7，但它却从第一次脱皮就有表达，暗示了其可能在家蚕的早期幼虫阶段也发挥作用。蚕的整个生命周期和培养细胞系的详细表达谱与脱皮激素脉冲和多种生物学过程有明显的关联。

农业部蚕桑学重点实验室，生物技术学院，西南大学，重庆

Identification and characteristics of microRNAs from *Bombyx mori*

He PA[1,2#] Nie ZM[1#] Chen JQ[1] Chen J[1] Lü ZB[1] Sheng Q[1]
Zhou SP[2] Gao XL[1,3] Kong LY[1] Wu XF[1] Jin YF[4*] Zhang YZ[1*]

Abstract:

Background: MicroRNAs (miRNAs) are small RNA molecules that regulate gene expression by targeting messenger RNAs (mRNAs) and causing mRNA cleavage or translation blockage. Of the 355 Arthropod miRNAs that have been identified, only 21 are *B. mori* miRNAs that were predicted computationally; of these, only *let-7* has been confirmed by Northern blotting.

Results: Combining a computational method based on sequence homology searches with experimental identification based on microarray assays and Northern blotting, we identified 46 miRNAs, an additional 21 plausible miRNAs, and a novel small RNA in *B. mori*. The latter, *bmo-miR-100-like*, was identified using the known miRNA *aga-miR-100* as a probe; *bmo-miR-100-like* was detected by microarray assay and Northern blotting, but its precursor sequences did not fold into a hairpin structure. Among these identified miRNAs, we found 12 pairs of miRNAs and miRNA*s. Northern blotting revealed that some *B. mori* miRNA genes were expressed only during specific stages, indicating that *B. mori* miRNA genes (e. g., *bmo-miR-277*) have developmentally regulated patterns of expression. We identified two miRNA gene clusters in the *B. mori* genome. *bmo-miR-2b*, which is found in the gene cluster *bmo-miR-2a-1/bmo-miR-2a-1*/bmo-miR-2a-2/bmo-miR-2b/bmo-miR-13a*/bmo-miR-13b*, encodes a newly identified member of the mir-2 family. Moreover, we found that methylation can increase the sensitivity of a DNA probe used to detect a miRNA by Northern blotting. Functional analysis revealed that 11 miRNAs may regulate 13 *B. mori* orthologs of the 25 known *Drosophila* miRNA-targeted genes according to the functional conservation. We predicted the binding sites on the 1 671 3'UTR of *B. mori* genes; 547 targeted genes, including 986 target sites, were predicted. Of these target sites, 338 had perfect base pairing to the seed region of 43 miRNAs. From the predicted genes, 61 genes, each of them with multiple predicted target sites, should be considered excellent candidates for future functional studies. Biological classification of predicted miRNA targets showed that "binding", "catalytic activity" and "physiological process" were over-represented for the predicted genes.

Conclusion: Combining computational predictions with microarray assays, we identified 46 *B. mori* miRNAs, 13 of which were miRNA*s. We identified a novel small RNA and 21 plausible *B. mori* miRNAs that could not be located in the available *B. mori* genome, but which could be detected by microarray. Thirteen and 547 target genes were predicted according to the functional conservation and binding sites, respectively. Identification of miRNAs in *B. mori*, particularly those that are developmentally regulated, provides a foundation for subsequent functional studies.

Published On: BMC Genomics, 2008, 9, 248.

1 Institute of Biochemistry, Zhejiang Sci-Tech University, Hangzhou 310018, China.

2 College of Science, Zhejiang Sci-Tech University, Hangzhou 310018, China.

3 Department of Biology and Biochemistry, University of Houston, Houston 77204-5001, USA.

4 Institute of Biochemistry, Zhejiang University, Hangzhou 310029, China.

[#]These authors contributed equally.

[*]Corresponding author E-mail: jinyf@zju.edu.cn; yaozhou@zstu.edu.cn.

家蚕microRNA的鉴定与特征

贺平安[1,2#]　聂作明[1#]　陈剑清[1]　陈　健[1]　吕正兵[1]　盛　清[1]
周颂平[2]　高晓莲[1,3]　孔令印[1]　吴祥甫[1]　金勇丰[4*]　张耀洲[1*]

摘要:背景——MicroRNA是一种小RNA分子,其可以通过与靶基因的mRNA结合,使mRNA剪切或者翻译阻断。在已经确定了的355个节肢动物的microRNA中,仅有21个可以通过计算预测在家蚕中存在。在这21个microRNA当中,只有*let-7*被Northern blot检测到。结果——基于序列同源性搜索计算方法、微芯片分析和Northern blot实验,我们在家蚕中鉴定了除之前的21个以外的46个microRNA和一个新的小RNA。利用*aga-miR-100*探针鉴定出*bmo-miR-100-like*小RNA;利用微芯片和Northern blot检测到*bmo-miR-100-like*小RNA,但其前体序列并没有折叠成发夹结构。在这些鉴定出来的microRNA中我们发现了12对miRNA和miRNA*。Northern blot实验表明,家蚕中一些miRNA只在生长的特定时期才表达,表明家蚕miRNA(如*bmo-miR-277*)具有发育阶段的表达模式。我们在家蚕基因组中鉴定出两个miRNA基因簇。*bmo-miR-2b*是我们在基因簇*bmo-miR-2a-1/bmo-miR-2a-1*/bmo-miR-2a-2/bmo-miR-2b/bmo-miR-13a*/bmo-miR-13b*中发现的,其编码mir-2家族一个最新发现的成员。此外,我们还发现甲基化可以提高Northern blot鉴定时的DNA探针灵敏度。功能分析表明,有11个miRNA可能通过保守结构直接调控与已知的25个果蝇基因同源的13个家蚕基因。我们在1 671个家蚕基因的3′UTR区预测miRNA结合位点,共预测到547个靶基因,包含986个靶位点。在这些靶位点中,338个能与43个miRNA的种子区域完美配对。在预测的基因中,有61个含有多个可能的结合位点,这些基因需要后续的深入研究。这些miRNA靶基因的生物学分类分析表明,"结合","催化活性"和"生理过程"代表了其主要的生物学功能。结论——通过计算机预测和微芯片分析,我们鉴定出了46个家蚕miRNA,其中13个是miRNA*。我们在家蚕中发现了一个新的miRNA和21个可能的miRNA,其无法在目前的家蚕基因组中定位,但是却可以通过微芯片检测出来。通过功能保守性分析和结合位点分析,分别预测到13个和547个靶基因。家蚕miRNA的鉴定,特别是特定发育阶段的miRNA鉴定为后续的功能研究奠定了基础。

1　生物化学研究所,浙江理工大学,杭州
2　理学院,浙江理工大学,杭州
3　生物学与生物化学系,休斯敦大学,美国
4　生物化学研究所,浙江大学,杭州

Insect-specific microRNA involved in the development of the silkworm *Bombyx mori*

Zhang Y[1#] Zhou X[2#] Ge X[1] Jiang JH[1] Li MW[3] Jia SH[1]
Yang XN[1] Kan YC[4] Miao XX[1] Zhao GP[1] Li F[2*] Huang YP[1*]

Abstract: MicroRNAs (miRNAs) are endogenous non-coding genes that participate in post-transcription regulation by either degrading mRNA or blocking its translation. It is considered to be very important in regulating insect development and metamorphosis. We conducted a large-scale screening for miRNA genes in the silkworm *Bombyx mori* using sequence-by-synthesis (SBS) deep sequencing of mixed RNAs from egg, larval, pupal, and adult stages. Of 2 227 930 SBS tags, 1 144 485 ranged from 17 to 25 nt, corresponding to 256 604 unique tags. Among these non-redundant tags, 95 184 were matched to the silkworm genome. We identified 3 750 miRNA candidate genes using a computational pipeline combining RNAfold and TripletSVM algorithms. We confirmed 354 miRNA genes using miRNA microarrays and then performed expression profile analysis on these miRNAs for all developmental stages. While 106 miRNAs were expressed in all stages, 248 miRNAs were egg- and pupa-specific, suggesting that insect miRNAs play a significant role in embryogenesis and metamorphosis. We selected eight miRNAs for quantitative RT-PCR analysis; six of these were consistent with our microarray results. In addition, we searched for orthologous miRNA genes in mammals, a nematode, and other insects and found that most silkworm miRNAs are conserved in insects, whereas only a small number of silkworm miRNAs has orthologs in mammals and the nematode. These results suggest that there are many miRNAs unique to insects.

Published On: PLoS ONE, 2009, 4(3).

1 Shanghai Institute of Plant Physiology and Ecology, Chinese Academy of Sciences, Shanghai 200032, China.
2 Nanjing Agricultural University, Nanjing 210095, China.
3 Sericultural Research Institute, Chinese Academy of Agriculture Sciences, Zhenjiang 212018, China.
4 Nanyang Normal University, Nanyang 473007, China.
[#]These authors contributed equally.
[*]Corresponding author E-mail: lifei@njau.edu.cn ; yphuang@sibs.ac.cn

昆虫特异的microRNA调控家蚕的生长发育

张 勇[1#] 周 学[2#] 戈 榭[1] 蒋剑豪[1] 李木旺[3] 贾世海[1]
杨晓楠[1] 阚云超[4] 苗雪霞[1] 赵国屏[1] 李 飞[2*] 黄勇平[1*]

摘要：miRNA是生物体内源性的非编码序列，通过降解mRNA或阻断其翻译来参与转录后调控。它在调控昆虫的发育和变态过程中很重要。用SBS测序来自卵、幼虫、蛹和成虫阶段的混合RNA，对家蚕miRNA基因进行了大规模筛选。在2 227 930个SBS标签中，1 144 485个标签的长度在17—25 nt，对应于256 604个单一的标签。而这些非冗余的标签中，有95 184条能够匹配到家蚕基因组上。我们结合RNAfold和TripletSVM算法确认了3 750个候选的miRNA基因。我们使用基因芯片确认了354 miRNA基因，并检测了这些miRNA在不同发育阶段的表达模式。其中只有106个miRNA在所有的发育阶段都表达，248个在卵期和蛹期特异性表达，暗示着昆虫的miRNA在胚胎形成和变态发育中起着重要的作用。我们挑选了8个miRNA做RT-PCR定量分析，6个与基因芯片检测结果一致，此外，我们在哺乳类动物、线虫类和其他昆虫中寻找同源miRNA基因，发现大多数家蚕miRNAs是昆虫保守的，仅有一小部分的家蚕miRNA在哺乳动物和线虫中有同源miRNA的存在。我们的结果进一步揭示了昆虫特异的miRNA序列。

1 上海植物生理生态研究所，中国科学院，上海
2 南京农业大学，南京
3 蚕业研究所，中国农业科学院，镇江
4 南阳师范学院，南阳

Computational identification and characteristics of novel microRNAs from the silkworm (*Bombyx mori* L.)

Huang Y[1,2] Zou Q[3] Tang SM[1,2] Wang LG[1,2] Shen XJ[1,2*]

Abstract: MicroRNAs (miRNAs) are a class of non-protein coding small RNAs that regulate expression of genes at post-transcriptional levels. Increasing evidence has shown that miRNAs play multiple roles in biological processes, including development, cell proliferation and apoptosis. Based on the conservation of miRNAs sequence, using a computational homology search based on genomic survey sequence analysis, a total of 16 novel miRNAs were identified and characteristics such as family and evolutionary conservation have been described. By using these newly identified miRNAs, the mRNA database of silkworm was blasted and 21 potential targets of miRNAs were detected. Most of these miRNA targeted genes were predicted to encode transcription factors. The semi-stem loop RT-PCR based assays were performed and found that silkworm miRNAs have diverse expression patterns during development.

Published On: Molecular Biology Reports, 2010, 37(7), 3171-3176.

1 Jiangsu University of Science and Technology, Zhenjiang 212018, China.

2 Key Laboratory of Silkworm and Mulberry Genetic Improvement, Ministry of Agriculture, Sericultural Research Institute, Chinese Academy of Agricultural Sciences, Zhenjiang 212018, China.

3 School of Information Science and Technology of Xiamen University, Xiamen 361005, China.

*Corresponding author E-mail: shenxj63@yahoo.com.cn.

家蚕新型microRNA的计算机鉴定及其特征

黄　勇[1,2]　邹　权[3]　唐顺明[1,2]　王力刚[1,2]　沈兴家[1,2*]

摘要：MicroRNA（miRNAs）是一类不编码蛋白质的小RNAs，可调节基因转录后的表达水平。越来越多的证据表明，miRNAs在生物过程中起多重作用，包括在发育、细胞增殖和细胞凋亡中。基于miRNAs序列的保守性，利用基于基因组测序序列分析的计算同源性检索，共鉴定到16种新型miRNAs，并对其家族和进化保守性进行分析描述。使用这些新鉴定的miRNA在家蚕mRNA数据库进行检索，共检测到miRNA的21个潜在靶标基因。对这些miRNA的靶标基因进行预测后发现，它们中的大多数编码转录因子。半茎环RT-PCR实验检测发现miRNA在家蚕发育过程中具有不同的表达模式。

1　江苏科技大学，镇江

2　农业部蚕桑遗传改良重点实验室，蚕业研究所，中国农业科学院，镇江

3　信息科学与技术学院，厦门大学，厦门

MicroRNAs show diverse and dynamic expression patterns in multiple tissues of *Bombyx mori*

Liu SP[1] Gao S[1] Zhang DY[1]
Yin JY[1] Xiang ZH[1] Xia QY[1,2*]

Abstract:

Background: MicroRNAs (miRNAs) repress target genes at the post-transcriptional level, and function in the development and cell-lineage pathways of host species. Tissue-specific expression of miRNAs is highly relevant to their physiological roles in the corresponding tissues. However, to date, few miRNAs have been spatially identified in the silkworm.

Results: We establish for the first time the spatial expression patterns of nearly 100 miRNAs in multiple normal tissues (organs) of *Bombyx mori* females and males using microarray and Northern-blotting analyses. In all, only 10 miRNAs were universally distributed (including *bmo-let-7* and *bmo-bantam*), while the majority were expressed exclusively or preferentially in specific tissue types (e.g., *bmo-miR-275* and *bmo-miR-1*). Additionally, we examined the developmental patterns of miRNA expression during metamorphosis of the body wall, silk glands, midgut and fat body. In total, 63 miRNAs displayed significant alterations in abundance in at least 1 tissue during the developmental transition from larvae to pupae (e.g., *bmo-miR-263b* and *bmo-miR-124*). Expression patterns of five miRNAs were significantly increased during metamorphosis in all four tissues (e.g., *bmo-miR-275* and *bmo-miR-305*), and two miRNA pairs, *bmo-miR-10b-3p/5p* and *bmo-miR-281-3p/5p*, showed coordinate expression.

Conclusions: In this study, we conducted preliminary spatial measurements of several miRNAs in the silkworm. Periods of rapid morphological change were associated with alterations in miRNA expression patterns in the body wall, silk glands, midgut and fat body during metamorphosis. Accordingly, we propose that corresponding ubiquitous or tissue-specific expression of miRNAs supports their critical roles in tissue specification. These results should facilitate future functional analyses.

Published On: BMC Genomics, 2010, 11, 85.

1 Key Sericultural Laboratory of Agricultural Ministry, College of Biotechnology, Southwest University, Chongqing 400716, China.
2 Institute of Agricultural and Life Sciences, Chongqing University, Chongqing 400030, China.
*Corresponding author E-mail: xiaqy@swu.edu.cn.

MicroRNAs在家蚕多种组织中呈现多样性及动态表达模式

刘仕平[1] 高 颂[1] 张丹宇[1]
尹纪云[1] 向仲怀[1] 夏庆友[1,2*]

摘要：背景——小RNA（miRNAs）在转录后水平参与抑制靶基因的表达，从而在宿主的发育和细胞通路中发挥重要作用。并且miRNA的组织特异性表达与其在相应组织中的生理作用高度相关。然而，迄今为止，对家蚕中miRNA的鉴定工作报道较少。结果——我们首次使用芯片和Northern印迹分析建立了家蚕蛹期雌性和雄性的多个组织（器官）中近100种miRNA的空间表达模式。研究发现，总共只有10种miRNA是普遍分布（包括*bmo-let-7*和*bmo–bantam*），而大多数miRNA仅在特定组织中表达（例如*bmo-miR-275*和*bmo-miR-1*）。另外，研究了家蚕变态期间的体壁、丝腺、中肠和脂肪体中miRNA的表达模式，发现总共有63种miRNA从幼虫到蛹的转变期间在至少1个组织中表现出显著变化（例如*bmo-miR-263b*和*bmo-miR-124*），5种miRNA在变态发育期间的4个组织中的表达模式呈递增趋势（例如*bmo-miR-275*和*bmo-miR-305*），并且*bmo-miR-10b-3p/5p*和*bmo-miR-281-3p/5p*呈现相同的表达模式。结论——在本研究中，我们初步验证了家蚕中几种miRNA的时空表达模式。发现家蚕形态快速变化时期与该过程中体壁、丝腺、中肠和脂肪体中miRNA的表达变化息息相关。因此，认为普遍存在的或组织特异性的miRNA的表达支撑其在特定组织中的关键作用，这些结果将有助于今后的功能分析。

1 家蚕基因组生物学国家重点实验室，西南大学，重庆
2 农学及生命科学研究院，重庆大学，重庆

MicroRNAs of *Bombyx mori* identified by solexa sequencing

Liu SP[1] Li D[1] Li QB[2] Zhao P[1] Xiang ZH[1] Xia QY[1,3*]

Abstract:

Background: MicroRNA (miRNA) and other small regulatory RNAs contribute to the modulation of a large number of cellular processes. We sequenced three small RNA libraries prepared from the whole body, and the anterior-middle and posterior silk glands of *Bombyx mori*, with a view to expanding the repertoire of silkworm miRNAs and exploring transcriptional differences in miRNAs between segments of the silk gland.

Results: With the aid of large-scale Solexa sequencing technology, we validated 257 unique miRNA genes, including 202 novel and 55 previously reported genes, corresponding to 324 loci in the silkworm genome. Over 30 known silkworm miRNAs were further corrected in their sequence constitutes and length. A number of reads originated from the loop regions of the precursors of two previously reported miRNAs (*bmo-miR-1920* and *miR-1921*). Interestingly, the majority of the newly identified miRNAs were silkworm-specific, 23 unique miRNAs were widely conserved from invertebrates to vertebrates, 13 unique miRNAs were limited to invertebrates, and 32 were confined to insects. We identified 24 closely positioned clusters and 45 paralogs of miRNAs in the silkworm genome. However, sequence tags showed that paralogs or clusters were not prerequisites for coordinated transcription and accumulation. The majority of silkworm-specific miRNAs were located in transposable elements, and displayed significant differences in abundance between the anterior-middle and posterior silk gland.

Conclusions: Conservative analysis revealed that miRNAs can serve as phylogenetic markers and function in evolutionary signaling. The newly identified miRNAs greatly enrich the repertoire of insect miRNAs, and provide insights into miRNA evolution, biogenesis, and expression in insects. The differential expression of miRNAs in the anterior-middle and posterior silk glands supports their involvement as new levels in the regulation of the silkworm silk gland.

Published On: BMC Genomics, 2010, 11, 148.

1 Key Sericultural Laboratory of Agricultural Ministry, College of Biotechnology, Southwest University, Chongqing 400716, China.

2 Beijing Genomics Institute, Shenzhen 518083, China.

3 Institute of Agricultural and Life Sciences, Chongqing University, Chongqing 400030, China.

*Corresponding author E-mail: xiaqy@swu.edu.cn.

利用高通量测序鉴定家蚕中的MicroRNAs

刘仕平[1]　李　东[1]　李启斌[2]　赵　萍[1]　向仲怀[1]　夏庆友[1,3*]

摘要：背景——MicroRNA（miRNA）和其他小的调节RNA参与调节大多数的细胞过程。我们对家蚕整蚕以及家蚕的前中部和后部丝腺的3种小RNA文库进行测序，以扩大家蚕miRNAs的种类，并探索丝腺不同区段miRNAs的转录差异。结果——在大规模高通量测序技术的帮助下，我们验证了257种独特的miRNA基因，包括202个新的和55个先前报道的基因，对应于家蚕基因组中的324个位点。超过30种已知的家蚕miRNA需要进一步校正其序列组成和长度。许多的reads来自两个先前报道的miRNA（*bmo-miR-1920*和*miR-1921*）的前体的loop区。有趣的是，大多数新鉴定的miRNA是家蚕特异性的，23种独特的miRNA从无脊椎动物到脊椎动物广泛保守，13种是无脊椎动物独特的miRNA，32种是昆虫中独特的miRNA。我们在家蚕基因组中确定了24个紧密定位的簇和45个旁系同源基因。然而，序列标签旁系同源基因或簇不是协调转录和积累的先决条件。大多数蚕特异性miRNA位于转座子之间，且前中部和后部丝腺之间丰度差异显著。结论——保守性分析显示，miRNA可以作为系统发育的标志物在信号传递中起作用。新鉴定的miRNA大大丰富了昆虫miRNA的谱系，并为昆虫中的miRNA进化、生物发生和表达研究提供了新思路。前中部和后丝腺中miRNA的差异表达参加家蚕丝腺的调控。

1　家蚕基因组生物学国家重点实验室，生物技术学院，西南大学，重庆

2　北京基因组研究所，深圳

3　农学及生命科学研究院，重庆大学，重庆

Bioinformatics analysis on structural features of microRNA precursors in insects

Li JS[1,2] Fan W[1] You ZY[1] Zhong BX[1*]

Abstract: To date, thousands of microRNAs (miRNAs) and their precursors (pre-miRNAs) have been identified in insects and their nucleotide sequences deposited in the miRBase database. In the present work, we have systematically analyzed, utilizing bioinformatics tools, the featural differences between human and insect pre-miRNAs, as well as differences across 24 insect species. Results showed that the nucleotide composition, sequence length, nucleotides preference and secondary structure features between human and insects were different. Subsequently, with the aid of three available SVM-based prediction programs, pre-miRNA sequences were evaluated and given corresponding scores. Thus it was found that of 2 633 sequences from the 24 chosen insect species, 2 229 (84.7%) were successfully recognized by the Mirident classifier, higher than Triplet-SVM (72.5%) and PMirP (72.6%). In contrast, four species, including the domesticated silkworm, *Bombyx mori* L., the fruit fly, *Drosophila melanogaster* Meigen, the honeybee, *Apis mellifera* L. and the red flour beetle, *Tribolium castaneum* (Herbst), were found to be largely responsible for the poor performance of some sequence matching. Compared with other species, *B. mori* especially showed the worst performance with the lowest average MFE index (0.73). Collectively these results pave the way for understanding specificity and diversity of miRNA precursors in insects, and lay the foundation for the further development of more suitable algorisms for insects.

Published On: European Journal of Entomology, 2013, 110(1), 13-20.

1 College of Animal Sciences, Zijingang Campus, Zhejiang University, Hangzhou 310058, China.

2 Institute of Sericulture, Chengde Medical University, Chengde 067000, China.

*Corresponding author E-mail: bxzhong@zju.edu.cn.

昆虫中microRNA前体结构特征的生物信息学分析

李季生[1,2]　范　伟[1]　尤征英[1]　钟伯雄[1*]

摘要：迄今为止，已在昆虫中发现了数以千计的microRNA（miRNAs）及其前体（pre-miRNAs）和其在miRBase数据库中的核苷酸序列。在本文中，我们利用生物信息学工具，系统地分析了人类和昆虫pre-miRNAs的个体特征差异，以及24种昆虫物种之间的差异。结果表明，人和昆虫的核苷酸组成、序列长度、核苷酸偏好性和二级结构特征存在差异。随后，借助3种可用的基于SVM的预测程序，对前体miRNA序列进行了评估并给出相应的分数。因此，从24个选择的昆虫种类中发现2 633条序列，其中2 229条序列被Mirident分类器成功地识别，比例高达84.7%，高于Triplet-SVM（72.5%）和PMirP（72.6%）。相比之下，4个物种，包括家蚕、果蝇、蜜蜂和赤拟谷盗，一些序列匹配较差。与其他物种相比，家蚕平均MFE指数（0.73）最小。这些结果为我们了解昆虫miRNA前体的特异性和多样性提供了线索，也为进一步开发出针对昆虫前体特征的miRNA识别软件奠定了基础。

1　动物科学学院，紫金港校区，浙江大学，杭州

2　蚕业研究所，承德医学院，承德

Expression analysis of miRNAs in *BmN* cells

Yang LC[1] Lu X[2] Liu Y[2] Lü ZB[1]
Chen J[1] Yu W[1] Zhang YZ[1] Nie ZM[1*]

Abstract: MicroRNAs (miRNAs) are the family of noncoding single-strand RNA molecules of 21-25 nucleotides in length and play a broad and key regulation role in various physiological and pathological processes including differentiation, apoptosis, proliferation, and tumorigenesis. In *Bombyx mori*, a total of 487 pre-miRNAs and 562 mature miRNAs were identified by experimental or computational approaches, but their functions remain unknown. To carry out the research of gain-of-function of miRNAs in *BmN* cells, we firstly identified the endogenous expression of miRNAs in *BmN* cells by microarray and found that only 73 miRNAs could be detected by miRNA microarray. Then three low abundance or undetected miRNAs, *pri-mir-1a*, *pri-mir-8* and *pri-mir-133*, were selected to express in *BmN* cells. The eukaryotic expression vector pIEx-1 harboring baculovirus *ie1* promoter and *hr5* enhancer was screened and used for expressing miRNA in *BmN* cells. Three miRNA expression vectors pIEx-1-EGFP-pri-mir-1a/8/133 were constructed, which contained the three corresponding pri-miRNA sequences, respectively. The constructed miRNA vectors were successfully transfected into *BmN* cells and the qRT-PCR analysis showed that relative abundance of *bmo-mir-1a*, *bmo-mir-8* and *bmo-mir-133* in *BmN* cells transfected with the pIEx-1-EGFP-pri-mir-1a/8/133 is as 32, 4.4 and 904 times as that in *BmN* cells transfected with the control vector pIEx-1-EGFP, respectively. The present work lays a foundation for the further functional studies of miRNAs in silkworm.

Published On: Gene, 2012, 505(2), 240-245.

1 Institute of Biochemistry, Zhejiang Sci-Tech University, Hangzhou 310018, China.
2 Zhejiang Economic and Trade Polytechnic, Hangzhou 310018, China.
*Corresponding author E-mail: wuxinzm@126.com.

*BmN*细胞中miRNAs的表达分析

羊兰翠[1] 陆 旋[2] 刘 悦[2] 吕正兵[1]
陈 健[1] 于 威[1] 张耀洲[1] 聂作明[1*]

摘要:MicroRNA（miRNA）是一类大小为21—25 nt的非编码RNA,在各种生理和病理过程中起到广泛和关键的作用,包括在细胞分化、凋亡、繁殖和肿瘤发生等过程中。通过实验验证或计算机预测,家蚕中目前已经鉴定出487个前体miRNA和562个成熟miRNA,但是它们的功能还未知。为了在*BmN*细胞中开展miRNA的获得功能的研究,我们首次通过基因芯片鉴定了*BmN*细胞中的内源miRNA,共检测到73个miRNA的表达。随后,选择3个低丰度或未检出的细胞miRNA——*pri-mir-1a*,*pri-mir-8*和*pri-mir-133*,在*BmN*中表达。采用真核表达载体pIEx-1在*BmN*细胞中表达miRNA,该表达载体包含有杆状病毒*ie1*启动子和*hr5*增强子。我们构建了3个含有miRNA前体序列的重组表达载体pIEx-1-EGFP-pri-mir-1a/8/133,转染*BmN*细胞。荧光定量PCR分析显示,和转染pIEx-1-EGFP的对照细胞相比,转染pIEx-1-EGFP-pri-mir-1a/8/133的*BmN*细胞中*bmo-mir-1a*,*bmo-mir-8*和*bmo-mir-133*的相对表达丰度为32,4.4和904。本研究为进一步研究家蚕miRNA的功能奠定了基础。

1 生物化学研究所,浙江理工大学,杭州
2 浙江经贸职业技术学院,杭州

MicroRNA expression profiling of the fifth-instar posterior silk gland of *Bombyx mori*

Li JS[1,2#] Cai YM[3#] Ye LP[1] Wang SH[1]
Che JQ[1] You ZY[1] Yu J[3*] Zhong BX[1*]

Abstract:

Background: The growth and development of the posterior silk gland and the biosynthesis of the silk core protein at the fifth larval instar stage of *Bombyx mori* are of paramount importance for silk production.

Results: Here, aided by next-generation sequencing and microarray assay, we profile 1 229 microRNAs (miRNAs), including 728 novel miRNAs and 110 miRNA/miRNA* duplexes, of the posterior silk gland at the fifth larval instar. Target gene prediction yields 14 222 unique target genes from 1 195 miRNAs. Functional categorization classifies the targets into complex pathways that include both cellular and metabolic processes, especially protein synthesis and processing.

Conclusion: The enrichment of target genes in the ribosome-related pathway indicates that miRNAs may directly regulate translation. Our findings pave a way for further functional elucidation of these miRNAs and their targets in silk production.

Published On: BMC Genomics, 2014, 15(1),410.

1 College of Animal Sciences, Zhejiang University, Hangzhou 310058, China.

2 Institute of Sericulture, Chengde Medical University, Chengde 067000, China.

3 Key Laboratory of Genome Sciences and Information, Beijing Institute of Genomics, Chinese Academy of Sciences, Beijing 100029, China.

#These authors contributed equally.

*Corresponding author E-mail: junyu@big.ac.cn; bxzhong@zju.edu.cn.

五龄家蚕后部丝腺microRNA表达谱分析

李季生[1,2#]　蔡亦梅[3#]　叶露鹏[1]　王少华[1]
车家倩[1]　尤征英[1]　于　军[3*]　钟伯雄[1*]

摘要：背景——五龄家蚕后部丝腺的生长发育以及丝蛋白的合成对蚕丝生产至关重要。结果——本研究利用深度测序和芯片技术检测到1 229个microRNA，包括了728条新miRNA和110对miRNA/miRNA*。同时，通过软件预测到了14 222个靶基因对应1 195个miRNA。功能分析表明，这些靶基因涉及很多复杂通路，其中包括细胞和代谢过程，特别是蛋白质的合成和加工。结论——另外，核糖体相关通路中靶基因的富集表明miRNAs可能直接调控翻译。这些结果为进一步阐明miRNAs及其靶基因在蚕丝合成中的功能提供了参考。

1　动物科学学院，浙江大学，杭州
2　蚕业研究所，承德医学院，承德
3　基因组科学与信息重点实验室，北京基因组研究所，中国科学院，北京

Characterization and profiling of microRNAs in posterior silk gland of the silkworm (*Bombyx mori*)

Song F[1#] Wang X[1#] Chen C[1] Fan YY[1]
Tang SM[1,2] Huang JS[1,2] Guo XJ[1,2] Shen XJ[1,2*]

Abstract: MicroRNAs (miRNAs) regulate expression of genes at post-transcriptional level by binding on complementary sequences of target mRNAs and play multiple roles in biological processes. To investigate the differential expression of miRNAs in posterior silk gland (PSG) of silkworm (*Bombyx mori*) in different periods and regulation of miRNAs on the expression of fibroin genes, Solexa sequencing technology was used to detect miRNAs in PSGs of fourth-instar day-2 larvae and fifth-instar day-3 larvae, respectively. As a result, 466 previously reported miRNAs, and 35 novel miRNAs were detected, and 499 of these detected miRNAs are predicted to target 13 383 genes by target prediction softwares. Additionally, 29 miRNAs expressed differently between the PSG of fourth-instar day-2 larvae and fifth-instar day-3 larvae were found, and the differential expression of these miRNAs may play an important role in the expression of fibroin genes.

Published On: Genes & Genomics, 2015, 37(8), 703-712.

1 Jiangsu Key Laboratory of Sericultural Biology and Biotechnology, School of Biotechnology, Jiangsu University of Science and Technology, Zhenjiang 212018, China.

2 Key Laboratory of Silkworm and Mulberry Genetic Improvement, Ministry of Agriculture, Sericultural Research Institute, Chinese Academy of Agricultural Sciences, Zhenjiang 212018, China.

[#]These authors contributed equally.

[*]Corresponding author E-mail: shenxjsri@163.com.

家蚕幼虫后部丝腺microRNAs的鉴定与表达分析

宋 菲[1#] 王 欣[1#] 陈 晨[1] 范洋洋[1]
唐顺明[1,2] 黄金山[1,2] 郭锡杰[1,2] 沈兴家[1,2*]

摘要：MicroRNA（miRNA）通过序列互补配对的方式在转录后水平对靶标mRNA的表达进行调控，参与一系列重要的生命过程，有着众多的生物学功能。为了研究家蚕幼虫不同时期后部丝腺中miRNAs的表达差异，以及家蚕miRNAs对蚕丝蛋白基因表达的调控作用，利用Solexa测序技术对家蚕四龄第2天和五龄第3天幼虫后部丝腺进行了miRNA测序，检测到466个已知miRNAs，预测到35个新的miRNAs；对获得的499个miRNAs进行靶基因预测，获得13 383个靶基因；家蚕四龄第2天和五龄第3天幼虫后部丝腺样品中，有29个miRNAs存在表达差异，推测这些差异表达miRNAs对丝素蛋白基因表达可能存在调控作用。

1 江苏蚕桑生物学与生物技术重点实验室，生物技术学院，江苏科技大学，镇江
2 农业部蚕桑遗传改良重点实验室，蚕业研究所，中国农业科学院，镇江

BmNPV-miR-415 up-regulates the expression of *TOR2* via *Bmo-miR-5738*

Cao XL[1,3] Huang Y[2,4] Xia DG[1,4] Qiu ZY[1,4]
Shen XJ[1,4] Guo XJ[1,4] Zhao QL[1,4*]

Abstract: MicroRNAs (miRNAs) have emerged as key players in host-pathogen interaction and many virus-encoded miRNAs have been identified (computationally and / or experimentally) in a variety of organisms. A novel *Bombyx mori* nucleopolyhedrosis virus (*Bm*NPV) - encoded miRNA miR-415 was previously identified through high-throughput sequencing. In this study, a *BmNPV-miR-415* expression vector was constructed and transfected into *BmN* cells. The differentially expressed protein target of rapamycin isoform 2 (TOR2) was observed through two-dimensional gel electrophoresis and mass spectrometry. Results showed that *TOR2* is not directly a target gene of *BmNPV-miR-415*, but its expression is up-regulated by *BmNPV-miR-415* via *Bmo-miR-5738*, which could be induced by *Bm*NPV.

Published On: Saudi Journal of Biological Sciences, 2017, 24(7), 1614-1619.

1 Jiangsu University of Science and Technology, Zhenjiang 212018, China.
2 College of Animal Science and Technology, Henan University of Science and Technology, Luoyang 471003, China.
3 Dezhou College, Dezhou 253023, China.
4 Sericultural Research Institute, Chinese Academy of Agricultural Sciences, Zhenjiang 212018, China.
*Corresponding author E-mail: qlzhao302@126.com.

BmNPV-miR-415 通过 *Bmo-miR-5738* 上调表达 *TOR2* 基因

曹学亮[1,3]　黄　勇[2,4]　夏定国[1,4]　裘智勇[1,4]
沈兴家[1,4]　郭锡杰[1,4]　赵巧玲[1,4*]

摘要：MicroRNAs（miRNAs）已经成为宿主-病原体相互作用的关键因子，现已从多种生物体中通过理论预测或实验鉴定到大量病毒编码的miRNA，其中*miR-415*是通过高通量测序的方法从家蚕核型多角体病毒*Bm*NPV中鉴定到的miRNA。在本研究中，通过构建*BmNPV-miR-415*表达载体并转染家蚕*BmN*细胞系，利用双向电泳技术和质谱技术从转染的细胞中鉴定获得一个差异表达蛋白TOR2。分析结果表明，虽然*TOR2*并不是*BmNPV-miR-415*调控的靶基因，但是*BmNPV-miR-415*能通过*Bmo-miR-5738*使该基因表达量上调，而且这个过程是受*Bm*NPV诱导的。

1　江苏科技大学，镇江
2　动物科技学院，河南科技大学，洛阳
3　德州学院，德州
4　蚕业研究所，中国农业科学院，镇江

Bmo-miR-2758 targets *BmFMBP-1* (Lepidoptera: Bombycidae) and suppresses its expression in *BmN* cells

Wang X[1,2] Tang SM[1,2] Song F[1]
Chen C[1] Guo XJ[1,2] Shen XJ[1,2*]

Abstract: MicroRNAs (miRNAs) are an abundant family of endogenous non-coding small RNA molecules. They play crucial roles on regulation of life processes both in plants and animals. Fibroin modulator binding protein-1 (FMBP-1) is a silk gland transcription factor of *Bombyx mori*, which is considered as a trans-activator of fibroin genes. And bioinformatics prediction showed that at the 3′ untranslated region (3′UTR) of *BmFMBP-1* there were binding sites for three *bmo*-miRNAs, *bmo-miR-2b**, *bmo-miR-305*, and *bmo-miR-2758*, separately. In order to validate whether these *bmo*-miRNAs involved in the regulation of *BmFMBP-1* expression, the expression levels of three *bmo*-miRNAs and *BmFMBP-1* in the middle silk gland (MSG) and posterior silk gland (PSG) during the fourth- and fifth- larval stages of *B. mori* were measured by semi-quantitative reverse transcription polymerase chain reaction. The results revealed that the expression level of *bmo-miR-2758* was the highest in the three, and it expressed higher in the PSG than in the MSG with a similar expression pattern as *BmFMBP-1*, implying that *bmo-miR-2758* may involved in regulation of *BmFMBP-1*. To validate the regulation function of *bmo-miR-2758* on *BmFMBP-1*, recombinant plasmids pcDNA3 [*ie1-egfp-pri-bmo*-miR-2758-SV40] and pGL3 [*A3-luc-FMBP*-1 3′UTR-SV40] were constructed and co-transfected in *BmN* cells. The dual-luciferase reporter assay system was used for assay of transient expression. The results showed that the expression of the luciferase reporter was significantly decreased when pGL3 [*A3-luc-FMBP*-1 3′UTR-SV40] co-transfected with pcDNA3 [*ie1-egfp-pri*-bmo-miR-2758-SV40] ($P<0.01$). Furthermore, when the artificial antisense RNA of *bmo-miR-2758* (inhibitor) was added to the above co-transfection, the expression of the luciferase reporter was recovered significantly ($P<0.01$). These results suggest that *bmo-miR-2758* represses the expression of *BmFMBP-1 in vitro*.

Published On: Journal of Insect Science, 2016, 16(1), 28.

1 Jiangsu Key Laboratory of Sericultural Biology and Biotechnology, School of Biotechnology, Jiangsu University of Science and Technology, Zhenjiang 212003, China.

2 Key Laboratory of Silkworm and Mulberry Genetic Improvement, Ministry of Agriculture, Sericultural Research Institute, Chinese Academy of Agricultural Sciences, Zhenjiang 212018, China.

*Corresponding author E-mail: shenxjsri@163.com.

家蚕*miR-2758*体外下调丝腺转录因子*FMBP-1*的表达

王　欣[1,2]　唐顺明[1,2]　宋　菲[1]
陈　晨[1]　郭锡杰[1,2]　沈兴家[1,2*]

摘要：MicroRNA（miRNA）是一类丰富的内源性非编码小RNA分子家族，其在动植物生命过程调控中起到重要作用。家蚕丝腺调节结合因子1（FMBP-1）是一种家蚕丝腺转录因子，也被认为是丝素基因的反式激活因子。生物信息学预测显示*BmFMBP-1*的3′UTR上有3个家蚕miRNAs结合位点，分别是*bmo-miR-2b**、*bmo-miR-305*和*bmo-miR-2758*。为了验证这3个家蚕miRNAs对*BmFMBP-1*基因的表达调控，利用半定量RT-PCR对四龄和五龄幼虫中部丝腺（MSG）和后部丝腺（PSG）的3个miRNA和*BmFMBP-1*基因表达水平进行检测。结果表明*bmo-miR-2758*的表达量最高，且在后部丝腺中的表达量高于在中部丝腺中，这与*BmFMBP-1*基因的表达趋势保持一致，暗示*bmo-miR-2758*可能参与*BmFMBP-1*基因表达的调控。为了研究*bmo-miR-2758*对*BmFMBP-1*的调控作用，分别构建重组质粒pcDNA3 [*ie1-egfp-pri-bmo*-miR-2758-SV40]和pGL3 [*A3-luc-FMBP*-1 3′UTR-SV40]并共转染家蚕*BmN*细胞，双荧光素酶报告分析系统检测其瞬时表达，结果表明，当pcDNA3 [*ie1-egfp-pri-bmo*-miR-2758-SV40]与pGL3 [*A3-luc-FMBP*-1 3′UTR-SV40]共转染后，荧光素酶报告基因的表达显著降低（$P<0.01$）。进一步实验，当人工合成的*bmo-miR-2758*反义RNA（抑制剂）加入到上述共转染系统中，荧光素酶报告基因的表达明显恢复（$P<0.01$）。上述结果表明*bmo-miR-2758*在体外抑制*BmFMBP-1*的表达。

1　农业部蚕桑遗传改良重点实验室，江苏科技大学，镇江
2　蚕桑遗传改良重点实验室，蚕业研究所，中国农业科学院，镇江

RIP-seq of *Bm*Ago2-associated small RNAs reveal various types of small non-coding RNAs in the silkworm, *Bombyx mori*

Nie ZM[1] Zhou F[1] Li D[1] Lü ZB[1] Chen J[1] Liu Y[2] Shu JH[1] Sheng Q[1]
Yu W[1] Zhang WP[1] Jiang CY[1] Yao YH[1] Yao JM[3] Jin YF[4] Zhang YZ[1*]

Abstract:

Background: Small non-coding RNAs (ncRNAs) are important regulators of gene expression in eukaryotes. Previously, only microRNAs (miRNAs) and piRNAs have been identified in the silkworm, *Bombyx mori*. Furthermore, only ncRNAs (50-500 nt) of intermediate size have been systematically identified in the silkworm.

Results: Here, we performed a systematic identification and analysis of small RNAs (18-50 nt) associated with the *Bombyx mori* argonaute2 (*Bm*Ago2) protein. Using RIP-seq, we identified various types of small ncRNAs associated with BmAGO2. These ncRNAs showed a multimodal length distribution, with three peaks at ~20 nt, ~27 nt and ~33 nt, which included tRNA-, transposable element (TE)-, rRNA-, snoRNA- and snRNA-derived small RNAs as well as miRNAs and piRNAs. The tRNA-derived fragments (tRFs) were found at an extremely high abundance and accounted for 69.90% of the BmAgo2-associated small RNAs. Northern blotting confirmed that many tRFs were expressed or up-regulated only in the *Bm*NPV-infected cells, implying that the tRFs play a prominent role by binding to *Bm*Ago2 during *Bm*NPV infection. Additional evidence suggested that there are potential cleavage sites on the D, anti-codon and TψC loops of the tRNAs. TE-derived small RNAs and piRNAs also accounted for a significant proportion of the *Bm*Ago2-associated small RNAs, suggesting that *Bm*Ago2 could be involved in the maintenance of genome stability by suppressing the activities of transposons guided by these small RNAs. Finally, Northern blotting was also used to confirm the *Bombyx* 5.8 S rRNA-derived small RNAs, demonstrating that various novel small RNAs exist in the silkworm.

Conclusions: Using an RIP-seq method in combination with Northern blotting, we identified various types of small RNAs associated with the *Bm*Ago2 protein, including tRNA-, TE-, rRNA-, snoRNA- and snRNA-derived small RNAs as well as miRNAs and piRNAs. Our findings provide new clues for future functional studies of the role of small RNAs in insect development and evolution.

Published On: BMC Genomics, 2013, 14.

1 College of Life Sciences, Zhejiang Sci-Tech University, Hangzhou 310018, China.
2 Zhejiang Economic & Trade Polytechnic, Hangzhou 310018, China.
3 College of Materials and Textile, Zhejiang Sci-Tech University, Hangzhou 310018, China.
4 College of Life Sciences, Zhejiang University, Hangzhou 310058, China.
*Corresponding author E-mail: yaozhou@chinagene.com.

RIP-seq 法鉴定 *Bm*Ago2 结合小 RNA 揭示家蚕中多种不同类型的非编码小 RNA

聂作明[1] 周 芳[1] 李 丹[1] 吕正兵[1] 陈 健[1] 刘 悦[2] 舒建洪[1] 盛 清[1]
于 威[1] 张文平[1] 蒋彩英[1] 姚玉华[1] 姚菊明[3] 金勇丰[4] 张耀洲[1*]

摘要：背景——非编码小 RNA（ncRNAs）是真核生物中调控基因表达的重要分子。目前家蚕中仅仅鉴定了 microRNAs（miRNAs）和 piRNAs。同时，只有大小在 50—500 nt 的非编码 RNA 在家蚕中被系统性地鉴定了。结果——本研究系统分析和鉴定了大小为 18—50 nt 的家蚕 *Bm*Ago2 结合小 RNA。采用 RIP-seq 方法，我们鉴定出多种 *Bm*AGO2 结合非编码小 RNA。这些 ncRNA 大小不一，长度分布图显示 3 个峰值分别在~20 nt、~27 nt 和~33 nt，包括 tRNA-、转座元件（TE）-、rRNA-、snoRNA-、snRNA-派生而来的小 RNA（除开 miRNA 和 piRNA）等。*Bm*Ago2 蛋白结合小 RNA 中 tRNA 来源小 RNA（tRFs）丰度极高，占整个 *Bm*Ago2 结合小 RNA 的 69.90%。Northern blot 结果证实许多 tRF 小 RNA 仅仅只在病毒感染的家蚕细胞中表达或上调表达，暗示 tRF 可能在 *Bm*NPV 病毒感染过程中通过结合 *Bm*Ago2 蛋白行使作用。另有证据暗示 tRF 主要来源于 tRNA 的 5′或 3′末端序列，且通过对反密码子环区、D 环区和 TψC 环区特异性剪切产生。*Bm*Ago2 结合小 RNA 中，TE 来源小 RNA 和 piRNA 也占很大一部分，表明 *Bm*Ago2 蛋白可能在这些小 RNA 的介导下通过抑制转座子的活性从而参与基因组稳定性的调控。最后，Northern blot 还证实了家蚕 5.8S rRNA 来源小 RNA，表明家蚕中可能存在多种新型小 RNA 分子。结论——采用 RIP-seq 和 Northern blotting 方法，我们鉴定了多种 *Bm*Ago2 蛋白结合小 RNA，包括 tRNA-、TE-、rRNA-、snoRNA-、snRNA-derived 小 RNA（除开 miRNA 和 piRNA）等。本文为进一步研究这些小 RNA 在昆虫发育和进化中的功能提供了新的线索。

1 生命科学学院，浙江理工大学，杭州
2 浙江经贸职业技术学院，杭州
3 材料与纺织学院，浙江理工大学，杭州
4 生命科学学院，浙江大学，杭州

Characterization of Argonaute family members in the silkworm, *Bombyx mori*

Wang GH[1,2] Jiang L[1] Zhu L[1] Cheng TC[1,2]
Niu WH[2] Yan YF[2] Xia QY[1*]

Abstract: The Argonaute protein family is a highly conserved group of proteins, which have been implicated in RNA silencing in both plants and animals. Here, four members of the *Argonaute* family were systemically identified based on the genome sequence of *Bombyx mori*. Based on their sequence similarity, *BmAgo1* and *BmAgo2* belong to the *Ago* subfamily, while *BmAgo3* and *BmPiwi* are in the *Piwi* subfamily. Phylogenetic analysis reveals that silkworm *Argonaute* family members are conserved in insects. Conserved amino acid residues involved in recognition of the 5′ end of the small RNA guide strand and of the conserved (aspartate, aspartate and histidine [DDH]) motif present in their PIWI domains suggest that these four Argonaute family members may have conserved slicer activities. The results of microarray expression analysis show that there is a low expression level for *B. mori Argonaute* family members in different tissues and different developmental stages, except for *BmPiwi.* All four *B. mori Argonaute* family members are upregulated upon infection with *B. mori* nucleopolyhedrovirus. The complete coding sequence of *BmPiwi*, the homolog of *Drosophila piwi*, was cloned and its expression occurred mainly in the area where spermatogonia and spermatocytes appear. Our results provide an overview of the *B. mori Argonaute* family members and suggest that they may have multiple roles. In addition, this is also the first report, to our knowledge, of the response of RNA silencing machinery to DNA virus infection in insects.

Published On: Insect Science, 2013, 20(1), 78-91.

1 State Key Laboratory of Silkworm Genome Biology, School of Biotechnology, Southwest University, Chongqing 400716, China.
2 Institute of Agriculture and Life Sciences, Chongqing University, Chongqing 400044, China.
*Corresponding author E-mail: xiaqy@swu.edu.cn.

家蚕Argonaute蛋白家族成员表征

王根洪[1,2]　蒋　亮[1]　祝　力[1]　程廷才[1,2]
牛维环[2]　闫亚飞[2]　夏庆友[1*]

摘要：Argonaute蛋白家族是一类高度保守的蛋白，在植物和动物中参与RNA沉默。本研究基于家蚕基因组序列鉴定到4个*Argonaute*家族成员。基于序列相似性，*BmAgo1*和*BmAgo2*属于*Ago*亚家族，而*BmAgo3*和*BmPiwi*属于*Piwi*亚家族。系统发育分析表明，家蚕*Argonaute*家族成员在昆虫中保守。4个家族成员均含有识别小RNA引导链5′端的保守氨基酸残基和存在于PIWI功能域的保守基序［天冬氨酸、天冬氨酸和组氨酸(DDH)］，说明它们可能拥有保守的剪切活性。基因芯片表达谱分析的结果显示，除了*BmPiwi*，其他3个家蚕*Argonaute*家族成员在不同组织和不同发育阶段均呈现较低水平的表达。在感染家蚕核型多角体病毒后，家蚕4个*Argonautes*成员均被诱导上调表达。此外，本研究还克隆了果蝇*piwi*同源基因*Bmpiwi*基因的完整编码序列，RNA原位杂交显示其在精巢主要表达于精原细胞和精母细胞中。本研究系统鉴定分析了家蚕的*Argonaute*家族成员，现有结果初步说明它们可能在多个重要的生命活动中发挥着作用。此外，这也是第一次在昆虫中报道RNA沉默可能参与宿主对DNA病毒的感染应答。

1　家蚕基因组生物学国家重点实验室，生物技术学院，西南大学，重庆
2　农学与生命科学研究院，重庆大学，重庆

第八章

蚕的细胞生物学

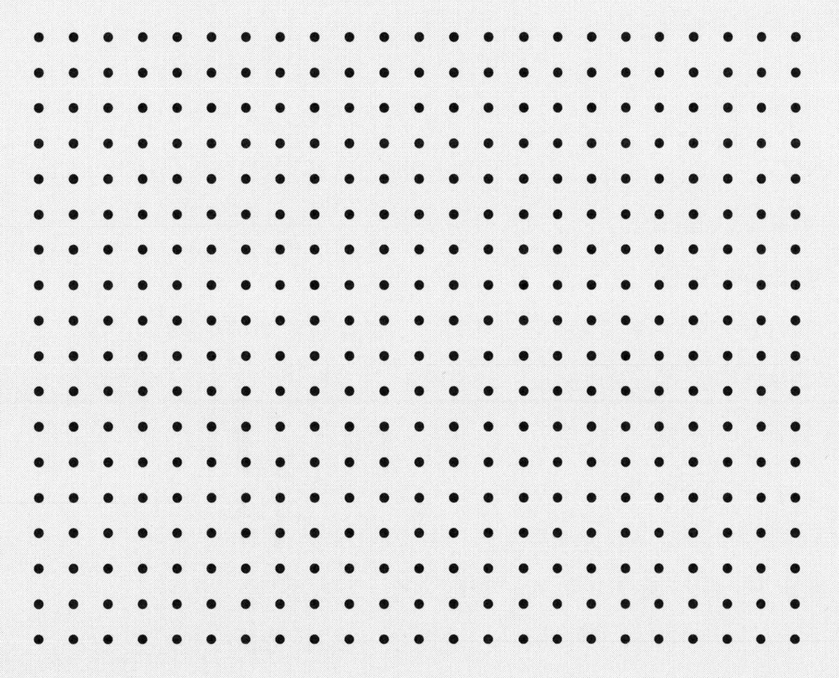

Chapter 8

昆虫细胞生物学属昆虫学前沿重大热点和难点领域，该领域研究是在分子细胞生物学层面阐明昆虫变态发育这一重大科学问题的关键之一，将有力推动家蚕变态发育分子机制以及重要经济性状基因功能、重大疾病发生机理与防治、突变系统的模式化等基础研究的发展。

1. 家蚕血细胞的发育分化调控研究

家蚕基因组生物学国家重点实验室家蚕干细胞与转化研究方向初步建立起家蚕干细胞研究技术平台，构建了适合家蚕等昆虫类的干细胞生物学研究体系，全面系统地对家蚕血细胞的发育调控分子机制进行研究。完成了对家蚕血液发育相关基因 *intergrin* 家族，*LZ*等基因的克隆鉴定及功能探究。

2. 家蚕中肠细胞的发育调控研究

建立家蚕中肠干细胞研究技术体系，初步定位家蚕中肠干细胞的位置位于家蚕幼虫中肠基底膜处。完成对家蚕中肠细胞发育调控机制初步研究。

3. 深化家蚕细胞的多元化利用研究

对家蚕不同组织进行了原代细胞培养，已成功建立 3 种家蚕胚胎细胞系，有望成为家蚕基因功能的研究和建立转基因工程细胞系的重要工具。并以原有的细胞系和新建的细胞系为研究对象，开展一系列的相关研究，如家蚕细胞凋亡、增殖、细胞周期调控等研究。

崔红娟、杨丽群

Isolation and characterization of a PUF-domain of *pumilio* gene from silkworm *Bombyx mori*

Tian BX Chen KP Yao Q*

Chen HQ Zhou Y Zhou J

Abstract: Pumilio is a sequence-specific RNA-binding protein that binds to target mRNA to repress its translation. The PUF-domain, the RNA-binding motif of pumilio, is highly conserved across species. In the present study, a partial *pumilio* gene with complete PUF-domain in *Bombyx mori* has been cloned using 3′ and 5′ RACE for the first time, designated as *BmPUM*. The sequence of *BmPUM* has been registered in GenBank under the accession number FJ461590. Comparative sequence analysis revealed that the deduced protein *Bm*PUM contains a PUF-domain and shares 83% identity with Drosophila pumilio, hence belongs to pumilio family. One encoding sequence of *BmPUM* fragment was successfully expressed in *Escherichia coli.* Western blot indicated that the anti-*Bm*PUM antibody and anti-*Drosophila* pumilio antibody all could specifically detect *BmPUM* expressed in prokaryotic cells. The tissue expression pattern performed by real-time PCR and Western blot demonstrated that *BmPUM* expressed in various tissues, especially in testis and ovary. Those data collectively indicated that *BmPUM* belongs to the extremely high conserved RNA-binding domain. Conserved function of pumilio protein in invertebrates and vertebrates suggested that *BmPUM* could also play an important role in the proliferation of silkworm germline stem cell.

Published On: African Journal of Biotechnology, 2009, 8（6）, 986-994.

Institute of Life Sciences, Jiangsu University, Zhenjiang 212013, China.

*Corresponding author E-mail: yaoqin@ujs.edu.cn.

家蚕*pumilio*基因PUF结构域的分离与鉴定

田宝霞　陈克平　姚　勤*
陈慧卿　周　阳　周　佳

摘要：Pumilio是一种序列特异性的RNA结合蛋白质，其结合靶mRNA从而抑制其翻译。PUF结构域作为Pumilio的RNA结合基序，在物种间是高度保守的。在本研究中，通过使用3′RACE和5′RACE技术首次克隆了在家蚕中具有完整PUF结构域的部分*pumilio*基因片段，并将其命名为*BmPUM*。*BmPUM*的序列已在GenBank中注册，登录号为FJ461590。序列比较分析显示，推导出的*BmPUM*蛋白质盒有一个PUF结构域，与果蝇的Pumilio序列具有83%的同源性，因此属于Pumilio家族。*BmPUM*片段的一个编码序列在大肠杆菌中成功表达。Western blot表明，*Bm*PUM抗体和果蝇Pumilio抗体均能够特异性检测到原核细胞中表达的*BmPUM*。通过实时PCR和Western blot进行组织表达模式分析，证明了*BmPUM*在各种组织中均有表达，特别是在精巢和卵巢 。这些数据表明*BmPUM*属于保守性极高的RNA结合域。无脊椎动物和脊椎动物中Pumilio蛋白质的保守功能表明，*BmPUM*也在家蚕生殖干细胞的增殖中起重要作用。

生命科学研究院，江苏大学，镇江

Transcriptome analysis of *BmN* cells following over-expression of *BmSTAT*

Hu XL[1,2#] Zhang X[1#] Wang J[1] Huang ML[1]
Xue RY[1,2] Cao GL[1,2] Gong CL[1,2*]

Abstract: The Janus kinase / signal transducers and activators of transcription (JAK / STAT) signaling pathway are involved in immune response, cell proliferation, differentiation, cell migration and apoptosis. In order to better understand the role of the JAK / STAT pathway in insects we chose *Bombyx mori* as an experimental model system. Over-expression of *BmSTAT* in a *BmN* cell line increased the number of cells in the G_2 phase of the cell cycle. Genome-wide transcriptome analysis was performed to identify genes that were differentially expressed following *BmSTAT* overexpression. Transcriptome data showed that 10 853 and 10 129 expressed genes were obtained from the normal *BmN* cells and transformed cells, respectively. A total of 800 differentially expressed genes (DEGs) were detected, of which 787 were up-regulated and 13 were down-regulated with T test. In case of FC-test, 252 DEGs were detected, and 123 were expressed in the transformed cells and remaining were in the normal cells. Gene ontology (GO) annotation predicted a functional role for DEGs in catalytic activity, binding, transport, biological regulation, cellular and metabolic processes and pigmentation, while Kyoto encyclopedia of genes and genomes (KEGG) analysis revealed the affected genes to be involved in a multitude of cell signaling pathways. Our findings implicate JAK / STAT signaling in regulating the cell cycle in *Bombyx mori*, probably in combination with other pathways. These findings justify further investigation into the functional role of the *BmSTAT* gene.

Published On: Molecular Genetics and Genomics, 2015, 290(6), 2137-2146.

1 School of Biology and Basic Medical Sciences ,Soochow University, Suzhou 215123, China.
2 National Engineering Laboratory for Modern Silk, Soochow University, Suzhou 215123, China.
#These authors contributed equally.
*Corresponding author E-mail: gongcl@suda.edu.cn

过表达*BmSTAT*的*BmN*细胞的转录组分析

胡小龙[1,2#] 张 曦[1#] 王 健[1] 黄茉莉[1]
薛仁宇[1,2] 曹广力[1,2] 贡成良[1,2*]

摘要：Janus激酶/信号转导和转录激活子(JAK / STAT)信号通路参与调节免疫应答、细胞增殖、细胞分化、细胞迁移和细胞凋亡。为了更好地了解JAK/STAT通路在昆虫中的作用，我们选择了家蚕作为实验模型系统。在*BmN*细胞中过表达*BmSTAT*增加了处于G_2期的细胞比例，进行基因组转录分析，以识别分析*BmSTAT*过表达后差异表达的基因。转录组测序结果显示，从正常*BmN*细胞、过表达*BmSTAT*细胞中分别检测到10 853个和10 129个基因表达，通过*t*检验共获得800个差异表达的基因，其中包括787个上调表达基因和13个下调表达基因；经过FC检验，共检测到252个差异表达的基因，其中123个基因在过表达*BmSTAT*细胞系中表达而其他基因在正常细胞中表达。GO富集分析显示，这些差异表达的基因在催化、结合、转运、生物调节、细胞代谢和色素沉着方面起作用；KEGG富集分析显示，这些差异表达基因参与调节多种细胞信号通路。我们的研究结果表明：JAK / STAT信号通路可能是通过与其他通路共同作用来参与调节家蚕细胞周期。这些发现也对*BmSTAT*基因的功能研究有很大帮助。

1 基础医学与生命科学学院，苏州大学，苏州
2 现代丝绸国家工程实验室，苏州大学，苏州

BmDredd regulates the apoptosis coordinating with *BmDaxx*, *BmCide-b*, *BmFadd* and *BmCreb* in *BmN* cells

Chen RT Jiao P Lu Y Xin HH
Zhang DP Wang MX Liang S Miao YG*

Abstract: The apoptosis mechanisms in mammals were investigated relatively clearly. However, little is known about how apoptosis is achieved at a molecular level in silkworm cells. We cloned a caspase homologous gene named *BmDredd* (Where *Bm* is *Bombyx mori* and *Dredd* is death-related ced-3/Nedd2-like Caspase) in *BmN* cells from the ovary of *Bombyx mori* and analyzed its biological information. We constructed the N-terminal, C-terminal and overexpression vector of *BmDredd* respectively. Our results showed that the transcriptional expression level of *BmDredd* was increased in the apoptotic *BmN* cells. Furthermore, overexpression of *BmDredd* increased the caspase-3 / 7 activity. Simultaneously, RNAi of *BmDredd* could save *BmN* cells from apoptosis. The immunofluorescence study showed that *BmDredd* located at the cytoplasm in normal cell, otherwise be found at the nucleus when cells undergo apoptosis. Moreover, we quantified the transcriptional expressions of apoptosis-related genes including *BmDredd*, *BmDaxx*(where Daxx is death-domain associated protein), *BmCide-b*(where Cide-b is cell death-domain associatal protein), *BmFadd*(Fadd is fas-associated via death domain) and *BmCreb*(where Creb is cAmp-response element binding protein) in *BmN* cells with dsRNA interferences to detect the molecular mechanism of apoptosis. In conclusion, *BmDredd* may function for promoting apoptosis and there are various regulatory interactions among these apoptosis-related genes.

Published On: Archives of Insect Biochemistry and Physiology, 2016, 93(3), 160 - 173.

Institute of Sericulture and Apiculture, College of Animal Sciences, Zhejiang University, Hangzhou 310058, China.
*Corresponding author E-mail: miaoyg@zju.edu.cn.

*BmDredd*与*BmDaxx*、*BmCide-b*、*BmFadd*和*BmCreb* 协同调控家蚕细胞的凋亡

陈瑞婷　矫　鹏　陆　岳　辛虎虎
张登攀　王梅仙　梁　爽　缪云根*

摘要：哺乳动物中的凋亡机制研究得相对清楚，然而对于家蚕细胞分子水平的凋亡我们知之甚少。我们在*BmN*细胞中克隆了家蚕*caspase*同源基因*BmDredd*（*Bm*为*Bombyx mori*，*Dredd*为死亡相关的ced-3/Nedd2样Caspase）并且进行了生物信息学分析。分别构建了*BmDredd N*端、C端以及ORF全长的过表达载体。研究结果表明，在发生凋亡的*BmN*细胞中*BmDredd*的表达量上升。另外，过表达*BmDredd*使得caspase-3/7活性升高。同时，RNA干扰*BmDredd*基因的表达可以减少细胞凋亡的发生。免疫荧光显示*Bm*Dredd定位在正常细胞的细胞质内，当细胞发生凋亡时，会转移到细胞核。最后，我们分别使用dsRNA干扰不同凋亡相关基因表达，检测了干扰后*BmDredd*，*BmDaxx*（其中*Daxx*是死亡域相关蛋白），*BmCide-b*（其中Cide-b是诱导细胞死亡的DFF45样效因子），*BmFadd*（Fas相关死亡域蛋白）和*BmCreb*（Creb是cAMP-response元件结合蛋白）的表达量，初步探索了它们的分子凋亡调控机理。研究结果表明，*BmDredd*有着类似caspase作用，可以引起细胞凋亡，并且与凋亡相关基因*BmDaxx*，*BmCide-b*，*BmFadd*和*BmCreb*有着复杂的调控关系。

蚕蜂研究所，动物科学学院，浙江大学，杭州

The protein import pore Tom40 in the microsporidian *Nosema bombycis*

Lin LP[1] Pan GQ[1] Li T[1] Dang XQ[1] Deng YH[1]
Ma C[1] Chen J[1] Luo J[1] Zhou ZY[1,2*]

Abstract: Microsporidia, an unusual group of unicellular parasites related to fungi, possess a highly reduced mitochondrion known as the mitosome. Since mitosomes lack an organellar genome, their proteins must be translated in the cytosol before being imported into the mitosome via translocases. We have identified a *Tom40* gene (*NbTom40*), the main component of the translocase of the outer mitochondrial membrane, in the genome of the microsporidian *Nosema bombycis*. *Nb*Tom40 is reduced in size, but it is predicted to form a β-barrel structure composed of 19 β-strands. Phylogenetic analysis confirms that *Nb*Tom40 forms a clade with Tom40 sequences from other species, distinct from a related clade of voltage-dependent anion channels (VDACs). The *Nb*Tom40 contains a β - signal motif that the polar residue is substituted by glycine. Furthermore, we show that expression of *Nb*Tom40, as a GFP fusion protein within yeast cells, directs GFP to mitochondria of yeast. These findings suggest that *Nb*Tom40 may serve as an import channel of the microsporidian mitosome and facilitate protein translocation into this organelle.

Published On: Journal of Eukaryotic Microbiology, 2012, 59(3), 251-257.

1 State Key Laboratory of Silkworm Genome Biology, Southwest University, Chongqing 400716, China.
2 Laboratory of Animal Biology, Chongqing Normal University, Chongqing 400047, China.
*Corresponding author E-mail: zyzhou@swu.edu.cn.

家蚕微孢子虫纺锤剩体转运蛋白*Nb*Tom40的鉴定

林立鹏[1] 潘国庆[1] 李 田[1] 党晓群[1] 邓远洪[1]
马 成[1] 陈 洁[1] 罗 洁[1] 周泽扬[1,2*]

摘要：微孢子虫是一种与真菌有关的不常见的单细胞寄生虫，具有高度减缩的线粒体源细胞器即纺锤剩体。由于纺锤剩体自身缺乏细胞器基因组，纺锤剩体蛋白需先在胞质溶胶中翻译，然后才能通过转运酶导入纺锤剩体。我们在家蚕微孢子虫基因组中鉴定了一个线粒体外膜转运系统的核心组分*NbTom40*。*Nb*Tom40在长度有所缩短，但经三维结构预测其19个β片层仍可折叠形成β桶状拓扑结构。系统发育树显示，*Nb*Tom40与其他物种的Tom40聚为一类，不属于相关的电压依赖阴离子通道（VDACs）亚家族。相关的*Nb*Tom40含有一个β信号基序，其极性氨基酸残基位点被甘氨酸替代。将*Nb*Tom40与GFP融合表达后发现该蛋白可以靶向酿酒酵母的线粒体。这些结果暗示，*Nb*Tom40作为家蚕微孢子虫纺锤剩体主要的转运通道蛋白，有利于蛋白转运至纺锤剩体中。

1 家蚕基因组生物学国家重点实验室，西南大学，重庆
2 动物学重点实验室，重庆师范大学，重庆

Hemocytes and hematopoiesis in the silkworm, *Bombyx mori*

Liu F[1,2] Xu QY[2] Zhang QL[2]
Lu AR[2] BT Beerntsen[3] Ling EJ[2*]

Abstract: The silkworm, *Bombyx mori*, is a typical Lepidopteran insect. In the silkworm hemolymph, there are 5 types of circulating hemocytes that are classified as prohemocytes, granulocytes, plasmatocytes, spherulocytes and oenocytoids. All of them are involved in humoral and cellular immunity either directly or indirectly. Insect hematopoietic organs can produce hemocytes that are continuously released into the circulation. Recent studies indicate that in the hematopoietic organs of silkworm larvae, there are mainly prohemocytes and oenocytoids. Based on *in vitro* observations, silkworm prohemocytes can differentiate into plasmatocytes and granulocytes, and granulocytes can differentiate into spherulocytes. The silkworm also has a novel type of hematopoiesis. When its hematopoietic organs are extirpated through a surgical operation, circulating hemocytes can still remain at a high level through the wandering stage due to an increase in the level of cell division. Previously, oenocytoids have been considered as the only source of prophenoloxidase (PPO) which is an important immunity protein in insects. However, recent studies in different insect species, as well as in the silkworm, show that additional hemocyte types contain PPO. Furthermore, PPO can be produced by epidermal cells in the hindgut of the silkworm. Consequently, the silkworm is a valuable model to study hemocyte development and cellular and humoral immune responses.

Published On: Invertebrate Survival Journal, 2013, 10(1), 102-109.

1 Department of Biological Sciences and Technology, Shanxi Xueqian Normal University, Xi'an 710100, China.

2 Key Laboratory of Insect Developmental and Evolutionary Biology, Institute of Plant Physiology and Ecology, Shanghai Institutes for Biological Sciences, Chinese Academy of Sciences, Shanghai 200032, China.

3 Department of Veterinary Pathobiology, University of Missouri, MO 65211, USA.

*Corresponding author E-mail: erjunling@sippe.ac.cn.

家蚕血细胞与造血功能

刘　菲[1,2]　徐秋云[2]　张巧利[2]
路岸瑞[2]　BT Beerntsen[3]　凌尔军[2*]

摘要：家蚕是一种典型的鳞翅目昆虫。在家蚕血淋巴中共有5种循环的血细胞，分别是原血细胞、粒细胞、浆细胞、珠血细胞、类绛色细胞。这些细胞均直接或者间接参与体液免疫和细胞免疫。昆虫造血器官能够产生血细胞连续释放进入血液循环中。研究结果显示家蚕幼虫的主要造血细胞是原血细胞以及类绛色细胞。基于体外的观察，家蚕原血细胞能够分化成粒细胞和浆细胞，而粒细胞可以再分化成珠血细胞。家蚕还有一种新的造血功能。当使用外科手术摘除其造血器官后，在游走期时血液循环中血细胞仍能维持在一个较高的水平，这是细胞分裂的水平增加所致。此前，类绛色细胞被认为是昆虫重要的免疫蛋白酚氧化酶(PPO)的唯一来源。然而，对包括家蚕在内的不同昆虫的研究表明其他类型的血细胞中也含有PPO。此外，PPO可在家蚕后肠肠道表皮细胞中产生。因此，家蚕是研究血细胞发育、细胞免疫和体液免疫的一种有价值的模式生物。

1　生物科学与技术系，陕西学前师范学院，西安
2　昆虫发育与进化生物学重点实验室，植物生理生态研究所，上海生命科学研究院，中国科学院，上海
3　兽医病理学系，密苏里大学，密苏里州，美国

A novel granulocyte-specific α integrin is essential for cellular immunity in the silkworm *Bombyx mori*

Zhang K[1#] Tan J[1,2#] Xu M[1] Su JJ[1] Hu RJ[1]
Chen YB[1] Xuan F[1] Yang R[1] Cui HJ[1*]

Abstract: Haemocytes play crucial roles in immune responses and survival in insects. Specific cell markers have proven effective in clarifying the function and haematopoiesis of haemocytes. The silkworm *Bombyx mori* is a good model for studying insect haemocytes; however, little is known about haemocyte-specific markers or their functions in silkworm. In this study, we identified the α subunit of integrin, *BmintegrinαPS3*, as being specifically and highly expressed in silkworm haemocytes. Immunofluorescence analysis validated the specificity of *Bm*integrinαPS3 in larval granulocytes. Further analyses indicated that haemocytes dispersed from haematopoietic organs (HPOs) into the circulating haemolymph could differentiate into granulocytes. In addition, the processes of encapsulation and phagocytosis were controlled by larval granulocytes. Our work demonstrated that *Bm*integrinαPS3 could be used as a specific marker for granulocytes and could be applied to future molecular cell biology studies.

Published On: Journal of Insect Physiology, 2014, 71, 61-67.

State Key Laboratory of Silkworm Genome Biology, Southwest University, Chongqing 400716, China.
[#]These authors contributed equally.
[*]Corresponding author E-mail: hcui@swu.edu.cn; hongjuan.cui@gmail.com.

一种新的粒细胞特定整合素在家蚕细胞免疫功能中是必不可少的

张　奎[1#]　谈　娟[1,2#]　徐　曼[1]　苏晶晶[1]
胡仁建[1]　陈毅彪　禤　凡[1]　杨　睿[1]　崔红娟[1*]

摘要：血细胞在昆虫的免疫反应和存活中起关键作用。特异性细胞标记物已经被证明在阐明血细胞的功能和造血作用方面是有效的。家蚕是研究昆虫血细胞的一个很好的模型；然而，关于血细胞特异性标记物和其在家蚕中的功能知之甚少。在这项研究中，我们确定了整合素α亚基（*BmintegrinαPS3*）在蚕血细胞中特异性高度表达。免疫荧光分析证实了*Bm*integrinαPS3在幼虫粒细胞中的特异性。进一步的分析表明，从造血器官（HPOs）分散到循环血淋巴中的血细胞可以分化成粒细胞。此外，包囊和吞噬过程由幼虫粒细胞控制。本研究的工作表明，*Bm*integrinαPS3可以用作粒细胞的特异性标记，可用于未来的分子细胞生物学研究。

家蚕基因组生物学国家重点实验室，西南大学，重庆

A novel *Lozenge* gene in silkworm, *Bombyx mori* regulates the melanization response of hemolymph

Xu M[1#] Wang X[1#] Tan J[1] Zhang K[1] Guan X[1]
Laurence H. Patterson[2] Ding HF[3] Cui HJ[1*]

Abstract: Runt-related (RUNX) transcription factors are evolutionarily conserved either in vertebrate or inverte-brate. Lozenge (Lz), a members of RUNX family as well as homologue of AML-1, functions as an important transcription factor regulating the hemocytes differentiation. In this paper, we identified and characterized RUNX family especially *Lz* in silkworm, which is a lepidopteran model insect. The gene expression analysis illustrated that *BmLz* was highly expressed in hemocytes throughout the whole development period, and reached a peak in glutonous stage. Over-expression of *BmLz* in silkworm accelerated the melanization process of hemolymph, and led to instantaneously up-regulation of prophenoloxidases (PPOs), which were key enzymes in the melanization process. Further down-regulation of *BmLz* expression by RNA interference resulted in the significant delay of melanization reaction of hemolymph. These findings suggested that *BmLz* regulated the melanization process of hemolymph by inducing PPOs expression, and played a critical role in innate immunity defense in silkworm.

Published On: Developmental and Comparative Immunology, 2015, 53(1), 191-198.

1 State Key Laboratory of Silkworm Genome Biology, Southwest University, Chongqing 400716, China.
2 Institute of Cancer Therapeutics, University of Bradford, West Yorkshire BD7 1DP, UK.
3 Cancer Center, Medical College of Georgia, Georgia Regents University, Augusta, GA 30912, USA.
[#]These authors contributed equally.
[*]Corresponding author E-mail: hcui@swu.edu.cn, hongjuan.cui@gmail.com.

家蚕中一种新的调节血淋巴黑化反应的*Lozenge*基因

徐　曼[1#]　王　雪[1#]　谈　娟[1]　张　奎[1]　关　熙[1]
Laurence H. Patterson[2]　丁寒飞[3]　崔红娟[1*]

摘要：Runt 相关（RUNX）转录因子无论是在脊椎动物还是无脊椎动物中进化上都很保守。Lozenge（Lz）作为RUNX家族的成员之一以及AML-1的同系物，是调节血细胞分化的重要转录因子。在本研究中，对鳞翅目模型昆虫家蚕RUNX家族中的*Lz*基因进行鉴定和表征。基因表达分析表明，在整个发育期间，*BmLz*在血细胞中高度表达，并在盛食期血细胞中达到峰值。家蚕*BmLz*基因的过度表达加速了血淋巴的黑化过程，并导致了黑化过程中的关键酶原酚氧化酶（PPO）的瞬时上调。通过RNA干扰进一步下调*BmLz*基因表达导致血淋巴黑化反应明显的延迟。这些发现表明*BmLz*基因通过诱导PPOs表达来调节家蚕血淋巴的黑化过程，并在蚕的先天免疫防御中发挥了关键作用。

1　家蚕基因组生物学国家重点实验室，西南大学，重庆
2　癌症治疗研究所，布拉德福德大学，西约克郡，英国
3　佐治亚州医学院，佐治亚摄政大学，奥古斯塔，美国

Identification and characterization of three novel hemocyte-specific promoters in silkworm *Bombyx mori*

Zhang K[#] Yu S[#] Su JJ Xu M
Tan P Zhang YJ Xiang ZH Cui HJ[*]

Abstract: Insect hemocytes play essential roles in the metabolism, metamorphosis and immunity, which are closely related events of growth and development. Here, four novel hemocyte-specific genes were obtained and conformed in our study, namely, *Bmintβ2, Bmintβ3, BmCatO, and BmSw04862*, respectively. Subsequently, their promoter sequences were cloned, and their activity in hemocytes, fat body, and silk gland were analyzed using recombinant *Ac*NPV vector system *in vivo*. Our results showed that *Bmintβ2*, *Bmintβ3*, and *BmCatO* were hemocyte-specific promoters in the silkworm, *Bombyx mori*. Interestingly, *Bmintβ2*, and *Bmintβ3* promoter regions were both located in their first intron. Further analysis of a series of *BmCatO* promoter truncations showed that a 254 bp region could function as a promoter element in the tissue-specificity expression. In summary, the results of this study revealed that we have identified three hemocyte-specific promoters in silkworm that will not only great significance for better understanding of hemocyte-specific gene, but also has potential applications in insect hematopoiesis and innate immunity research.

Published On: Biochemical and Biophysical Research Communications, 2015, 461(1), 102-108.

State Key Laboratory of Silkworm Genome Biology, Southwest University, Chongqing 400716, China.
[#]These authors contributed equally.
[*]Corresponding author E-mail: hcui@swu.edu.cn.

家蚕中三种新型血细胞特异性启动子的鉴定与表征

张　奎[#]　余　霜[#]　苏晶晶　徐　曼
谭　鹏　张亚军　向仲怀　崔红娟[*]

摘要：昆虫血液细胞在代谢、变态和免疫中起重要作用，而这些都与生长发育密切相关。在本研究中获得了4种新型血细胞特异性基因，分别为*Bmintβ2*，*Bmintβ3*，*BmCatO*和*BmSw04862*。随后，克隆了它们的启动子序列，其在血细胞、脂肪体和丝腺中的活性分别用体内重组*Ac*NPV载体系统进行分析。结果表明，*Bmintβ2*，*Bmintβ3*和*BmCatO*是家蚕（*Bombyx mori*）血细胞特异性启动子。有趣的是，*Bmintβ2*和*Bmintβ3*启动子区都位于其第一个内含子。对一系列*BmCatO*启动子截短的实验进一步表明，254 bp的区域可作为组织特异表达的启动子元件。综上所述，本研究的结果表明，我们已经在蚕中鉴定了3种血细胞特异性启动子，不仅对于更好地理解血细胞特异性基因具有重要意义，还在昆虫造血和先天免疫研究中具有潜在的应用。

家蚕基因组生物学国家重点实验室，西南大学，重庆

Characterization of hemocytes proliferation in larval silkworm, *Bombyx mori*

Tan J　Xu M　Zhang K　Wang X

Chen SY　Li T　Xiang ZH　Cui HJ*

Abstract: Hemocytes play multiple important roles during insect growth and development. Five types of hemocytes have been identified in the silkworm, *Bombyx mori*: prohemocyte, plasmatocyte, granulocyte, spherulocyte, and oenocytoid. We used the S-phase marker bromodeoxyuridine (BrdU) antibody along with the mitosis marker phosphohistone H3 (PHH3) antibody to monitor proliferation of hemocytes *in vivo*. The results indicate that silkworm hematopoiesis not only occurs in the circulatory system but also in hematopoietic organs (HPOs). During the 5th instar, the hemocyte proliferation in the circulatory system reaches a peak at the pre-wandering stage. Following infection by *Escherichia coli*, circulating hemocytes increase their cell divisions as demanded by the cellular immune response. All hemocytes, except spherulocytes, have the capacity to multiply *in vivo*. The BrdU label-retaining assay shows that a small portion of cells from the circulatory system and the HPOs are continuously labelled up to 9 days and 4 days respectively. A small number of long-term label retaining cells (LRCs) quiescently locate in circulatory system. All results indicate that there are a few quiescent stem cells or some progenitors in the larval circulatory system and HPO that produce new hemocytes and continuously release them into the circulating system.

Published On: Journal of Insect Physiology, 2013, 59(6), 595-603.

State Key Laboratory of Silkworm Genome Biology, Southwest University, Chongqing 400716, China.

*Corresponding author E-mail: hongjuan. cui@gmail.com; hcui@swu.edu.cn.

家蚕幼虫血细胞增殖特性的研究

谈　娟　徐　曼　张　奎　王　雪
陈思源　李　泰　向仲怀　崔红娟*

摘要：血细胞在昆虫生长发育过程中起着多重重要作用。家蚕中已发现5种血细胞类型：原血细胞、浆细胞、粒细胞、珠血细胞和类绛色细胞。我们使用S期标记物溴脱氧尿苷（BrdU）抗体，连同有丝分裂标记磷酸化组蛋白（PHH3）抗体监测血细胞在体内的增殖。结果表明，家蚕造血系统不仅发生在循环系统，而且发生在造血器官（HPOs）中。在五龄期间，循环系统中的血细胞增殖在游走期达到峰值。在大肠杆菌感染后，循环中的血细胞按照细胞免疫反应的要求，加强其细胞分裂程度。所有血细胞，除了珠血细胞，均有能力在体内繁殖。BrdU滞留实验表明，来自HPOs和循环系统中的一小部分细胞分别被连续标记长达9 d和4 d。少量长期标签保持细胞（LRCs）静止地分布在循环系统中。结果表明，幼虫循环系统中存在少量静止的干细胞或一些祖细胞，HPO产生新的血细胞并不断将它们释放到循环系统中。

家蚕基因组生物学国家重点实验室，西南大学，重庆

Proteomics identification and annotation of proteins of a cell line of *Bombyx mori*, *BmN* cells.

Yao HP Chen L Xiang XW

Guo AQ Lu XM Wu XF*

Abstract: A cell line is an important experimental platform for biological sciences as it can basically reflect the biology of its original organism. In this study, we firstly characterized the proteome of cultured *BmN* cells, derived from *Bombyx mori*. Total 1 478 proteins were identified with two or more peptides by using 1D (one-dimensional) SDS/PAGE and LTQ-Orbitrap. According to the gene ontology annotation, these proteins presented diverse pI values and molecular masses, involved in various molecular functions, including catalytic activity, binding, molecular transducer activity, motor activity, transcription regulator activity, enzyme regulator activity and antioxidant activity. Some proteins related to virus infection were also identified. These results provided us with useful information to understand the molecular mechanism of *B. mori* as well as antiviral immunity.

Published On: Bioscience Reports, 2010, 30(3), 209-215.

College of Animal Science, Zhejiang University, Hangzhou 310029, China.

*Corresponding author E-mail: wuxiaofeng@zju.edu.cn.

家蚕*BmN*细胞系的蛋白质组学鉴定及蛋白功能注释

姚慧鹏　陈　琳　相兴伟
郭爱芹　鲁兴萌　吴小锋*

摘要：细胞系是生物科学的重要实验平台，因为它可以基本反映原始生物体内的生物学特征。在本研究中，首先对来自家蚕的*BmN*细胞系进行了蛋白质组学鉴定。通过使用一维（1D）SDS/PAGE和LTQ-Orbitrap技术，鉴定了共1 478个蛋白质，这些蛋白质至少被鉴定到两个或更多个肽段。根据基因本体注释，这些蛋白质呈现了不同的等电点和分子质量，涉及包括催化活性、结合活性、分子转导活性、肌动活性、转录调节活性、酶调节活性和抗氧化活性等多种分子功能。另外还鉴定到一些与病毒感染相关的蛋白质。这些结果为我们了解家蚕抗病毒免疫的分子机制提供了有用信息。

动物科学学院，浙江大学，杭州

Establishment and characterization of two embryonic cell lines of *Bombyx mori*

Pan MH Xiao SQ Chen M Hong XJ Lu C*

Abstract: Two cell lines, i.e., *BmE-SWU1* and *BmE-SWU2*, were established from silkworm embryonic tissues of the reversion phase through primary culture in Grace's medium supplemented with 20% fetal bovine serum. The *BmE-SWU1* cell line mainly included diploid spindle cells and round cells, which were large and had severe heteroploidy karyotypes. The population doubling time of the 30th passage of the cell line was 58.7 hr. *BmE-SWU2* cells were oblong or round, and small. The population doubling time for the 30th passage of the cell line was 46.6 hr. Of *BmE-SWU2* cells 89.9% were diploid (2*n*=56). Both strains were attached to epithelial-like cell lines and were susceptible to *Bombyx mori* nucleopolyhedroviruse (*Bm*NPV). Inter simple sequence repeat (ISSR) fingerprinting of silkworm embryonic cell line was obtained.

Published On: Vitro Cellular & Developmental Biology Animal, 2007, 43(2), 101-104.

Key Sericultural Laboratory of Ministry of Agriculture, Southwest University, Chongqing 400716, China.
*Corresponding author E-mail: lucheng@swau.cq.cn.

家蚕两个胚胎细胞系的建立和表征

潘敏慧　肖仕全　陈　敏　洪锡钧　鲁　成*

摘要：通过使用含有20%胎牛血清的Grace培养基对家蚕反转期胚胎组织进行原代培养，从蚕胚胎组织中建立了两个细胞系，即*BmE-SWU1*和*BmE-SWU2*。*BmE-SWU1*细胞系以二倍体梭形细胞和圆形细胞为主，体积大并且核型异倍化严重，第30代细胞群体的倍增时间为58.7 h。*BmE-SWU2*细胞呈椭圆形或圆形，体积小，第30代细胞群体的倍增时间是46.6 h，*BmE-SWU2*细胞89.9%为二倍体（$2n$=56）。这两个细胞系都附着于类上皮细胞系并且都对家蚕多角体病毒（*Bm*NPV）敏感。获得了家蚕胚胎细胞系的简单序列重复ISSR指纹图谱。

农业部蚕桑学重点实验室，西南大学，重庆

Establishment and characterization of an ovarian cell line of the silkworm, *Bombyx mori*

Pan MH Cai XJ Liu M Lü J Tang H Tan J Lu C*

Abstract: A cell line *BmN-SWU1* was established from the ovarian tissues of 3-day-old fourth instar *Bombyx mori* larvae of the *21-872nlw* variety by performing primary cultures in Grace' s medium supplemented with 20% fetal bovine serum (FBS). The cell line primarily consisted of short spindle cells and round cells. The frequency of cells with chromosome number $2n = 56$ was 80.5%; therefore, the cell line was considered to be a diploid cell line. The population-doubling time (PDT) at 45th passage line was 57.7 h. This cell line was susceptible to the *B. mori* nuclear polyhedrovirus (*Bm*NPV), and the median tissue culture infective dose ($TCID_{50}$) at a cell density of 10^5 cells/mL was 16.3 OBs/mL. The transient expression efficiency of the *green fluorescent protein* (*GFP*) gene in this cell line was 54.8%. We used the *BmN-SWU1* cell line to select and establish a GFP transgenic cell line.

Published On: Tissue and Cell, 2010, 42(1), 42-46.

Key Sericultural Laboratory of Agricultural Ministry, Southwest University, Chongqing 400716, China.
*Corresponding author E-mail: lucheng@swu.edu.cn.

家蚕卵巢细胞系的建立与鉴定

潘敏慧　蔡秀娟　刘　敏　吕　军
唐　辉　谈　娟　鲁　成*

摘要：通过在添加20%胎牛血清（FBS）的Grace培养基中进行原代培养，从四龄第3天*21-872nlw*品种的卵巢组织建立细胞系*BmN-SWU1*。细胞系主要由短梭形细胞和圆形细胞组成。染色体数$2n=56$的细胞占80.5%；因此，该细胞系被认为是二倍体细胞系。45号染色体倍增时间（PDT）为57.7 h。该细胞系对蚕核型多角体病毒（*Bm*NPV）敏感，细胞密度为10^5个/mL的中位数组织培养感染剂量（$TCID_{50}$）为16.3 OBs/mL。绿色荧光蛋白（GFP）基因在该细胞系中的瞬时表达效率为54.8%。我们使用*BmN-SWU1*细胞系选择和建立GFP转基因细胞系。

农业部蚕桑学重点实验室，西南大学，重庆

Existence of prophenoloxidase in wing discs: a source of plasma prophenoloxidase in the silkworm, *Bombyx mori*

Diao YP[1] Lu AR[1] Yang B[1] Hu WL[1] Peng Q[1]
Ling QZ[2] Brenda T. Beerntsen[3] Kenneth Söderhäll[4] Ling EJ[1*]

Abstract: In insects, hemocytes are considered as the only source of plasma prophenoloxidase (PPO). PPO also exists in the hemocytes of the hematopoietic organ that is connected to the wing disc of *Bombyx mori*. It is unknown whether there are other cells or tissues that can produce PPO and release it into the hemolymph besides circulating hemocytes. In this study, we use the silkworm as a model to explore this possibility. Through tissue staining and biochemical assays, we found that wing discs contain PPO that can be released into the culture medium *in vitro*. An *in situ* assay showed that some cells in the cavity of wing discs have *PPO1* and *PPO2* mRNA. We conclude that the hematopoietic organ may wrongly release hemocytes into wing discs since they are connected together through many tubes as repost in previous paper. In wing discs, the infiltrating hemocytes produce and release PPO through cell lysis in the wing disc and the PPO is later transported into the hemolymph. Therefore, this might be another source of plasma PPO in the silkworm: some infiltrated hemocytes sourced from the hematopoietic organ release PPO via wing discs.

Published On: PLoS ONE, 2012, 7(7).

1 Key Laboratory of Insect Developmental and Evolutionary Biology, Shanghai Institute of Plant Physiology and Ecology, Shanghai Institutes for Biological Sciences, Chinese Academy of Sciences, Shanghai 200032, China.
2 Department of Applied Biology, Zhejiang Pharmaceutical College, Ningbo 315100, China.
3 Department of Veterinary Pathobiology, University of Missouri, Columbia, MO 65211, USA.
4 Department of Comparative Physiology, Uppsala University, Uppsala 75236, Sweden.
*Corresponding author E-mail: erjunling@sippe.ac.cn.

翅原基中存在酚氧化酶：家蚕血淋巴中酚氧化酶的来源

刁玉璞[1] 路岸瑞[1] 杨 兵[1] 胡文利[1] 彭 琴[1]
凌庆枝[2] Brenda T. Beerntsen[3] Kenneth Söderhäll[4] 凌尔军[1*]

摘要：在昆虫中，血细胞被认为是血淋巴多酚氧化酶（PPO）的唯一来源。连接家蚕翅原基造血器官的血细胞中也存在PPO。目前尚不清楚除了循环血细胞外，是否还有其他细胞或组织可以产生PPO并将其释放到血淋巴中。在本研究中，我们使用家蚕作为模型来探究这种可能性。通过组织染色和生化分析，我们发现翅原基含有PPO而且在体外可以释放到培养基中。原位杂交实验表明翅原基腔的一些细胞含有*PPO1*和*PPO2*的mRNA。我们认为造血器官可能会错误地将血细胞释放到翅原基中，因为它们通过许多管道连接在一起，就像以前的文章所述。在翅原基中，浸润的血细胞可能通过细胞裂解产生并释放PPO，随后PPO被转运到血淋巴中。因此，这可能是家蚕血淋巴PPO的另一个来源：一些来自造血器官的浸润血细胞通过翅原基释放PPO。

1 昆虫发育与进化生物学重点实验室，上海植物生理生态研究所，上海生命科学研究院，中国科学院，上海
2 应用生物学系，浙江医药高等专科学校，宁波
3 兽医病理学系，密苏里大学，哥伦比亚，美国
4 比较生理学系，乌普萨拉大学，乌普萨拉，瑞典

Characterization and identification of the integrin family in silkworm, *Bombyx mori*

Zhang K[1] Xu M[1] Su JJ[1] Yu S[1] Sun ZF[1]
Li YT[1] Zhang WB[1] Hou JB[1] Shang LJ[2] Cui HJ[1*]

Abstract: As an important economic insect, *Bombyx mori* is also a useful model organism for lepidopteran insect. Integrins are evolutionarily conserved from sponges to humans, and play vital roles in many physiological and pathological processes. To explore their diverse functions of integrins in insect, eleven integrins including six α and five β subunits were cloned and characterized from silkworm. Our results showed that integrins from silkworm own more family members compared to other invertebrates. Among those α subunits, integrins α1, α2, and the other four subunits belong to PS1, PS2, and PS3 groups, respectively. The β subunits mainly gather in the insect βv group except the β1 subunit which belongs to the insect β group. Expression profiles demonstrated that the integrins exhibited distinct patterns, but were mainly expressed in hemocytes. α1 and β2 subunits are the predominant ones either in the embryogenesis or larva stages. Interestingly, integrins were significantly up-regulated after stimulated by 20-hydroxyecdysone (20-E) *in vivo*. These results indicate that integrins perform diverse functions in hemocytes of silkworm. Overall, our results provide a new insight into the functional and evolutionary features of integrins.

Published On: Gene, 2014, 549(1), 149-155.

1 State Key Laboratory of Silkworm Genome Biology, Southwest University, Chongqing 400716, China.
2 School of Life Sciences, University of Bradford, Bradford, BD7 1DP, UK.
*Corresponding author E-mail: hcui@swu.edu.cn; hongjuan.cui@gmail.com.

家蚕整合素家族的表征及鉴定

张　奎[1]　徐　曼[1]　苏晶晶[1]　余　霜[1]　孙中锋[1]
李钰添[1]　张维博[1]　侯建兵[1]　商立军[2]　崔红娟[1*]

摘要:家蚕作为一种重要的经济昆虫,也是鳞翅目昆虫的模式生物。从海绵动物到人类,整合素在进化上是保守的,并在许多生理和病理过程中发挥重要作用。为了探讨整合素在昆虫中的不同功能,我们鉴定和克隆得到11个整合素成员,包括6个α亚基和5个β亚基。结果表明,与其他无脊椎动物相比,家蚕中的整合素拥有更多家庭成员。在这些α亚基中,整合素α1、α2和其他4种亚基分别属于PS1、PS2和PS3组。除了属于昆虫β组的β1亚基,β亚基主要聚集在昆虫βν组。表达谱显示,整合素表现出不同的表达模式,但主要在血细胞中表达。α1亚基和β2亚基主要在胚胎或幼虫阶段表达。有趣的是,在体内受蜕皮激素(20-E)的刺激后,整合素表达明显上调。这些结果表明整合素在家蚕血淋巴中执行不同的功能。综上,我们的结果为整合素的功能和进化特性的研究提供了新的思路。

1　家蚕基因组生物学国家重点实验室,西南大学,重庆
2　生命科学学院,布拉德福德大学,布拉德福德,英国

Cloning and analysis of *DnaJ* family members in the silkworm, *Bombyx mori*

Li YN[1] Bu CY[1] Li TT[1] Wang SB[1]
Jiang F[1] Yi YZ[2] Yang HP[1] Zhang ZF[1*]

Abstract: Heat shock proteins (Hsps) are involved in a variety of critical biological functions, including protein folding, degradation, and translocation and macromolecule assembly, act as molecular chaperones during periods of stress by binding to other proteins. Using expressed sequence tag (EST) and silkworm (*Bombyx mori*) transcriptome databases, we identified 27 cDNA sequences encoding the conserved J domain, which is found in DnaJ-type Hsps. Of the 27 J domain-containing sequences, 25 were complete cDNA sequences. We divided them into three types according to the number and presence of conserved domains. By analyzing the gene structures, intron numbers, and conserved domains and constructing a phylogenetic tree, we found that the *DnaJ* family had undergone convergent evolution, obtaining new domains to expand the diversity of its family members. The acquisition of the new DnaJ domains most likely occurred prior to the evolutionary divergence of prokaryotes and eukaryotes. The expression of *DnaJ* genes in the silkworm was generally higher in the fat body. The tissue distribution of DnaJ1 protein was detected by Western blot, demonstrating that in the fifth-instar larvae, the DnaJ1 protein was expressed at their highest levels in hemocytes, followed by the fat body and head. We also found that the *DnaJ1* transcripts were likely differentially translated in different tissues. Using immunofluorescence cytochemistry, we revealed that in the blood cells, DnaJ1 was mainly localized in the cytoplasm.

Published On: Gene, 2016, 576(1), 88-98.

1 Biotechnology Research Institute, Chinese Academy of Agricultural Sciences, Beijing 100081, China.
2 Sericultural Research Institute, Chinese Academy of Agricultural Sciences, Zhenjiang 212018, China.
*Corresponding author E-mail: zhifangzhang@yahoo.com.

家蚕*DnaJ*家族成员的克隆与分析

李铁女[1] 步翠玉[1] 李田田[1] 王石宝[1]
江 峰[1] 易咏竹[2] 杨慧鹏[1] 张志芳[1*]

摘要：热激蛋白(Hsps)与蛋白折叠、降解、迁移和大分子组装等一系列关键的生物学功能相关，并在特殊时期与其他蛋白相互结合行使着分子伴侣的功能。通过基因表达序列标签和家蚕转录数据，我们鉴定到27个编码保守J结构域的cDNA的序列，该序列位于DnaJ型Hsps中。在27个含J结构域的序列中，有25个是完整的cDNA序列。我们根据保守结构域的数量将其分成3种类型。通过分析基因结构、内含子数和保守结构域，并构建系统进化发育树，我们发现*DnaJ*家族经历了趋同进化，获得了新的结构域来拓展家族成员的多样性，且新DnaJ结构域的获得很可能发生在原核生物和真核生物进化之前。*DnaJ*基因在家蚕脂肪体中的表达量普遍较高。通过蛋白质免疫印迹分析DnaJ1蛋白在各组织的分布，表明在五龄幼虫中，DnaJ1蛋白在血淋巴中的表达量最高，其次是在脂肪体和头部。我们发现*DnaJ1*可能在不同的组织中有不同的转录本。通过免疫组织化学实验，我们发现DnaJ1蛋白主要定位在血细胞的细胞质中。

1 生物技术研究所，中国农业科学院，北京
2 蚕业研究所，中国农业科学院，镇江

Establishment and characterization of a new embryonic cell line from the silkworm, *Bombyx mori*

Xu M# Tan J# Wang X Zhong XX Cui HJ*

Abstract: Insect cell lines are widely used for basic and applied research in the fields of insect pathology, genetics, and molecular biology. In the present study, a new continuous cell line designated *BmE-SWU3* was established from blastokinesis-stage embryos of the silkworm *Bombyx mori* (*Furong* strain). The primary culture was initially performed using Grace's medium supplemented with 20% foetal bovine serum (FBS) at a constant temperature of 27 °C. The dominant cell type was round and spindle-shaped. Thus far, this cell line has been cultured continuously for 60 passages. The cell doubling time was approximately 3.0 days. The SSR profile of *BmE-SWU3* differs from those of the silkworm *BmE* and *BmN-SWU1* cell lines and from those of the *Spodoptera frugiperda* cell line *Sf9* and the *Drosophila* cell line *S2*. However, the SSR profiles among the various passages of *BmE-SWU3* were stable and identical. This new cell line was highly susceptible to *Bombyx mori* nucleopolyhedrovirus (*Bm*NPV). Semi-quantitative RT-PCR indicated that the tissue-specific gene expression patterns were completely distinct from those of *BmE* and *BmN-SWU1*.

Published On: Invertebrate Survival Journal, 2015, 12, 13-18.

State Key Laboratory of Silkworm Genome Biology, Southwest University, Chongqing 400716, China.
#These authors contributed equally.
*Corresponding author E-mail: hongjuan.cui@gmail.com.

一个新的家蚕胚胎细胞系的建立与鉴定

徐　曼[#]　谈　娟[#]　王　雪　钟晓霞　崔红娟[*]

摘要：昆虫细胞系广泛应用于昆虫病理学基础研究、应用研究以及遗传学与分子生物学。在本研究中，用家蚕（芙蓉品系）胚动期的胚胎建立了称为*BmE-SWU3*的新的连续细胞系。最初使用Grace培养基进行原代培养，该培养基补充有20%胎牛血清（FBS），温度恒定在27 ℃。主要细胞类型为圆形和纺锤形。到目前为止，该细胞系已连续培养60代。细胞倍增时间约为3 d。*BmE-SWU3*的SSR曲线不同于家蚕*BmE*和*BmN-SWU1*细胞系以及草地夜蛾细胞系*Sf9*和果蝇细胞系*S2*。然而，*BmE-SWU3*各代间的SSR分布稳定且一致。这种新的细胞系对家蚕核型多角体病毒（*Bm*NPV）高度敏感。半定量RT-PCR表明，*BmE*-SWU3的组织特异性基因的表达模式与*BmE*和*BmN-SWU1*完全不同。

家蚕基因组生物学国家重点实验室，西南大学，重庆

Agkistin-s, a disintegrin domain, inhibits angiogenesis and induces BAECs apoptosis

Ren AX[1,2#] Wang SH[2#] Cai WJ[4] Yang GZ[2]

Zhu YC[4] Wu XF[2,3*] Zhang YZ [3*]

Abstract: Previous work in our laboratory has shown agkistin, a snake venom metalloproteases (SVMPs) from the venom of *Agkistrodon halys*, possesses antiplatelet aggregation activity. In this study, we further examined the antiangiogenic activity of agkistin-s, the disintegrin domain of agkistin. Recombinant agkistin-s was produced in *Escherichia coli* by subcloning its cDNA into pET28a vector, and the effect of purified agkistin-s was evaluated. At the concentration of 0.5-1.5 μM, the recombinant agkistin-s exhibited inhibitory activities on the bovine aortic endothelial cells (BAECs) migration and proliferation in a dose-dependent manner. In addition, it exhibited an effective antiangiogenic effect when assayed by using the 10-day-old embryo chick CAM model and effectively inhibits the tube-like structure formation. Furthermore, it potently induced BAECs apoptosis as examined by flow cytometric assays.

Published On: Journal of Cellular Biochemistry, 2006, 99(6), 1517-1523.

1 College of Life Sciences, Zhejiang University, Hangzhou 310029, China.

2 Shanghai Institute of Biochemistry and Cell Biology, Chinese Academy of Sciences, Shanghai 200031, China.

3 College of Life Sciences, Zhejiang Sci-Tech University, Hangzhou 310018, China.

4 Department of Physiology and Pathophysiology, Fudan University Shanghai Medical College, Shanghai 200032, China.

#These authors contributed equally.

*Corresponding author E-mail: xfwu@sunm.shcnc.ac.cn; yaozhou@chinagene.com.

蛇毒金属蛋白酶去整合素结构域agkistin-s抑制血管生成并诱导BAECs凋亡

任爱霞[1,2#] 王韶辉[2#] 蔡文杰[4] 杨冠珍[2]
朱依纯[4] 吴祥甫[2,3*] 张耀洲[3*]

摘要:本实验室前期研究表明,蝮蛇蛇毒源的金属蛋白酶(SVMPs)agkistin具有抗血小板凝集活性。在这项研究中,我们进一步研究agkistin-s(去整合素结构域)的抗血管生成活性。将agkistin-s cDNA亚克隆到原核表达载体pET28a,在大肠杆菌中获得重组agkistin-s,并对纯化的agkistin-s效果进行评价。在0.5—1.5 μM浓度,重组agkistin-s以剂量依赖性抑制牛主动脉内皮细胞(BAECs)的迁移和增殖。当使用10 d大的鸡胚CAM模型分析时,它表现出有效的抗血管生成作用并抑制管状结构的形成。此外,流式细胞检测分析表明它还可以诱导BAEC细胞凋亡。

1 生命科学学院,浙江大学,杭州
2 上海生物化学与细胞生物学研究所,中国科学院,上海
3 生命科学学院,浙江理工大学,杭州
4 生理学与病理生理学系,复旦大学上海医学院,上海

第九章

家蚕突变体分子机理研究

Chapter 9

家蚕在超过5000年的人工驯化过程中积累了大量突变体，这些突变体为遗传学和分子生物学的研究提供了宝贵的资源。随着家蚕基因组精细图、SSR及SNP图谱的完成，定位克隆技术已在家蚕突变基因的研究中被广泛地应用，目前为止，有40余个突变基因被成功分离克隆。这些突变体主要包括家蚕体色、斑纹、血色、茧色、眠性及抗病等性状。这些突变基因的克隆及突变机理的阐释为家蚕的生理生化、发育变态和病理研究提供了宝贵的资源，同时这些研究也为昆虫的相关领域提供了理论基础。在10年期间，项目组围绕家蚕的关键性状的突变基因开展了精细定位及定位克隆研究，取得的研究成果主要集中在三个方面。一是利用组学方法对突变体进行差异基因的表达分析，初步获得一些基因的表达信息；二是对一些突变基因开展了精细定位；三是成功分离克隆突变基因并对其进行了功能研究及突变机理的阐释。

突变体相关基因的表达分析及组学分析：项目执行期间，对一些突变体的相关基因的表达进行了研究，同时利用组学方法对其差异表达基因进行了全面分析。如利用蛋白质组学对无鳞毛(*sl*)突变进行了分析；利用转录组学对体色突变体鹑斑(*q*)、黑色蚕(*bd*)表皮及行为突变体缩蚕蚁蚕等突变体开展了研究。由于通过组学都会获得大量的差异基因，很难通过差异基因分析成功鉴定突变基因，还需结合连锁定位分析及功能研究。通过组学研究中所发现的差异表达基因及蛋白所涉及的通路及功能对突变机制的阐释提供了线索。

突变基因的精细定位：项目执行期间，对抗病基因、体色、黄茧、附肢突变基因进行了精细定位。在抗病方面，主要对浓核病抗性基因(*nsd-Z*)进行了精细定位。在体色研究方法，通过连锁分析研究发现，分别位于24连锁群0.0座位的黄体色突变*Sel*和位于27连锁群0.0座位的*Xan*突变基因事实上为24染色体的等位基因，其与分子标记S2404的遗传距离分别为13cM和13.7cM。通过对这两个基因的精细定位，更正了这两个基因的遗传定位，为后期这两个突变基因的成功克隆奠定了基础。在附肢突变基因精细定位方面，项目组主要对*Ekp*，*Ecs-1*及*ap*三个突变基因进行了精细定位。在项目执行早期，由于基因组数据不够完善，只有框架图，开发的可用标记数量不够，而且品种间的标记差异还比较大，因此大部分定位克隆工作都只进行到精细图谱的定位。在家蚕精细图谱和遗传图谱完成后，大量的突变基因被成功克隆，但也还有少数最后无法锁定突变基因，主要是锁定区域内没有编码基因，而一些调控序列如启动子、内含子或者基因间区发生突变等等，如E群突变体。这些突变基因较难进行相应的功能验证，但随着家蚕基因功能研究手段的完善，如CRISPR/Cas9技术的建立，相信家蚕越来越多的突变基因会被逐一克隆。

突变基因的定位克隆及功能验证：项目执行期间，特别是家蚕精细图谱和遗传图谱完成后，利用定位克隆方法成功分离了近10例突变基因，并利用基因功能研究方法进行了功能验证，主要包括家蚕卵子发生突变体（*vit*），“Ming”致死卵突变体（$l\text{-}e^{m}$），蚁蚕体色突变体-伴性赤蚁（*sch*），幼虫和成虫体色突变体黑化型（*mln*），行为突变体缩蚕（*cot*），石蚕（*stony*）等突变基因。在这些成功的定位克隆工作中，突变基因的突变类型主要是由转座元件介导的基因调控序列的缺失，如$l\text{-}e^{m}$，*sch*。另一种主要是编码序列的异常，产生功能缺失或者异常的蛋白。如*vit*，*mln*，*cot*，*stony*等。不难发现，这几个突变体都是隐性突变体。在已经克隆成功的突变体中，绝大部分的都是隐性突变体，显性突变基因被成功克隆的比较少，如L、P^{s}、C，M3等。这些突变位点都主要集中在基因的调控区域，导致基因的异常表达从而导致其突变表型。这些突变体的定位克隆周期通常比较长，需要的群体数量也比较多，在候选基因的确定及功能验证上较隐性突变体更具挑战性。

刘春

The scaleless wings mutant in *Bombyx mori* is associated with a lack of scale precursor cell differentiation followed by excessive apoptosis

Zhou QX[1,3] Li YN[1] Shen XJ[3] Yi YZ[2] Zhang YZ [2*] Zhang ZF[1*]

Abstract: The scaleless wings mutant in *Bombyx mori* (*scaleless, sl*) was previously reported morphologically. In the present study, we give data to clarify the mechanism of the mutation at the developmental level. Programmed cell death participates in the wing scale development during early pupal stage, and there are significant differences between that of *sl* and the wild type (*WT*) at each phase. Well-differentiated scale precursor cells do not form in *sl* when they have formed in *WT*. The peak of Caspase-3/7 activity in *sl* occurs 1 day later than and ten times as much as that in *WT*. Apoptotic bodies and DNA ladder studies also show that there is excessive apoptosis in *sl* early pupal wing. In addition, we have studied *Bm-ASH1*, an *achaete-scute* homolog in *B. mori*, which is thought to play a key role during the development of wing scales, and have found that the gene structure and expression levels of *Bm-ASH1* in *sl* and *WT* are identical. All the data indicate that the wing scale precursor differentiation mechanism is abnormal in *sl*, which causes failing determination of scale cells and the downstream symptom of excessive apoptosis. But some of the elements to the scale differentiation circuit, such as *Bm-ASH1*, still operate in *sl*.

Published On: Development Genes and Evolution, 2006, 216(1), 721-726.

1 Biotechnology Research Institute, National Key Facility for Crop Gene Resources and Genetic Improvement, Chinese Academy of Agricultural Sciences, Beijing 100081, China.

2 Institute of Biochemistry, Zhejiang Sci-Tech University, Hangzhou 310018, China.

3 Sericultural Research Institute, Chinese Academy of Agricultural Sciences, Zhenjiang 212018, China.

*Corresponding author E-mail: yaozhou@chinagene.com; zhifangzhang@yahoo.com.

家蚕翅无鳞毛突变体与鳞片前体细胞分化不足及随后的过度凋亡有关

周庆祥[1,3] 李轶女[1] 沈兴家[3] 易咏竹[2] 张耀洲[2*] 张志芳[1*]

摘要：家蚕翅无鳞毛突变体(*sl*)在形态上已有报道，本研究在发育水平阐明了它的突变机制。细胞程序性死亡参与蛹早期的翅鳞毛发育，且*sl*和野生型(*WT*)在各阶段存在显著差异。在*sl*细胞中，分化良好的鳞片前体细胞并未形成，而在野生型中却已形成。*sl*中Capases-3/7活性峰值出现时间比*WT*晚一天，是*WT*的10倍。凋亡小体与DNA梯形条带分析结果也表明，在*sl*蛹翅早期的发育过程中存在过度凋亡的现象。此外，我们在家蚕中研究了一个无鳞毛同源基因*Bm-ASH1*，该基因在翅鳞毛的发育过程中发挥重要作用，并发现*Bm-ASH1*在*sl*和*WT*中的基因结构和表达水平都是相似的。所有数据都表明，*sl*中翅鳞毛前体分化机制的异常导致鳞毛细胞的分化失败和下游过度凋亡症状发生。但是一些鳞翅细胞分化相关基因，如*Bm-ASH1*在*sl*体内仍然在起作用。

1 生物技术研究所，中国农业科学院，北京
2 生物化学研究所，浙江理工大学，杭州
3 蚕业研究所，中国农业科学院，镇江

Proteomic analysis of the phenotype of the scaleless wings mutant in the silkworm, *Bombyx mori*

Shi XF[1#] Han B[1,2#] Li YN[1] Yi YZ[2]
Li XM[3] Shen XJ[2*] Zhang ZF[1*]

Abstract: A scaleless wing mutant of silkworm, *Bombyx mori*, has much fewer scales than wild type (WT). The scaleless phenotype was associated with tracheal system developmental deficiency and excessive apoptosis of scale cells. In this study, the wing discs proteins of WT and scaleless during pupation were studied using 2-DE and mass spectrometry. Of the 99 identified protein spots, four critical differentially expressed proteins between WT and scaleless were further verified using Q-PCR. At the first day of pupation (P0) in WT, imaginal disk growth factor (IDGF) was upregulated, whereas actin-depolymerizing factor 1 (ADF1) and profilin (PFN), which associated with cellular motility and cytoplasmic extension, were downregulated. We speculated their coaction counteracts the correct organization of the tracheal system in wing disc. Thiol peroxiredoxin (TPx) was upregulated in scaleless at P0, but its mRNA higher expression occurred in the day before pupation (S4). TPx could inhibit the formation of hydrogen peroxide, preventing the release of cytochrome C and activation of the caspase family protease. Its higher expression in scaleless was responsible for the apoptosis of scale cells delayed. The results provide further evidence that the scale less phenotype was related to the tracheal system developmental deficiency and excessive apoptosis of scale cells.

Published On: Journal of Proteomics, 2013, 78, 15-25.

1 Biotechnology Research Institute, Chinese Academy of Agricultural Sciences, Beijing 100081, China.
2 Sericultural Research Institute, Chinese Academy of Agricultural Sciences, Zhenjiang 212018, China.
3 State Key Laboratory of Biomembrane and Membrane Biotechnology, Institute of Zoology, Chinese Academy of Sciences, Beijing 100101, China.
[#]These authors contributed equally.
[*]Corresponding author E-mail: shenxj63@yahoo.com.cn; zhifangzhang@yahoo.com.

家蚕翅无鳞毛突变体表型的蛋白质组学分析

石小峰[1#] 韩 宾[1,2#] 李轶女[1] 易咏竹[2]
李晓明[3] 沈兴家[2*] 张志芳[1*]

摘要：家蚕翅无鳞毛突变体与野生型家蚕相比有更少的鳞毛。无鳞毛表型与气管系统发育缺陷和鳞毛细胞过度凋亡有关。在本研究中，使用二维凝胶电泳和质谱法研究了蛹化过程中野生型和翅无鳞毛突变体的翅原基蛋白。在野生型和翅无鳞毛突变体中鉴定到99个蛋白点，其中4个关键的差异表达蛋白通过Q-PCR进行进一步的验证。在野生型家蚕蛹期第一天（P0），成虫原基生长因子（IDGF）上调，而与细胞迁移和细胞质扩张相关的肌动蛋白解聚因子1（ADF1）和前纤维蛋白（PFN）则下调。我们推测它们的相互作用阻碍了翅原基中气管系统的正确组装。在翅无鳞毛突变体P0时期，硫醇过氧化酶（TPx）上调，但它的信使RNA在化蛹前有较高表达。TPx能够抑制过氧化氢的形成，阻止细胞色素C的释放和caspase家族蛋白酶的激活。TPx在无鳞毛突变体中的高表达是造成鳞毛细胞凋亡延迟的原因。我们的结果为翅无鳞毛表型与气管系统发育缺陷和鳞毛细胞的过度凋亡有关提供了进一步的证据。

1 生物技术研究所，中国农业科学院，北京
2 蚕业研究所，中国农业科学院，镇江
3 生物膜和膜生物工程国家重点实验室，动物所，中国科学院，北京

Cloning, expression and enzymatic properties analysis of dihydrofolate reductase gene from the silkworm, *Bombyx mori*

Wang WJ[#] Gao JS[#] Wang J Liu CL Meng Y[*]

Abstract: Tetrahydrobiopterin (BH4) is an essential cofactor for aromatic acid hydroxylases, which control the levels of monoamine neurotransmitters. BH4 deficiency has been associated with many neuropsychological disorders. Dihydrofolate reductase (DHFR) can catalyze 7,8-dihydrobiopterin to 5,6,7,8-tetrahydrobiopterin (BH4) in the salvage pathway of BH4 synthesis from sepiapterin (SP), a major pigment component contained in the integument of silkworm *Bombyx mori* mutant *lemon* (*lem*) in high concentration. In this study, we report the cloning of *DHFR* gene from the silkworm *B. mori* (*BmDhfr*) and identification of enzymatic properties of *Bm*DHFR. *BmDhfr* is located on scaffold *Bm*_199 with a predicted gene model BGIBMGA013340, which encodes a 185-aa polypeptide with a predicted molecular mass of about 21 kDa. Biochemical analyses showed that the recombinant *Bm*DHFR protein exhibited high enzymatic activity and suitable parameters to substrate. Together with our previous studies on SP reductase of *B. mori* (*Bm*SPR) and the *lem* mutant, it may be an effective way to industrially extract SP from the *lem* silkworms in large scale to produce BH4 *in vitro* by co-expressing *Bm*SPR and *Bm*DHFR and using the extracted SP as a substrate in the future.

Published On: Molecular Biology Reports, 2012, 39(12), 10285-10291.

School of Life Sciences, Anhui Agricultural University, Hefei 230036, China.

[#]These authors contributed equally.

[*]Corresponding author E-mail: mengyan@ahau.edu.cn.

家蚕二氢叶酸还原酶基因的克隆、表达及酶学特征分析

王文静[#] 高俊山[#] 王 敬 刘朝良 孟 艳[*]

摘要：四氢生物蝶呤(BH4)是芳香酸羟化酶的必需辅助因子，其控制单胺类神经递质的水平。BH4缺乏与许多神经心理失调有关。二氢叶酸还原酶(DHFR)可以将来自墨蝶呤(SP)的BH4合成的补救途径中的7,8-二氢生物蝶呤催化为5,6,7,8-四氢生物蝶呤(BH4)，墨蝶呤在柠檬突变蚕(*lem*)体壁中浓度高，为主要色素成分。在本研究中，我们克隆了家蚕中的*DHFR*基因(*BmDhfr*)及鉴定了*Bm*DHFR的酶学特征。*BmDhfr*位于编号*Bm*_199的Scaffold上，在上面预测到一个典型的基因——BGIBMGA013340，该基因编码具有185个氨基酸的多肽，预测蛋白质分子质量为21 kDa。生物化学分析表明重组*Bm*DHFR蛋白具有较高的酶活性和对底物的合适参数。结合我们以前关于家蚕和*lem*突变体的SP还原酶(*Bm*SPR)的研究，在未来通过同时表达SP还原酶和二氢叶酸还原酶，并以*lem*突变体提取的墨蝶呤为底物，在体外工业化生产四氢生物蝶呤是一个有效的途径。

生命科学学院，安徽农业大学，合肥

Effects of altered catecholamine metabolism on pigmentation and physical properties of sclerotized regions in the silkworm melanism mutant

Qiao L[1#] Li YH[1#] Xiong G[1#] Liu XF[1] He SZ[1] Tong XL[1,2] Wu SY[1] Hu H[1]
Wang RX[1] Hu HW[3] Chen LS[1] Zhang L[1] Wu J[1] Dai FY[1,3*] Lu C[1*] Xiang ZH[1]

Abstract: Catecholamine metabolism plays an important role in the determination of insect body color and cuticle sclerotization. To date, limited research has focused on these processes in silkworm. In the current study, we analyzed the interactions between catecholamines and melanin genes and their effects on the pigmentation patterns and physical properties of sclerotized regions in silkworm, using the melanic mutant melanism (*mln*) silkworm strain as a model. Injection of β-alanine into *mln* mutant silkworm induced a change in catecholamine metabolism and turned its body color yellow. Further investigation of the catecholamine content and expression levels of the corresponding *melanin* genes from different developmental stages of *Dazao-mln* (mutant) and *Dazao* (wild-type) silkworm revealed that at the larval and adult stages, the expression patterns of melanin genes precipitated dopamine accumulation corresponding to functional loss of *Bm-iAANAT*, a repressive effect of excess NBAD on *ebony*, and upregulation of *tan* in the *Dazao-mln* strain. During the early pupal stage, dopamine did not accumulate in *Dazao-mln*, since upregulation of *ebony* and *black* genes led to conversion of high amounts of dopamine into NBAD, resulting in deep yellow cuticles. Scanning electron microscope analysis of a cross-section of adult dorsal plates from both wild-type and mutant silkworm disclosed the formation of different layers in *Dazao-mln* owing to lack of NADA, compared to even and dense layers in *Dazao*. Analysis of the mechanical properties of the anterior wings revealed higher storage modulus and lower loss tangent in *Dazao-mln*, which was closely associated with the altered catecholamine metabolism in the mutant strain. Based on these findings, we conclude that catecholamine metabolism is crucial for the color pattern and physical properties of cuticles in silkworm. Our results should provide a significant contribution to Lepidoptera cuticle tanning research.

Published On: PLoS ONE, 2012, 7(8).

1 State Key Laboratory of Silkworm Genome Biology, Southwest University, Institute of Sericulture and Systems Biology, Southwest University, Chongqing 400716, China.
2 Institute of Agriculture and Life Science, Chongqing University, Chongqing 40030, China.
3 College of Biotechnology, Southwest University, Chongqing 400716, China.
[#]These authors contributed equally.
[*]Corresponding author E-mail: fydai@swu.edu.cn; lucheng@swu.edu.cn.

改变儿茶酚胺代谢对蚕黑色素突变体的色素沉积和物理性质的影响

乔 梁[1#] 李元浩[1#] 熊 高[1#] 刘小凡[1] 何松真[1] 童晓玲[1,2] 吴松原[1] 胡 海[1]
王日欣[1] 胡红伟[3] 陈璐诗[1] 张 丽[1] 伍 杰[1] 代方银[1,3*] 鲁 成[1*] 向仲怀[1]

摘要: 儿茶酚胺代谢在昆虫体色决定及长皮硬化方面发挥重要作用。迄今为止,家蚕中有关这些过程的研究非常有限。在本研究中,以黑化突变体(*mln*)家蚕为研究对象,我们分析了儿茶酚胺与黑色素基因之间的相互作用及其对家蚕硬化区色素沉积模式和物理性质的影响。将β-丙氨酸注射到*mln*突变体家蚕体内,会诱导儿茶酚胺代谢发生改变,体色变黄。对*mln*突变体大造及野生型大造家蚕不同发育阶段儿茶酚胺含量及相应黑色素基因表达水平进一步研究表明,在幼虫和成虫期,家蚕黑色素期的表达促进黑色素的沉积导致*Bm-iAANAT*基因功能丧失,过量NBAD对*ebony*基因的抑制,以及*mln*大造突变体中*tam*基因的上调表达。在中间早期,多巴胺没有在*mln*突变体中积累,因为上调的*ebony*和*black*基因导致多巴胺转化为NBAD,形成深黄色的表皮。对野生型及突变体家蚕背板横截面进行扫描电镜观察,发现由于缺少NADA,*mln*突变体大造形成了不同层次的结构,而野生型大造层厚较均匀。对*mln*突变体大造的前翅力学性能进行分析,发现其具有较高的存储模量和较低的损失模量,这与突变体中儿茶酚胺代谢的改变密切相关。基于以上发现,我们认为儿茶酚胺代谢对家蚕角质层的颜色和物理特性至关重要。本研究结果对鳞翅目昆虫表皮鞣制的研究具有重要意义。

1 家蚕基因组生物学国家重点实验室,蚕学与系统生物学研究所,西南大学,重庆

2 农学及生命科学研究院,重庆大学,重庆

3 生物技术学院,西南大学,重庆

Isolation, purification, and identification of an important pigment, sepiapterin, from integument of the *lemon* mutant of the silkworm, *Bombyx mori*

Gao JS   Wang J   Wang WJ   Liu CL*   Meng Y*

Abstract: Sepiapterin is the precursor of tetrahydrobiopterin, an important coenzyme of aromatic amino acid hydroxylases, the lack of which leads to a variety of physiological metabolic diseases or neurological syndromes in humans. Sepiapterin is a main pigment component in the integument of the *lemon* mutant of the silkworm, *Bombyx mori* (L.) (Lepidoptera: Bombycidae), and is present there in extremely high content, so *lemon* is a valuable genetic resource to extract sepiapterin. In this study, an effective experimental system was set up for isolation and purification of sepiapterin from *lemon* silkworms by optimizing homogenization solvent, elution buffer, and separation chromatographic column. The results showed that ethanol was the most suitable solvent to homogenize the integument, with a concentration of 50% and solid:liquid ratio of 1∶20 (g/mL). Sepiapterin was purified successively by column chromatography of cellulose Ecteola, sephadex G-25-150, and cellulose phosphate, and was identified by ultraviolet-visible absorption spectrometry. A stable and accurate high performance liquid chromatography method was constructed to identify sepiapterin and conduct qualitative and quantitative analyses. Sepiapterin of high purity was achieved, and the harvest reached about 40 μg/g of integument in the experiments. This work helps to obtaining natural sepiapterin in large amounts in order to use the *lemon B. mori* mutant to produce BH4 *in vitro*.

Published On: Journal of Insect Science, 2013, 13.

School of Life Sciences, Anhui Agricultural University, Hefei 230036, China.
*Corresponding author E-mail: cyschx@163.com; mengyan@ahau.edu.cn.

家蚕*lemon*突变体表皮中重要色素墨蝶呤的分离、纯化和鉴定

高俊山　王　敬　王文静　刘朝良*　孟　艳*

摘要：墨蝶呤是四氢生物蝶呤的前体，四氢生物蝶呤是芳香族氨基酸羟化酶的重要辅酶，缺乏四氢生物蝶呤可导致人类多种生理代谢疾病或神经性综合征。墨蝶呤是家蚕*lemon*突变体表皮内的主要色素成分，含量极高，使得*lemon*突变体可以作为提取墨蝶呤的宝贵遗传资源。在本研究中，我们建立了一个有效的实验系统用于从家蚕*lemon*突变体中分离和纯化墨蝶呤，主要包括最适匀浆溶剂、洗脱缓冲液和分离色谱柱。结果显示乙醇为最佳的体壁匀浆溶剂，其最适体积分数为50%，最佳固液比例为1∶20（g/mL）。墨蝶呤相继通过交联醇胺纤维素色谱柱、交联葡聚糖G-25-150色谱柱和磷酸纤维素色谱柱纯化，并通过紫外—可见吸收光谱法进行鉴定。同时建立了一个稳定和精确的高效液相色谱系统对墨蝶呤进行定性和定量分析。通过以上方法，在家蚕表皮内获得高纯度的墨蝶呤色素，产量可达40 μg/g。该研究为通过家蚕*lemon*突变体体外获得大量天然墨蝶呤从而生产四氢生物蝶呤提供了帮助。

生命科学学院，安徽农业大学，合肥

Transcriptome analysis of neonatal larvae after hyperthermia-induced seizures in the contractile silkworm, *Bombyx mori*

Nie HY[1] Liu C[1,2] Zhang YX[1] Zhou MT[1]
Huang XF[1] Peng L[1] Xia QY[1,2*]

Abstract: The ability to respond quickly and efficiently to transient extreme environmental conditions is an important property of all biota. However, the physiological basis of thermotolerance in different species is still unclear. Here, we found that the *cot* mutant showed a seizure phenotype including contraction of the body, rolling, vomiting gut juice and a momentary cessation of movement, and the heartbeat rhythm of the dorsal vessel significantly increases after hyperthermia. To comprehensively understand this process at the molecular level, the transcriptomic profile of *cot* mutant, which is a behavior mutant that exhibits a seizure phenotype, was investigated after hyperthermia (42℃) that was induced for 5 min. By digital gene expression profiling, we determined the gene expression profile of three strains (*cot/cot ok/ok*, +/+ *ok/ok* and +/+ +/+) under hyperthermia (42℃) and normal (25℃) conditions. A Venn diagram showed that the most common differentially expressed genes (DEGs, FDR<0.01 and log2 Ratio≥1) were up-regulated and annotated with the heat shock proteins (HSPs) in 3 strains after treatment with hyperthermia, suggesting that HSPs rapidly increased in response to high temperature; 110 unique DEGs, could be identified in the *cot* mutant after inducing hyperthermia when compared to the control strains. Of these 110 unique DEGs, 98.18% (108 genes) were up-regulated and 1.82% (two genes) were downregulated in the *cot* mutant. KEGG pathways analysis of these unique DEGs suggested that the top three KEGG pathways were "Biotin metabolism," "Fatty acid biosynthesis" and "Purine metabolism," implying that diverse metabolic processes are active in *cot* mutant induced - hyperthermia. Unique DEGs of interest were mainly involved in the ubiquitin system, nicotinic acetylcholine receptor genes, cardiac excitation-contraction coupling or the Notch signaling pathway. Insights into hyperthermia-induced alterations in gene expression and related pathways could yield hints for understanding the relationship between behaviors and environmental stimuli (hyperthermia) in insects.

Published On: PLoS ONE, 2014, 9(11).

1 State Key Laboratory of Silkworm Genome Biology, Southwest University, Chongqing 400716, China.
2 Key Sericultural Laboratory of the Ministry of Agriculture, Southwest University, Chongqing 400716, China.
*Corresponding author E-mail: xiaqy@swu.edu.cn.

高温诱导癫痫缩蚕的初生幼虫的转录组分析

聂红毅[1]　刘　春[1,2]　张银霞[1]　周梦婷[1]
黄小凤[1]　彭　莉[1]　夏庆友[1,2*]

摘要：快速响应极端瞬时环境变化是所有生物体的重要能力。然而，不同物种的耐热性生理基础仍然不清楚。本文中，我们发现家蚕*cot*突变种会在高温处理后出现癫痫的表征，包括体节收缩、翻滚、吐液和短暂停止运动，背血管心跳节律也显著变快。为了全面地了解这一过程的分子基础，我们对42 ℃热处理5 min后出现癫痫表征的*cot*家蚕进行了转录组分析。通过数字基因表达谱，我们检测了3个品系家蚕（*cot*/*cot ok*/*ok*，+/+ *ok*/*ok*和+/+ +/+）在42 ℃和25 ℃处理下的基因表达模式。韦恩图显示高温处理后大多数差异基因都表现出了上调趋势（FDR<0.01和log2 Ratio≥1），并且被注释为热休克蛋白（HSPs），表明热休克蛋白含量在高温处理后会迅速升高。相比对照组，*cot*品系在高温处理后鉴定到了110个差异基因。在这110个差异基因中，98.18%（108个）基因上调，1.82%（2个）基因下调。对这些差异基因进行KEGG通路分析，发现富集度最高的3个通路为“生物素代谢”、“脂肪酸生物合成”和“嘌呤代谢”，暗示这些生物过程可能在高温处理后的*cot*品系中较为活跃。这些差异基因主要涉及泛素系统、烟碱乙酰胆碱受体基因、心脏收缩耦合或者Notch信号通路。对高温处理引起的基因变化的研究，可以为理解昆虫行为与环境刺激之间的关系提供新视点。

1　家蚕基因组生物学国家重点实验室，西南大学，重庆

2　农业部蚕桑学重点实验室，西南大学，重庆

Transcriptome analysis of integument differentially expressed genes in the pigment mutant (*quail*) during molting of silkworm, *Bombyx mori*

Nie HY[#] Liu C[#] Cheng TC Li QY
Wu YQ Zhou MT Zhang YX Xia QY[*]

Abstract: In the silkworm *Bombyx mori*, pigment mutants with diverse body colors have been maintained throughout domestication for about 5 000 years. The silkworm larval body color is formed through the mutual interaction of melanin, ommochromes, pteridines and uric acid. These pigments/compounds are synthesized by the cooperative action of various genes and enzymes. Previous reports showed that melanin, ommochrome and pteridine are increased in silkworm *quail* (*q*) mutants. To understand the pigment increase and alterations in pigment synthesis in *q* mutant, transcriptome profiles of the silkworm integument were investigated at 16 h after head capsule slippage in the fourth molt in *q* mutants and wild-type (*Dazao*). Compared to the wild-type, 1 161 genes were differentially expressed in the *q* mutant. Of these modulated genes, 62.4% (725 genes) were upregulated and 37.6% (436 genes) were downregulated in the *q* mutant. The molecular function of differently expressed genes was analyzed by Blast2GO. The results showed that upregulated genes were mainly involved in protein binding, small molecule binding, transferase activity, nucleic acid binding, specific DNA-binding transcription factor activity and chromatin binding, while exclusively down-expressed genes functioned in oxidoreductase activity, cofactor binding, tetrapyrrole binding, peroxidase activity and pigment binding. We focused on genes related to melanin, pteridine and ommochrome biosynthesis; transport of uric acid; and juvenile hormone metabolism because of their importance in integument coloration during molting. This study identified differently expressed genes implicated in silkworm integument formation and pigmentation using silkworm *q* mutant. The results estimated the number and types of genes that drive new integument formation.

Published On: PLoS ONE, 2014, 9(4).

State Key Laboratory of Silkworm Genome Biology, Southwest University, Chongqing 400716, China.
[#]These authors contributed equally.
[*]Corresponding author E-mail: xiaqy@swu.edu.cn.

家蚕蜕皮过程中色素突变体(*quail*)表皮差异表达基因的转录组分析

聂红毅[#] 刘 春[#] 程廷才 李琼艳
吴玉乾 周梦婷 张银霞 夏庆友[*]

摘要:在家蚕5 000年的驯养过程中,保存了不同体色的突变品系。家蚕幼虫体色是由黑色素、眼色素、蝶呤和尿酸之间的相互作用决定的,而多种基因和酶之间的相互作用合成了这些色素复合物。之前有报道表明在家蚕*quail*(*q*)突变品系中的黑色素、眼色素和蝶呤的含量多。为了解析*q*突变品系中色素增加和色素合成的变化,我们进行了*Dazao*与*q*突变品系4眠头壳脱落16 h后的表皮转录组分析。与野生型相比,*q*突变体中的差异基因有1 161个。在这些差异基因中,62.4%(725个)在*q*突变体中上调,37.6%(436个)下调。用Blast2GO分析了不同表达基因的分子功能。结果表明,上调基因主要参与蛋白质结合、小分子结合、转移酶活性、核酸结合、特异的DNA结合转录因子活性和染色质结合,而仅下调表达的基因在氧化还原酶活性、辅因子结合、四吡啶结合、过氧化物酶活性和色素结合中起作用。我们重点研究了与黑色素、蝶呤和眼色素生物合成相关的基因,尿酸转运,以及保幼激素的代谢,因为它们在蜕皮过程中对表皮着色有重要影响。本研究利用家蚕*q*突变体鉴定了家蚕表皮形成和色素沉着过程中的不同表达基因,并评估了驱动新表皮形成的基因数量和类型。

家蚕基因组生物学国家重点实验室,西南大学,重庆

Molecular and enzymatic characterization of two enzymes *Bm*PCD and *Bm*DHPR involving in the regeneration pathway of tetrahydrobiopterin from the silkworm *Bombyx mori*

Li WT Gong MX Shu R

Li X Gao JS Meng Y*

Abstract: Tetrahydrobiopterin (BH4) is an essential cofactor of aromatic amino acid hydroxylases and nitric oxide synthase so that BH4 plays a key role in many biological processes. BH4 deficiency is associated with numerous metabolic syndromes and neuropsychological disorders. BH4 concentration in mammals is maintained through a *de novo* synthesis pathway and a regeneration pathway. Previous studies showed that the *de novo* pathway of BH4 is similar between insects and mammals. However, knowledge about the regeneration pathway of BH4 (RPB) is very limited in insects. Several mutants in the silkworm *Bombyx mori* have been approved to be associated with BH4 deficiency, which are good models to research on the RPB in insects. In this study, homologous genes encoding two enzymes, pterin-4a-carbinolamine dehydratase (PCD) and dihydropteridine reductase (DHPR) involving in RPB have been cloned and identified from *B. mori*. Enzymatic activity of DHPR was found in the fat body of wild type silkworm larvae. Together with the transcription profiles, it was indicated that *BmPcd* and *BmDhpr* might normally act in the RPB of *B. mori* and the expression of *BmDhpr* was activated in the brain and sexual glands while *BmPcd* was expressed in a wider special pattern when the *de novo* pathway of BH4 was lacked in *lemon*. Biochemical analyses showed that the recombinant *Bm*DHPR exhibited high enzymatic activity and more suitable parameters to the coenzyme of NADH *in vitro*. The results in this report give new information about the RPB in *B. mori* and help in better understanding insect BH4 biosynthetic networks.

Published On: Comparative Biochemistry and Physiology Part B—Biochemistry and Molecular Biology, 2015, 186, 20-27.

School of Life Sciences, Anhui Agricultural University, Hefei 230036, China.

*Corresponding author E-mail: mengyan@ahau.edu.cn.

家蚕四氢生物蝶呤再生途径中的两种酶*Bm*PCD和*Bm*DHPR的分子及酶学特征分析

李闻天　龚美霞　舒　蕊
李　鑫　高俊山　孟　艳*

摘要：四氢生物蝶呤（BH4）是芳香族氨基酸羟化酶和一氧化氮合酶的重要辅因子，因此BH4在许多生物过程中起关键作用。BH4缺乏与许多代谢综合征和神经心理障碍有关。哺乳动物中通过从头合成途径和再生途径维持BH4浓度。以前的研究表明，BH4的从头合成途径在昆虫和哺乳动物之间是相似的。然而，在昆虫中关于BH4的再生途径（RPB）的认知非常有限。家蚕中的几种突变体已经被验证与BH4缺陷相关，这是很好的研究昆虫中RPB的模型。在本研究中，我们克隆并鉴定了RPB途径两种酶：蝶呤-4a-甲醇胺脱水酶（PCD）和二氢蝶啶还原酶（DHPR）的同源基因。并且在野生型家蚕幼虫的脂肪体中发现DHPR的酶活性。结合转录水平检测，表明*BmPcd*和*BmDhpr*一般在家蚕RPB途径发挥作用。当*lemon*突变体中缺少BH4的从头合成途径时，*BmDhpr*的表达在脑和生殖腺中被激活，而*BmPcd*则表现出更广泛的特殊表达模式。生物化学分析显示，重组*Bm*DHPR在体外对NADH辅酶具有较高的酶促活性和更合适的参数。本研究的结果提供了家蚕RPB途径的新信息，有助于更好地了解昆虫BH4生物合成网络。

生命科学学院，安徽农业大学，合肥

Comparative analysis of the integument transcriptomes of the *black dilute* mutant and the wild-type silkworm *Bombyx mori*

Wu SY[1] Tong XL[1] Peng CX[1] Xiong G[1] Lu KP[1] Hu H[1]
Tan D[1] Li CL[1] Han MJ[1] Lu C[1,2*] Dai FY[1,2*]

Abstract: The insect cuticle is a critical protective shell that is composed predominantly of chitin and various cuticular proteins and pigments. Indeed, insects often change their surface pigment patterns in response to selective pressures, such as threats from predators, sexual selection and environmental changes. However, the molecular mechanisms underlying the construction of the epidermis and its pigmentation patterns are not fully understood. Among Lepidoptera, the silkworm is a favorable model for color pattern research. The *black dilute* (*bd*) mutant of silkworm is the result of a spontaneous mutation; the larval body color is notably melanized. We performed integument transcriptome sequencing of the wild-type strain *Dazao* and the mutant strains *+/bd* and *bd/bd*. In these experiments, during an early stage of the fourth molt, a stage at which approximately 51% of genes were expressed genome wide (RPKM ≥1) in each strain. A total of 254 novel transcripts were characterized using Cuffcompare and BLAST analyses. Comparison of the transcriptome data revealed 28 differentially expressed genes (*DEGs*) that may contribute to *bd* larval melanism, including 15 cuticular protein genes that were remarkably highly expressed in the *bd/bd* mutant. We suggest that these significantly up-regulated cuticular proteins may promote melanism in silkworm larvae.

Published On: Scientific Reports, 2016, 6.

1 State Key Laboratory of Silkworm Genome Biology, Southwest University, Chongqing 400716, China.
2 Key Laboratory for Sericulture Functional Genomics and Biotechnology of Agricultural Ministry, Southwest University, Chongqing 400716, China.
*Corresponding author E-mail: lucheng@swu.edu.cn; fydai@swu.edu.cn.

淡墨突变体和野生型家蚕的表皮转录组比较分析

吴松原[1] 童晓玲[1] 彭晨星[1] 熊 高[1] 陆昆鹏[1] 胡 海[1]
谭 端[1] 李春林[1] 韩民锦[1] 鲁 成[1,2*] 代方银[1,2*]

摘要：昆虫表皮主要由几丁质、表皮蛋白和色素组成，是重要的保护壳。事实上，昆虫通常改变它们的体色以应对选择压力，如来自捕食者的威胁、性别选择和环境变化等。然而，表皮的构造及其色素沉着的分子机制尚未得到完全解析。在鳞翅目昆虫中，家蚕是体色研究的理想模式生物。家蚕的淡墨突变体 *bd* 是自发突变的结果，其幼虫体色显著变黑。我们对野生型 *Dazao* 和突变品系 *+/bd* 和 *bd/bd* 的外表皮进行转录组测序。通过分析我们发现，在家蚕幼虫四眠的早期阶段，野生型 *Dazao* 和突变品系的基因组上约有 51% 的基因在表达（RPKM≥1）。利用 Cuffcompare 和 BLAST 分析对总共 254 个新的差异转录本进行表征。转录组数据的比较分析显示，有 28 个差异表达基因（*DEGs*）可能导致 *bd* 幼虫黑化，其中 15 个表皮蛋白基因在 *bd/bd* 突变体中显著高表达。我们认为，这些显著上调的表皮蛋白可能促进家蚕幼虫黑化。

1 家蚕基因组生物学国家重点实验室，西南大学，重庆
2 农业部蚕桑功能基因组学与生物技术重点实验室，西南大学，重庆

Linkage and mapping analyses of the densonucleosis non-susceptible gene *nsd-Z* in the silkworm *Bombyx mori* using SSR markers

Li MW[1,2] Guo QH[2] Hou CX[1] Miao XX[2]
Xu AY[1] Guo XJ[1] Huang YP[2*]

Abstract: In the silkworm *Bombyx mori*, non-susceptibility to the *Zhenjiang* (China) strain of the densonucleosis virus (DNV-Z) is controlled by the recessive gene *nsd-Z* (non-susceptible to DNV-Z), which is located on chromosome 15. Owing to a lack of crossing over in females, reciprocal backcrossed F_1 (BC_1) progeny were used for linkage analysis and mapping of the *nsd-Z* gene using silkworm strains *Js* and *L10*, which are classified as being highly susceptible and non-susceptible to DNV-Z, respectively. BC_1 larvae were inoculated with the DNV-Z virus at the first instar, and DNA was extracted from the individual surviving pupae and analyzed for simple sequence repeat (SSR) markers. The *nsd-Z* gene was found to be linked to 7 SSR markers, as all the surviving larvae in the BC_1♀ (F_1♀ × L10♂) showed the homozygous profile of strain *L10*, and the sick larvae in the BC_1♀ (F_1♀ × L10♂) showed the heterozygous profile of *Js* × *L10* F_1 hybrids. Using a reciprocal BC_1♀ (F_1♀ × L10♂) cross, we constructed a linkage map of 80.6 cM, with *nsd-Z* mapped at 30 cM and the closest SSR marker at a distance of 4.4 cM.

Published On: Genome, 2006, 49(4), 397-402.

1 Sericultural Research Institute, Chinese Academy of Agricultural Sciences, Zhenjiang 212018, China.
2 Shanghai Institute of Plant Physiology and Ecology, Shanghai Institutes for Biological Sciences, Chinese Academy of Sciences, Shanghai 200032, China.
*Corresponding author E-mail: yongping@sippe.ac.cn.

利用SSR标记对导致家蚕浓核病的非易感基因*nsd-Z*进行连锁图谱分析

李木旺[1,2]　郭秋红[2]　侯成香[1]　苗雪霞[2]
徐安英[1]　郭锡杰[1]　黄勇平[2*]

摘要： 隐性基因*nsd-Z*（DNV-Z非易感性）控制镇江（中国）家蚕品系对浓核病毒株（DNV-Z）的非易感性，它位于15号染色体上。由于雌性家蚕不发生重组交换，因此用F_1回交代BC_1♀（F_1♀×L10♂）来进行连锁定位分析，选择DNV-Z高度易感的家蚕*Js*品系和非易感性的品系*L10*对*nsd-Z*基因进行定位分析。BC_1一龄幼虫添食DNV-Z病毒，饲喂至蛹期，提取存活蛹的DNA，进行SSR标记分析。最终发现*nsd-Z*基因与7个SSR标签连锁，BC_1品系中存活的幼虫都是*L10*品系的纯合型，而BC_1品系中感病幼虫显示出*Js*×*L10*的杂合型。利用BC_1回交构建了80.6 cM的连锁图，发现*nsd-Z*定位在30 cM，与最近的一个SSR标记相距4.4 cM。

1　蚕业研究所，中国农业科学院，镇江
2　上海植物生理生态研究所，上海生命科学研究院，中国科学院，上海

Detection of homozygosity in near isogenic lines of non-susceptible to *Zhenjiang* strain of densonucleosis virus in silkworm

Li MW[1*] Hou CX[1] Zhao YP[3] Xu AY[1] Guo XJ[1,3] Huang YP[2]

Abstract: The near isogenic lines (NILs) of non-susceptible to DNV-Z were bred through backcrossing successively using *L10* as donor parent and *Js* as recurrent parent. The homozygosity of the NILs was detected using the SSR markers from the SSR linkage map. The results showed that the ratio of the linkage groups from the recurrent parent in the NILs increased rapidly along with the increase of the backcrossing generations. After 6 generations of backcrossing, it was selfed and was fed with DNV-Z. The linkage groups of 9 of the 5 BC_6 pairs, whose 1/4 offspring were non-susceptible to DNV-Z, were all replaced by the recurrent parent except the linkage group that held *nsd-Z*, and only part of one linkage group from the other one individual had not been replaced. According to the markers linked to *nsd-Z*, Fl0316 had not been replaced in all of the 10 individuals, but crossover happened in 2 individuals between Fl0568 and *nsd-Z*.

Published On: African Journal of Biotechnology, 2017, 6(14), 1629-1633.

1 Sericultural Research Institute, Chinese Academy of Agricultural Sciences, Zhenjiang 212018, China.

2 Shanghai Institute of Plant Physiology and Ecology, Shanghai Institutes for Biological Sciences, Chinese Academy of Sciences, Shanghai 200032, China.

3 College of Biotechnology, Jiangsu University of Science and Technology, Zhenjiang 212018, China.

*Corresponding author E-mail: cangene@pub.zj.jsinfo.net.

家蚕浓核病毒(镇江株)不敏感近等基因系的纯合性检测

李木旺[1*] 侯成香[1] 赵云坡[3] 徐安英[1] 郭锡杰[1,3] 黄勇平[2]

摘要：以*L10*作为供体亲本，以*Js*作为轮回亲本，通过回交先后选育对DNV-Z不易感染的近等基因系(NILs)。用SSR连锁图中的SSR标记检测NILs的纯合性。结果表明，随着回交世代的增加，NILs中轮回亲本的连锁群比例迅速增加。经6代回交后，自交并饲喂DNV-Z检测抗病性。结果显示，5对BC_6个体中1/4的后代对DNV-Z不敏感，而它们的9个连锁群中除含有*nsd-Z*连锁群未被回交亲本所取代外，其他均被回交亲本替代，且只有来自另一个个体的一个连锁群的一部分没有被替换。根据*nsd-Z*的关联标记，Fl0316在这10个个体中均未被替换，但在其中两个个体中，Fl0568和*nsd-Z*之间发生了交换。

1 蚕业研究所，中国农业科学院，镇江
2 上海植物生理生态研究所，上海生命科学研究院，中国科学院，上海
3 生物技术学院，江苏科技大学，镇江

Linkage analysis of the visible mutations *Sel* and *Xan* of *Bombyx mori* (Lepidoptera: Bombycidae) using SSR markers

Miao XX[1] Li MW[2] Dai FY[3] Lu C[3]

Marian R. Goldsmith[4] Huang YP[1*]

Abstract: Wild type silkworm larvae have opaque white skin, whereas the mutants *Sel* (Sepialumazine) and *Xan* (Xanthous) are yellow-skinned. Previous genetic analysis indicated that *Sel* and *Xan* are on established linkage groups 24 (0.0) and 27 (0.0), respectively. However, in constructing a molecular linkage map using simple sequence repeat (SSR) loci, we found that the two mutations were linked. To confirm this finding, we developed a set of SSR markers and used them to score reciprocal backcross populations. Taking advantage of the lack of crossing-over in female silkworms, we found that the progeny of backcrosses between F_1 females and males of the parental strains (BC_1F) of the two visible mutations had the same inheritance patterns linked to the same SSR markers. This indicated that the two visible mutations belonged to the same chromosome. To confirm this finding, we tested for independent assortment by crossing *Sel* and *Xan* marker strains with each other to obtain F_1 and F_2 populations. Absence of the expected wild type class among 5 000 F_2 progeny indicated that the two visible mutations were located on the same linkage group. We carried out recombination analysis for each mutation by scoring 190 progeny of backcrosses between F_1 males and parental females (BC_1M) and constructed a linkage map for each strain. The results indicated that the *Sel* gene was 12 cM from SSR marker S2404, and the *Xan* gene was 7.03 cM from SSR marker S2407. To construct a combined SSR map and to avoid having to discriminate the two similar dominant mutations in heterozygotes, we carried out recombination analysis by scoring recessive wild type segregants of F_2 populations for each mutation. The results showed that the *Sel* and *Xan* genes were 13 cM and 13.7 cM from the S2404 marker, respectively, consistent with the possibility that they are alleles of the same locus, which we provisionally assigned to SSR linkage group 24. We also used the F_2 recessive populations to construct two linkage groups for the *Sel* and *Xan* genes.

Published On: European Journal of Entomology, 2007, 104(4), 647-652.

1 Shanghai Institute of Plant Physiology and Ecology, Shanghai Institutes for Biological Sciences, Chinese Academy of Sciences, Shanghai 200032, China.

2 Sericultural Research Institute, Chinese Academy of Agricultural Sciences, Zhenjiang 212018, China.

3 Southwest University, Chongqing 400716, China.

4 Biological Sciences Department, University of Rhode Island, Kingston, RI 02881, USA.

*Corresponding author E-mail: yphuang@sibs.ac.cn.

利用SSR标记对家蚕（鳞翅目：蚕蛾科）*Sel*和*Xan*基因进行连锁分析

苗雪霞[1] 李木旺[2] 代方银[3]
鲁 成[3] Marian R. Goldsmith[4] 黄勇平[1*]

摘要：野生型家蚕幼虫的表皮是白色、不透明的，而突变型*Sel*和*Xan*家蚕品系的表皮为黄色。先前的遗传分析表明*Sel*和*Xan*基因分别定位在24（0.0）和27（0.0）两个连锁群。然而，在用简单重复序列（SSR）位点分析构建分子标记连锁图谱的过程中，我们发现这两种突变体存在着连锁关系。为了证实这一发现，我们构建了一组SSR分子标记，用它们分析相互的回交群体。利用雌性家蚕减数分裂过程中染色体内基因不发生重组交换的特点，我们发现两个突变品系F_1代的雌性家蚕和亲本的雄性家蚕回交的后代中两个突变基因具有和SSR分子标记相同的遗传模式，表明这两种突变基因位于同一条染色体。为了证实这个发现，我们将*Sel*和*Xan*突变品系杂交，得到F_1和F_2代，在5 000个F_2代中没有发现预期的野生型个体，说明这两个基因位于同一条染色体上。通过对F_1代的雄性家蚕和亲本的雌性家蚕进行重组分析，并对每一个品系构建连锁图谱。结果发现*Sel*基因距离SSR分子标记S2404有12 cM，*Xan*基因距离SSR分子标记S2407有7.03 cM。为了构建联合的SSR图谱并避免在杂合子中区分这两种显性突变，我们对每个突变F_2代个体隔离的隐性野生型进行了重组分析，结果显示，*Sel*和*Xan*基因分别距离S2404分子标记13 cM和13.7 cM，这与它们是同一基因座的等位基因的可能性一致，我们将它们暂时分配到SSR连锁组24。此外，我们还利用F_2代隐性个体构建了*Sel*和*Xan*基因的两个连锁图谱。

1 上海植物生理生态研究所，上海生命科学研究院，中国科学院，上海
2 蚕业研究所，中国农业科学研究院，镇江
3 西南大学，重庆
4 生物科学系，罗德岛大学，金斯顿，美国

SSR based linkage and mapping analysis of *C*, a yellow cocoon gene in the silkworm, *Bombyx mori*

Zhao YP[1,2] Li MW[1*] Xu AY[1] Hou CX[1]
Li MH[3] Guo QH[3] Huang YP[3] Guo XJ[1,2]

Abstract: The yellow color of the cocoon of the silkworm *Bombyx mori* is controlled by three genes, *Y (Yellow haemolymph)*, *I (Yellow inhibitor)* and *C (Outer-layer yellow cocoon)*, which are located on linkage groups 2, 9 and 12, respectively. Taking advantage of a lack of crossing over in females, reciprocal backcrossed F_1 (BC_1) progeny were used for linkage analysis and mapping of the *C* gene using silkworm strains *C108* and *KY*, which spin white and yellow cocoons, respectively. DNA was extracted from individual pupae and analyzed for simple sequence repeat (SSR) markers. The *C* gene was found to be linked to seven SSR markers. All the yellow cocoon individuals from a female heterozygous backcross (BC_1F) showed a heterozygous profile for SSR markers on linkage group 12, whereas individuals with light yellow cocoons showed the homozygous profile of the strain *C108*. Using a reciprocal heterozygous male backcross (BC_1M), we constructed a linkage map of 36.4 cM with the *C* gene located at the distal end, and the closest SSR marker at a distance of 13.9 cM.

Published On: Insect Science, 2008, 15(5), 399-404.

1 Sericultural Research Institute, Chinese Academy of Agricultural Sciences, Zhenjiang 212000, China.
2 College of Biotechnology and Environmental Engineering, Jiangsu University of Science and Technology, Zhenjiang 212013, China.
3 Shanghai Institute of Plant Physiology and Ecology, Shanghai Institutes for Biological Sciences, Chinese Academy of Sciences, Shanghai 200032, China.
*Corresponding author E-mail: cangene@pub.zj.jsinfo.net.

利用SSR法对家蚕黄茧基因C进行连锁定位分析

赵云坡[1,2]　李木旺[1*]　徐安英[1]　侯成香[1]
李明辉[3]　郭秋红[3]　黄勇平[3]　郭锡杰[1,2]

摘要：家蚕(*Bombyx mori*)黄茧表型主要由*Y*(黄血)、*I*(黄色抑制剂)和*C*(外层黄茧)3个基因来控制，其分别定位在家蚕的2号、9号和12号连锁群上。由于在雌性中缺乏姐妹染色单体的交叉互换，利用结白色茧的*C108*和黄色茧的*KY*品系相互回交的后代$F_1(BC_1)$，对*C*基因进行定位。从单个蛹中提取DNA，分析简单序列重复(SSR)标记，发现*C*基因与7个SSR标记有关。雌性杂合子回交(BC_1F)的所有黄茧个体在12号连锁群SSR标记上均表现为杂合，而淡黄色茧个体则表现为*C108*品系纯合。我们使用一个互异杂合子雄性回交(BC_1M)，构建了一个36.4 cM的连锁图谱，其中*C*基因位于远端，最近的SSR标记距离为13.9 cM。

1　蚕业研究所，中国农业科学院，镇江
2　生物与环境工程学院，江苏科技大学，镇江
3　上海植物生理生态研究所，上海生命科学研究院，中国科学院，上海

Fine mapping of E^{kp}-*1*, a locus associated with silkworm (*Bombyx mori*) proleg development

Xiang H[1] Li MW[1,2] Yang F[1] Guo QH[1]
Zhan S[1] Lin HX[1] Miao XX[1] Huang YP[1*]

Abstract: The silkworm homeotic mutant E^{kp} has a pair of rudimentary abdominal legs, called prolegs, in its A2 segment. This phenotype is caused by a single dominant mutation at the E^{kp}-*1* locus, which was previously mapped to chromosome 6. To explore the possible association of *Hox* genes with proleg development in the silkworm, a map-based cloning strategy was used to isolate the E^{kp}-*1* locus. Five E^{kp}-*1*-linked simple sequence repeat markers on chromosome 6 were used to generate a low-resolution map with a total genetic distance of 39.5 cM. Four additional cleaved amplified polymorphic sequence markers were developed based on the initial map. The closest marker to E^{kp}-*1* was at a genetic distance of 2.7 cM. A high-resolution genetic map was constructed using nine BC_1 segregating populations consisting of 2 396 individuals. Recombination suppression was observed in the vicinity of E^{kp}-*1*. Four molecular markers were tightly linked to E^{kp}-*1*, and three were clustered with it. These markers were used to screen a BAC library. A single bacterial artificial chromosome (BAC) clone spanning the E^{kp}-*1* locus was identified, and E^{kp}-*1* was delimited to a region less than 220 kb long that included the *Hox* gene *abdominal-A* and a non-coding locus, *iab-4*. These results provide essential information for the isolation of this locus, which may shed light on the mechanism of proleg development in the silkworm and possibly in Lepidoptera.

Published On: Heredity, 2008, 100(5), 533-540.

1 Shanghai Institute of Plant Physiology and Ecology, Shanghai Institutes for Biological Sciences, Chinese Academy of Sciences, Shanghai 200032, China.

2 Sericultural Research Institute, Chinese Academy of Agricultural Sciences, Zhenjiang 212000, China.

*Corresponding author E-mail: yphuang@sibs.ac.cn.

家蚕腹足发育相关基因座E^{kp}-*1*的精细定位

相　辉[1]　李木旺[1,2]　杨　斐[1]　郭秋红[1]
詹　帅[1]　林鸿宣[1]　苗雪霞[1]　黄勇平[1*]

摘要：家蚕同源异型突变体E^{kp}的A2腹节具有一对退化的腹足，称之为原足。这种表型是由E^{kp}-*1*基因座上的单一显性突变引起的，该位点先前已定位在6号染色体上。为了探索*Hox*基因与家蚕原足发育的相关性，利用定位克隆法分离E^{kp}-*1*基因座。使用6号染色体上E^{kp}-*1*连锁的5个简单重复序列标记，绘制了总遗传距离为39.5 cM的低分辨率连锁图谱。基于初始图谱增加了4个酶切扩增多态性序列标记。最接近E^{kp}-*1*的标记位于遗传距离为2.7 cM的位点。利用2 396个个体组成的9个BC_1分离群体构建了高分辨率遗传图谱。在E^{kp}-*1*邻近位点观察到重组抑制。4个分子标记与E^{kp}-*1*紧密相关，其中有3个分子标记与E^{kp}-*1*聚类。这些标记用于筛选BAC文库。鉴定到一个跨越E^{kp}-*1*基因座的单个细菌人工染色体（BAC）克隆，并且E^{kp}-*1*被限定在小于220 kb的区域内，其中还包括*Hox*基因*abdominal-A*和一个非编码位点*iab-4*。这些结果为分离E^{kp}-*1*基因座提供了重要信息，为进一步揭示家蚕和鳞翅目中原足发育的机制提供了可能。

1　上海植物生理生态研究所，上海生命科学研究院，中国科学院，上海
2　蚕业研究所，中国农业科学院，镇江

Fine mapping of a supernumerary proleg mutant (E^{Cs}-*l*) and comparative expression analysis of the *abdominal-A* gene in silkworm, *Bombyx mori*

Chen P[1,2] Tong XL[1] Li DD[1] Liang PF[1] Fu MY[1]
Li CF[1] Hu H[1] Xiang ZH[1] Lu C[1*] Dai FY[1,2*]

Abstract: Patterning and phenotypic variations of appendages in insects provide important clues on developmental genetics. In the silkworm *Bombyx mori*, morphological variations associated with the E complex, an analogue of the *Drosophila melanogaster* bithorax complex, mainly determine the shape and number of prolegs on abdominal segments. Here, we report the identification and characterization of the allele responsible for the supernumerary crescents and legs-like (E^{Cs}-*l*) mutant, a model derived from spontaneous mutation of the E complex, with supernumerary legs and extra crescents. Fine mapping with 1 605 individuals revealed a ~68 kb sequence in the upstream intergenic region of *B. mori abdominal-A* (*Bmabd-A*) clustered with the E^{Cs}-*l* locus. Quantitative real-time PCR (qRT-PCR) and Western blotting analyses disclosed a marked increase in *Bmabd-A* expression in the E^{Cs}-*l* mutant at both the transcriptional and translational levels, compared to wild-type *Dazao*. Furthermore, we observed ectopic expression of the *Bm*abd-A protein in the second abdominal segment (A2) of the E^{Cs}-*l* mutant. Our results collectively suggest that the 68 kb region contains important regulatory elements of the *Bmabd-A* gene, and provide evidence that the gene is required for limb development in abdominal segments in the silkworm.

Published On: Insect Molecular Biology, 2013, 22(5), 497-504.

1 State Key Laboratory of Silkworm Genome Biology, Southwest University, Chongqing 400716, China.
2 Key Laboratory for Sericulture Functional Genomics and Biotechnology of Agricultural Ministry, Southwest University, Chongqing 400716, China.
*Corresponding author E-mail: lucheng@swu.edu.cn; fydai@swu.edu.cn.

家蚕 E^{Cs}-*l* 突变体的精细定位以及家蚕 *abd-A* 基因的比较表达分析

陈　鹏[1,2]　童晓玲[1]　李丹丹[1]　梁平凤[1]　付明月[1]
李春峰[1]　胡　海[1]　向仲怀[1]　鲁　成[1*]　代方银[1,2*]

摘要：昆虫附肢的形态和表型变异为发育遗传学提供了重要线索。在家蚕中，形态变异与E复合物有关，黑腹果蝇中该复合物的一个类似物决定腹部节段的形状和数量。在本研究中，我们鉴定并分析了家蚕 E^{Cs}-*l* 突变体的关键基因，该突变体来源于E复合体的自发突变，并产生了额外的足和半月纹的表型。对1 605个个体进行精细定位的结果表明，*Bmabd-A* 的上游基因间的一段68 kb序列与 E^{Cs}-*l* 基因座成簇。qPCR和Western blot实验分析揭示了转录和翻译水平的 E^{Cs}-*l* 突变体中 *Bmabd-A* 的表达显著增加。此外，我们观察到 *B*mabd-A 蛋白在 E^{Cs}-*l* 突变体的第二腹足（A2）中异位表达。我们的研究结果表明，68 kb区域含有 *Bmabd-A* 基因的重要调控元件，且该基因是家蚕腹部肢节发育所必需的。

1　家蚕基因组生物学国家重点实验室，西南大学，重庆
2　农业部蚕桑功能基因组与生物技术重点实验室，西南大学，重庆

Molecular mapping and characterization of the silkworm *apodal* mutant

Chen P[1,2#] Tong XL[1#] Fu MY[1] Hu H[1] Song JB[1]
He SZ[1] Gai TT[1] Dai FY[1,2*] Lu C[1,2*]

Abstract: The morphological diversity of insects is important for their survival; in essence, it results from the differential expression of genes during development of the insect body. The silkworm *apodal* (*ap*) mutant has degraded thoracic legs making crawling and eating difficult and the female is sterile, which is an ideal subject for studying the molecular mechanisms of morphogenesis. Here, we confirmed that the infertility of *ap* female moths is a result of the degradation of the bursa copulatrix. Positional cloning of *ap* locus and expression analyses reveal that the *Bombyx mori* sister of *odd* and *bowl* (*Bmsob*) gene is a strong candidate for the *ap* mutant. The expression of *Bmsob* is down-regulated, while the corresponding *Hox* genes are up-regulated in the *ap* mutant compared to the wild type. Analyses with the dual luciferase assay present a declined activity of the *Bmsob* promoter in the *ap* mutant. Furthermore, we demonstrate that *Bmsob* can inhibit *Hox* gene expression directly and by suppressing the expression of other genes, including the *BmDsp* gene. The results of this study are an important contribution to our understanding of the diversification of insect body plan.

Published On: Scientific Reports, 2016, 6.

1 State Key Laboratory of Silkworm Genome Biology, Southwest University, Chongqing 400716, China.
2 Key Laboratory for Sericulture Functional Genomics and Biotechnology of Agricultural Ministry, Southwest University, Chongqing 400716, China.
#These authors contributed equally.
*Corresponding author E-mail: fydai@swu.edu.cn; lucheng@swu.edu.cn.

家蚕*apodal*突变体的分子定位与鉴定

陈　鹏[1,2#]　童晓玲[1#]　付明月[1]　胡　海[1]　宋江波[1]
何松真[1]　盖停停[1]　代方银[1,2*]　鲁　成[1,2*]

摘要：昆虫形态的多样性是昆虫发育过程中的基因差异表达引起的，对其生存至关重要。家蚕*apodal*（*ap*）突变体由于胸足退化，致使爬行和进食困难，且雌性不育，是研究形态发生分子机制的理想对象。在这里，我们证实了雌蛾的不育症是交配囊退化的结果。定位克隆显示*Bmsob*基因是家蚕*ap*突变体最佳的候选基因。与野生型相比，*Bmsob*在*ap*突变体中下调，而相应的*Hox*基因表达上调。双荧光素酶实验表明，在*ap*突变体中，*Bmsob*启动子的活性下降。此外，我们证明*Bmsob*可以通过抑制包括*BmDsp*基因在内的其他基因的表达，直接抑制*Hox*基因的表达。这些研究结果对我们了解昆虫体态特征的多样化有重要意义。

1　家蚕基因组生物学国家重点实验室，西南大学，重庆
2　农业部蚕桑功能基因组学与生物技术重点实验室，西南大学，重庆

A composite method for mapping quantitative trait loci without interference of female achiasmatic and gender effects in silkworm, *Bombyx mori*

Li CL[1,2] Zuo WD[1,2] Tong XL[1,2] Hu H[1,2] Qiao L[3]
Song JB[1,2] Xiong G[1,2] Gao R[1,2] Dai FY[1,2*] Lu C[1,2]

Abstract: The silkworm, *Bombyx mori*, is an economically important insect that was domesticated more than 5 000 years ago. Its major economic traits focused on by breeders are quantitative traits, and an accurate and efficient QTL mapping method is necessary to explore their genetic architecture. However, current widely used QTL mapping models are not well suited for silkworm because they ignore female achiasmate and gender effects. In this study, we propose a composite method combining rational population selection and special mapping methods to map QTL in silkworm. By determining QTL for cocoon shell weight (CSW), we demonstrated the effectiveness of this method. In the CSW mapping process, only 56 markers were used and five loci or chromosomes were detected, more than in previous studies. Additionally, loci on chromosomes 1 and 11 dominated and accounted for 35.10% and 15.03% of the phenotypic variance respectively. Unlike previous studies, epistasis was detected between loci on chromosomes 11 and 22. These mapping results demonstrate the power and convenience of this method for QTL mapping in silkworm, and this method may inspire the development of similar approaches for other species with special genetic characteristics.

Published On: Animal Genetics, 2015, 46(4), 426-432.

1 State Key Laboratory of Silkworm Genome Biology, Southwest University, Chongqing 400716, China.
2 Key Laboratory for Sericulture Functional Genomics and Biotechnology of Agricultural Ministry, Southwest University, Chongqing, 400716, China.
3 Institute of Entomology and Molecular Biology, College of Life Sciences, Chongqing Normal University, Chongqing 401331, China.
*Corresponding author E-mail: fydai@swu.edu.cn.

一种不影响蚕蛾雌性性别的用于定位数量性状基因座的复合方法

李春林[1,2] 左伟东[1,2] 童晓玲[1,2] 胡 海[1,2] 乔 梁[3]
宋江波[1,2] 熊 高[1,2] 高 瑞[1,2] 代方银[1,2*] 鲁 成[1,2]

摘要：家蚕是一种重要的经济昆虫，5 000多年前就已被驯化。育种学家关注的主要经济性状是数量性状，准确有效的QTL定位方法是探索遗传结构所必需的。然而，目前广泛使用的QTL定位模型忽视了雌蚕无重组互换和性别效应，因此并不适用对家蚕进行分析。在本研究中，我们提出了一种综合方法，结合合理种群选择和特殊映射方法对家蚕中的QTL进行定位。通过确定茧重（CSW）的QTL，我们证明了该方法的有效性。在CSW定位过程中，仅使用56个标记，就能检测到5个位点或染色体，比以前的研究获得的结果更多。另外，1号和11号染色体上的位点分别占表型变异的35.10%和15.03%。与以前研究不同的是，在11号染色体和22号染色体的基因座之间检测到上位性。这些定位结果证明了该方法在家蚕QTL定位中的有效性和便利性，并且该方法可能对具有特殊遗传特征的其他物种开发相似定位方法具有一定的启发作用。

1 家蚕基因组生物学国家重点实验室，西南大学，重庆
2 农业部蚕桑功能基因组学与生物技术重点实验室，西南大学，重庆
3 昆虫与分子生物学研究所，生命科学学院，重庆师范大学，重庆

Mutations of an Arylalkylamine-*N*-acetyl transferase, *Bm–iAANAT*, are responsible for silkworm melanism mutant

Dai FY[1#] Qiao L[1#] Tong XL[1,2#] Cao C[1]
Chen P[1] Chen J[1] Lu C[1*] Xiang ZH[1]

Abstract: Coloration is one of the most variable characters in animals and provides rich material for studying the developmental genetic basis of pigment patterns. In the silkworm, more than 100 genes mutation systems are related to aberrant color patterns. The melanism(*mln*) is a rare body color mutant that exhibits an easily distinguishable phenotype in both larval and adult silkworms. By positional cloning, we identified the candidate gene of the *mln* locus, *Bm-iAANAT*, whose homologous gene (*Dat*) converts dopamine into *N*-acetyldopamine, a precursor for *N*-acetyldopamine sclerotin in *Drosophila*. In the *mln* mutant, two types of abnormal *Bm-iAANAT* transcripts were identified, whose expression levels are markedly lower than the wild type (*WT*). Moreover, dopamine content was approximately twice as high in the sclerified tissues(head, thoracic legs, and anal plate) of the mutant as in *WT*, resulting in phenotypic differences between the two. Quantitative reverse transcription PCR analyses showed that other genes involved in the melanin metabolism pathway were regulated by the aberrant *Bm-iAANAT* activity in *mln* mutant in different ways and degrees. We therefore propose that greater accumulation of dopamine results from the functional deficiency of *BmiAANAT* in the mutant, causing a darker pattern in the sclerified regions than in the *WT*. In summary, our results indicate that *Bm-iAANAT* is responsible for the color pattern of the silkworm mutant, *mln*. To our knowledge, this is the first report showing a role for arylalkylamine-*N*-acetyltransferases in color pattern mutation in Lepidoptera.

Published On: Journal of Biological Chemistry, 2010, 285(25), 19553-19560.

1 College of Biotechnology, Institute of Sericulture and Systems Biology, Southwest University, Chongqing 400716, China.
2 Institute of Agriculture and Life Science, Chongqing University, Chongqing 400030, China.
[#]These authors contributed equally.
[*]Corresponding author E-mail: lucheng@swu.edu.cn.

芳烷基胺-*N*-乙酰转移酶基因 *Bm-iAANAT* 突变导致家蚕黑化突变体的产生

代方银[1#] 乔 梁[1#] 童晓玲[1,2#] 曹 存[1]
陈 鹏[1] 陈 军[1] 鲁 成[1*] 向仲怀[1]

摘要：体色是动物变化最大的特征之一，为研究色素模式的发育遗传基础提供了丰富的素材。在家蚕中，超过100种基因突变与异常体色模式有关。黑化型（*mln*）是一种罕见的体色突变体，在家蚕幼虫和成虫中都表现出易于识别的表型。通过定位克隆，我们确定了*mln*突变位点的候选基因为*Bm-iAANAT*，其同源基因（*Dat*）将多巴胺转化为*N*-乙酰多巴胺。在果蝇中，*N*-乙酰多巴胺是*N*-乙酰多巴胺硬化蛋白的前体。在*mln*突变体中，鉴定到两种异常的*Bm-iAATAT*转录本，其表达水平明显低于野生型（*WT*）。此外，该突变体硬化组织（头、胸足和肛板）中多巴胺含量约为野生型的两倍，导致两者出现表型差异。实时荧光定量PCR分析显示，突变体*mln*中异常的*Bm-iAATAT*活性以不同方式和程度调节黑色素代谢途径相关基因。因此，我们认为多巴胺的大量积累是由突变体中*Bm-iAANAT*的功能缺陷引起的，进而导致硬化区域的颜色比*WT*更暗。总而言之，我们的研究结果表明，*Bm-iAANAT*调控家蚕突变体*mln*的色型。据我们所知，这是首次报道芳烷基胺-*N*-乙酰转移酶在鳞翅目体色突变中发挥作用。

1 生物技术学院，蚕学与系统生物学研究所，西南大学，重庆
2 农学及生命科学研究院，重庆大学，重庆

Disruption of an *N-acetyltransferase* gene in the silkworm reveals a novel role in pigmentation

Zhan S[1,2#] Guo QH[1#] Li MH[1] Li MW[3]
Li JY[4] Miao XX[1] Huang YP[1*]

Abstract: The pigmentation of insects has served as an excellent model for the study of morphological trait evolution and developmental biology. The *melanism* (*mln*) mutant of the silkworm *Bombyx mori* is notable for its strong black coloration, phenotypic differences between larval and adult stages, and its widespread use in strain selection. Here, we report the genetic and molecular bases for the formation of the *mln* morphological trait. Fine mapping revealed that an *arylalkylamine N-acetyltransferase* (*AANAT*) gene co-segregates with the black coloration patterns. Coding sequence variations and expression profiles of *AANAT* are also associated with the melanic phenotypes. A 126 bp deletion in the *mln* genome causes two alternatively spliced transcripts with premature terminations. An enzymatic assay demonstrated the absolute loss of *AANAT* activity in the mutant proteins. We also performed RNA interference of *AANAT* in wild-type pupae and observed a significant proportion of adults with ectopic black coloration. These findings indicate that functional deletion of this *AANAT* gene accounts for the *mln* mutation in silkworm. *AANAT* is also involved in a parallel melanin synthesis pathway in which *ebony* plays a role, whereas no pigmentation defect has been reported in the *Drosophila* model or in other insects to date. To the best of our knowledge, the *mln* mutation is the first characterized mutant phenotype of insects with *AANAT*, and this result contributes to our understanding of dopamine metabolism and melanin pattern polymorphisms.

Published On: Development, 2010, 137(23), 4083-4090.

1 Shanghai Institute of Plant Physiology and Ecology, Shanghai Institutes for Biological Sciences, Chinese Academy of Sciences, Shanghai, 200032, China.
2 Graduate School of Chinese Academy of Sciences, Shanghai 200031, China.
3 Sericultural Research Institute, Chinese Academy of Agriculture Sciences, Zhenjiang 212018, China.
4 Virginia Polytechnic Institute and State University, Blacksburg, VA 24061, USA.
[#]These authors contributed equally.
[*]Corresponding author E-mail: yphuang@sibs.ac.cn.

家蚕*N*–乙酰转移酶基因的功能缺失揭示其在色素沉积中的新作用

詹 帅[1,2#] 郭秋红[1#] 李明辉[1] 李木旺[3]
李建勇[4] 苗雪霞[1] 黄勇平[1*]

摘要：昆虫的色素沉着是形态特征进化和发育生物学研究的一个理想模型，家蚕的黑化突变体（*mln*）因其具有明显的黑化表型、幼虫和成虫阶段的表型差异，以及在品种选择中广泛应用而闻名。这里我们揭示了家蚕*mln*表型性状形成的遗传学和分子机理。精细定位结果表明*N*–乙酰转移酶（*AANAT*）基因和黑化表型呈现共分离的特征。*AANAT*基因编码序列差异和表达特征也与黑化表型有关。*mln*基因组中126 bp的缺失导致两个选择性剪接的转录物提前终止。酶促实验证实*mln*突变体中*AANAT*的活性完全丧失。我们在野生型蚕蛹中进行*AANAT* RNAi，观察到有相当比例的成虫出现异位黑色。这些发现表明*AANAT*基因的功能缺失导致家蚕中的*mln*突变。*AANAT*同时参与*ebony*基因起重要作用的黑色素合成途径，但在果蝇及其他昆虫中并没有黑色素形成缺陷的报道。据我们所知，*mln*突变是具有*AANAT*的昆虫的第一个特征性突变表型，这一结果有助于我们对多巴胺代谢和黑色素模式多态性的理解。

1 上海植物生理生态研究所，上海生命科学研究院，中国科学院，上海
2 研究生院，中国科学院，上海
3 蚕业研究所，中国农业科学研究院，镇江
4 弗吉尼亚理工学院暨州立大学，弗吉尼亚州，美国

Repression of tyrosine hydroxylase is responsible for the sex-linked chocolate mutation of the silkworm, *Bombyx mori*

Liu C[1] Kimiko Yamamoto[2] Cheng TC[3] Keiko Kadono Okuda[2] Junko Narukawa[2] Liu SP[1] Han Y[1] Ryo Futahashi[4] Kurako Kidokoro[2] Hiroaki Noda[2] Isao Kobayashi[2] Toshiki Tamura[2] Akio Ohnuma[5] Yutaka Banno[6] Dai FY[1] Xiang ZH[1] Marian R. Goldsmith[7] Kazuei Mita[2*] Xia QY[1,3*]

Abstract: Pigmentation patterning has long interested biologists, integrating topics in ecology, development, genetics, and physiology. Wild-type neonatal larvae of the silkworm, *Bombyx mori*, are completely black. By contrast, the epidermis and head of larvae of the homozygous recessive sex-linked chocolate (*sch*) mutant are reddish brown. When incubated at 30 °C, mutants with the *sch* allele fail to hatch; moreover, homozygous mutants carrying the allele *sch* lethal (sch^l) do not hatch even at room temperature (25 ℃). By positional cloning, we narrowed a region containing sch to 239 622 bp on chromosome 1 using 4 501 backcross (BC_1) individuals. Based on expression analyses, the best sch candidate gene was shown to be tyrosine hydroxylase (*BmTh*). *BmTh* coding sequences were identical among *sch*, sch^l, and wild-type. However, in *sch* the ~70 kb sequence was replaced with ~4.6 kb of a *Tc1-mariner* type transposon located ~6 kb upstream of *BmTh*, and in sch^l, a large fragment of an *L1Bm* retrotransposon was inserted just in front of the transcription start site of *BmTh*. In both cases, we observed a drastic reduction of *BmTh* expression. Use of RNAi with *BmTh* prevented pigmentation and hatching, and feeding of a tyrosine hydroxylase inhibitor also suppressed larval pigmentation in the wild-type strain, pnd^+ and in a *pS* (black-striped) heterozygote. Feeding L-dopa to *sch* neonate larvae rescued the mutant phenotype from chocolate to black. Our results indicate the *BmTh* gene is responsible for the *sch* mutation, which plays an important role in melanin synthesis producing neonatal larval color.

Published On: Proceeding of the National Academy of Sciences of the United States of America, 2010, 107(29), 12980-12985.

1 Key Sericultural Laboratory of Agricultural Ministry, Institute of Sericulture and Systems Biology, Southwest University, Chongqing 400716, China.

2 National Institute of Agrobiological Sciences, Owashi, Tsukuba, Ibaraki 305-8634, Japan.

3 Institute of Agronomy and Life Science, Chongqing University, Chongqing 400044, China.

4 Research Institute of Genome-based Biofactory, National Institute of Advanced Industrial Science and Technology, Tsukuba, Ibaraki 305-8566, Japan.

5 Institute of Sericulture, Ami, Ibaraki 300-0324, Japan.

6 Institute of Genetic Resources, Graduate School of Bioresource and Bioenvironmental Science, Kyushu University, Fukuoka 812-8581, Japan.

7 Biological Sciences Department, University of Rhode Island, Kingston, RI 02881, USA.

[*]Corresponding author E-mail: kmita@nias.affrc.go.jp; xiaqy@swu.edu.cn.

抑制酪氨酸羟化酶引起家蚕伴性遗传巧克力色突变体产生

刘 春[1] Kimiko Yamamoto[2] 程廷才[3] Keiko Kadono Okuda[2] Junko Narukawa[2]
刘仕平[1] 韩 宇[1] Ryo Futahashi[4] Kurako Kidokoro[2] Hiroaki Noda[2] Isao Kobayashi[2]
Toshiki Tamura[2] Akio Ohnuma[5] Yutaka Banno[6] 代方银[1] 向仲怀[1]
Marian R. Goldsmith[7] Kazuei Mita[2*] 夏庆友[1,3*]

摘要：长期以来科学家们对色素形成的模式非常感兴趣，这种模式将生态学、发育学、遗传学和生理学整合在一起。野生型家蚕的蚁蚕体色为黑色。相比之下，一种纯合的伴性赤蚁突变体（*sch*）的表皮和头部为赤褐色。在30℃孵育时，具有*sch*等位基因的突变体不能孵化；此外，携带*sch*致死等位基因的纯合突变体（sch^{l}）即使在室温（25℃）下也不孵化。通过定位克隆，我们使用4 501个回交（BC_1）个体，将*sch*突变基因缩小到1号染色体上239 622 bp区域。基于表达分析，发现*sch*突变体的最佳候选基因为酪氨酸羟化酶（*BmTh*）。*BmTh*编码序列在*sch*、sch^{l}和野生型之间是相同的。然而，在*sch*突变体*BmTh*上游约6 kb处，约70 kb的序列被替换为约4.6 kb的*Tc1-mariner*类型转座子，并且在sch^{l}突变体中，*L1Bm*反转录转座子的一个大片段插入*BmTh*的转录起始位点前面。在这两种情况下，我们观察到*BmTh*的表达量急剧下降。使用RNAi干扰*BmTh*能阻止色素沉着和孵化，饲喂酪氨酸羟化酶抑制剂也能抑制野生型、pnd^{+}型和杂合体*pS*（黑条纹）型幼虫色素沉着。对*sch*突变体蚁蚕添食L-多巴能把突变体的表型从巧克力色挽救成黑色。我们的研究结果表明，*BmTh*基因控制*sch*突变，其通过在黑色素合成中发挥作用，进而调控蚁蚕体色的形成。

1 农业部蚕桑学重点实验室，蚕学与系统生物学研究所，西南大学，重庆
2 国立农业生物科学研究所，茨城，日本
3 农学与生命科学研究院，重庆大学，重庆
4 基因生物工厂研究院，国家先进工业科学技术研究所，茨城，日本
5 蚕业研究所，茨城，日本
6 遗传资源研究所，生物资源和生物环境科学研究生院，九州大学，福冈，日本
7 生物科学系，罗德岛大学，金斯顿，美国

Vitellogenin receptor mutation leads to the oogenesis mutant phenotype "*scanty vitellin*" of the silkworm, *Bombyx mori*

Lin Y[1#] Meng Y[2,3#] Wang YX[1#] Luo J[1] Susumu Katsuma[3]
Yang CW[1] Yutaka Banno[4] Takahiro Kusakabe[4] Toru Shimada[3*] Xia QY[1*]

Abstract: In insects, the vitellogenin receptor (VgR) mediates the uptake of vitellogenin (Vg) from the hemolymph by developing oocytes. The oogenesis mutant *scanty vitellin* (*vit*) of *Bombyx mori* (*Bm*) lacks vitellin and 30-kDa proteins, but *B. mori* egg-specific protein and *Bm*Vg are normal. The *vit* eggs are white and smaller compared with the pale yellow eggs of the wild type and are embryonic lethal. This study found that a mutation in the *B. mori VgR* gene (*BmVgR*) is responsible for the *vit* phenotype. We cloned the cDNA sequences encoding *WT* and *vit Bm*VgR. The functional domains of *Bm*VgR are similar to those of other low-density lipoprotein receptors. When compared with the wild type, a 235-bp genomic sequence in *vit BmVgR* is substituted for a 7-bp sequence. This mutation has resulted in a 50-amino acid deletion in the third Class B region of the first epidermal growth factor (EGF1) domain. *BmVgR* is expressed specifically in oocytes, and the transcriptional level is changed dramatically and consistently with maturation of oocytes during the previtellogenic periods. Linkage analysis confirmed that *BmVgR* is mutated in the *vit* mutant. The coimmunoprecipitation assay confirmed that mutated *Bm*VgR is able to bind *Bm*Vg but that *Bm*Vg cannot be dissociated under acidic conditions. The *WT* phenotype determined by RNA interference was similar to that of the *vit* phenotype for nutritional deficiency, such as *Bm*Vg and 30-kDa proteins. These results showed that *Bm*VgR has an important role in transporting proteins for egg formation and embryonic development in *B. mori*.

Published On: Journal of Biological Chemistry, 2013, 288(19), 13345-13355.

1 State Key Laboratory of Silkworm Genome Biology, Southwest University, Chongqing 400716, China.

2 School of Life Sciences, Anhui Agricultural University, Hefei 230036, China.

3 Department of Agricultural and Environmental Biology, Graduate School of Agricultural and Life Sciences, University of Tokyo, Tokyo 113-8657, Japan.

4 Laboratory of Silkworm Science, Kyushu University Graduate School of Bioresource and Bioenvironmental Sciences, Fukuoka 812-8581, Japan.

[#]These authors contributed equally.

[*]Corresponding author E-mail: shimada@ss.ab.a.u-tokyo.ac.jp; xiaqy@swu.edu.cn.

家蚕卵黄原蛋白受体(*Bm*VgR)突变导致卵子发生突变体"*scanty vitellin*"(*vit*)的形成

林　英[1#]　孟　艳[2,3#]　王艳霞[1#]　罗　娟[1]　Susumu Katsuma[3]
杨从文[1]　Yutaka Banno[4]　Takahiro Kusakabe[4]　Toru Shimada[3]　夏庆友[1*]

摘要：在昆虫中，卵黄原蛋白受体(VgR)通过发育中的卵母细胞介导血淋巴中卵黄原蛋白(Vg)的吸收。家蚕卵子发生突变体“*scanty vitellin*”(*vit*)缺少卵黄磷蛋白和30-kDa蛋白，但卵特异蛋白(*Bm*ESP)和*Bm*Vg是正常的。*vit*的卵较正常淡黄色野生型卵偏小偏白，并且胚胎致死。本研究发现*BmVgR*基因的突变是形成*vit*表型的主要原因。我们首先克隆了野生型(*WT*)和*vit*的*BmVgR*基因，其功能域类似低密度脂蛋白受体。并且发现野生型的*BmVgR*基因组序列中的一段235 bp的序列在*vit BmVgR*上被替换成了只含7 bp的序列，该突变直接导致了*vit BmVgR*上的第一个表皮生长因子结构域(EGF1)上的第三个Class B区域缺失50个氨基酸。*BmVgR*在卵母细胞中特异表达，并且其转录水平随着卵黄发生前期卵母细胞的成熟显著提高。连锁分析表明*vit*突变品种的*BmVgR*基因发生突变，并且免疫共沉淀实验也证实*vit*突变的*Bm*VgR能够结合*Bm*Vg，但*Bm*Vg在酸性条件下不能解离。通过RNAi干涉野生型*BmVgR*后产生了类似的*vit*表型(如*Bm*Vg和30-kDa蛋白等营养物质的缺失)。这些结果均表明*Bm*VgR在家蚕卵的成熟及胚胎发育过程中的蛋白运输方面扮演着重要角色。

1　家蚕基因组生物学国家重点实验室，西南大学，重庆
2　生命科学学院，安徽农业大学，合肥
3　农业与环境生物学部门，东京大学，东京，日本
4　生物环境科学研究院，九州大学，福冈，日本

Mutation of a vitelline membrane protein, *Bm*EP80, is responsible for the silkworm "Ming" lethal egg mutant

Chen AL[1,2#] Gao P[1,2#] Zhao QL[1,2*] Tang SM[1,2] Shen XJ[1,2] Zhang GZ[1,2]
Qiu ZY[1,2] Xia DG[1,2] Huang YP[3] Xu YM[4] He NJ[4]

Abstract: The egg stage is an important stage in the silkworm (*Bombyx mori*) life cycle. Normal silkworm eggs are usually short, elliptical, and laterally flattened, with a sometimes hollowed surface on the lateral side. However, the eggs laid by homozygous recessive "Ming" lethal egg mutants (*l-e^m^*) lose water and become concaved around 1 h, ultimately exhibiting a triangular shape on the egg surfaces. We performed positional cloning, and narrowed down the region containing the gene responsible for the *l-e^m^* mutant to 360 kb on chromosome 10 using 2 287 F_2 individuals. Using expression analysis and RNA interference, the best *l-e^m^* candidate gene was shown to be *BmEP80*. The results of the inverse polymerase chain reaction showed that an ~1.9 kb region from the 3′ untranslated region of *BmVMP23* to the forepart of *BmEP80* was replaced by a >100 kb DNA fragment in the *l-e^m^* mutant. Several eggs laid by the normal moths injected with *BmEP80* small interfering RNAs were evidently depressed and exhibited a triangular shape on the surface. The phenotype exhibited was consistent with the eggs laid by the *l-e^m^* mutant. Moreover, two-dimensional gel electrophoresis showed that the *Bm*EP80 protein was expressed in the ovary from the 9th day of the pupa stage to eclosion in the wild-type silkworm, but was absent in the *l-e^m^* mutant. These results indicate that *BmEP80* is responsible for the *l-e^m^* mutation.

Published On: Gene, 2013, 515(2), 313-319.

1 Sericultural Research Institute, Jiangsu University of Science and Technology, Zhenjiang 212018, China.
2 Sericultural Research Institute, Chinese Academy of Agricultural Sciences, Zhenjiang 212018, China.
3 Shanghai Institute of Plant Physiology and Ecology, Shanghai Institutes for Biological Sciences, Chinese Academy of Sciences, Shanghai 200032, China.
4 Institute of Sericulture and Systems Biology, Southwest University, Chongqing 400716, China.
[#]These authors contributed equally.
[*]Corresponding author E-mail: qlzhao302@126.com.

卵黄膜蛋白*Bm*EP80突变是导致家蚕“Ming”卵致死的原因

陈安利[1,2#]　高　鹏[1,2#]　赵巧玲[1,2*]　唐顺明[1,2]　沈兴家[1,2]　张国政[1,2]
裘智勇[1,2]　夏定国[1,2]　黄勇平[3]　徐云敏[4]　何宁佳[4]

摘要：卵的发育是家蚕（*Bombyx mori*）生命周期的重要阶段。正常蚕卵通常是呈短、扁平椭圆状，外侧表面有时是空心的。然而，由纯合隐性“Ming”致死卵突变体（$l\text{-}e^{m}$）产下的卵，会快速失去水分，并在约1h出现表面凹陷，最终呈三角状卵。我们利用2 287个F_2代个体进行定位克隆，将负责该突的基因缩小到10号染色体上360 kb区域。通过表达分析和RNA干涉实验，我们发现最佳的$l\text{–}e^{m}$候选基因为*BmEP80*。反向PCR结果表明，在$l\text{–}e^{m}$突变体中，从*BmVMP23*的3′非翻译区到*BmEP80*前部的1.9 kb区域被替换为大于100 kb的DNA片段。对正常蛾子注射*BmEP80* siRNA，发现其所产的卵明显凹陷，呈三角状，该表型与$l\text{–}e^{m}$突变体的卵一致。此外，双向凝胶电泳显示，*Bm*EP80蛋白在蛹期第9天至羽化期的卵巢中表达，但在$l\text{–}e^{m}$突变体中不表达。这些结果表明$l\text{–}e^{m}$突变体的关键基因为*BmEP80*。

1　蚕业研究所，江苏科技大学，镇江
2　蚕业研究所，中国农业科学院，镇江
3　上海植物生理生态研究所，上海生命科学研究院，中国科学院，上海
4　蚕学与系统生物学研究所，西南大学，重庆

Mutation of a cuticular protein, *Bmor*CPR2, alters larval body shape and adaptability in silkworm, *Bombyx mori*

Qiao L[1,2#] Xiong G[1#] Wang RX[1#] He SZ[1] Chen J[1] Tong XL[1] Hu H[1] Li CL[1]
Gai TT[1] Xin YQ[1] Liu XF[1] Chen B[2] Xiang ZH[1] Lu C[1*] Dai FY[1*]

Abstract: Cuticular proteins (CPs) are crucial components of the insect cuticle. Although numerous genes encoding cuticular proteins have been identified in known insect genomes to date, their functions in maintaining insect body shape and adaptability remain largely unknown. In the current study, positional cloning led to the identification of a gene encoding an RR1-type cuticular protein, *BmorCPR2*, highly expressed in larval chitin-rich tissues and at the mulberry leaf-eating stages, which is responsible for the silkworm stony mutant. In the *Dazao-stony* strain, the *BmorCPR2* allele is a deletion mutation with significantly lower expression, compared to the wild-type *Dazao* strain. Dysfunctional *BmorCPR2* in the *stony* mutant lost chitin binding ability, leading to reduced chitin content in larval cuticle, limitation of cuticle extension, abatement of cuticle tensile properties, and aberrant ratio between internodes and intersegmental folds. These variations induce a significant decrease in cuticle capacity to hold the growing internal organs in the larval development process, resulting in whole-body stiffness, tightness, and hardness, bulging intersegmental folds, and serious defects in larval adaptability. To our knowledge, this is the first study to report the corresponding phenotype of stony in insects caused by mutation of RR1-type cuticular protein. Our findings collectively shed light on the specific role of cuticular proteins in maintaining normal larval body shape and will aid in the development of pest control strategies for the management of Lepidoptera.

Published On: Genetics, 2014, 196(4), 1103-1115.

1 State Key Laboratory of Silkworm Genome Biology, Southwest University, Chongqing 400716, China.
2 Institute of Entomology and Molecular Biology, College of Life Sciences, Chongqing Normal University, Chongqing 401331, China.
[#]These authors contributed equally.
[*]Corresponding author E-mail: lucheng@swu.edu.cn; fydai@swu.edu.cn.

家蚕表皮蛋白*Bmor*CPR2的突变改变家蚕幼虫的体形及适应性

乔 梁[1,2#] 熊 高[1#] 王日欣[1#] 何松真[1] 陈 洁[1] 童晓玲[1] 胡 海[1] 李春林[1]
盖停停[1] 辛亚群[1] 刘小凡[1] 陈 斌[2] 向仲怀[1] 鲁 成[1*] 代方银[1*]

摘要：表皮蛋白(CPs)是昆虫表皮的重要组成部分。尽管迄今在昆虫基因组中已经鉴定了许多编码表皮蛋白的基因，但它们在维持昆虫体形和适应性方面的功能在很大程度上仍然是未知的。本研究在石蚕突变体定位克隆鉴定了一个编码RR1型表皮蛋白基因*BmorCPR2*，*BmorCPR2*在幼虫富含几丁质的组织中以及食桑期高度表达。*Dazao-stony*突变品系与野生型相比，*BmorCPR2*等位基因有一个缺失突变，且显著低表达。石蚕突变体中，*BmorCPR2*的功能障碍导致其失去几丁质结合能力，进而导致幼虫表皮中几丁质含量降低，表皮延伸受到限制，表皮拉伸性能减弱，以及节间和节间褶皱之间比例异常。这些变化导致在幼虫发育过程中，表皮维持生长中的内部器官的能力显著下降，导致全身僵硬、紧致和坚硬、隆起节段间褶皱以及幼虫适应性的严重缺陷。据我们所知，这是第一个报道由RR1型表皮蛋白突变引起昆虫表皮石化的研究。我们的研究结果阐明了表皮蛋白在维持正常幼虫体型中的特殊作用，并将有助于鳞翅目的害虫防治策略的发展。

1 家蚕基因组生物学国家重点实验室，西南大学，重庆
2 昆虫与分子生物学研究所，生命科学学院，重庆师范大学，重庆

Functional loss of *Bmsei* causes thermosensitive epilepsy in contractile mutant silkworm, *Bombyx mori*

Nie HY[1,2,3,4] Cheng TC[1,2,3] Huang XF[1] Zhou MT[1] Zhang YX[1]
Dai FY[1,2,3] Kazuei Mita[1] Xia QY[1,2,3*] Liu C[1,2,3*]

Abstract: The thermoprotective mechanisms of insects remain largely unknown. We reported the *Bombyx mori* contractile (*cot*) behavioral mutant with thermo-sensitive seizures phenotype. At elevated temperatures, the *cot* mutant exhibit seizures associated with strong contractions, rolling, vomiting, and a temporary lack of movement. We narrowed a region containing *cot* to ~268 kb by positional cloning and identified the mutant gene as *Bmsei* which encoded a potassium channel protein. *Bmsei* was present in both the cell membrane and cytoplasm in wild-type ganglia but faint in *cot*. Furthermore, *Bmsei* was markedly decreased upon high temperature treatment in *cot* mutant. With the RNAi method and injecting potassium channel blockers, the wild type silkworm was induced the *cot* phenotype. These results demonstrated that *Bmsei* was responsible for the cot mutant phenotype and played an important role in thermoprotection in silkworm. Meanwhile, comparative proteomic approach was used to investigate the proteomic differences. The results showed that the protein of Hsp-1 and Tn1 were significantly decreased and increased on protein level in *cot* mutant after thermo-stimulus, respectively. Our data provide insights into the mechanism of thermoprotection in insect. As *cot* phenotype closely resembles human epilepsy, *cot* might be a potential model for the mechanism of epilepsy in future.

Published On: Scientific Reports, 2015, 5.

1 State Key Laboratory of Silkworm Genome Biology, Southwest University, Chongqing 400716, China.
2 Key Sericultural Laboratory of the Ministry of Agriculture, Southwest University, Chongqing 400716, China.
3 College of Biotechnology, Southwest University, Chongqing 400716, China.
4 College of Bee Science, Fujian Agriculture and Forestry University, Fuzhou 350002, China.
*Corresponding author E-mail: xiaqy@swu.edu.cn; mlliuchun@163.com.

家蚕缩蚕突变体中*Bmsei*功能缺失引起热敏性癫痫

聂红毅[1,2,3,4] 程廷才[1,2,3] 黄小凤[1] 周梦婷[1] 张银霞[1]
代方银[1,2,3] Kazuei Mita[1] 夏庆友[1,2,3*] 刘 春[1,2,3*]

摘要：昆虫的热保护机制在很多方面仍是未知的。我们报道了具有热敏性癫痫发作表型的家蚕收缩行为突变体(*cot*)。在高温条件下，*cot*突变体表现出强烈收缩、滚动、呕吐和暂时停止运动等癫痫相关表征。我们通过定位克隆将*cot*突变区域缩小到约268 kb区域，并将突变基因锁定为钾离子通道蛋白*Bmsei*。*Bmsei*存在于野生型家蚕神经节的细胞膜和细胞质中，但在*cot*突变体中表达较微弱。此外，经过高温处理后，*cot*突变体中*Bmsei*显著下调。通过RNAi和注射钾离子通道阻断剂发现，野生型家蚕被诱导出*cot*突变体表型。这些结果表明，*Bmsei*是引起*cot*突变体表型的主要原因，并在家蚕热保护机制中扮演重要角色。同时我们采用比较蛋白质组学方法来研究蛋白质组上的差异，结果表明，热刺激后，*cot*突变体中Hsp-1和Tn1蛋白分别显著降低和升高。我们的数据为研究昆虫的热保护机制提供可靠参考。同时由于*cot*突变体表型与人类癫痫病非常相似，*cot*突变体也可能作为未来研究癫痫病机制的潜在模型。

1 家蚕基因组生物学国家重点实验室，西南大学，重庆
2 农业部蚕桑学重点实验室，西南大学，重庆
3 生物技术学院，西南大学，重庆
4 蜂学学院，福建农林大学，福州

Silkworms can be used as an animal model to screen and evaluate gouty therapeutic drugs

Zhang XL[1#] Xue RY[1,2#] Cao GL[1,2]
Pan ZH[1,2] Zheng XJ[1,2] Gong CL[1,2*]

ABSTRACT: In the past few decades, the mouse has been used as a mammalian model for hyperuricemia and gout, which has increased not only in prevalence, but also in clinical complexity, accentuated in part by a dearth of novel advances in treatments for hyperuricemia and gouty arthritis. However, the use of mice for the development of gouty therapeutic drugs creates a number of problems. Thus, identification and evaluation of the therapeutic effects of chemicals in an alternative animal model is desirable. In the present study, the effects of gouty therapeutic drugs on lowering the content of uric acid and inhibiting activity of xanthine oxidase were evaluated by using a silkworm model, *Bombyx mori* L. (Lepidoptera: Bombycidae). The results showed that the effectiveness of oral administration of various gouty therapeutic drugs to 5th instar silkworms is consistent with results for human. The activity of xanthine oxidase of silkworm treated with allopurinol was lower, and declined in a dose-dependent manner compared with control silkworms, while sodium bicarbonate failed at inhibiting the activity of xanthine oxidase. The concentration of uric acid in the both hemolymph and fat body declined by 90% and 95% at six days post-administration with 25 mg/mL of allopurinol, respectively ($P < 0.01$), while the concentration of uric acid in both the hemolymph and fat body also declined by 81 and 95% at six days post-administration with 25 mg/mL of sodium bicarbonate, respectively ($P < 0.01$). Moreover, the epidermis of silkworm treated with allopurinol or sodium bicarbonate became transparent compared with the negative control group. These results suggest that silkworm larva can be used as an animal model for screening and evaluation of gouty therapeutic drugs.

Published On: Journal of Insect Science, 2012, 12, 4.

1 Pre-clinical Medical and Biological Science College, Soochow University, Suzhou 215123, China.
2 National Engineering Laboratory for Modern Silk, Soochow University, Suzhou 215123, China.
#These authors contributed equally.
*Corresponding author E-mail: gongcl@suda.edu.cn.

家蚕可作为筛选和评估痛风治疗药物的动物模型

张晓丽[1#]　薛仁宇[1,2#]　曹广力[1,2]
潘中华[1,2]　郑小坚[1,2]　贡成良[1,2*]

摘要：过去的十几年里，小鼠被广泛地用作研究高尿酸血症和痛风的哺乳动物模型。然而，高尿酸血症和痛风患病率及临床复杂性的不断增加，让其缺乏新的治疗手段的现状变得更加严重。因此有必要利用其他动物模型鉴定和评估药物的疗效。在本研究中，我们利用了一种昆虫模式动物——家蚕（鳞翅目，蚕蛾科）来评估痛风治疗药物在降低尿酸含量和抑制黄嘌呤氧化酶活性中的作用。结果显示，五龄家蚕口服多种痛风治疗药物效果与人体实验结果一致。别嘌呤醇处理过的家蚕的黄嘌呤氧化酶活性与对照相比较低，并且呈现剂量依赖性，而碳酸氢钠则不能抑制黄嘌呤氧化酶活性。口服25 mg/mL别嘌呤醇6d之后，家蚕淋巴系统和脂肪体内的尿酸浓度分别减少了90%和95%（$P < 0.01$），口服25 mg/mL碳酸氢钠6d后，血淋巴系统和脂肪体内的尿酸浓度同样减少了81%和95%（$P < 0.01$）。此外，用别嘌呤醇或者碳酸氢钠处理后，家蚕表皮与对照相比更加透亮。这些结果表明家蚕可作为筛选和评估痛风治疗药物的动物模型。

1　基础医学与生物科学学院，苏州大学，苏州
2　现代丝绸国家工程实验室，苏州大学，苏州

第十章

家蚕蛋白质组研究

Chapter 10

家蚕是鳞翅目昆虫的模式生物，其生长、发育、变态、免疫、丝蛋白合成等重要的生命过程均由各个组织器官中的蛋白质所承担。为了揭示家蚕生命活动的规律和本质，项目组在两轮“973”项目的支持上，对家蚕多个组织器官开展了蛋白质组学研究。

家蚕多组织蛋白质组学研究。蛋白质组学技术较早便被应用于家蚕研究，研究者利用双向电泳、鸟枪法与质谱联用技术、液相色谱-串联质谱技术等，成功对家蚕多个组织时期的样品进行了蛋白质组分析，其中包括胚胎、血淋巴、中肠、脂肪体、脑、咽下神经节、前胸腺、马氏管、头、翅原肌、围食膜等。以上蛋白质组学分析获得的数据和结果，为深入了解这些相关组织的生物学功能奠定了基础。

病毒感染家蚕比较蛋白质组研究。家蚕核型多角体病毒（*Bm*NPV）是家蚕最常见且严重的病原体之一，对经NPV感染后家蚕进行比较蛋白质组学分析有助于揭示家蚕抵御病毒入侵、免疫、病毒与宿主相互作用等机制，对蚕业发展具有重大意义。基于此，项目组采用二维电泳和质谱法鉴定了存在于ODV病毒粒子内或与之相关的蛋白质，包括16个来源于*Bm*NPV的蛋白质和4个来源于家蚕的蛋白质。对具有不同NPV抗性的家蚕进行比较蛋白质组学分析，发现半胱天冬酶-1和丝氨酸蛋白酶仅在*Bm*NPV抗性家蚕中表达，而在*Bm*NPV易感的家蚕中不表达，暗示其在家蚕抗*Bm*NPV感染中可能起关键作用。

比较蛋白质组学在家蚕杂交育种中的应用。项目组前期育成改良杂交种“金秋”，为了揭示“金秋”与其亲本之间蛋白质表达的差异，探讨其基因构建，阐明家蚕成功杂交育种的分子机制，项目组分别对“金秋”与其亲本丝腺、血淋巴和中肠进行蛋白质组学分析。发现“金秋”与亲本的匹配蛋白点约为70%，剩余约30%是各个品种中特异的蛋白点。以上结果说明新品种的育成，除汇集亲本的优良基因外，还有赖于基因互作产生的新功能蛋白质的作用。

家蚕蛋白质组学新技术的开发与应用。针对家蚕组织样品在进行蛋白质组学分析中产生的普遍性及特殊性问题，项目组通过大量实验，进行了广泛的技术革新。项目组针对蚕卵蛋白提取中高丰度蛋白如ESP和30K蛋白的干扰问题，开发出将富含ESP和30K蛋白的提取物分级分离的方法。该方法可将蚕卵蛋白点的检测的数量提升三分之一以上。针对双向电泳中蛋白点样过程造成的蛋白损失、分辨率低等缺陷，项目组开发出droplet-tap点样新方法，即将样品液滴均匀地滴入到再水化缓冲液中，然后将胶条降低到溶液表面来加载样品。实验表明这种新方法获得的蛋白点数量可增加约1/3，蛋白质的分辨率显著提高，更有利于蛋白质的回收和提取。

王鑫

Identification of the proteome of the midgut of silkworm, *Bombyx mori* L., by multidimensional liquid chromatography (MDLC) LTQ-Orbitrap MS

Yao HP Xiang XW Chen L Guo AQ He FQ
Lan LP Lu XM Wu XF*

Abstract: The midgut is the digestive apparatus of the silkworm and its proteome was studied by using nano-LC (liquid chromatography) electrospray ionization MS/MS (tandem MS). MS data were analysed by using X! Tandem searching software using different parameters and validated by using the Poisson model. A total of 90 proteins were identified and 79 proteins were described for the first time. Among the new proteins, (i) 22 proteins were closely related to the digestive function of the midgut, including 11 proteins of digestive enzymes secreted by the epithelium, eight proteins of intestine wall muscle and mechanical digestion and three proteins of peritrophic membrane that could prevent the epithelium from being mechanically rubbed; (ii) 44 proteins were involved in metabolism of substance and energy; and (iii) 11 proteins were associated with signal transduction, substance transport and cell skeleton.

Published On: Bioscience Reports, 2009, 29(6), 363-373.

College of Animal Science, Zhejiang University, Hangzhou 310029, China.
*Corresponding author E-mail: wuxiaofeng@zju.edu.cn.

利用多维液相色谱(MDLC)LTQ-Orbitrap质谱鉴定家蚕中肠蛋白质组

姚慧鹏　相兴伟　陈　琳　郭爱芹
何芳青　兰丽盼　鲁兴萌　吴小锋*

摘要:中肠是家蚕的消化器官,我们采用液相色谱LC和串联质谱MS/MS的方法研究其蛋白质组。通过使用不同参数的X!Tandem搜索软件分析MS数据,并使用泊松模型进行验证。共鉴定出90种蛋白质,其中有79种蛋白质是首次鉴定到的。这些首次鉴定到的蛋白质包括:(i)22种与中肠消化功能密切相关的蛋白质,包括由上皮组织分泌的11种消化酶、8种肠壁肌肉和机械消化相关蛋白质,以及可以减轻围食膜上机械摩擦的3种蛋白质;(ii)44种参与物质和能量代谢的蛋白质;(iii)11种与信号转导、物质运输和细胞骨架相关的蛋白质。

动物科学学院,浙江大学,杭州

Shotgun proteomics approach to characterizing the embryonic proteome of the silkworm, *Bombyx mori*, at labrum appearance stage

Li JY[1] Chen X[2] S. Hossein Hosseini Moghaddam[1,3]
Chen M[2] Wei H[1] Zhong BX[1*]

Abstract: The shotgun approach has gained considerable acknowledgement in recent years as a dominant strategy in proteomics. We observed a dramatic increase of specific protein spots in two-dimensional electrophoresis (2-DE) gels of the silkworm (*Bombyx mori*) embryo at labrum appearance, a characteristic stage during embryonic development of silkworm which is involved with temperature increase by silkworm raiser. We employed shotgun liquid chromatography tandem mass spectrometry (LC-MS/MS) technology to analyse the proteome of *B. mori* embryos at this stage. A total of 2 168 proteins were identified with an in-house database. Approximately 47% of them had isoelectric point (pI) values distributed theoretically in the range pI 5-7 and approximately 60% of them had molecular weights of 15-45 kDa. Furthermore, 111 proteins had a pI greater than 10 and were difficult to separate by 2-DE. Many important functional proteins related to embryonic development, stress response, DNA transcription / translation, cell growth, proliferation and differentiation, organogenesis and reproduction were identified. Among them proteins related to nervous system development were noticeable. All known heat shock proteins (HSPs) were detected in this developmental stage of *B. mori* embryo. In addition, Gene Ontology (GO) and Kyoto Encyclopedia of Genes and Genomes (KEGG) pathway analysis showed energetic metabolism at this stage. These results were expected to provide more information for proteomic monitoring of the insect embryo and better understanding of the spatiotemporal expression of genes during embryonic developmental processes.

Published On: Insect Molecular Biology, 2009, 18(5), 649-660.

1 College of Animal Sciences, Zhejiang University, Hangzhou 310029, China.
2 College of Life Sciences, Zhejiang University, Hangzhou 310058, China.
3 Agriculture Faculty, University of Guilan, Rasht 41457, Iran.
*Corresponding author E-mail: bxzhong@zju.edu.cn.

鸟枪法蛋白质组学方法分析上唇突显期的家蚕胚胎

李建营[1] 陈 祥[2] S. Hossein Hosseini Moghaddam[1,3]
陈 铭[2] 危 浩[1] 钟伯雄[1*]

摘要：鸟枪法在最近几年作为一种主导的蛋白质组学分析策略得到了广泛的认可。我们利用双向电泳发现在家蚕胚胎发育的上唇突显期特异蛋白点数量急剧增加，这是一个需要提高培养温度的关键时期。利用鸟枪法液相色谱串联质谱技术分析这一时期的家蚕胚胎蛋白质组，利用内部数据库共鉴定到2 168种蛋白质，其中约47%的蛋白质等电点在5—7之间，约60%的蛋白质分子质量为15—45 kDa。而且有111种蛋白质的等电点大于10，这些蛋白是难以被双向电泳分离的。此外，还鉴定到很多涉及胚胎发育、应激反应、DNA转录/翻译、细胞生长、增殖和分化、器官发生和生殖等过程的重要蛋白质。在这些蛋白质中，涉及神经系统发育的蛋白质尤其值得注意。所有已知的热休克蛋白在家蚕胚胎发育时期均被检测到，另外，GO分类和KEGG通路分析发现这一时期的代谢很旺盛。这些结果为在蛋白质组学水平上监测昆虫胚胎发育提供了更多的信息，而且有助于更好地了解胚胎发育期基因的时空表达。

1 动物科学学院，浙江大学，杭州
2 生命科学学院，浙江大学，杭州
3 农学院，桂兰大学，拉什特，伊朗

Shotgun proteomic analysis of the fat body during metamorphosis of domesticated silkworm (*Bombyx mori*)

Yang HJ[1] Zhou ZH[1] Zhang HR[2] Chen M[3]
Li JY[1] Ma YY[2] Zhong BX[1*]

Abstract: Protein expression profiles in the fat bodies of larval, pupal, and moth stages of silkworm were determined using shotgun proteomics and MS sequencing. We identified 138, 217, and 86 proteins from the larval, pupal and moth stages, respectively, of which 12 were shared by the 3 stages. There were 92, 150, and 45 specific proteins identified in the larval, pupal and moth stages, respectively, of which 17, 68, and 9 had functional annotations. Among the specific proteins identified in moth fat body, sex-specific storage-protein1 precursor and chorion protein B8 were unique to the moth stage, indicating that the moth stage fat body is more important for adult sexual characteristics. Many ribosomal proteins (L23, L4, L5, P2, S10, S11, S15A and S3) were found in pupal fat bodies, whereas only three (L14, S20, and S7) and none were identified in larval and moth fat bodies, respectively. Twenty-three metabolic enzymes were identified in the pupal stage, while only four and two were identified in the larval and moth stages, respectively. In addition, an important protein, gloverin2, was only identified in larval fat bodies. Gene ontology (GO) analysis of the proteins specific to the three stages linked them to the cellular component, molecular function, and biological process categories. The most diverse GO functional classes were involved by the relatively less specific proteins identi fied in larva. GO analysis of the proteins shared among the three stages showed that the pupa and moth stages shared the most similar protein functions in the fat body.

Published On: Amino Acids, 2010, 38(5), 1333-1342.

1 College of Animal Sciences, Zhejiang University, Hangzhou 310029, China.
2 Zhejiang California International NanoSystems Institute, Zhejiang University, Hangzhou 310029, China.
3 College of Life Sciences, Zhejiang University, Hangzhou 310029, China.
*Corresponding author E-mail: bxzhong@zju.edu.cn.

家蚕变态期脂肪体的蛋白质组分析

杨惠娟[1] 周仲华[1] 张华蓉[2] 陈 铭[3]
李建营[1] 马莹莹[2] 钟伯雄[1*]

摘要：通过鸟枪法蛋白质组学和质谱测序方法对家蚕幼虫、蛹期和蛾期的脂肪体蛋白质表达情况进行研究。在家蚕幼虫、蛹期和蛾期分别鉴定到138、217和86种蛋白质，其中有12种蛋白质为3个时期所共有。在这些蛋白质中，有92、150和45种蛋白质分别为幼虫、蛹期和蛾期3个时段的特异蛋白质，其中17、68和9种蛋白质具有功能注释。在蛾期脂肪体鉴定的特异蛋白质中，性别特异性贮藏蛋白1前体和卵壳蛋白B8两个蛋白在蛾期特异性表达，表明蛾期脂肪体对成虫性别特征具有较为重要的作用。在蛹期脂肪体中发现许多核糖体蛋白（L23、L4、L5、P2、S10、S11、S15A和S3），而在幼虫期脂肪体中仅发现3种（L14、S20和S7），在蛾期脂肪体中未发现。在蛹期鉴定到23种代谢酶类蛋白质，而在幼虫和蛾期只分别鉴定到4种和2种代谢酶。此外，一个重要的蛋白gloverin2只在幼虫的脂肪体中被发现。本研究对3个时期的特异蛋白质按照细胞组分、分子功能、生物学进程进行GO（Gene ontology）分析。结果表明，幼虫期特异蛋白质虽相对较少，但其涉及的GO功能分类却最为多样。对3个阶段共有的蛋白质进行GO分析，结果表明蛹期和蛾期阶段的脂肪体具有最相似的蛋白质功能。

1 动物科学学院，浙江大学，杭州
2 浙江加州国际纳米技术研究院，浙江大学，杭州
3 生命科学学院，浙江大学，杭州

Shotgun proteomic analysis on the embryos of silkworm *Bombyx mori* at the end of organogenesis

Li JY[1] S.Hossein Hosseini Moghaddam[1,3] Chen JE[1] Chen M[2] Zhong BX[1*]

Abstract: Embryonic development of silkworm, *Bombyx mori* is a process of systematical expression of genes and proteins which is dominated by complex regulatory networks. To gain comprehensive insight into the molecular basis of embryonic development and its regulation mechanisms, the proteome profile of the *B. mori* embryos at the end of organogenesis (tubercle appearance stage, TA) was characterized using LTQ-Orbitrap mass spectrometer. Totally 963 proteins were identified with a false discovery rate (FDR) of 0.12%. They were involved in embryonic development, chemoreception, and stimuli response and so forth. The proteins with the largest number of identified unique peptides, implying their possibly higher abundance, were involved in heat shock response, lipid transport and metabolism, and apoptosis. It was consistent with the physiological status of embryo at the end of organogenesis. Many functionally important proteins were identified for the first time in *B. mori* embryo such as the progesterone receptor membrane component 2, antennal binding protein, sericotropin, and molting fluid carboxypeptidase A (MF-CPA). 253 (26.27%) specific proteins in TA versus labrum appearance stage (LA, four days before TA) embryos were identified, which were mainly associated with musculature, nervous system, and chemoreception system. They disclosed the differential temporal and spatial expression of proteins in the process of organogenesis. The relative mRNA levels of fifteen identified proteins in the two experimented stages were also compared using quantitative reverse transcription PCR (qRT-PCR) and showed some inconsistencies with protein expression. Gene Ontology (GO) annotation of the identified proteins showed that the most proteome representations were in the categories of "binding" and "catalytic" in molecular function, and "cellular process" and "metabolic process" in biological process.

Published On: Insect Biochemistry and Molecular Biology, 2010, 40(4), 293-302.

1 College of Animal Sciences, Zhejiang University, Hangzhou 310029, China.

2 College of Life Sciences, Zhejiang University, Hangzhou 310058, China.

3 Agriculture Faculty, University of Guilan, Rasht 41457, Iran.

*Corresponding author E-mail: bxzhong@zju.edu.cn.

鸟枪法分析组织发生末期的家蚕胚胎蛋白质组

李建营[1] S.Hossein Hosseini Moghaddam[1,3] 陈金娥[1] 陈 铭[2] 钟伯雄[1*]

摘要：家蚕胚胎发育过程是一个以复杂调控网络为主导的基因和蛋白系统表达的过程。为了更全面地了解胚胎发育的分子基础及其调控机制，我们利用LTQ-Orbitrap质谱对家蚕胚胎组织发生末期（结节突显期）的蛋白质特征进行分析。共鉴定到963个蛋白质，其错误发现率为0.12%。这些蛋白质涉及胚胎发育、化学感应和刺激反应等。丰度比较高的蛋白质涉及到热休克反应、脂质运输和代谢以及细胞凋亡。这与组织发生末期胚胎的生理状态是一致的。许多重要的功能蛋白质在胚胎里首次被发现，例如孕酮受体膜组分2、触角结合蛋白、sericotropin和蜕皮液羧肽酶A（MF-CPA）等。与上唇突显期胚胎相比，结节突显期鉴定到253种特异蛋白质，这些蛋白质主要与肌肉组织、神经系统和化学感应系统有关。这揭示了在组织发生过程中的蛋白质时空差异表达。我们选择了15种在这两个时期存在差异的蛋白质，利用定量PCR比较了其相对mRNA水平，发现与蛋白质表达存在一定差异。GO分析发现大部分蛋白质在分子功能上表现为“结合”和“催化”，在生物学进程中表现为“细胞进程”和“代谢进程”。

1 动物科学学院，浙江大学，杭州
2 生命科学学院，浙江大学，杭州
3 农学院，桂兰大学，拉什特，伊朗

Shotgun proteomic analysis of wing discs from the domesticated silkworm (*Bombyx mori*) during metamorphosis

Zhang YL[1] Xue RY[1,2] Cao GL[1,2] Zhu YX[1] Pan ZH[1,2] Gong CL[1,2*]

Abstract: Proteomic profiles from the wing discs of silkworms at the larval, pupal, and adult moth stages were determined using shotgun proteomics and MS sequencing. We identified 241, 218 and 223 proteins from the larval, pupal and adult moth stages, respectively, of which 139 were shared by all three stages. In addition, there were 55, 37 and 43 specific proteins identified at the larval, pupal and adult moth stages, respectively. More metabolic enzymes were identified among the specific proteins expressed in the wing disc of larvae compared with pupae and moths. The identification of FKBP45 and the chitinase-like protein EN03 as two proteins solely expressed at the larval stage indicate these two proteins may be involved in the immunological functions of larvae. The myosin heavy chain was identified in the pupal wing disc, suggesting its involvement in the formation of wing muscle. Some proteins, such as proteasome alpha 3 subunits and ribosomal proteins, specifically identified from the moth stage may be involved in the degradation of old cuticle proteins and new cuticle protein synthesis. Gene ontology analysis of proteins specific to each of these three stages enabled their association with cellular component, molecular function, and biological process categories. The analysis of similarities and differences in these identified proteins will greatly further our understanding of wing disc development in silkworm and other insects.

Published On: Amino Acids, 2013, 45(5), 1231-1241.

1 School of Biology and Basic Medical Sciences, Soochow University, Suzhou 215123, China.

2 National Engineering Laboratory for Modern Silk, Soochow University, Suzhou 215123, China.

*Corresponding author E-mail: gongcl@suda.edu.cn.

家蚕变态发育过程中翅原基的蛋白质组学分析

张铁岭[1]　薛仁宇[1,2]　曹广力[1,2]
朱越雄[1]　潘中华[1,2]　贡成良[1,2*]

摘要：本研究利用鸟枪法蛋白质组学方法和质谱测序技术进行了家蚕幼虫、蛹和成虫阶段翅原基的蛋白组学分析。我们分别从家蚕幼虫、蛹和成虫3个阶段鉴定到241、218和223种蛋白质，其中139种蛋白质为3个阶段共同表达。此外，在3个阶段还分别鉴定到55、37和43种特异表达的蛋白质。相对蛹期和成虫期，幼虫期鉴定到更多的代谢酶类。FKBP45和几丁质酶样蛋白EN03仅在幼虫期表达，说明这两种蛋白质可能参与幼虫的免疫功能。在蛹期翅原基中鉴定到的肌球蛋白重链可能参与翅膀肌肉的形成。成虫期特异的一些蛋白质，例如蛋白酶体α3亚基和核糖体蛋白可能参与旧表皮蛋白的退化和新表皮蛋白的形成。对3个时期鉴定到的蛋白质进行的GO分析，使其与细胞组分、分子功能和生物学进程相关联。这些鉴定到的蛋白质的相似性和差异性分析，进一步加深了我们对家蚕和其他昆虫翅原基发育过程的理解。

1　基础医学与生物科学学院，苏州大学，苏州
2　现代丝绸国家工程实验室，苏州大学，苏州

Proteome analysis on differentially expressed proteins of the fat body of two silkworm breeds, *Bombyx mori*, exposed to heat shock exposure

S.Hossein Hosseini Moghaddam[1,2] Du X[1] Li J[1]

Cao JR[3] Zhong BX[1] Chen YY[1*]

Abstract: Proteomes of heat tolerant (multivoltine) and heat susceptible (bivoltine) silkworms (*Bombyx mori*) in response to heat shock were studied. Detected proteins from fat body were identified by using MALDI-TOF / TOF spectrometer, MS / MS, and MS analysis. Eight proteins, including small heat shock proteins (sHSPs) and HSP70, were expressed similarly in both breeds, while 4 protein spots were expressed specifically in the bivoltine breed and 12 protein spots were expressed specifically in the multivoltine breed. In the present proteomics approach, 5 separate spots of sHSP proteins (HSP19.9, HSP20.1, HSP20.4, HSP20.8, and HSP21.4) were identified. Protein spot intensity of sHSPs was lower in the multivoltine breed than in the bivoltine breed after the 45°C heat shock treatment, while the difference between two breeds was not significant after the 41°C heat shock treatment. These results indicated that some other mechanisms might be engaged in thermal tolerance of multivotine breed except for the expression of sHSP and HSP70. There were visible differences in the intensity of heat shock protein expression between male and female, however, differences were not statistically significant.

Published On: Biotechnology and Bioprocess Engineering, 2008, 13(5), 624-631.

1 College of Animal Sciences, Zhejiang University, Hangzhou 310029, China.

2 Agriculture Faculty, University of Guilan, Rasht 41457, Iran.

3 Sericultural Research Institute, Zhejiang Academy of Agricultural Sciences, Hangzhou 310021, China.

*Corresponding author E-mail: chenyy@zju.edu.cn.

家蚕不同品种在热激处理下的脂肪体蛋白质组学分析

S. Hossein Hosseini Moghaddam[1,2] 杜 鑫[1] 李 军[1]
曹锦如[3] 钟伯雄[1] 陈玉银[1*]

摘要：本文对热耐受（多化性）与热敏感（二化性）家蚕对热休克反应的蛋白质组学进行研究。利用MALDI-TOF/TOF，MS/MS和MS等技术和设备对热激处理后脂肪体的蛋白质进行鉴定。有8个蛋白质在两个品种中表达量相似，包括小热激蛋白（sHSPs）和HSP70。另外，有4个蛋白质在二化性品种中特异表达，12个蛋白质在多化性品种中特异表达。通过蛋白质组学技术，5个sHSP蛋白（包括HSP19.9、HSP20.1、HSP20.4、HSP20.8和HSP21.4）被鉴定出来。蛋白质丰度结果显示，45 ℃热处理后，多化性品种中sHSPs的表达量低于二化性品种，而41 °C处理后则没有明显差异。这些结果暗示多化性品种的热耐受机制可能不仅仅与sHSP和HSP70的表达量相关。同时，这些热激蛋白在雌雄家蚕中的表达量也有所不同，但是没有统计学意义的差异。

1 动物科学学院，浙江大学，杭州
2 农学院，桂兰大学，拉什特，伊朗
3 蚕桑研究所，浙江省农业科学院，杭州

Application of proteomic technology in silkworm research

Li XH Wu XF Yue WF Liu JM
Li GL Zhong BX Miao YG*

Abstract: From structural genome to functional one, the proteomic research has been an important content in the post-genome times. This paper presents the concept of proteome, and introduces its research technology and status in silkworm research.

Published On: Oriental Insects, 2007, 41, 453-458.

College of Animal Sciences, Zhejiang University, Hangzhou 310029, China.
*Corresponding author E-mail: miaoyg@zju.edu.cn.

蛋白质组学在家蚕研究中的应用

李兴华　吴小锋　岳万福　刘剑梅
李广立　钟伯雄　缪云根*

摘要：从结构基因组到功能基因组，蛋白质组学的研究是后基因时代的重要内容。本文介绍了蛋白质组学的概念，并介绍其在家蚕研究中的研究技术和现状。

动物科学学院，浙江大学，杭州

Proteomic and bioinformatic analysis on endocrine organs of domesticated silkworm, *Bombyx mori* L. for a comprehensive understanding of their roles and relations

Li JY[1] Chen X[2] Fan W[1] Hosseini Moghaddam SH[1]
Chen M[2] Zhou ZH[1] Yang HJ[1] Chen JE[1] Zhong BX[1*]

Abstract: Three organs of silkworm larva endocrine system, including brain (Br), subesophageal ganglion (SG) and prothoracic glands (PG), were studied employing shotgun LC-MS / MS combined with bioinformatic analysis to comprehensively understand their roles and relations. Totally, 3 430, 2 683, and 3 395 proteins were identified including 1 885 common and 652, 253, and 790 organ-specific ones in Br, SG, and PG, respectively. Identified common-expressed proteins indicated the existence of intrinsic complex interactions among these parts of endocrine system. Most of the reputed organs-specific proteins were identified by this approach. KEGG pathway analysis showed 162 same pathways among the 169, 164, and 171 relating Br, SG, and PG. This analysis revealed functional similarities with exceptional resemblance in their metabolism and signaling pathways of the three organs. On the other hand, 70, 57, and 114 organ-specific enzymes related pathways were detected for Br, SG, and PG confirming their functional differences. These results reveal a cooperative mechanism among the three endocrine organs in regulating various physiological and developmental events, and also suggest that the organ-specific proteins might be the fundamental factors responsible for the functional differentiation of these organs.

Published On: Journal of Proteome Research, 2009, 8(6), 2620-2632.

1 College of Animal Sciences, Zhejiang University, Hangzhou 310029, China.
2 College of Life Sciences, Zhejiang University, Hangzhou 310058, China.
*Corresponding author E-mail: bxzhong@zju.edu.cn.

蛋白质组学和生物信息学分析家蚕内分泌器官的功能及其关系

李建营[1] 陈 祥[2] 范 伟[1] Hosseini Moghaddam SH[1]
陈 铭[2] 周仲华[1] 杨惠娟[1] 陈金娥[1] 钟伯雄[1*]

摘要：家蚕幼虫的内分泌系统包括脑、咽下神经节和前胸腺。利用鸟枪法-液质联用质谱以及生物信息学分析来全面了解这3个器官的功能及关系。我们在这3个器官中分别鉴定到了3 430、2 683和3 395种蛋白质，其中包括1 885种共有蛋白质，以及652、253和790种器官特异蛋白质。鉴定的共有蛋白质表明这些内分泌器官之间存在内在的复杂关系。利用这种方法，我们鉴定到了大部分已知的器官特异蛋白质。KEGG通路分析发现这3个器官的蛋白质分别涉及到169、164和171个通路，其中162个通路是相同的。这3个器官在代谢和信号转导通路的相似性暗示了它们功能的相似性。另一方面，我们在这3个器官中分别发现了40、57和114个器官特异的酶相关通路，暗示它们可能存在着功能差异。这些结果表明这3个内分泌器官在调节生理和发育过程中可能存在合作机制，而器官特异蛋白质则与它们的功能差异有关。

1 动物科学学院，浙江大学，杭州
2 生命科学学院，浙江大学，杭州

Proteomic identification of the silkworm (*Bombyx mori* L.) prothoracic glands during the fifth instar stage

Wu XF[1] Li XH[1] Yue WF[1] Bhaskar Roy[1] Li GL[1]
Liu JM[1] Zhong BX[1] Gao QK[2] Wan Chi Cheong David[3] Miao YG[1*]

Abstract: Although the ecdysteroid of the silkworm had been studied for decades, the proteome of the prothoracic gland, the primary source of ecdysteroid hormones, has not been studied previously. In the present paper, we utilized a proteomic approach to investigate the fifth instar prothoracic gland during the growth and development of the silkworm, *Bombyx mori* L. The two-dimensional electrophoresis results showed that the majority of proteins were acidic proteins, especially concentrated in the area of 25-65 kDa, with pI values of between 4 and 7, and the difference was not distinct. When compared with *Qiufeng* (Japanese strain), the interspecific distinction was larger than the intraspecific distinction, and 19 particular spots, excized from the third, fifth and ninth days of *p50* (Chinese strain) and *Qiufeng* were subjected to MALDI-TOF-MS (matrix-assisted laser-desorption ionization-time-of-flight MS) analysis. We sorted them into seven catagories: energetics and/or metabolism, storage proteins, protection, lipid metabolism, signal transduction, cell function and unknown function proteins. Of these proteins, arginine methyltransferase is discussed as playing an important role in regulating the activation of ecdysteroidogenesis via transcription or translation.

Published On: Bioscience Reports, 2009, 29(2), 121-129.

1 Key Laboratory of Animal Epidemic Etiology and Immunological Prevention of Ministry of Agriculture, College of Animal Sciences, Zhejiang University, Hangzhou 310029, China.
2 Center of Analysis and Measurement, Zhejiang University, Hangzhou 310029, China.
3 Department of Biochemistry, The Chinese University of Hong Kong, Hong Kong SAR 999077, China.
*Corresponding author E-mail: miaoyg@zju.edu.cn.

家蚕五龄幼虫前胸腺蛋白组鉴定

吴小锋[1]　李兴华[1]　岳万福[1]　Bhaskar Roy[1]　李广立[1]
刘剑梅[1]　钟伯雄[1]　高其康[2]　Wan Chi Cheong David[3]　缪云根[1*]

摘要：虽然家蚕蜕皮激素的作用机制已被研究了数十年，但作为蜕皮激素主要来源的前胸腺蛋白质组尚未进行研究。在本文中，我们利用蛋白质组学方法对家蚕生长发育过程中的五龄幼虫的前胸腺进行了研究。双向电泳结果表明，前胸腺中大多数蛋白质是酸性蛋白质，分子质量集中在25—65 kDa，等电点介于4和7之间，差异不明显。与秋丰蚕品种（日本品系）相比，种间差异大于种内差异，从*p50*（中国品系）和秋丰蚕的五龄第3天、第5天和第9天分别取19个特定的点进行MALDI-TOF-MS（基质辅助激光解析串联飞行时间质谱）分析。将这些蛋白质分为7类：能量和/或代谢蛋白、储存蛋白、保护蛋白、脂质代谢蛋白、信号转导蛋白、细胞功能蛋白和未知功能蛋白质。在这些蛋白质中，精氨酸甲基转移酶被鉴定为通过转录或翻译在调节蜕皮激素的激活中发挥关键作用。

1　农业部动物疫病病原学与免疫控制重点开放实验室，动物科学学院，浙江大学，杭州
2　分析测试中心，浙江大学，杭州
3　生物化学系，香港中文大学，香港

Shotgun strategy-based proteome profiling analysis on the head of silkworm *Bombyx mori*

Li JY[1] Hosseini Moghaddam SH[1,2] Chen X[3]
Chen M[3] Zhong BX[1*]

Abstract: Insect head is comprised of important sensory systems to communicate with internal and external environment and endocrine organs such as brain and corpus allatum to regulate insect growth and development. To comprehensively understand how all these components act and interact within the head, it is necessary to investigate their molecular basis at protein level. Here, the spectra of peptides digested from silkworm larval heads were obtained from liquid chromatography tandem mass spectrometry (LC-MS/MS) and were analyzed by bioinformatics methods. Totally, 539 proteins with a low false discovery rate (FDR) were identified by searching against an in-house database with SEQUEST and X!Tandem algorithms followed by trans-proteomic pipeline (TPP) validation. Forty-three proteins had the theoretical isoelectric point (pI) greater than 10 which were too difficult to separate by two-dimensional gel electrophoresis (2-DE). Four chemosensory proteins, one odorant-binding protein, two diapause-related proteins, and a lot of cuticle proteins, interestingly including pupal cuticle proteins were identified. The proteins involved in nervous system development, stress response, apoptosis and so forth were related to the physiological status of head. Pathway analysis revealed that many proteins were highly homologous with the human proteins which involved in human neurodegenerative disease pathways, probably implying a symptom of the forthcoming metamorphosis of silkworm. These data and the analysis methods were expected to be of benefit to the proteomics research of silkworm and other insects.

Published On: Amino Acids, 2010, 39(3), 751-761.

1 College of Animal Sciences, Zhejiang University, Hangzhou 310029, China.
2 Agriculture Faculty, University of Guilan, Rasht 41457, Iran.
3 College of Life Sciences, Zhejiang University, Hangzhou 310058, China.
*Corresponding author E-mail: bxzhong@zju.edu.cn.

基于鸟枪法策略的家蚕头部蛋白质组分析

李建营[1] Hosseini Moghaddam SH[1,2] 陈 祥[3]
陈 铭[3] 钟伯雄[1*]

摘要：昆虫头部包含不同的感觉系统，可以与内外环境和大脑、咽侧体等内分泌器官进行交流，对昆虫的生长与发育进行调节。为了了解这些元件在头部是如何起作用以及相互作用的，需要在蛋白质水平研究它们的分子基础。在本研究中，我们利用液相色谱串联质谱（LC-MS/MS）获得了家蚕幼虫头部蛋白质的多肽图谱信息，并运用生物信息学的方法进行分析。利用SEQUEST和X!Tandem两种算法检索自建数据库，并用TPP的方法进行验证，共鉴定到539个低错误发现率（FDR）的蛋白质。其中包含了43个难以被双向电泳（2-DE）分离检测、理论等电点（pI）大于10的蛋白质。该方法还同时检测到了4种化学感受器蛋白质、1种气味结合蛋白质、2种滞育相关蛋白质和大量的表皮蛋白，有趣的是其中含有两个蛹表皮蛋白。与神经系统发育、应激反应和细胞凋亡等相关的蛋白与脑的生理状态关系密切。

通路分析发现其中许多蛋白质与人类神经退行性疾病通路中的蛋白质高度同源，这可能是家蚕变态发育将要到来的征兆，这些数据与分析方法有望为家蚕和其他昆虫的蛋白质组学研究提供参考。

1 动物科学学院，浙江大学，杭州
2 农学院，桂兰大学，拉什特，伊朗
3 生命科学学院，浙江大学，杭州

Proteomic profiling of the hemolymph at the fifth instar of the silkworm *Bombyx mori*

Li JY Li JS Zhong BX*

Abstract: Two-dimensional gel electrophoresis (2-DE) followed by matrix-assisted laser desorption ionization-time-of-flight / time-of-flight mass spectrometry (MS) analysis were used to characterize the hemolymph proteomic profiles of the silkworm, *Bombyx mori*. At days 4 (V4) and 5 (V5) of the fifth (final) instar, when the larvae were at the fast-growing stage, we found dramatic changes in spots representing proteins having an approximate molecular weight (MW) of 30 kDa. Of these spots, four 30K proteins were highly upregulated, implying a close association with the growth and development of *B. mori* larvae. To understand the molecular basis and underlying mechanisms involved in development and metamorphosis, the proteome of whole hemolymph at V5 was analyzed using shotgun liquid chromatography tandem mass spectrometry with an LTQ-Orbitrap. A total of 108 proteins were identified without any false discovery hits. These proteins were involved in a variety of cellular functions, including metabolism, development, nutrient transport and reserve, and defense response. Gene ontology analysis showed that 3.4% of these proteins had nutrient reservoir activities and 5.7% were involved in the response to stimulus. Pathway analysis revealed that 22 proteins with common targets were involved in various cellular processes such as immunity, differentiation, proliferation and metamorphosis. These results suggested that some key factors such as the 30K proteins in hemolymph play important roles in *B. mori* growth and development. Moreover, the multiple functions of hemolymph may be operated by a complex biological network.

Published On: Insect Science, 2012, 19(4), 441-454.

College of Animal Sciences, Zhejiang University, Hangzhou 310058, China.
*Corresponding author E-mail: bxzhong@zju.edu.cn.

家蚕五龄期血淋巴蛋白质组分析

李建营　李季生　钟伯雄*

摘要：本研究利用双向凝胶电泳结合基质辅助激光解吸电离质谱分析了家蚕血淋巴蛋白质组。在五龄第4天和第5天这个幼虫快速生长的阶段，我们发现分子质量为30 kDa的蛋白点发生了巨大的变化。在这些蛋白点中，4种30K蛋白显著上调，表明这些蛋白质与家蚕幼虫的生长和发育有密切关系。为了解析发育和变态的分子基础和机制，我们利用LTQ-Orbitrap质谱进行了液质联用质谱对五龄第5天的血淋巴进行了蛋白质组分析，共鉴定到108种蛋白质，无假阳性。这些蛋白质涉及一系列的分子功能，包括代谢、发育、营养运输和保存，以及防御反应。GO分类分析表明，3.4%的蛋白质具有营养贮存活性，5.7%的蛋白质与刺激反应有关。通路分析揭示了22种具有共同靶标的蛋白质涉及多种细胞过程，比如免疫、分化、增殖和变态。这些结果暗示在血淋巴中有许多关键因子比如30K蛋白在家蚕生长和发育过程中发挥了重要功能。而且，血淋巴的多重功能可能被一个复杂的生物网络所调控。

动物科学学院，浙江大学，杭州

Proteomic analysis of peritrophic membrane (PM) from the midgut of fifth-instar larvae, *Bombyx mori*

Hu XL Chen L Xiang XW
Yang R Yu SF Wu XF*

Abstract: The insect peritrophic membrane (PM), separating midgut epithelium and intestinal contents, is protective lining for the epithelium and plays the important role in absorption of nutrients, and also is the first barrier to the pathogens ingested through oral feeding. In order to understand the biological function of silkworm larval PM, shotgun liquid chromatography tandem mass spectrometry (LC-MS/MS) approach was applied to investigate its protein composition. Total 47 proteins were identified, of which 51.1% of the proteins had the isoelectric point (pI) within the range of 5-7, and 53.2% had molecular weights within the range 15-45 kDa. Most of them were found to be closely related to larval nutrients metabolism and innate immunity. Furthermore, these identified proteins were annotated according to Gene Ontology Annotation in terms of molecular function, biological process and cell localization. Most of the proteins had catalytic activity, binding activity and transport function. The knowledge obtained from this study will favour us to well understand the role of larval PM in larval physiological activities, and also help us to find the potential target and design better biopesticides to control pest, particularly the Lepidoptera insect.

Published On: Molecular Biology Reports, 2012, 39(4), 3427-3434.

College of Animal Science, Zhejiang University, Hangzhou 310029, China.
*Corresponding author E-mail: wuxiaofeng@zju.edu.cn.

家蚕五龄幼虫中肠围食膜的蛋白质组分析

胡小龙　陈　琳　相兴伟
杨　锐　于少芳　吴小锋*

摘要：昆虫中肠围食膜(peritrophic membrane，PM)将上皮细胞和肠道内容物隔离，保护着肠道内的上皮组织并且对于吸收营养物质发挥着重要作用，同时它也是防御经口饲喂摄入的病原体的第一道屏障。为了研究家蚕幼虫围食膜的生物学功能，我们利用液相色谱串联质谱(LC-MS/MS)的方法研究其蛋白质组成。共鉴定到47种蛋白质，约51.1%的蛋白质等电点为5—7，53.2%的蛋白质分子质量为15—45 kDa。发现大多数的蛋白质与幼虫营养物质代谢和先天免疫相关。此外，对这些被鉴定到的蛋白质在分子功能、生化过程和细胞定位等方面进行了基因功能注释。大多数的蛋白质有催化、结合活性以及转运功能。本研究获得的信息让我们很好地了解到围食膜在幼虫生理活动中的作用，并且可以帮助我们找到潜在的靶标，以便设计更好的生物农药去控制害虫。

动物科学学院，浙江大学，杭州

Proteomic-based insight into malpighian tubules of silkworm *Bombyx mori*

Zhong XW[#] Zou Y[#] Liu SP Yi QY
Hu CM Wang C Xia QY Zhao P[*]

Abstract: Malpighian tubules (MTs) are highly specific organs of arthropods (Insecta, Myriapoda and Arachnida) for excretion and osmoregulation. In order to highlight the important genes and pathways involved in multi-functions of MTs, we performed a systematic proteomic analysis of silkworm MTs in the present work. Totally, 1 367 proteins were identified by one dimensional gel electrophoresis coupled with liquid chromatography-tandem mass spectrometry, and as well as by Trans Proteomic Pipeline (TPP) and Absolute protein expression (APEX) analyses. Forty-one proteins were further identified by two-dimensional gel electrophoresis. Some proteins were revealed to be significantly associated with various metabolic processes, organic solute transport, detoxification and innate immunity. Our results might lay a good foundation for future functional studies of MTs in silkworm and other lepidoptera.

Published On: PLoS ONE, 2013, 8(9).

State Key Laboratory of Silkworm Genome Biology, Southwest University, Chongqing 400716, China.
[#]These authors contributed equally.
[*]Corresponding author E-mail: zhaop@swu.edu.cn.

家蚕马氏管蛋白质组学研究

钟晓武[#] 邹 勇[#] 刘仕平 衣启营
胡翠美 王 晨 夏庆友 赵 萍[*]

摘要：马氏管是节肢动物(昆虫纲、多足纲、蛛形纲)负责排泄和渗透调节的高度特异的器官。为了研究参与马氏管多项功能的重要基因及其通路，我们对家蚕马氏管进行了系统的蛋白质组学分析。最后，通过单向凝胶电泳结合液相串联质谱的方法和TPP及APEX蛋白表达分析，我们鉴定到1 367种蛋白质。其中，41种蛋白质能够进一步被双向电泳鉴定到。研究结果表明，某些蛋白质与多种代谢过程密切相关，如溶质转运、解毒作用和先天免疫等，本研究结果为进一步研究马氏管在家蚕及其他节肢动物中的功能奠定了基础。

家蚕基因组生物学国家重点实验室，西南大学，重庆

Comparative proteomic analysis reveals the suppressive effects of dietary high glucose on the midgut growth of silkworm

Feng F[1,2] Chen L[2] Lian CQ[1,2] Xia HC[2]
Zhou Y[2] Yao Q[2] Chen KP[1,2*]

Abstract: The silkworm, *Bombyx mori*, is an important model of lepidoptera insect, and it has been used for several models of human diseases. In human being, long-term high-sugar diet can induce the occurrence of diabetes and other related diseases. Interestingly, our experiments revealed the high glucose diet also has a suppressive effect on the development of silkworms. To investigate the molecular mechanism by which high-glucose diet inhibited the midgut growth in silkworms, we employed comparative proteomic analysis to globally identify proteins differentially expressed in normal and high-glucose diet group silkworms. In all, 28 differently proteins were suppressed and 5 proteins induced in high-glucose diet group. Gene ontology analysis showed that most of these differently proteins are mainly involved in metabolic process, catalytic and cellular process. A development related protein, imaginal disk growth factor (IDGF), was further confirmed by western blot exclusively expressing in the normal diet group silkworms. Taken together, our data suggests that IDGF plays a critical role in impairing the development of silkworms by a high-glucose diet.

Biological significance: Glucose has been thought to play essential roles in growth and development of silkworm. In this paper, we certified firstly that high-glucose diet can suppress the growth of silkworm, and comparative proteomic was employed to reveal the inhibition mechanism. Moreover, an important regulation related protein (IDGF) was found to involve in this inhibition process. These results will help us get a deeper understanding of the relationship between diet and healthy. Furthermore, IDGF may be the critical protein for reducing the blood sugar in silkworm, and it may be used for screening human hypoglycemic drug. The work has not been submitted elsewhere for publication, in whole or in part, and all the authors have approved the manuscript.

Published On: Journal of Proteomics, 2014, 108, 124-132.

1 School of Food and Biological Engineering, Jiangsu University, Zhenjiang 212013, China.
2 Institute of Life Sciences, Jiangsu University, Zhenjiang 212013, China.
*Corresponding author E-mail: kpchen@ujs.edu.cn.

比较蛋白质组学分析揭示摄食高葡萄糖对蚕中肠生长的抑制作用

冯　凡[1,2]　陈　亮[2]　连超群[1,2]　夏恒传[2]
周　阳[2]　姚　勤[2]　陈克平[1,2*]

摘要:家蚕是鳞翅目昆虫的重要模型,已被广泛应用于几种人类疾病模型。在人类身上,长期高糖饮食可诱发糖尿病及其他相关疾病的发生。有趣的是,我们的实验显示,高糖饮食也对蚕的发育具有抑制作用。为了研究高葡萄糖饮食抑制家蚕中肠生长的分子机制,我们采用比较蛋白质组学全面分析鉴定了正常和高糖饮食组蚕差异表达的蛋白质。总共有28种不同的蛋白质被抑制,5种蛋白质在高葡萄糖饮食组中诱导。GO富集分析显示,这些差异表达蛋白质大多数主要参与代谢过程、催化和细胞过程。利用蛋白质免疫印迹(Western blot)方法证实发育相关蛋白质——成虫原基生长因子(IDGF)只在正常饮食组家蚕中表达。综合起来,我们的数据表明IDGF在高葡萄糖饮食阻碍蚕的发育过程中起关键作用。

生物学意义:葡萄糖被认为在蚕的生长和发育中发挥重要作用。在本文中,我们首先证明高糖饮食可以抑制蚕的生长,采用比较蛋白质组来揭示其抑制机制。此外,发现重要的调节相关蛋白IDGF参与该抑制过程。这些结果将有助于我们更深入地了解饮食与健康之间的关系。此外,IDGF可能是降低蚕血糖的关键蛋白质,可用于筛选人类降血糖药物。

1　食品与生物工程学院,江苏大学,镇江
2　生命科学研究院,江苏大学,镇江

Development of an effective sample preparation approach for proteomic analysis of silkworm eggs using two-dimensional gel electrophoresis and mass spectrometry

Long XH[1] Zhu JW[2] Mo ZH[3] Feng S[1]

Cheng G[4] Zhou XW[4] Zhang YZ[5*] Yang PY[4*]

Abstract: Sample preparation is still the first and important step toward successful two-dimensional gel electrophoresis (2-DE) and identification in proteomics study. The 2-DE profiling of eggs of silkworm species by using conventional one-step extraction, however, is unsatisfactory because high-abundance proteins such as egg-specific protein (ESP) and No 30 family (30KP) in the extract lead to difficulties in detecting most of biologically relevant proteins. Based on the tendency of these abundant proteins to be soluble in Tris-HCl buffer, we report herein a robust approach in which the extract enriched in ESP and 30KP was fractionationed and mixed with the re-extract of residual pellet in an optimal proportion. In comparison with the one-step method, the 2-DE pattern was improved by this new method with over one-third enhancement in spots. A total of 48 unique proteins obtained have been furthermore identified by mass spectrometry (MS) and MS/MS. The identified proteins are found to include heat shock proteins families, ribosomal proteins, disulfide isomerase proteins, Glutathione S-transferase, and elongation factor, etc., which are mainly involved in some important processes. To our knowledge, this is the first time that the several proteins have been detected in silkworm eggs by proteomics means. This simple and reproducible approach would raise the opportunity of discovering and identifying more biomarkers and determining their possible roles in further studies.

Published On: Journal of Chromatography A, 2006, 1128(1-2), 133-137.

1 College of Life Science, Zhejiang University, Hangzhou 310029, China.

2 College of Agricultural and Biotechnology, Zhejiang University, Hangzhou 310029, China.

3 College of Chemistry and Chemical Engineering, Chongqing University, Chongqing 400044, China.

4 Department of Chemistry, Fudan University, Shanghai 200433, China.

5 College of Life Science, Zhejiang University of Sci-Tech, Hangzhou 310018, China.

*Corresponding author E-mail: yaozhou@chinagene.com; pyyang@fudan.edu.cn.

用于双向电泳和质谱检测的蚕卵蛋白质高效样品制备方法开发

龙晓辉[1] 朱金文[2] 莫志宏[3] 冯 珊[1]
程 刚[4] 周新文[4] 张耀洲[5*] 杨芃原[4*]

摘要：样品制备仍是蛋白质组学研究中成功进行双向凝胶电泳（2-DE）和鉴定的第一步，并且是重要的一步。蚕卵的2-DE电泳样品制备一般是使用传统的一步提取法提取，但是，由于存在高丰度蛋白（ESP）和30K蛋白家族（30KP），导致多数蛋白质的检测十分困难，效果并不令人满意。鉴于这些高丰度蛋白在Tris-HCl溶液中是可溶的，我们提出了一个有效方法，就是将富含ESP和30KP的提取物分级分离并以最佳比例与剩余颗粒的再提取物混合。与一步法相比，采用这种新方法的2-DE将蛋白点的检测的数量提升了三分之一以上。通过质谱分析（MS）和MS/MS进一步鉴定了总共48种独特的蛋白质，这些鉴定到的蛋白质包括热休克蛋白家族、核糖体蛋白、二硫键异构酶蛋白、谷胱甘肽S-转移酶和延伸因子等主要涉及重要生理过程的蛋白质。据我们所知，这是首次通过蛋白质组学手段在家蚕卵中检测到多种蛋白质。这种简单可重复的方法为将来发现和识别更多的生物标志物并确定其可能的作用提供了研究基础。

1 生命科学学院，浙江大学，杭州
2 农业与生物技术学院，浙江大学，杭州
3 化学化工学院，重庆大学，重庆
4 化学系，复旦大学，上海
5 生命科学学院，浙江理工大学，杭州

Shotgun analysis on the peritrophic membrane of the silkworm *Bombyx mori*

Zhong XW Zhang LP Zou Y Yi QY
Zhao P* Xia QY Xiang ZH

Abstract: The insect midgut epithelium is generally lined with a unique chitin and protein structure, the peritrophic membrane (PM), which facilitates food digestion and protects the gut epithelium. We used gel electrophoresis and mass spectrometry to identify the extracted proteins from the silkworm PM to obtain an in-depth understanding of the biological function of the silkworm PM components. A total of 305 proteins, with molecular weights ranging from 8.02 kDa to 788.52 kDa and the isoelectric points ranging from 3.39 to 12.91, were successfully identified. We also found several major classes of PM proteins, i. e. PM chitin-binding protein, invertebrate intestinal mucin, and chitin deacetylase. The protein profile provides a basis for further study of the physiological events in the PM of *Bombyx mori*.

Published On: BMB Reports, 2012, 45(11), 665-670.

State Key Laboratory of Silkworm Genome Biology, Southwest University, Chongqing 400716, China.
*Corresponding author E-mail: zhaop@swu.edu.cn.

家蚕围食膜“鸟枪法”蛋白质组学分析

钟晓武　张利平　邹　勇　衣启营
赵　萍*　夏庆友　向仲怀

摘要：昆虫的中肠上皮内表面有一层由几丁质和蛋白质构成的特殊结构——围食膜(PM)，它有利于食物的消化，同时也可以保护中肠上皮细胞。我们用凝胶电泳和质谱分析法鉴定并分离出家蚕围食膜中的蛋白质，以对家蚕围食膜中化合物的生物学功能有更深层的理解。成功鉴定出305种蛋白质，其分子质量从8.02 kDa到788.52 kDa不等，等电点介于3.39到12.91之间。同时，我们发现了几种主要的围食膜蛋白类型，比如，围食膜几丁质结合蛋白、无脊椎动物肠道黏液素和几丁质脱乙酰酶。这些蛋白质表达谱为对家蚕围食膜中生理过程的进一步研究奠定了基础。

家蚕基因组生物学国家重点实验室，西南大学，重庆

Comparative proteomic analysis between the domesticated silkworm (*Bombyx mori*) reared on fresh mulberry leaves and on artificial diet

Zhou ZH[1] Yang HJ[1] Chen M[2] Lou CF[1] Zhang YZ[3] Chen KP[4]
Wang Y[4] Yu ML[3] Yu F[1] Li JY[1] Zhong BX[1*]

Abstract: To gain an insight into the effects of different diets on growth and development of the domesticated silkworm at protein level, we employed comparative proteomic approach to investigate the proteomic differences of midgut, hemolymph, fat body and posterior silk gland of the silkworms reared on fresh mulberry leaves and on artificial diet. Seventy-six differentially expressed proteins were identified by MALDI TOF / TOF MS, and among them, 41 proteins were up-regulated, and 35 proteins were down-regulated. Database searches, combined with GO analysis and KEGG pathway analysis revealed that some hemolymph proteins such as Nuecin, Gloverin-like proteins, PGRP, P50 and β-N-acetylglucosa-midase were related to innate immunity of the silkworm, and some proteins identified in silkworm midgut including Myosin 1 light chain, Tropomyosin 1, Profilin, Serpin-2 and GSH-Px were involved in digestion and nutrition absorption. Moreover, two up-regulated enzymes in fat body of larvae reared on artificial diet were identified as V-ATPase subunit B and Arginine kinase which participate in energy metabolism. Furthermore, 6 down-regulated proteins identified in posterior silk gland of silkworm larvae reared on artificial diet including Ribosomal protein SA, EF-2, EF-1γ, AspAT, ERp57 and PHB were related to silk synthesis. Our results suggested that the different diets could alter the expression of proteins related to immune system, digestion and absorption of nutrient, energy metabolism and silk synthesis poor nutrition and absorption of nutrition in silkworm. The results also confirmed that the poor nutrient absorption, weakened innate immunity, decreased energy metabolism and reduced silk synthesis are the main reasons for low cocoons yield, inferior filament quality, low survival rate of young larvae and insufficient resistance against specific pathogens in the silkworms fed on artificial diet.

Published On: Journal of Proteome Research, 2008, 7(2): 5103-5111.

1 College of Animal Sciences, Zhejiang University, Hangzhou 310029, China.
2 College of Life Sciences, Zhejiang University, Hangzhou 310058, China.
3 College of Life Sciences, Zhejiang Sci-Tech University, Hangzhou 310018, China.
4 Institute of Life Sciences, Jiangsu University, Zhenjiang 212013, China.
*Corresponding author E-mail: bxzhong@zju.edu.cn.

新鲜桑叶和人工饲料饲养的家蚕间的比较蛋白质组学研究

周仲华[1] 杨慧娟[1] 陈 铭[2] 楼程富[1] 张耀洲[3] 陈克平[4]
王 勇[4] 俞梅兰[2] 于 芳[1] 李建营[1] 钟伯雄[1*]

摘要：为了在蛋白质水平调查不同饲养方式对家蚕生长发育的影响，我们用比较蛋白质组学的方法调查了饲喂新鲜桑叶和人工饲料的家蚕中肠、血液、脂肪体和后部丝腺等组织中的蛋白质表达差异情况。利用MALDI TOF/TOF MS技术鉴定到76个差异表达的蛋白质，其中41个蛋白质上调表达，35个蛋白质下调表达。数据库检索结合GO分析以及KEGG通路分析显示：血淋巴蛋白，如Nuecin，Gloverin-like，PGRP，P50和β-N-acetylglucosa-midase，与家蚕的先天性免疫相关；而一些中肠蛋白，包括Myosin 1轻链，Tropomyosin 1，Profilin，Serpin-2和GSH-Px，与家蚕的消化和营养吸收相关。另外有2个蛋白质，即V-ATPase B亚基和精氨酸激酶，在人工饲料饲养的幼虫的脂肪体内上调表达，可能参与了能量代谢过程。此外，在人工饲料饲喂的家蚕后部丝腺中，有6个下调的蛋白质被鉴定到，包括核糖体蛋白SA，EF-2，EF-1γ，AspAT，ERp57和PHB，它们均涉及丝蛋白的合成过程。实验结果表明，不同的饮食对家蚕免疫系统相关蛋白的表达、营养物质的消化吸收、能量代谢和蚕丝合成均有影响。结果还说明，营养吸收不良，先天免疫弱，能量代谢降低和丝蛋白的合成减少等因素是造成人工饲料饲养的家蚕茧层量低、纤维质量差、幼虫存活率低和抵抗特异病原菌能力弱的主要原因。

1 动物科学学院，浙江大学，杭州
2 生命科学学院，浙江大学，杭州
3 生命科学学院，浙江理工大学，杭州
4 生命科学研究院，江苏大学，镇江

Shotgun proteomic analysis of *Bombyx mori* brain: emphasis on regulation of behavior and development of the nervous system

Wang GB　Zheng Q　Shen YW　Wu XF*

Abstract: The insect brain plays crucial roles in the regulation of growth and development and in all types of behavior. We used sodium dodecyl sulfate polyacrylamide gel electrophoresis and high-performance liquid chromatography electron spray ionization tandem mass spectrometry (ESI-MS/MS) shotgun to identify the proteome of the silkworm brain, to investigate its protein composition and to understand their biological functions. A total of 2 210 proteins with molecular weights in the range of 5.64-1 539.82 kDa and isoelectric points in the range of 3.78-12.55 were identified. These proteins were annotated according to Gene Ontology Annotation into the categories of molecular function, biological process and cellular component. We characterized two categories of proteins: one includes behavior-related proteins involved in the regulation of behaviors, such as locomotion, reproduction and learning; the other consists of proteins related to the development or function of the nervous system. The identified proteins were classified into 283 different pathways according to KEGG analysis, including the PI3K-Akt signaling pathway which plays a crucial role in mediating survival signals in a wide range of neuronal cell types. This extensive protein profile provides a basis for further understanding of the physiological functions in the silkworm brain.

Published On:Insect Science, 2016, 23(1), 15 - 27.

College of Animal Science, Zhejiang University, Hangzhou 310029, China.
*Corresponding author E-mail: wuxiaofeng@zju.edu.cn.

鸟枪法蛋白质组分析家蚕脑组织：着重于行为的调控以及神经系统发育

王国宝　郑　秦　沈运旺　吴小锋*

摘要：昆虫脑组织在调控生长发育中发挥着重要作用并且操纵着昆虫的所有行为。为了研究家蚕脑组织的蛋白质组成和了解其生物学功能，我们利用聚丙烯酰胺凝胶电泳和高效液相色谱法——电喷雾串联质谱（ESI-MS/MS）鸟枪法鉴定家蚕脑组织的蛋白质组。共鉴定到2 210种蛋白质，分子质量为5.64—1 539.82 kDa，等电点为3.78—12.55。将这些蛋白质根据GO（Gene Ontology）对分子功能、生化过程和细胞组分等方面生物学功能进行注释。我们鉴定了两类蛋白：一类是行为相关蛋白，主要参与了行为的调控，例如运动能力等；另一类是神经系统发育或功能相关蛋白。KEGG分析将鉴定到的蛋白质归类到283种不同的通路中，其中包括了在神经类细胞中广泛存在的并对生存信号的调控发挥重要功能的PI3K-Akt信号通路。这种广泛的蛋白质谱为深入了解家蚕脑组织的生理功能提供了良好的基础。

动物科学学院，浙江大学，杭州

Proteomic analysis of the silkworm (*Bombyx mori* L.) hemolymph during developmental stage

Li XH　Wu XF　Yue WF
Liu JM　Li GL　Miao YG*

Abstract: We utilized the proteomic approach to investigate the proteome of the fifth instar hemolymph during growth and development, and to improve the understanding of this important bioprocess and gene expression situation. A total of 25 μL of hemolymph was used for 2D analysis, and the separated proteins were visualized by silver staining and analyzed using the ImageMaster 2D software. The report showed as many as 241 of protein spots were expressed in the beginning of the fifth instar. Among them, most were concentrated in pI 3.5-6.5, which reached 76% of the total protein spots. As for the protein molecular sizes, 182 protein spots concentrated between 35 and 90 kDa, which comes to 75% of the total spots. When the larvae grow to the seventh day (total fifth instar duration was 9 days), 298 protein spots were visualized through 2D electrophoresis. Fifty-seven spots were newly expressed compared to the image of the first day in fifth instar. The results implied that these proteins are related to biosynthesis of silk protein and metamorphosis preparation from larva to pupa. In total, 19 protein spots including 6 special spots expressed in seventh day were analyzed through MALDI-TOF-MS. The relations between proteins and growth and development of silkworm were discussed.

Published On: Journal of Proteome Research, 2006, 5(10), 2809-2814.

College of Animal Sciences, Zhejiang University, Hangzhou 310029, China.
*Corresponding author E-mail: miaoyg@zju.edu.cn.

家蚕不同发育阶段血淋巴蛋白质组学分析

李兴华　吴小锋　岳万福
刘剑梅　李广立　缪云根*

摘要：我们利用蛋白质组学方法研究了五龄家蚕生长发育过程中血淋巴的蛋白质组，以促进我们对家蚕生长发育中的生物过程和基因表达情况的理解。本研究提取五龄家蚕幼虫25 μL血淋巴进行双向电泳，通过双向电泳对蛋白质进行分离并银染，使用ImageMaster 2D软件进行分析。结果显示，在五龄起蚕，分离出多达241个蛋白点。其中大部分蛋白等电点集中在3.5—6.5，占总蛋白点的76%。在蛋白质分子质量上，182个蛋白点集中在35—90 kDa，达到总蛋白点的75%。当幼虫生长到五龄第7天（总共五龄持续时间为9d）时，通过双向电泳可分离出298个蛋白点，与五龄第1天相比，新发现了57个蛋白点，暗示这些蛋白质与丝蛋白的生物合成和从幼虫到蛹的变态反应有关。通过MALDI-TOF-MS分析了五龄第7天表达的19个蛋白点，包括6个只在五龄第7天特异表达的蛋白点，并讨论了这些蛋白质与蚕生长发育的关系。

动物科学学院，浙江大学，杭州

Proteomics of larval hemolymph in *Bombyx mori* reveals various nutrient–storage and immunity–related proteins

Zhang Y　Dong ZM　Wang DD　Wu Y
Song QR　Gu PM　Zhao P*　Xia QY

Abstract: The silkworm, *Bombyx mori*, is an important economic insect for its production of silk. The larvae of many lepidopteran insects are major agricultural pests and often silkworm is explored as a model organism for other lepidopteran pest species. The hemolymph of caterpillars contains a lot of nutrient and immune components. In this study, we applied liquid chromatography-tandem mass spectrometry to gain a better understanding of the larval hemolymph proteomics in *B. mori.* We identified 752 proteins in hemolymph collected from day-4 fourth instar and day-7 fifth instar. Nearly half the identified proteins (49%) were predicted to function as binding proteins and 46% were predicted to have catalytic activities. Apolipophorins, storage proteins, and 30K proteins constituted the most abundant groups of nutrient-storage proteins. Of them, 30K proteins showed large differences between fourth instar larvae and fifth instar larvae. Besides nutrient-storage proteins, protease inhibitors are also expressed very highly in hemolymph. The analysis also revealed lots of immunity-related proteins, including recognition, signaling, effectors and other proteins, comprising multiple immunity pathways in hemolymph. Our data provide an exhaustive research of nutrientstorage proteins and immunity-related proteins in larval hemolymph, and will pave the way for future physiological and pathological studies of caterpillars.

Published On: Amino Acids, 2014, 46, 1021-1031.

State Key Laboratory of Silkworm Genome Biology, Southwest University, Chongqing 400716, China.
*Corresponding author E-mail: zhaop@swu.edu.cn.

家蚕幼虫血淋巴蛋白质组学揭示各种营养储存和免疫相关蛋白

张　艳　董照明　王丹丹　吴　用
宋倩茹　顾培明　赵　萍*　夏庆友

摘要：家蚕是一种重要的蚕丝生产经济昆虫。许多鳞翅目昆虫的幼虫是主要的农业害虫，而家蚕经常被用作其他鳞翅目害虫研究的模式生物。鳞翅目昆虫幼虫的血淋巴中含有大量营养成分及免疫相关成分。在本研究中，我们采用液相色谱-串联质谱法，对家蚕幼虫血淋巴蛋白质组学进行了深入的研究。我们收集家蚕四龄第4天和五龄第7天的血淋巴，并鉴定到752个蛋白质。近一半(49%)的蛋白质被预测是结合蛋白，46%的蛋白质被预测具有催化活性。载脂蛋白、贮藏蛋白和30K蛋白是营养储存最丰富的蛋白质。其中30K蛋白在四龄幼虫与五龄幼虫之间显示出较大的差异。除了营养储存蛋白，蛋白酶抑制剂在血淋巴中也有高丰度的表达。该分析还发现了许多与免疫相关的蛋白质，包括识别、信号传导蛋白，效应子和其他蛋白质，构成了血淋巴中的多条免疫途径。本研究结果对幼虫血淋巴中的营养储存蛋白和免疫相关蛋白进行了详细的研究，为未来鳞翅目昆虫幼虫的生理和病理学研究奠定了一定基础。

家蚕基因组生物学国家重点实验室，西南大学，重庆

Determination of protein composition and host-derived proteins of *Bombyx mori* nucleopolyhedrovirus by 2-dimensional electrophoresis and mass spectrometry

Liu XY Chen KP* Cai KY Yao Q

Abstract:

Objective: The occlusion-derived form of baculovirus is specially adapted for primary infection of host midgut epithelial cells. The virion contains proteins essential for host-range determination and initiation of infection. Determination of protein composition of the occlusion-derived virus (ODV) will help us to understand this virion's infection mechanism.

Methods: We obtained the ODV virion of *Bombyx mori* nucleopolyhedrovirus (*Bm*NPV) by sucrose density gradient centrifugation, and used SDS-PAGE, 2-dimensional electrophoresis and mass spectrometry to identify for the first time proteins present within or associated with the ODV virion.

Results: 20 proteins, including 16 major proteins of the virion encoded by *Bm*NPV and 4 proteins of the silkworm, were identified with confidence.

Conclusion: Three *Bm*NPV proteins (orf40, P95 and protein tyrosine phosphatase), which had not been detected as components of ODV previously, were identified in this study. We propose that the host-derived proteins may be important for the assembly and maturation of the ODV capsid.

Published On: Intervirology, 2009, 51(5), 369-376.

Institute of Life Science, Jiangsu University, Zhenjiang 212013, China.

*Corresponding author E-mail:kpchen@ujs.edu.cn.

通过二维电泳和质谱法检测家蚕核型多角体病毒的蛋白质组成及宿主来源的蛋白质的分析

刘晓勇　陈克平*　蔡克亚　姚　勤

摘要：目的——杆状病毒的包涵体形式特别适应于宿主中肠上皮细胞的原发感染。病毒粒子含有确定宿主范围和启动感染所必需的蛋白质。测定包涵体病毒（ODV）的蛋白质组成将有助于我们了解病毒的感染机制。方法——我们通过蔗糖密度梯度离心法获得家蚕核型多角体病毒（*Bm*NPV）的ODV病毒粒子，并采用SDS-PAGE、二维电泳和质谱法首次鉴定了存在于ODV病毒粒子内或与之相关的蛋白质。结果——鉴定获得20种蛋白质，包括16个来源于*Bm*NPV的蛋白质和4个来源于家蚕的蛋白质。结论——本研究鉴定了3种此前未被鉴定为ODV成分的*Bm*NPV蛋白（orf40，P95和蛋白酪氨酸磷酸酶）。我们初步认为鉴定获得的来源于宿主的蛋白质可能对ODV衣壳的组装和成熟具有重要的作用。

生命科学研究院，江苏大学，镇江

Comparative proteomic analysis reveals that caspase-1 and serine protease may be involved in silkworm resistance to *Bombyx mori* nuclear polyhedrosis virus

Qin LG[1,2#] Xia HC[2#] Shi HF[2] Zhou YJ[2] Chen L[3]
Yao Q[2] Liu XY[2] Feng F[2] Yuan Y[2] Chen KP[1,2*]

Abstract: The silkworm *Bombyx mori* is of great economic value. The *B. mori* nuclear polyhedrosis virus (*Bm*NPV) is one of the most common and severe pathogens for silkworm. Although certain immune mechanisms exist in silkworms, most silkworms are still susceptible to *Bm*NPV infection. Interestingly, *Bm*NPV infection resistance in some silkworm strains is varied and naturally existing. We have previously established a silkworm strain *NB* by genetic cross, which is highly resistant to *Bm*NPV invasion. To investigate the molecular mechanism of silkworm resistance to *Bm*NPV infection, we employed proteomic approach and genetic cross to globally identify proteins differentially expressed in parental silkworms *NB* and *306*, a *Bm*NPV-susceptible strain, and their F_1 hybrids. In all, 53 different proteins were found in direct cross group (*NB*♀, *306*♂, F_1 hybrid) and 21 in reciprocal cross group (*306*♀, *NB*♂, F_1 hybrid). Gene ontology and KEGG pathway analyses showed that most of these different proteins are located in cytoplasm and are involved in many important metabolisms. Caspase-1 and serine protease expressed only in *Bm*NPV-resistant silkworms, but not in BmNPV-susceptible silkworms, which was further confirmed by Western blot. Taken together, our data suggests that both caspase-1 and serine protease play a critical role in silkworm resistance against *Bm*NPV infection.

Published On:Journal of Proteomics, 2012, 75(12), 3630-3638.

1 School of Food and Biological Engineering, Jiangsu University, Zhenjiang 212013, China.
2 Institute of Life Sciences, Jiangsu University, Zhenjiang 212013, China.
3 School of Life Science, Sichuan University, Chengdu 10065, China.
#These authors contributed equally.
*Corresponding author E-mail: kpchen@ujs.edu.cn.

比较蛋白质组学分析显示半胱天冬酶-1和丝氨酸蛋白酶可能参与了家蚕对家蚕核型多角体病毒的抗性

覃吕高[1,2#] 夏恒传[2#] 施海峰[2] 周亚竞[2] 陈 亮[3]
姚 勤[2] 刘晓勇[2] 冯 凡[2] 袁 弋[2] 陈克平[1,2*]

摘要：家蚕具有很高的经济价值。家蚕核型多角体病毒（*Bm*NPV）是家蚕最常见且严重的病原体之一。虽然家蚕存在一定的免疫机制，但大多数家蚕仍然对*Bm*NPV的感染很易感。有趣的是，一些家蚕品系中对*Bm*NPV感染存在多样性和天然的抗性。我们之前已经通过遗传杂交获得了一种家蚕品系*NB*，它对*Bm*NPV的入侵具有高度的抗性。为了研究家蚕抗*Bm*NPV感染的分子机制，我们采用蛋白质组学方法和遗传杂交来全面鉴定亲本蚕*NB*和*306*（一种*Bm*NPV易感品系）及其F_1杂种的差异表达蛋白质。在直接杂交组（*NB*♀，*306*♂，F_1杂交）中发现了53种不同的蛋白质，在互交组（*306*♀，*NB*♂，F_1杂交）中发现了21种不同的蛋白质。GO分析和KEGG通路分析显示，这些不同的蛋白质大多数位于细胞质中，并且参与许多重要的代谢过程。半胱天冬酶-1和丝氨酸蛋白酶仅在*Bm*NPV抗性家蚕中表达，而在*Bm*NPV易感的家蚕中不表达，Western blot进一步证实了此结果。总之，我们的数据表明，半胱天冬酶-1和丝氨酸蛋白酶在蚕抗*Bm*NPV感染中起关键作用。

1 食品与生物工程学院，江苏大学，镇江
2 生命科学研究院，江苏大学，镇江
3 生命科学学院，四川大学，成都

Proteomics analysis of digestive juice from silkworm during *Bombyx mori* nucleopolyhedrovirus infection

Hu XL[1,2#] Zhu M[1#] Wang SM[1] Zhu LY[1]
Xue RY[1,2] Cao GL[1,2] Gong CL[1,2*]

Abstract: Previous studies have analyzed the midgut transcriptome and proteome after challenge with *Bombyx mori* nucleopolyhedrovirus (*Bm*NPV), however little information is available on the digestive juice proteome after *Bm*NPV challenge. This study investigated *Bm*NPV infection-induced protein changes in the digestive juice of silkworms using shotgun proteomics and MS sequencing. From the digestive juice of normal third-day, fifth-instar silkworm larvae, 75 proteins were identified, 44 of which were unknown; from larvae 6 h after inoculation with *Bm*NPV, 106 proteins were identified, of which 39 were unknown. After *Bm*NPV challenge, more secreted proteins appeared that had antiviral and digestive features. GO annotation analysis clustered most proteins in the lumen into catalytic, binding, and metabolic processes. Numerous proteins were reported to have *Bm*NPV interactions. *Hsp70 protein cognate*, *lipase-1*, and *chlorophyllide A-binding protein precursor* were upregulated significantly after *Bm*NPV challenge. Levels of *trypsin-like serine protease*, *beta-1, 3-glucanase*, *catalase*, and *serine protease* transcripts decreased or were not significantly change after *Bm*NPV challenge. Taken together, these findings provided insights into the interaction between host and *Bm*NPV and revealed potential functions of digestive juice after per os *Bm*NPV infection.

Published On: Proteomics, 2015, 15(15), 2691-2700

1 School of Biology and Basic Medical Sciences ,Soochow University, Suzhou 215123, China.
2 National Engineering Laboratory for Modern Silk, Soochow University, Suzhou 215123, China.
#These authors contributed equally.
*Corresponding author E-mail: gongcl@suda.edu.cn

感染核型多角体病毒后家蚕消化液的蛋白质组学分析

胡小龙[1,2#]　朱　敏[1#]　王四妹[1]　朱丽媛[1]
薛仁宇[1,2]　曹广力[1,2]　贡成良[1,2*]

摘要：先前的研究已经对家蚕感染核型多角体病毒(*Bm*NPV)后的中肠进行了转录组和蛋白质组分析，但关于感染后消化液蛋白质组变化的报道却很少。本研究采用鸟枪法和MS测序法对*Bm*NPV感染后家蚕消化液的蛋白质组学进行比较。结果显示，在正常五龄第3天家蚕幼虫的消化液中鉴定到75种蛋白质，其中44种为未知功能蛋白质；经*Bm*NPV感染6 h后的幼虫消化液中鉴定到106种蛋白质，其中39种为未知功能蛋白质。*Bm*NPV感染后，鉴定到更多的分泌蛋白，包括抗病毒和消化相关的蛋白质。GO富集分析发现这些蛋白质大多参与了催化、结合和代谢过程，多数蛋白质与*Bm*NPV具有相互作用。*Bm*NPV感染后，Hsp70蛋白同源物，脂肪酶-1和叶绿素A结合蛋白前体均明显上调；胰蛋白酶类丝氨酸蛋白酶，β-1，3-葡聚糖酶、过氧化氢酶和丝氨酸蛋白酶下调表达或没有显著变化。综上，这些发现为研究宿主与*Bm*NPV之间相互作用提供了思路，揭示了*Bm*NPV感染后消化液的潜在功能。

1　基础医学与生物科学学院，苏州大学，苏州
2　现代丝绸国家工程实验室，苏州大学，苏州

Analysis of protein expression patterns of silkworm *Jinqiu* and its cross parents

Yu F[1,3] Yang HJ[1] Li JY[1] Ding N[2] Zhou ZH[1]
Ye J[1] Zhang JW[2] Duan JL[3] Zhong BX[1*]

Abstract: The differences of protein expression between the improved cross breeding race *Jinqiu* and its parents were analyzed to discuss the gene construction, and to form a base for illuminating the molecular mechanisms of successful cross breeding in silkworm. Protein samples from silk gland, hemolymph, and midgut were separated by 2-dimensional gel electrophoresis (2-DE). In the three tissues the matched protein spots between *Jinqiu* and its cross parents were approximately 70% with approximately 30% specific protein spots. In the matched protein spots, 9%-24% was differentially expressed representing up- and down-regulated expression. These specific protein spots might be either the newly appeared, which were produced from the genic interaction of cross parents' genes in cross breeding, or posttranscriptionally modified, which were produced from the different modifications on the same original proteins. These results indicate that it is important for a new successful breed, by cross breeding, relying on the actions of some newly produced functional proteins from genic interaction, in addition to marshaling excellent genes of cross parents.

Published On: Agricultural Science in China, 2009, 8(9), 1130-1137.

1 College of Animal Sciences, Zhejiang University, Hangzhou 310029, China.
2 Huzhou Academy of Agriculture Sciences, Huzhou 313000, China.
3 College of Life Sciences, Anhui University, Hefei 230036, China.
*Corresponding author E-mail: bxzhong@zju.edu.cn.

家蚕“金秋”及其亲本之间的蛋白表达模式分析

余　芳[1,3]　杨惠娟[1]　李建营[1]　丁　农[2]　周仲华[1]
叶　键[1]　张金卫[2]　段家龙[3]　钟伯雄[1*]

摘要：本研究分析了改良杂交种“金秋”与其亲本之间蛋白质表达的差异，探讨了其基因构建，为阐明家蚕成功杂交育种的分子机制奠定了基础。我们采用二维凝胶电泳技术(2-DE)分离了丝腺、血淋巴和中肠中的蛋白质。结果显示：3种组织器官中育成品种“金秋”与两个亲本的匹配蛋白点约为70%，剩余约30%是各个品种中特异的蛋白点；在匹配蛋白点中，还有9%—24%的蛋白质出现上调或下调变化。这些特异蛋白质可能是两个亲本的基因在杂交育成品种中产生了互作而形成的新蛋白质，也可能是对相同的蛋白质进行了不同的修饰而获得的修饰蛋白质。这些结果表明，一个新品种的育成，除了汇集亲本的优良基因外，还有赖于基因互作产生的新功能蛋白质的作用。

1　动物科学学院，浙江大学，杭州
2　湖州市农业科学院，湖州
3　生命科学学院，安徽大学，合肥

A novel droplet-tap sample-loading method for two-dimensional gel electrophoresis

Long XH[1] Lin JF[2] Zhang YZ[3*]

Abstract: A novel procedure, droplet-tap mode, has been devised for sample application for two-dimensional gel electrophoresis (2DE) expression profiles. The sample was loaded by evenly distributed tapping of droplets of the sample on to the rehydration buffer (RB) and then lowering the strip on to the solution surface. At normal loading concentrations, the number of spots obtained was increased by approximately one-third by this new approach compared with the rehydration loading procedure. The method also resulted in significantly improved resolution compared with cup loading when high concentrations of proteins were present, indicating its potential usefulness in micro-preparative separation. In addition, recovery of the proteins confirmed that protein uptake was enhanced by use of this method. By enabling improved performances in 2DE, the proposed procedure has much potential for sample loading to meet the requirements of global proteome analysis.

Published On: Analytical and Bioanalytical Chemistry, 2006, 384(7-8), 1578-1583.

1 College of Life Science, Zhejiang University, Hangzhou 310029, China.

2 Key Laboratory of Proteomics, Shanghai Institute of Biochemistry and Cell Biology, Shanghai Institutes for Biological Sciences, Chinese Academy of Sciences, Shanghai 200031, China.

3 College of Life Science, Zhejiang Sci-Tech University, Hangzhou 310018, China.

*Corresponding author E-mail: yaozhou@chinagene.com.

一种新的二维凝胶电泳液滴取样加载方法

龙晓辉[1]　林建峰[2]　张耀洲[3*]

摘要：我们设计出一种用于二维凝胶电泳（2DE）表达谱的点样新方法——droplet-tap模式。即将样品的液滴均匀地滴入到再水化缓冲液（RB）中，然后将胶条降低到溶液表面来加载样品。在正常加样浓度下，与再水化加载程序相比，通过这种新方法获得的斑点数量增加了约1/3。当存在高浓度蛋白质时，相较于杯装法该方法也显著地提高了蛋白质的分辨率，表明其在微制备分离中具有潜在的应用价值。此外，蛋白质回收实验证明，使用该方法更有利于蛋白质的提取。通过对2DE电泳性能的改进，该方法在加样上具有很大的潜力，可以满足对整体蛋白质组学分析的要求。

1　生命科学学院，浙江大学，杭州

2　蛋白质组学重点实验室，上海生物化学与细胞生物学研究所，上海生命科学研究院，中国科学院，上海

3　生命科学学院，浙江理工大学，杭州

第十一章

蚕的进化与比较基因组

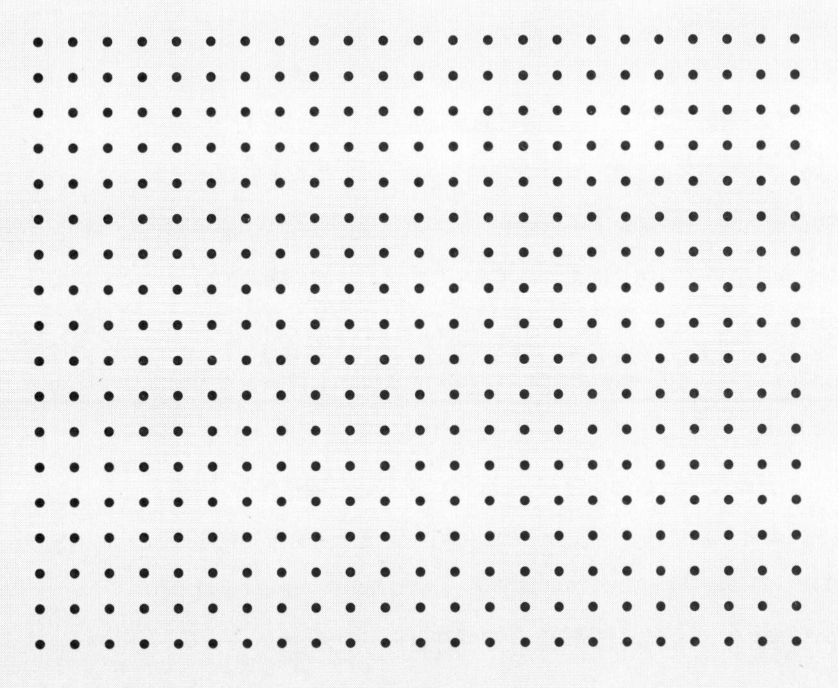

Chapter 11

将野桑蚕驯化为家蚕，在生物学史上具有重要意义，人类历史上只有家蚕和蜜蜂等为数不多的昆虫被驯化。而家蚕吐的丝主要成分就是蛋白质，蛋白质是生命的存在和表现形式，因此研究家蚕的起源进化具有特别重要的意义。

随着家蚕基因组计划的实施，在家蚕基因组框架图的基础上，中日双方合作完成了家蚕基因组精细图谱的构建，家蚕分子生物学研究已步入全基因组水平研究阶段，也为鳞翅目昆虫比较基因组学研究奠定了数据基础。昆虫重要基因家族的多物种比较研究进入了全盛时期，如核受体、细胞凋亡相关、激素信号通路、蛋白酶及其抑制剂、免疫球蛋白、同源异形盒（Hox）、基因组重复序列等比较进化研究。在这一时期，设计并制备了家蚕全基因组水平的表达谱芯片，同时完成了家蚕全组织基因表达谱图谱构建。基于表达谱芯片，开展了家蚕病源微生物诱导表达图谱构建工作，为揭示宿主响应病源的分子机制研究奠定了基础。

高通量测序技术在家蚕基因组重测序的应用。绘制完成了世界上第一张基因组水平上的蚕类单碱基遗传变异图谱，是世界上首次报道的昆虫基因组变异图。获得了40个家蚕突变品系和中国野桑蚕的全基因组序列，共获得632.5亿对碱基序列，覆盖了99.8%的基因组区域。有助于从全基因组范围研究驯化和人工选择对家蚕生物学的影响，阐释家蚕及野桑蚕之间生物学差异和线粒体基因组的进化的遗传基础，从全基因组水平上揭示了家蚕的起源进化，为家蚕的起源和驯化研究开启了一扇大门。

程廷才

Complete resequencing of 40 genomes reveals domestication events and genes in silkworm (*Bombyx mori* L)

Xia QY[1,2#] Guo YR[3#] Zhang Z[1,2#] Li D[1,3#] Xuan ZL[3#] Li Z[3#] Dai FY[1] Li YR[3] Cheng DJ[1] Li RQ[3,4] Cheng TC[1,2] Jiang T[3] Celine Becquet[5,6] Xu X[3] Liu C[1] Zha XF[1] Fan W[3] Lin Y[1] Shen YH[1] Jiang L[3] Jeffrey Jensen[5,6] Ines Hellmann[5,6] Tang S[5,6] Zhao P[1] Xu HF[1] Yu C[3] Zhang GJ[3] Li J[3] Cao JJ[3] Liu SP[1] He NJ[1] Zhou Y[3] Liu H[3] Zhao J[3] Ye C[3] Du ZH[1] Pan GQ[1] Zhao AC[1] Shao HJ[3,8] Zeng W[3] Wu P[3] Li CF[1] Pan MH[1] Li JJ[3] Yin XY[3] Li DW[3] Wang J[3] Zheng HS[3] Wang W[3] Zhang XQ[3] Li SG[3] Yang HM[3] Lu C[1] Rasmus Nielsen[4,5,6] Zhou ZY[1,7] Wang J[3] Xiang ZH[1*] Wang J[3,4*]

Abstract: A single-base pair resolution silkworm genetic variation map was constructed from 40 domesticated and wild silkworms, each sequenced to approximately threefold coverage, representing 99.88% of the genome. We identified ~16 million single-nucleotide polymorphisms, many indels, and structural variations. We find that the domesticated silkworms are clearly genetically differentiated from the wild ones, but they have maintained large levels of genetic variability, suggesting a short domestication event involving a large number of individuals. We also identified signals of selection at 354 candidate genes that may have been important during domestication, some of which have enriched expression in the silk gland, midgut, and testis. These data add to our understanding of the domestication processes and may have applications in devising pest control strategies and advancing the use of silkworms as efficient bioreactors.

Published On: Science, 2009, 326(5951), 433-436.

1 Key Sericultural Laboratory of Agricultural Ministry, Southwest University, Chongqing 400716, China.
2 Institute of Agronomy and Life Sciences, Chongqing University, Chongqing 400044, China.
3 BGIShenzhen, Shenzhen 518083, China.
4 Department of Biology, University of Copenhagen, DK-2100kbh O, Denmark.
5 Departments of Integrative Biology and Statistics, University of California Berkeley, Berkeley CA 94720, USA.
6 Department of Statistics, University of California Berkeley, Berkeley CA94720, USA.
7 Chongqing Normal University, Chongqing 400047, China.
8 Innovative Program for Undergraduate Students, South China University of Technology, Guangzhou 510006, China.
#These authors contributed equally.
*Corresponding author E-mail: wangj@genomics.org.cn (J.W.); xbxzh@swu.edu.cn (Z.X.).

40个基因组完全重测序揭示蚕的驯化事件及相关基因

夏庆友[1,2#] 郭一然[3#] 张 泽[1,2#] 李 东[1,3#] 玄兆伶[3#] 李 卓[3#] 代方银[1]
李英睿[3] 程道军[1] 李瑞强[3,4] 程廷才[1,2] 蒋 涛[3] Celine Becquet[5,6] 徐 讯[3]
刘 春[1] 查幸福[1] 樊 伟[3] 林 英[1] 沈以红[1] 蒋 岚[3] Jeffrey Jensen[5,6]
Ines Hellmann[5,6] 唐 思[5,6] 赵 萍[1] 徐汉福[1] 余 昶[3] 张国捷[3] 李 俊[3] 曹建军[3]
刘仕平[1] 何宁佳[1] 周 妍[3] 刘 慧[3] 赵 静[3] 叶 辰[3] 杜周和[1] 潘国庆[1]
赵爱春[1] 邵浩靖[3,8] 曾 巍[3] 吴 平[3] 李春峰[1] 潘敏慧[1] 李晶晶[3] 殷旭阳[3]
李大为[3] 王 娟[3] 郑会松[3] 王 文[3] 张秀清[3] 李松岗[3] 杨焕明[3] 鲁 成[1]
Rasmus Nielsen[4,5,6] 周泽扬[1,7] 汪 建[3] 向仲怀[1*] 王 俊[3,4*]

摘要：利用40个家蚕品种和野蚕构建了单碱基分辨率家蚕遗传变异图谱，每一个品种的测序达到3倍覆盖度，覆盖基因组的99.88%。我们鉴定了约1 600万个SNP，及一些插入缺失和结构变异，发现家蚕与野蚕在遗传上具有显著差异，但其自身保留了很高的遗传变异水平，暗示着大量的个体参与了一个较短的驯化过程。我们还鉴定了354个候选基因的选择信号，这些候选基因可能在驯化过程中起着重要作用，其中一些基因在丝腺、中肠、精巢中高量表达。这些数据加深了我们对家蚕驯化过程的理解，并可能促进家蚕在害虫控制和生物反应器方面的应用。

1 农业部蚕桑学重点实验室，西南大学，重庆
2 农学与生命科学研究院，重庆大学，重庆
3 北京基因组研究所（深圳），深圳
4 生物学院，哥本哈根大学，丹麦
5 综合生物学与统计学系，加州大学伯克利分校，伯克利
6 统计学系，加州大学伯克利分校，伯克利
7 重庆师范大学，重庆
8 华南理工大学，广州

Advances in silkworm studies accelerated by the genome sequencing of *Bombyx mori*

Xia QY[1*] Li S[2] Feng QL[3]

Abstract: Significant progress has been achieved in silkworm (*Bombyx mori*) research since the last review on *this insect* was published in this journal in 2005. In this article, we review the new and exciting progress and discoveries that have been made in *B. mori* during the past 10 years, which include the construction of a fine genome sequence and a genetic variation map, the evolution of genomes, the advent of functional genomics, the genetic basis of silk production, metamorphic development, immune response, and the advances in genetic manipulation. These advances, which were accelerated by the genome sequencing project, have promoted *B. mori* as a model organism not only for lepidopterans but also for general biology.

Published On: Annual Review of Entomology, 2014, 59(1), 513-536.

1 State Key Laboratory of Silkworm Genome Biology, Southwest University, Chongqing 400716, China.

2 Key Laboratory of Insect Developmental and Evolutionary Biology, Shanghai Institute of Plant Physiology and Ecology, Shanghai Institutes for Biological Sciences, Chinese Academy of Sciences, Shanghai 500032, China.

3 Guangdong Provincial Key Lab of Biotechnology for Plant Development, School of Life Sciences, South China Normal University, Guangzhou 510631, China.

*Corresponding author E-mail: xiaqy@swu.edu.cn.

家蚕基因组测序加速了家蚕研究的进展

夏庆友[1*]　李　胜[2]　冯启理[3]

摘要：自从关于家蚕的评述于2005年发表以来，家蚕的研究已经取得了重大进展。本论文综述了2005—2014年家蚕领域取得的令人振奋的进展和发现，包括：家蚕全基因组精细图和遗传变异图的完成、基因组的进化、功能基因组学研究的出现、丝合成的遗传基础、变态发育、免疫应答以及遗传操作的进展。全基因组测序计划推动家蚕研究向前发展，这也让家蚕成为了鳞翅目乃至普通生物学的模式生物。

1　家蚕基因组生物学国家重点实验室，西南大学，重庆

2　昆虫发育和进化生物学重点实验室，上海植物生理生态研究所，上海生命科学研究所，中国科学院，上海

3　广东省植物发育生物工程重点实验室，生命科学学院，华南师范大学，广州

Genetic diversity, molecular phylogeny and selection evidence of the silkworm mitochondria implicated by complete resequencing of 41 genomes

Li D[1,2] Guo YR[2] Shao H[2] Laurent C Tellier[2,3]
Wang J[2,3] Xiang ZH[1] Xia QY[1,4*]

Abstract:

Background: Mitochondria are a valuable resource for studying the evolutionary process and deducing phylogeny. A few mitochondria genomes have been sequenced, but a comprehensive picture of the domestication event for silkworm mitochondria remains to be established. In this study, we integrate the extant data, and perform a whole genome resequencing of Japanese wild silkworm to obtain breakthrough results in silkworm mitochondrial (*mt*) population, and finally use these to deduce a more comprehensive phylogeny of the Bombycidae.

Results: We identified 347 single nucleotide polymorphisms (SNPs) in the *mt* genome, but found no past recombination event to have occurred in the silkworm progenitor. A phylogeny inferred from these whole genome SNPs resulted in a well-classified tree, confirming that the domesticated silkworm, *Bombyxmori,* most recently diverged from the Chinese wild silkworm, rather than from the Japanese wild silkworm. We showed that the population sizes of the domesticated and Chinese wild silkworms both experience neither expansion nor contraction. We also discovered that one *mt* gene, named *cytochrome* b, shows a strong signal of positive selection in the domesticated clade. This gene is related to energy metabolism, and may have played an important role during silkworm domestication.

Conclusions: We present a comparative analysis on 41 *mt* genomes of *B. mori* and *B. mandarina* from China and Japan. With these, we obtain a much clearer picture of the evolution history of the silkworm. The data and analyses presented here aid our understanding of the silkworm in general, and provide a crucial insight into silkworm phylogeny.

Published On: BMC Evolutionary Biology, 2010, 10.

1 Key Sericultural Laboratory of Agricultural Ministry, College of Biotechnology, Southwest University, Chongqing 400716, China.
2 BGI-Shenzhen, Shenzhen 518083, China.
3 Department of Biology, University of Copenhagen, DK-2100 Kbh O, Denmark.
4 Institute of Agronomy and Life Sciences, Chongqing University, Chongqing 400030, China.
*Corresponding author E-mail: xiaqy@swu.edu.cn.

通过41个线粒体基因组的完全重测序对家蚕线粒体进行遗传多样性、分子系统发生和选择进化分析

李 东[1,2] 郭一然[2] 邵 浩[2] Laurent C Tellier[2,3]
王 俊[2,3] 向仲怀[1] 夏庆友[1,4*]

摘要：背景——线粒体是研究进化过程和推导系统发育的宝贵资源。一些线粒体基因组已被测序，但是我们对家蚕线粒体的驯化事件的全貌尚不清楚。在本研究中，我们整合了现有的数据，并对日本野蚕进行了全基因组重新测序，以获得家蚕线粒体（*mt*）种群的突破性结果，并以此推断家蚕更全面的系统发育。结果——我们在*mt*基因组中鉴定了347个单核苷酸多态性（SNPs），但发现在家蚕祖先中没有发生过重组事件。从这些全基因组单核苷酸多态性推断的系统发育形成了一个分类良好的树，证实家蚕最近与中国野蚕分离，而不是来自日本野蚕。我们发现驯养的家蚕和中国野蚕的种群规模既没有扩张也没有收缩。我们还发现一个名为细胞色素b的*mt*基因在驯化的分支中显示出强烈的阳性选择信号。该基因与能量代谢有关，可能在家蚕驯化过程中发挥了重要作用。结论——我们对来自中国和日本的家蚕、野桑蚕的41个*mt*基因组进行了比较分析。通过这些，我们可以更清楚地了解家蚕的进化史。本研究提供的数据和分析有助于我们对家蚕的整体认识，并为家蚕的系统发育提供了重要的见解。

1 农业部蚕桑学重点实验室，生物技术学院，西南大学，重庆
2 北京基因组研究所（深圳），深圳
3 生物系，哥本哈根大学，丹麦
4 农学与生命科学研究院，重庆大学，重庆

Genome-wide patterns of genetic variation among silkworms

Zhang XT[1,2] Nie MY[1] Zhao Q[3]
Wu YQ[1,2] Wang GH[1] Xia QY[1*]

Abstract: Although the draft genome sequence of silkworm is available for a decade, its genetic variations, especially structural variations, are far from well explored. In this study, we identified 1 298 659 SNPs and 9 731 indels, of which 32 % of SNPs and 92.2 % of indels were novel compared to previous silkworm re-sequencing analysis. In addition, we applied a read depth-based approach to investigate copy number variations among 21 silkworm strains at genome-wide level. This effort resulted in 562 duplicated and 41 deleted CNV regions, and among them 442 CNV were newly identified. Functional annotation of genes affected by these genetic variations reveal that these genes include a wide spectrum of molecular functions, such as immunity and drug detoxification, which are important for the adaptive evolution of silkworms. We further validated the predicted CNV regions using q-PCR. 94.7 % (36 / 38) of the selected regions show divergent copy numbers compared to a single-copy gene *OR2*. In addition, potential presence / absence variations are also observed in our study: 11 genes are present in the reference genome, but absent in other strains. Overall, we draw an integrative map of silkworm genetic variation at genome-wide level. The identification of genetic variations in this study improves our understanding that these variants play important roles in shaping phenotypic variations between wild and domesticated silkworms.

Published On: Molecular Genetics and Genomics, 2015, 290(4), 1575-1587.

1 State Key Laboratory of Silkworm Genome Biology, Southwest University, Chongqing 400716, China.
2 School of Life Sciences, Chongqing University, Chongqing 400044, China.
3 Department of Biology, Syracuse University, Syracuse NY13210, USA.
*Corresponding author E-mail: xiaqy@swu.edu.cn.

家蚕全基因组遗传变异模式

张兴坦[1,2] 聂梦云[1] 赵 乾[3]
吴玉乾[1,2] 王根洪[1] 夏庆友[1*]

摘要：虽然家蚕基因组序列草图已经有十几年历史了，但是其遗传变异，尤其是结构变异还远未得到充分的研究。在本研究中，我们鉴定了1 298 659个SNPs和9 731个插入缺失，其中32%的SNPs和92.2%的插入缺失与原来家蚕基因序列重测序分析相比是新发现的。此外，我们采用了基于深度的阅读方法，在全基因组水平上研究了21个家蚕品系拷贝数变异。这一工作产生了562个重复和41个删除的CNV区域，并且其中有442个CNV是新鉴定的。受这些基因变异影响的基因的功能注释表明这些基因包括广泛的分子功能，如免疫和药物解毒作用，这对蚕的适应性进化很重要。我们通过qPCR进一步验证这些预测到的CNV区域。与单拷贝基因*OR2*相比，94.7%（36/38）的所选区域的基因显示出不同的拷贝数。此外，在我们的研究中也观察到潜在存在/缺失的变化：有11个基因在参照基因组中存在，但却在其他变异体中缺失。总之，我们在全基因组水平上绘制了家蚕遗传变异的综合图谱。本研究增强了我们对遗传变异的认识，即这些变异在野蚕和驯养家蚕的表型变异中扮演重要角色。

1 家蚕基因组生物学国家重点实验室，西南大学，重庆
2 生命科学学院，重庆大学，重庆
3 生物系，锡拉丘兹大学，锡拉丘兹，美国

The complete mitogenome and phylogenetic analysis of *Bombyx mandarina* strain *Qingzhou*

Hu XL[#] Cao GL[#] Xue RY Zheng XJ

Zhang X Duan HR Gong CL[*]

Abstract: The complete nucleotide sequence of the mitogenome of *Bombyx mandarina* strain *Qingzhou* was determined. The circular genome is 15 717 bp long and has the typical gene organization and order of lepidopteran mitogenomes. All protein-coding sequences are initiated with a typical ATN codon, except the *COI* gene, which has a 4 bp TTAG putative initiator codon. Eleven of the 13 protein-coding gene have a complete termination codon (all TAA), but the remaining two genes terminate with incomplete codons. All transfer RNAs (tRNAs) have a clover-leaf structure typical of the mitochondrial tRNAs, and some of them have a mismatch in the four-stem-and-loop structure. The length of the A + T rich region of *B. mandarina* strain *Qingzhou* is 495 bp, shorter than that of *B. mandarina* strain *Tsukuba* (747 bp) but similar to that of *Bombyx mori.* Phylogenetic analysis based on the whole mitochondrial genome sequences of the available sequenced species (*B. mori* strains *C-108*, *Aojuku*, *Backokjam*, and *Xiafang*, *B. mandarina* strains *Tsukuba*, *Ankang*, and *Qingzhou*, and *Antheraea pernyi*) shows the origin of the domesticated silkmoth *B. mori* to be the Chinese *B. mandarina.* Nuclear mitochondrial pseudogene sequences were detected in the nuclear genome of *B. mori* with the MEGA BLAST search program. A phylogenetic analysis of these nuclear mitochondrial pseudogene sequences suggests that *B. mori* was domesticated independently in different areas and periods.

Published On: Molecular Biology Reports, 2010, 37(6), 2599-2608.

Medical College, Soochow University, Suzhou 215123, China.

[#]These authors contributed equally.

[*]Corresponding author E-mail: gongcl@suda.edu.cn.

野桑蚕青州品系的线粒体基因组序列和系统发育分析

胡小龙[#] 曹广力[#] 薛仁宇 郑小坚
张 星 段海蓉 贡成良[*]

摘要:我们对青州野桑蚕线粒体全基因组序列进行了测定和分析,结果显示青州野桑蚕线粒体DNA长度为15 717 bp,具有鳞翅目昆虫线粒体典型的基因组成和基因排列方向。除了*COI*基因具有一个4 bp TTAG的假定起始密码子,其他所有蛋白质编码序列均由ATN密码子启动。在13个蛋白编码基因中,11个具有完整的终止密码子(TAA),另外2个基因的终止密码子是不完整的。所有的转运RNA(tRNA)具有典型的线粒体tRNA的三叶草结构,并且有些tRNA在四茎环结构中存在错配。青州野桑蚕的A+T丰富区长度为495 bp,比日本野蚕(747 bp)短,但与家蚕的长度相当。基于4种家蚕、3种野桑蚕和柞蚕的线粒体全基因组序列的系统进化的分析表明,驯化的家蚕起源于中国野桑蚕。通过MEGA BLAST搜索程序在家蚕核基因组中搜索到核线粒体假基因序列。对这些核线粒体假基因进行系统进化分析,表明家蚕是在不同的时期和区域被独立驯化的。

医学部,苏州大学,苏州

A genetic diversity study of silkworm using cleaved amplified polymorphic sequence (CAPS) markers

Huang JH[1,2] Li MW[1,2] Zhang Y[1,2] Liu WB[1,2]
Li MH[1,2] Miao XX[1] Huang YP[1*]

Abstract: The silkworm, *Bombyx mori* is a beneficial insect of great economic importance in China for its silk production. In this study, we obtained 11 cleaved amplified polymorphic sequence (CAPS) markers and one PCR polymorphism marker from the genes of the silkworm, *B. mori.* A backcross progeny analysis showed that all these molecular markers were segregated in a Mendelian fashion and that polymorphisms were co-dominant. These markers were used to investigate the genetic diversity among 29 strains of *B. mori* from China, Japan and Europe. Cluster analysis, based on the genetic similarities calculated from CAPS data, grouped these strains roughly according to their geographical origin. One group contained silkworm strains from Europe and some of the Japanese strains were interspersed into the Chinese groups, whereas other Japanese strains clustered together.

Published On: Biochemical Systematics and Ecology, 2006, 34(12), 868-874.

1 Institute of Plant Physiology and Ecology, Chinese Academy of Sciences, Shanghai 200032, China.
2 Graduate School of Chinese Academy of Sciences, Beijing 100039, China.
*Corresponding author E-mail: yphuang@sibs.ac.cn.

利用酶切扩增多态序列(CAPS)进行家蚕遗传多样性分析

黄健华[1,2]　李木旺[1,2]　张　勇[1,2]　刘文彬[1,2]
李明辉[1,2]　苗雪霞[1]　黄勇平[1*]

摘要：家蚕具有吐丝能力，是一种非常重要的经济昆虫。在本研究中，我们从家蚕基因组里寻找到11对酶切扩增多态序列(cleaved amplified polymorphic sequence，CAPS)标记和一个PCR多态性标记，通过回交后代分析，我们发现这些分子标记是互相分离的共显性标记，符合孟德尔遗传规律。我们利用这些分子标记分析了来自中国、日本和欧洲的29个家蚕品系之间的遗传多样性。根据CAPS标记的数据，基于遗传相似形计算，我们进行了聚类分析，参照这些家蚕品系的原产地，我们对它们进行了初步的分组。其中一组来自欧洲的族群和一些来自日本的品系被聚类到中国族群中，其他的日本品系归类为一个族群。

1　植物生理生态研究所，中国科学院，上海
2　研究生院，中国科学院，北京

Evolutionarily conserved repulsive guidance role of Slit in the silkworm *Bombyx mori*

Yu Q[1#] Li XT[1#] Liu C[2] Cui WZ[1]
Mu ZM[1] Zhao X[1] Liu QX[1*]

Abstract: Axon guidance molecule Slit is critical for the axon repulsion in neural tissues, which is evolutionarily conserved from planarians to humans. However, the function of Slit in the silkworm *Bombyx mori* was unknown. Here we showed that the structure of *Bombyx mori* Slit (*Bm*Slit) was different from that in most other species in its C-terminal sequence. *Bm*Slit was localized in the midline glial cell, the neuropil, the tendon cell, the muscle and the silk gland and colocalized with *Bm*Robo1 in the neuropil, the muscle and the silk gland. Knock-down of *Bmslit* by RNA interference (RNAi) resulted in abnormal development of axons and muscles. Our results suggest that *Bm*Slit has a repulsive role in axon guidance and muscle migration. Moreover, the localization of *Bm*Slit in the silk gland argues for its important function in the development of the silk gland.

Published On :PLoS ONE, 2014, 9（10）.

1 Laboratory of Developmental Genetics, Shandong Agricultural University, Taian 271018, China.
2 State Key Laboratory of Silkworm Genome Biology, Southwest University, Chongqing 400716, China.
#These authors contributed equally.
*Corresponding author E-mail: liuqingxin@sdau.edu.cn.

家蚕Slit在进化上保守的排斥导向作用

于　奇[1#]　李晓童[1#]　刘　春[2]　崔为正[1]
牟志美[1]　赵　晓[1]　刘庆信[1*]

摘要：轴突指导分子Slit对神经组织中的轴突排斥至关重要，Slit从涡虫到人类在进化上相对保守。然而，家蚕中Slit的功能尚不清楚。本研究表明，家蚕Slit（*Bm*Slit）的结构与大多数其他物种的C-末端序列不同。*Bm*Slit定位在中线神经胶质细胞、神经纤维网、肌腱细胞、肌肉和丝腺中，并在神经纤维网、肌肉和丝腺中能够与*Bm*Robo1进行共定位。RNA干扰（RNAi）敲低*Bmslit*导致轴突和肌肉的发育异常。我们的研究结果表明，*Bm*Slit在轴突导向和肌肉迁移中具有排斥作用。此外，*Bm*Slit在丝腺中的表达定位表明其在丝腺发育中具有重要功能。

1　发育遗传学实验室，山东农业大学，泰安
2　家蚕基因组生物学国家重点实验室，西南大学，重庆

Analyzing genetic relationships in *Bombyx mori* using intersimple sequence repeat amplification

Li MW[1,2] Hou CX[2] Miao XX[1] Xu AY[2] Huang YP[1*]

Abstract: Intersimple sequence repeat (ISSR) amplification was used to analyze genetic relationships among silkworm, *Bombyx mori* L., strains. Nineteen primers containing simple sequence repeat (SSR) motifs were tested for amplification on a panel of 42 strains, representative of the diversity of silkworm germplasm; 12 of the primers amplified distinct, reproducible bands. The primers amplified a total of 108 bands, of which 85 (78.7%) were polymorphic. The ISSR results suggested that within the dinucleotide class, the poly (CA) motif was more common than the poly (CT) motif. The ISSR amplification pattern was used to group the silkworm strains into seven subclusters based on their origin in an unweighted pair-group method with arithmetic average cluster analysis by using Nei's genetic distance. Seven major ecotypic silkworm groups were analyzed. Principal component analysis of the ISSR data supported the unweighted pair-group method with arithmetic average clustering. Therefore, ISSR amplification is a valuable method for determining genetic variability among silkworm varieties. This efficient genetic fingerprinting technique should be useful for characterizing the large numbers of silkworm strains held in national and international germplasm centers.

Published On: Journal of Economic Entomology, 2007, 100 (1), 202-208.

1 Institute of Plant Physiology and Ecology, Chinese Academy of Sciences, Shanghai 200032, China.

2 Sericultural Research Institute, Chinese Academy of Agricultural Sciences, Zhenjiang 212018, China.

*Corresponding author E-mail: yphuang@sibs.ac.cn.

利用简单重复扩增序列分析家蚕不同品系之间的遗传关系

李木旺[1,2]　侯成香[2]　苗雪霞[1]　徐安英[2]　黄勇平[1*]

摘要：重复扩增序列被用来分析家蚕不同品系中的基因关系。利用包括简单重复序列(SSR)在内的19对引物对代表性的42个家蚕品系进行了研究；其中有12对引物扩增出了明显可区分和可重现的108条条带，其中有85条(78.7%)具有多态型。ISSR结果显示，在二核苷酸类，多聚CA探针要比多聚CT探针更普遍。利用ISSR扩增模式和Nei遗传距离分析法(基于算术平均聚类非加权配对法，UPGMA法)将家蚕品兼按来源分为7个亚类。所获得的ISSR数据经主成分分析(Principal components analysis，PCA)，支持UPGMA方法的聚类结果。因此，ISSR扩增序列是测定家蚕遗传变异的一种有价值的方法，这种高效的遗传指纹分析技术将有助于国内外种质资源中心保存的大量家蚕品系的鉴定。

1　植物生理生态研究所，中国科学院，上海

2　蚕业研究所，中国农业科学研究院，镇江

Gene analysis of an antiviral protein SP-2 from Chinese wild silkworm, *Bombyx mandarina* Moore and its bioactivity assay

Yao HP　He FQ　Guo AQ
Cao CP　Lu XM　Wu XF*

Abstract: The cDNA encoding an antiviral protein SP-2 against *Bm*NPV was cloned from the midgut of Chinese wild silkworm, *Bombyx mandarina* Moore (GenBank access AY945210) based on the available information of the domesticated silkworm. Its cDNA was 855 bp encoding 284 amino acids with predicted molecular weight of 29.6 kDa. Its full length in genomics was 1 376 bp, including 5 exons and 4 introns. The expression analysis indicated that it was only expressed in midgut, and its expression level was higher during feeding stage of larval instars while very lower during the moltism and mature stages. The deduced amino acid sequence of this protein showed eight-amino-acid variation compared with the counterpart of domesticated silkworm. Its antiviral activity was assayed through *in vitro* test. The results indicated that it showed strong bioactivity against *Bm*NPV, and its activity was 1.6 fold higher that the counterpart of domesticated silkworm.

Published On: Science in China Series C—Life Sciences, 2008, 51(10), 879-884.

College of Animal Science, Zhejiang University, Hangzhou 310029, China.
*Corresponding author E-mail: wuxiaofeng@zju.edu.cn.

中国野桑蚕抗病毒蛋白SP-2的序列分析及其生物活性测定

姚慧鹏　何芳青　郭爱芹
曹翠萍　鲁兴萌　吴小锋*

摘要：基于家蚕的序列信息，以中国野桑蚕的中肠cDNA为模板，克隆获得一条编码抗病毒蛋白SP-2的cDNA（GenBank access AY945210）。野桑蚕SP-2的cDNA长度为855 bp，编码284个氨基酸，预测分子质量为29.6 kDa。其基因组序列全长为1 376 bp，包含5个外显子和4个内含子。表达谱分析表明，SP-2仅在中肠表达，且在幼虫期的饲养阶段表达水平较高，而在蜕皮期和成熟期表达水平则较低。该蛋白预测的氨基酸序列与家蚕的相比，存在8个氨基酸变异。通过体外试验测定其抗病毒活性，结果表明，野蚕的SP-2具有更高的抗*Bm*NPV病毒活性，其活性是家蚕的1.6倍。

动物科学学院，浙江大学，杭州

Comparative genomics of parasitic silkworm microsporidia reveal an association between genome expansion and host adaptation

Pan GQ[1#] Xu JS[2,3#] Li T[1#] Xia QY[1] Liu SL[3] Zhang GJ[4] Li SG[4] Li CF[1] Liu HD[1] Yang L[1] Liu T[1] Zhang X[2] Wu ZL[1] Fan W[4] Dang XQ[1] Xiang H[1] Tao ML[1] Li YH[1] Hu JH[1] Li Z[1,2] Lin LP[1] Luo J[1] Geng LN[1] Wang LL[2] Long MX[1] Wan YJ[1] He NJ[1] Zhang Z[1] Lu C[1] Patrick J Keeling[3] Wang J[4] Xiang ZH[1] Zhou ZY[1,2*]

Abstract:

Background: Microsporidian *Nosema bombycis* has received much attention because the pébrine disease of domesticated silkworms results in great economic losses in the silkworm industry. So far, no effective treatment could be found for pébrine. Compared to other known *Nosema parasites, N. bombycis* can unusually parasitize a broad range of hosts. To gain some insights into the underlying genetic mechanism of pathological ability and host range expansion in this parasite, a comparative genomic approach is conducted. The genome of two *Nosema parasites*, *N. bombycis* and *N. antheraeae* (an obligatory parasite to undomesticated silkworms *Antheraea pernyi*), were sequenced and compared with their distantly related species, *N. ceranae* (an obligatory parasite to honey bees).

Results: Our comparative genomics analysis show that the *N. bombycis* genome has greatly expanded due to the following three molecular mechanisms: 1) the proliferation of host-derived transposable elements, 2) the acquisition of many horizontally transferred genes from bacteria, and 3) the production of abundant gene duplications. To our knowledge, duplicated genes derived not only from small-scale events (e. g., tandem duplications) but also from largescale events (e.g., segmental duplications) have never been seen so abundant in any reported microsporidia genomes. Our relative dating analysis further indicated that these duplication events have arisen recently over very short evolutionary time. Furthermore, several duplicated genes involving in the cytotoxic metabolic pathway were found to undergo positive selection, suggestive of the role of duplicated genes on the adaptive evolution of pathogenic ability.

Conclusions: Genome expansion is rarely considered as the evolutionary outcome acting on those highly reduced and compact parasitic microsporidian genomes. This study, for the first time, demonstrates that the parasitic genomes can expand, instead of shrink, through several common molecular mechanisms such as gene duplication, horizontal gene transfer, and transposable element expansion. We also showed that the duplicated genes can serve as raw materials for evolutionary innovations possibly contributing to the increase of pathologenic ability. Based on our research, we propose that duplicated genes of *N. bombycis* should be treated as primary targets for treatment designs against pébrine.

Published On: BMC Genomics, 2013, 14, 186.

1 State Key Laboratory of Silkworm Genome Biology, Southwest University, Chongqing 400716, China.

2 College of Life Sciences, Chongqing Normal University, Chongqing 400047, China.

3 Department of Botany, University of British Columbia, Vancouver, British Columbia V6T 1Z4, Canada.

4 Beijing Genomics Institute at Shenzhen, Shenzhen 518000, China.

[#]These authors contributed equally.

[*]Corresponding author E-mail: zyzhou@swu.edu.cn.

家蚕微孢子虫的比较基因组学显示基因组扩增与宿主适应之间的关联

潘国庆[1#] 许金山[2,3#] 李田[1#] 夏庆友[1] 刘绍伦[3] 张国捷[4] 李松岗[4] 李春峰[1] 刘含登[1]
杨柳[1] 刘铁[1] 张玺[2] 吴正理[1] 樊伟[4] 党晓群[1] 向恒[1] 陶美林[1] 李艳红[1]
胡军华[1] 李治[1,2] 林立鹏[1] 罗洁[1] 耿莉娜[1] 王林玲[2] 龙梦娴[1] 万永继[1] 何宁佳[1]
张泽[1] 鲁成[1] Patrick J Keeling[3] 王俊[4] 向仲怀[1] 周泽扬[1,2*]

摘要：背景—— 家蚕微孢子虫(*Nosema bombycis*)受到广泛关注,因为家蚕的微粒子病在蚕业中造成了巨大的经济损失。到目前为止,还没有找到有效的治疗方法。与其他已知的寄生虫微孢子虫相比,*N. bombycis* 寄生范围异常广泛。为了获得对这种微孢子虫病理能力和宿主范围扩大的潜在遗传机制的一些了解,本文进行了比较基因组分析。将两种微孢子虫,*N. bombycis* 和 *N.antheraeae*(半驯化柞蚕 *Antheraea pernyi* 特有微孢子虫)的基因组进行测序,并与其远源相关物种 *N. ceranae*(东方蜜蜂专有微孢子虫)进行比较。结果——我们的比较基因组学分析表明,由于以下3种分子机制,*N. bombycis* 基因组已经大大扩展:1)宿主衍生的转座因子的增殖,2)从细菌获得许多水平转移的基因,3)产生了丰富的基因重复。据我们所知,来自小规模事件(例如,串联重复)和来自大规模事件(例如,分段重复)的重复基因的丰富程度,在任何报道的微孢子虫基因组中从未有过。我们的分析进一步表明,这些重复事件出现在很短的进化时间内。此外,一些参与细胞毒性代谢途径的重复基因被发现经历正向筛选,暗示了重复基因在致病能力适应性进化中的作用。结论——基因组扩增很少被认为是作用于那些高度减少和紧凑的寄生微孢子虫基因组的进化结果。这项研究首次证明寄生基因组可以通过几种常见的分子机制(如基因重复、基因水平转移和转座因子扩增)扩大而不是收缩。结果还表明,重复的基因可以作为进化创新的原料,可能有助于微孢子虫致病能力的增加。基于我们的研究,我们建议将 *N. bombycis* 的重复基因作为治理家蚕微粒子病的主要靶标。

1 家蚕基因组生物学国家重点实验室,西南大学,重庆
2 生命科学学院,重庆师范大学,重庆
3 植物学院,英属哥伦比亚大学,温哥华,加拿大
4 北京基因组研究所(深圳),深圳

The varying microsporidian genome: existence of long-terminal repeat retrotransposon in domesticated silkworm parasite *Nosema bombycis*

Xu JS[1] Pan GQ[1] Fang L[2] Li J[2]
Tian XJ[2] Li T[1] Zhou ZY[1,3*] Xiang ZH[1]

Abstract: Microsporidia are a group of intracellular parasites with an extremely compact genome and there is no confirmed evidence that retroelementsare parasitised in these organisms. Using the dataset of 200 000 genomic shotgun reads of the silkworm pebrine *Nosemabombycis*, we have identified the eight complete *N. bombycis* long-terminal repeat retrotransposon (*Nbr*) elements. All of the *Nbr* elements are Ty3/gypsy members and have close relationships to *Saccharomycetes* long-terminal repeat retrotransposons identified previously, providing further evidence of their relationship to fungi. To explore the effect of retrotransposons in microsporidian genome evolution, their distribution was characterised by comparisons between two *N. bombycis* contigs containing the *Nbr* elements with the completed genome of the human parasite *Encephalitozoon cuniculi*, which is closely related to *N. bombycis*. The *Nbr* elements locate between or beside syntenic blocks, which are often clustered with other transposable-like sequences, indicating that they are associated with genomesize variation and syntenic discontinuities. The ratios of the number of non-synonymous substitutions per non-synonymous site to the number of synonymous substitutions per synonymous site of the open reading frames among members of each of the eight *Nbr* families were estimated, which reveal the purifying selection acted on the *N. bombycis* long-terminal repeat retrotransposons. These results strongly suggest that retrotransposons play a major role in reorganization of the microsporidian genome and they might be active.The present study presents an initial characterization of some transposable elements in the *N. bombycis* genome and provides some insight into the evolutionary mechanism of microsporidian genomes.

Published On: International Journal for Parasitology, 2006, 36(9), 1049-1056.

1 Key Sericultural Laboratory of Agricultural Ministry of China, Southwest University, Chongqing 400716, China.
2 Beijing Institute of Genomics, Chinese Academy of Sciences, Beijing 101300, China.
3 Laboratory of Animal Biology, Chongqing Normal University, Chongqing 400047, China.
*Corresponding author E-mail: zyzhou@swau.cq.cn.

不同微孢子虫基因组：家蚕微孢子虫中存在长末端重复逆转录转座子

许金山[1]　潘国庆[1]　方　林[2]　李　军[2]
田相军[2]　李　田[1]　周泽扬[1,3*]　向仲怀[1]

摘要：微孢子虫是一类具有致密基因组的细胞内寄生生物，目前为止尚未有证据表明微孢子虫具有逆转录元件。利用200 000个家蚕微孢子虫基因组鸟枪法读长数据，我们鉴定到了8个完整的家蚕微孢子虫长末端重复逆转录转座子（*Nbr*）元件。鉴定到的*Nbr*元件全部属于Ty3/gypsy家族，并且与先前报道所鉴定到的酵母菌的长末端重复逆转录转座子高度相似，这为它们与真菌的关系提供了进一步的证据。为了探究逆转录转座子对微孢子虫基因组进化所起到的作用，我们选取了两个含有*Nbr*元件的微孢子虫重叠群与人类脑寄生虫脑管虫的完全基因组进行了对比。结果显示，*Nbr*元件通常位于共线块的中间或旁边，这些共线块通常与其他的转座类似序列成簇出现，这也暗示了它们可能与基因组的大小改变和合成的不连续性相关。对8个*Nbr*元件开放阅读框中每个非同义位点的非同义替换数目与同义位点的同义替换数目的比值进行了评估，揭示了家蚕微孢子虫的长末端重复逆转录转座子都是经过纯化选择的。这些结果强烈地暗示了逆转录转座子在微孢子虫基因组的重组过程中起着非常重要的作用。本研究初步描述了家蚕微孢子虫基因组中存在一些转座元件的特征，并且对微孢子虫基因组的进化机制提供了一些新的参考。

1　农业部蚕桑学重点实验室，西南大学，重庆
2　北京基因组研究所，中国科学院，北京
3　动物生物学实验室，重庆师范大学，重庆

Evolutionary analysis of the ubiquitin gene of baculovirus and insect hosts

Ma SS Zhang Z Xia HC Chen L
Yang YH Yao Q Chen KP*

Abstract: Baculovirus is the only virus that has been found to encode the ubiquitin protein. In this study, ubiquitin sequences from 16 insects and 49 viruses were collected and compared. The resulting sequences were aligned with virus genomes. Then MAGE 5.0, *k*-estimated software, as well as other software programs were used for systemic evolutionary, selection pressure, and evolutionary distance analysis. The results of the pairwise ratio of non-synonymous to synonymous substitution values and evolutionary distances showed that ubiquitin from baculovirus and insect hosts have been under purifying selection during evolution and are thus evolutionarily conserved. Moreover, genes from insect hosts were more conserved than those in baculovirus. Analysis of the non-synonymous to synonymous substitution rates at each site and entropy calculations revealed the evolutionary status of every site in the ubiquitin genes of baculovirus and their hosts. Genome locations and phylogenetic trees indicated that granuloviruses and non-photosynthetic vegetation evolved, and granulovirus evolution was more similar to that of insect hosts. Our results suggest that the ubiquitin gene in baculovirus may have been acquired through horizontal transfer from the host.

Published On: Genetics and Molecular Research, 2015, 14(3), 9963-9973.

Institute of Life Sciences, Jiangsu University, Zhenjiang 212013, China.
*Corresponding author E-mail: kpchen@ujs.edu.cn.

杆状病毒和昆虫宿主泛素基因的进化分析

马上上　张　朝　夏恒传　陈　亮
杨艳华　姚　勤　陈克平*

摘要：杆状病毒是唯一被发现编码泛素蛋白的病毒。在本研究中，收集并比较了来自16种昆虫和49种病毒的泛素序列。将所得序列与病毒基因组进行比对。然后使用MAGE 5.0，*k*-estimated软件以及其他软件程序进行系统进化、选择压力和进化距离的分析。从非同义替代值与同义替代值的两两比值和进化距离来看，来自杆状病毒和昆虫宿主的泛素在进化过程中一直处于纯化选择中，因此在进化上是保守的。此外，来自昆虫宿主的基因比杆状病毒中的基因更加保守。对每个位点非同义到同义替代率和熵计算的分析揭示了杆状病毒及其宿主的泛素基因中每个位点的进化状态。基因组位置和系统发育树表明了颗粒病毒和非光合植被的演化，并且颗粒体病毒进化与昆虫宿主更相似。我们的研究结果表明，杆状病毒中的泛素基因可能是从宿主水平转移获得的。

生命科学研究院，江苏大学，镇江

Characterization of mitochondrial genome of Chinese wild mulberry silkworm, *Bombyx mandarina* (Lepidoptera: Bombycidae)

Pan MH[1] Yu QY[1] Xia YL[1] Dai FY[1]
Liu YQ[2] Lu C[1*] Zhang Z[1] Xiang ZH[1]

Abstract: The complete mitochondrial genome of Chinese *Bombyx mandarina* (Ch*Bm*) was determined. The circular genome is 15 682 bp long, and contains a typical gene complement, order, and arrangement identical to that of *Bombyx mori* (*B. mori*) and Japanese *Bombyx mandarina* (Ja*Bm*) except for two additional tRNA-like structures: $tRNA^{Ser(TGA)}$-like and $tRNA^{Ile(TAT)}$-like. All protein-coding sequences are initiated with a typical ATN codon except for the *COI* gene, which has a 4-bp TTAG putative initiator codon. Eleven of 13 protein-coding genes (PCGs) have a complete termination codon (all TAA), but the remaining two genes terminate with incomplete codons. All tRNAs have the typical clover-leaf structures of mitochondrial tRNAs, with the exception of $tRNA^{Ser(TGA)}$-like, with a four stem-and-loop structure. The length of the A+T-rich region of Ch*Bm* is 484 bp, shorter than those of Ja*Bm* (747 bp) and *B. mori* (494-499 bp). Phylogenetic analysis among *B. mori*, Ch*Bm*, Ja*Bm*, and *Antheraea pernyi* (*Anpe*) showed that *B. mori* is more closely related to Ch*Bm* than Ja*Bm*. The earliest divergence time estimate for *B. mori*-Ch*Bm* and *B. mori*-Ja*Bm* is about(1.08±0.18)-(1.41±0.24) and (1.53±0.20)-(2.01±0.26) Mya, respectively. Ch*Bm* and Ja*Bm* diverged around(1.11±0.16)-(1.45±0.21) Mya.

Published On: Science In China Series C—Life Sciences, 2008, 51(8), 693-701.

1 Key Sericultural Laboratory of Agricultural Ministry, Southwest University, Chongqing 400716, China.
2 College of Bioscience and Biotechnology, Shenyang Agricultural University, Shenyang 110161,China.
*Corresponding author E-mail: lucheng@swu.edu.cn.

中国野桑蚕线粒体基因组特征

潘敏慧[1] 余泉友[1] 夏玉玲[1] 代方银[1]
刘彦群[12] 鲁 成[1*] 张 泽 向仲怀[1]

摘要：中国野桑蚕（Ch*Bm*）线粒体全基因组序列全长为15 682 bp，除了两个类似tRNA结构、tRNA$^{Ser(TGA)}$和tRNA$^{Ile(TAT)}$的结构外，还包含有与家蚕（*B. mori*）和野桑蚕（Ja*Bm*）相同的典型的基因互补、排序和排列。除*COI*基因外（含有一个TTAG假想起始密码子），所有蛋白编码基因含有典型的ATN起始密码子；13个蛋白编码基因中，有11个使用TAA作为终止密码子但其余两个基因的终止密码子不完整。除tRNA$^{Ser(TGA)}$-like具有四茎环结构外，其他的tRNA具有典型的三叶草结构。Ch*Bm*的AT富集区长度为484 bp，短于Ja*Bm*（747 bp）和家蚕（494—499 bp）。系统发育分析表明，家蚕与中国野桑蚕（Ch*Bm*）的亲缘关系较近，与Ja*Bm*的亲缘关系较远。Ch*Bm*与家蚕的分化时间约在（1.08±0.18）百万年至（1.41±0.24）百万年前，家蚕和Ja*Bm*的分化时间约在（1.53±0.20）百万年至（2.01±0.26）百万年前，Ch*Bm*和Ja*Bm*的分化时间约在（1.11±0.16）百万年至（1.45±0.21）百万年。

1 农业部蚕桑学重点实验室，西南大学，重庆
2 生物科学与生物技术学院，沈阳农业大学，沈阳

An integrated genetic linkage map for silkworms with three parental combinations and its application to the mapping of single genes and QTL

Zhan S[1#] Huang JH[1#] Guo QH[1] Zhao YP[1] Li WH[1]
Miao XX[1] Marian R Goldsmith[3] Li MW[2*] Huang YP[1*]

Abstract: *Bombyx mori*, the domesticated silkworm, is a well-studied model insect with great economic and scientific significance. Although more than 400 mutations have been described in silkworms, most have not been identified, especially those affecting economically-important traits. Simple sequence repeats (SSRs) are effective and economical tools for mapping traits and genetic improvement. The current SSR linkage map is of low density and contains few polymorphisms. The purpose of this work was to develop a dense and informative linkage map that would assist in the preliminary mapping and dissection of quantitative trait loci (QTL) in a variety of silkworm strains. Through an analysis of > 50 000 genotypes across new mapping populations, we constructed two new linkage maps covering 27 assigned chromosomes and merged the data with previously reported data sets. The integrated consensus map contains 692 unique SSR sites, improving the density from 6.3 cM in the previous map to 4.8 cM. We also developed 497 confirmed neighboring markers for corresponding low-polymorphism sites, with 244 having polymorphisms. Large-scale statistics on the SSR type were suggestive of highly efficient markers, based upon which we searched 16 462 available genomic scaffolds for SSR loci. With the newly constructed map, we mapped single-gene traits, the QTL of filaments, and a number of ribosomal protein genes. The integrated map produced in this study is a highly efficient genetic tool for the high-throughput mapping of single genes and QTL. Compared to previous maps, the current map offers a greater number of markers and polymorphisms; thus, it may be used as a resource for marker-assisted breeding.

Published On: BMC Genomics, 2009, 10, 389.

1 Shanghai Institute of Plant Physiology and Ecology, Shanghai Institutes for Biological Sciences, Chinese Academy of Sciences, Shanghai 200032, China.

2 Sericultural Research Institute, Chinese Academy of Agriculture Sciences, Zhenjiang 212018, China.

3 Biological Sciences Department, University of Rhode Island, Kingston, RI 02881, USA.

[#]These authors contributed equally.

[*]Corresponding author E-mail: yphuang@sibs.ac.cn.

家蚕三亲本组合遗传连锁图谱及其在单基因和QTL定位中的应用

詹　帅[1#]　黄建华[1#]　郭秋红[1]　赵云坡[1]　李伟华[1]
苗雪霞[1]　Marian R Goldsmith[3]　李木旺[2*]　黄勇平[1*]

摘要:家蚕是一种被广泛研究的模式昆虫,具有重要的经济和科学意义。尽管家蚕已有400多个突变体,但绝大多数尚未被鉴定,尤其是那些影响重要经济性状的突变品种。简单重复序列(SSR)是用于性状定位和遗传改良的有效而又经济的工具。目前的SSR连锁图谱密度低且具多态性少。本研究的目标是开发一个高密度和信息丰富的遗传连锁图谱,将有助于对各种家蚕品种中的数量性状基因座(QTL)的初步定位和分解。通过对新作图群体中>50 000个基因型的分析,我们构建了两个新的覆盖了27条指定染色体的连锁图谱,并将数据与之前报道的数据集合并。该图谱包含692个特异的SSR位点,将以前的图谱密度从6.3 cM提高到了4.8 cM。我们还为相应的低多态位点开发了497个确认的邻近标记,其中244个位点具有多态性。大规模的SSR类型的统计表现出高效标记,基于此我们搜索了16 462个可用于SSR位点的基因组框架。利用新构建的图谱,我们绘制了单基因的特征、丝的QTL和一些核糖体蛋白基因。这个整合图谱是一个高效的单基因定位和QTL高通量定位工具。与之前相比,它提供了更多的标记和多态性,因此,它可以用于分子标记辅助育种的资源。

1　上海植物生理生态研究所,上海生命科学研究院,中国科学院,上海
2　蚕业研究所,中国农业科学院,镇江
3　生物科学系,罗德岛大学,美国

The genome of a lepidopteran model insect, the silkworm *Bombyx mori*

The International Silkworm Genome Consortium

Abstract: *Bombyx mori*, the domesticated silkworm, is a major insect model for research, and the first lepidopteran for which draft genome sequences became available in 2004. Two independent data sets from whole genome shotgun sequencing were merged and assembled together with newly obtained fosmid- and BAC-end sequences. The remarkably improved new assembly is presented here. The 8.5-fold sequence coverage of an estimated 432 Mb genome was assembled into scaffolds with an N50 size of ~3.7 Mb; the largest scaffold was 14.5 million base pairs. With help of a high-density SNP linkage map, we anchored 87% of the scaffold sequences to all 28 chromosomes. A particular feature was the high repetitive sequence content estimated to be 43.6% and that consisted mainly of transposable elements. We predicted 14 623 gene models based on a GLEAN-based algorithm, a more accurate prediction than the previous gene models for this species. Over three thousand silkworm genes have no homologs in other insect or vertebrate genomes. Some insights into gene evolution and into characteristic biological processes are presented here and in other papers in this issue. The massive silk production correlates with the existence of specific tRNA clusters, and of several sericin genes assembled in a cluster. The silkworm's adaptation to feeding on mulberry leaves, which contain toxic alkaloids, is likely linked to the presence of new-type sucrase genes, apparently acquired from bacteria. The silkworm genome also revealed the cascade of genes involved in the juvenile hormone biosynthesis pathway, and a large number of cuticular protein genes.

Published On: Insect Biochemistry and Molecular Biology, 2008, 38(12), 1036-1045.

*Corresponding author E-mail: xbxzh@swu.edu.cn; kmita@nias.affrc.go.jp; xiaqy@swu.edu.cn; wangjian@genomics.org.cn; moris@cb.k.utokyo.ac.jp; shimada@ss.ab.a.u-tokyo.ac.jp.

鳞翅目模式昆虫家蚕的基因组

国际家蚕基因组联合会

摘要：家蚕是一个典型的昆虫研究模式生物，也是第一个完成基因组框架图序列的鳞翅目昆虫。整合中国与日本科研人员分别独立发表的2套家蚕全基因组鸟枪法测序数据，并结合fosmid和BAC末端序列，对家蚕基因组进行了重组装，基因组质量得到显著改进。结果表明，将估算的8.5倍覆盖率的432 Mb基因组组装成N50大小约为3.7 Mb的scaffold，其中最大的scaffold为14.5 Mb。利用高密度SNP连锁图谱，我们将87%的scaffold定位于家蚕全部28条染色体上。家蚕全基因组最大的特点是存在大量重复序列，大约占全基因组的43.6%，主要由转录元件组成。我们其于GLEAN算法在家蚕全基因组中共预测到14 623个基因，这个预测比之前针对这个物种的基因模型更为准确，其中3 000多个基因在其他昆虫或脊椎动物中均未发现同源序列。本文和本期的其他论文中，介绍一些对基因进化和特征生物学过程的见解。分析发现，蚕丝的生成与特殊的tRNA群及丝胶蛋白基因的成群分布相关。家蚕对以含有毒生物碱的桑叶为食的适应很可能与新型蔗糖酶基因的存在有关，这种基因显然是从细菌中获得的。此外，在家蚕全基因组中还发现了与保幼激素合成通路有关的串联分布基因及表皮蛋白相关基因。

Molecular determinants and evolutionary dynamics of wobble splicing

Lü JN[1#] Yang Y[1#] Yin H[1] Chu FJ[2]
Wang HT[1] Zhang WJ[1] Zhang YZ[2] Jin YF[1,2*]

Abstract: Alternative splicing at tandem splice sites (wobble splicing) is widespread in many species, but the mechanisms specifying the tandem sites remain poorly understood. Here, we used *synaptotagmin I* as a model to analyze the phylogeny of wobble splicing spanning more than 300 My of insect evolution. Phylogenetic analysis indicated that the occurrence of species-specific wobble splicing was related to synonymous variation at tandem splice sites. Further mutagenesis experiments demonstrated that wobble splicing could be lost by artificially induced synonymous point mutations due to destruction of splice acceptor sites. In contrast, wobble splicing could not be correctly restored through mimicking an ancestral tandem acceptor by artificial synonymous mutation *in vivo* splicing assays, which suggests that artificial tandem splice sites might be incompatible with normal wobble splicing. Moreover, combining comparative genomics with hybrid minigene analysis revealed that alternative splicing has evolved from the 3′tandem donor to the 5′ tandem acceptor in *Culex pipiens*, as a result of an evolutionary shift of *cis* element sequences from 3′ to 5′ splice sites. These data collectively suggest that the selection of tandem splice sites might not simply be an accident of history but rather in large part the result of coevolution between splice site and *cis* element sequences as a basis for wobble splicing. An evolutionary model of wobble splicingis proposed.

Published On: Molecular Biology and Evolution. 2009, 26(5), 1081-1092.

1 Institute of Biochemistry, College of Life Sciences, Zhejiang University, Hangzhou 310018, China.

2 College of Life Sciences, Zhejiang Sci-Tech University, Hangzhou 310018, China.

#These authors contributed equally.

*Corresponding author E-mail: jinyf@zju.edu.cn.

摆动剪接的分子决定因素和进化动力学

吕剑宁[1#] 杨 赟[1#] 尹 姮[1] 褚凤娇[2]
王海涛[1] 张文静[1] 张耀洲[2] 金勇丰[1,2*]

摘 要: 串联剪接位点的选择性剪接(摆动剪接)在许多物种中广泛存在,但是对指定串联位点的机制仍然知之甚少。在这里,我们使用突触结合蛋白作为模型来分析超过300种昆虫进化时的摆动剪接的系统发育。系统发育分析表明,物种特异性摆动剪接的发生与串联剪接位点的同义变异有关。进一步的诱变实验表明,由于人为诱导点突变导致剪接受体位点的破坏,致使摆动剪接丢失。相反,在体内剪接实验中,摆动剪接无法通过人为同义突变模拟原始串联受体进行正确修复,这表明人造串联剪接位点可能与正常摆动剪接不相容。此外,结合比较基因组学与混合微基因分析表明,在尖音库蚊中,由于顺式元件序列从3′到5′剪接位点的进化转变导致了选择性剪接从3′串联供体进化为5′串联受体。这些数据共同表明,串联剪接位点的选择可能不仅仅是历史的偶然,而在很大程度上是剪接位点和顺式元件序列之间作为摆动剪接基础的协同进化的结果。摆动剪接的演化模型便由此提出。

1 生物化学研究所,生命科学学院,浙江大学,杭州
2 生命科学学院,浙江理工大学,杭州

A-to-I editing sites are a genomically encoded G: Implications for the evolutionary significance and identification of novel editing sites

Tian N[1] Wu XJ[1] Zhang YZ[2] Jin YF[1,2*]

Abstract: Ribonucleic acid (RNA) editing can extend transcriptomic and proteomic diversity by changing the identity of a particular codon. Genetic recoding as a result of adenosine-to-inosine (A-to-I) RNA editing can alter highly conserved or invariant coding positions in proteins. Interestingly, examples exist in which A - to-I editing sites in one species are fixed genomically as a G in a closely related species. Phylogenetic analysis indicates that G-to-A mutations at the DNA level may be corrected by posttranscriptional A-to-I RNA editing, while in turn, the edited I (G) may be hardwired into the genome, resulting in an A-to-G mutation. We propose a model in which nuclear A-to-I RNA editing acts as an evolutionary intermediate of genetic variation. We not only provide information on the mechanism behind the evolutionary acquisition of an A-to-I RNA editing site but also demonstrate how to predict nuclear A-to-I editing sites by identifying positions where an RNA editing event would maintain the conservation of a protein relative to its homologs in other species. We identified a novel edited site in the fourth exon of the *cacophony* transcript coding calcium channel α1 and verified it experimentally.

Published On: RNA, 2008, 14(2), 211-216.

1 Institute of Biochemistry, College of Life Sciences, Zhejiang University, Hangzhou 310058, China.

2 Institute of Biochemistry, Zhejiang Sci-Tech University, Hangzhou 310018, China.

*Corresponding author E-mail: jinyf@zju.edu.cn.

A-to-I编辑位点是基因组编码的G：对新的编辑位点和进化意义的启示

田　男[1]　吴晓杰[1]　张耀洲[2]　金勇丰[1,2*]

摘要：核糖核酸（RNA）编辑可以通过改变特定密码子的属性从而拓展转录组和蛋白质组的多样性。A-to-I的RNA编辑所致的遗传编码可以改变蛋白质高度保守或不变的编码位置。有趣的是，在基因组学上一个物种及相近物种中A-to-I编辑位点是固定的G。进化学分析表明DNA水平上的G-to-A突变或许可以被转录后的A-to-I的RNA编辑修正，反过来，编辑的I（G）或许可以进入基因组中造成A-to-G的突变。我们提出一个RNA编辑的模型，即细胞核A-to-I的RNA编辑扮演着进化中遗传变异的媒介。我们不仅阐释了A-to-I的RNA编辑位点在进化上产生的机制，也展示了如何通过鉴定维持蛋白保守性的RNA编辑来预测细胞核中A-to-I编辑位点。我们在分析钙通道蛋白α1编码基因*cacophony*的第4个外显子时发现了一个新的编辑位点并得到了实验验证。

1　生物化学研究所，生命科学学院，浙江大学，杭州

2　生物化学研究所，浙江理工大学，杭州

A-to-I RNA editing alters less-conserved residues of highly conserved coding regions: Implications for dual functions in evolution

Yang Y[1] Lü JN[1] Gui B[1] Yin H[1]
Wu XJ[1] Zhang YZ[2*] Jin YF[1,2*]

Abstract: The molecular mechanism and physiological function of recoding by A-to-I RNA editing is well known, but its evolutionary significance remains a mystery. We analyzed the RNA editing of the Kv2 K^+ channel from different insects spanning more than 300 million years of evolution: *Drosophila melanogaster*, *Culex pipiens* (Diptera), *Pulex irritans* (Siphonaptera), *Bombyx mori* (Lepidoptera), *Tribolium castaneum* (Coleoptera), *Apis mellifera* (Hymenoptera), *Pediculus humanus* (Phthiraptera), and *Myzus persicae* (Homoptera). RNA editing was detected across all Kv2 orthologs, representing the most highly conserved RNA editing event yet reported in invertebrates. Surprisingly, five of these editing sites were conserved in squid (Mollusca) and were possibly of independent origin, suggesting phylogenetic conservation of editing between mollusks and insects. Based on this result, we predicted and experimentally verified two novel A-to-I editing sites in squid *synaptotagmin I* transcript. In addition, comparative analysis indicated that RNA editing usually occurred within highly conserved coding regions, but mostly altered less-conserved coding positions of these regions. Moreover, more than half of these edited amino acids are genomically encoded in the orthologs of other species; an example of a conversion model of the nonconservative edited site is addressed. Therefore, these data imply that RNA editing might play dual roles in evolution by extending protein diversity and maintaining phylogenetic conservation.

Published On: RNA, 2008, 14(8), 1516-1525.

1 Institute of Biochemistry, College of Life Sciences, Zhejiang University, Hangzhou 310058, China.
2 Institute of Biochemistry, Zhejiang Sci-Tech University, Hangzhou 310018, China.
*Corresponding author E-mail: yaozhou@chinagene.com; jinyf@zju.edu.cn.

A-to-I RNA编辑改变高度保守编码区的不保守氨基酸残基:在进化中的双重功能提示

杨　赟[1]　吕建宁[1]　桂　宾[1]　尹　姮[1]
吴晓杰[1]　张耀洲[2*]　金勇丰[1,2*]

摘要:A-to-I RNA编辑的分子机制和生理功能是众所周知的,但其进化意义仍然是一个谜。我们分析了超过3亿年的不同昆虫的Kv2 K^+通道蛋白的RNA编辑,这些昆虫包括:果蝇、淡色库蚊(双翅目)、人蚤(蚤目)、家蚕(鳞翅目)、赤拟谷盗(鞘翅目)、蜜蜂(膜翅目)、人虱(虱目)和桃蚜(同翅目)。检测的所有Kv2同源基因均发现有RNA编辑事件,说明RNA编辑在无脊椎动物中是高度保守的。令人惊讶的是,5个RNA编辑位点在鱿鱼(软体动物)中保守,可能是独立起源,提示软体动物和昆虫之间编辑的进化保守。在此基础上,我们预测和实验验证了两种新的鱿鱼突触蛋白I基因转录本的A-to-I编辑位点。此外,比较分析表明,RNA编辑通常发生在高度保守的编码区域,但主要改变这些区域的非保守编码位点。此外,这些编码氨基酸有一半以上都是由其他物种中的同系物所编码。因此,这些数据表明RNA编辑在扩大蛋白质多样性和维持系统进化保守性这两个方面发挥着双重作用。

1　生物化学研究所,生命科学学院,浙江大学,杭州
2　生物化学研究所,浙江理工大学,杭州

Complete sequence and organization of *Antheraea pernyi* nucleopolyhedrovirus, a dr-rich baculovirus

Nie ZM[1#] Zhang ZF[2#] Wang D[1#] He PA[1] Jiang CY[1] Song L[1] Chen F[1] Xu J[1]
Yang L[2] Yu LL[2] Chen J[1] Lü ZB[1] Lu JJ[1] Wu XF[1] Zhang YZ[1*]

Abstract:

Background: The completion and reporting of baculovirus genomes is extremely important as it advances our understanding of gene function and evolution. Due to the large number of viral genomes now sequenced it is very important that authors present significantly detailed analyses to advance the understanding of the viral genomes. However, there is no report of the *Antheraea pernyi* nucleopolyhedrovirus (*Anpe*NPV) genome.

Results: The genome of *Anpe*NPV, which infects Chinese tussah silkworm (*Antheraea pernyi*), was sequenced and analyzed. The genome was 126 629 bp in size. The G+C content of the genome, 53.4%, was higher than that of most of the sequenced baculoviruses. 147 open reading frames (ORFs) that putatively encode proteins of 50 or more amino acid residues with minimal overlap were determined. Of the 147 ORFs, 143 appeared to be homologous to other baculovirus genes, and 4 were unique to *Anpe*NPV. Furthermore, there are still 29 and 33 conserved genes present in all baculoviruses and all lepidopteran baculoviruses respectively. In addition, the total number of genes common to all lepidopteran NPVs is still 74, however the 74 genes are somewhat different from the 74 genes identified before because of some new sequenced NPVs. Only 6 genes were found exclusively in all lepidopteran NPVs and 12 genes were found exclusively in all Group 1 NPVs. *Anpe*NPV encodes *v-trex*(Anpe115, a 3′ to 5′ repair exonuclease), which was observed only in *CfM*NPV and *CfDEF*NPV in Group 1 NPVs. This gene potentially originated by horizontal gene transfer from an ancestral host. In addition, *Anpe*NPV encodes two *conotoxin*-like gene homologues (*ctls*), *ctl1* and *ctl2*, which were observed only in *Hycu*NPV, *OpM*NPV and *LdM*NPV. Unlike other baculoviruses, only 3 typical homologous regions (*hrs*)were identified containing 2~9 repeats of a 30 bp-long palindromic core. However, 24 perfect or imperfect direct repeats(*drs*) with a high degree of AT content were found within the intergenic spacer regions that may function as *non-hr*, *ori-like* regions found in *Grle*GV, *Cp*GV and *Ador*GV. 9 *drs* were also found in intragenic spacer regions of *Anpe*NPV.

Conclusion: AnpeNPV belongs to Group 1 NPVs and is most similar to *Hycu*NPV, *Eppo*NPV, *OpM*NPV and *CfM*NPV based on gene content, genome arrangement, and amino acid identity. In addition, analysis of genes that flank *hrs* supported the argument that these regions are involved in the transfer of sequences between the virus and host.

Published On: BMC Genomics, 2007, 8.

1 Institute of Biochemistry, Zhejiang Sci-Tech University, Hangzhou 310018, China.

2 Biotechnology Research Institute, Chinese Academy of Agricultural Sciences, Beijing 100081, China.

[#]These authors contributed equally.

[*]Corresponding author E-mail: yaozhou@zstu.edu.cn.

柞蚕核型多角体病毒的全基因组序列和组成，一种富含dr区的杆状病毒

聂作明[1#] 张志芳[2#] 王 丹[1#] 贺平安[1] 蒋彩英[1] 宋 力[1] 陈 芳[1] 徐 杰[1]
杨 灵[2] 于琳琳[2] 陈 健[1] 吕正兵[1] 陆晶晶[1] 吴祥甫[1] 张耀洲[1*]

摘要：背景——杆状病毒基因组的完成和报道非常重要，因为它促进了我们对基因功能和进化的理解。现在对大量的病毒基因组进行了测序，并提供了非常重要的详细分析，以提高对病毒基因组的理解。然而，目前还没有柞蚕核型多角体病毒（*Anpe*NPV）基因组的报道。结果——本论文对中国柞蚕（*Antheraea pernyi*）感染的*Anpe*NPV的基因组进行了测序和分析。基因组大小为126 629 bp。基因组的G + C含量高达53.4%，高于大多数测序的杆状病毒。在基因组上鉴定出编码50个或更多氨基酸残基的蛋白质的开放阅读框（ORFs）147个。在147个ORF中，143个可能与其他杆状病毒基因同源，4个是*Anpe*NPV独有的。此外，仍有29个和33个保守基因分别存在于所有杆状病毒和所有鳞翅目杆状病毒中。所有鳞翅目杆状病毒共有的基因总数为74个，一些新的杆状病毒被测序，74个共有基因与之前鉴定的74个基因有一定的差异。只有6个基因在所有鳞翅目NPV中被完全发现。12个基因仅在所有1组NPV中发现。*Anpe*NPV编码*v-trex*（Anpe115，3′至5′修复核酸外切酶），其仅在1组NPV的*CfM*NPV和*CfDEF*NPV中发现，该基因可能起源于宿主基因的水平转移。此外，*Anpe*NPV编码两种类毒素样基因同源物（*ctl*），*ctl1*和*ctl2*，仅在*Hycu*NPV、*OpM*NPV和*LdM*NPV中发现。与其他杆状病毒不同的是，*Anpe*NPV只存在3个典型的同源区段（*hrs*），其被鉴定含有2—9个重复的30 bp长的回文核心序列。然而，在基因间隔区内发现了具有高度AT含量的24个完全或不完全直接重复区（*drs*），其具有潜在的*non-hr*，*ori*-like功能，且只在*Grle*GV、*Cp*GV和*Ador*GV中发现。另外，在*Anpe*NPV的基因内部也发现了9 *drs*。结论——基于基因内容、基因组排列和氨基酸鉴定发现，*Anpe*NPV属于1组NPVs，与*Hycu*NPV、*Eppo*NPV、*OpM*NPV和*CfM*NPV最相似。此外，分析*hrs*区侧翼基因发现，这些基因片段可能涉及病毒和宿主之间的序列转移。

1 生物化学研究所，浙江理工大学，杭州
2 生物技术研究所，中国农业科学院，北京

Microarray-based gene expression profiles in multiple tissues of the domesticated silkworm, *Bombyx mori*

Xia QY[1] Cheng DJ[1] Duan J[1] Wang GH[1] Cheng TC[1] Zha XF[1]
Liu C[1] Zhao P[1] Dai FY[1] Zhang Z[1] He NJ[1] Zhang L[2*] Xiang ZH[1*]

Abstract: We designed and constructed a genome-wide microarray with 22 987 70-mer oligonucleotides covering the presently known and predicted genes in the silkworm genome, and surveyed the gene expression in multiple silkworm tissues on day 3 of the fifth instar. Clusters of tissue-prevalent and tissue-specific genes and genes that are differentially expressed in different tissues were identified, and they reflect well major tissue-specific functions on the molecular level. The data presented in this study provide a new resource for annotating the silkworm genome.

Published On: Genome Biology, 2007, 8(8).

1 Key Sericultural Laboratory of Agricultural Ministry, Key Laboratory for Sericultural Sciences and Genomics of the Ministry of Education, College of Biotechnology, Southwest University, Chongqing 400716, China.

2 National Engineering Center for Beijing Biochip Technology, Life Science Parkway, Beijing 102206, China.

*Corresponding author E-mail: lzhang@capitalbio.com; xbxzh@swu.edu.cn

家蚕不同组织中基因表达特征的生物芯片研究

夏庆友[1] 程道军[1] 段 军[1] 王根洪[1] 程廷才[1] 查幸福[1]
刘 春[1] 赵 萍[1] 代方银[1] 张 泽[1] 何宁佳[1] 张 亮[2*] 向仲怀[1*]

摘要：本研究设计并构建了覆盖家蚕基因组中已知和预测基因的全基因组芯片，共计22 987个70-mer寡核苷酸，并在5龄第3天对家蚕多个组织中的基因表达进行了检测。鉴定了组织普遍和组织特异表达的基因，以及在不同组织中差异表达基因，它们在分子水平上很好地反映了主要的组织特异性功能。研究提供的数据为家蚕基因组注释提供了新的资源。

1 农业部蚕桑学重点实验室，教育部蚕学与基因组学重点实验室，生物技术学院，西南大学，重庆
2 国家生物芯片技术中心，生命科学园，北京

From genome to proteome: great progress in the domesticated silkworm (*Bombyx mori* L.)

Zhou ZH Yang HJ Zhong BX*

Abstract: As the only truly domesticated insect, the silkworm not only has great economic value, but it also has value as a model for genetics and molecular biology research. Genomics and proteomics have recently shown vast potential to be essential tools in domesticated silkworm research, especially after the completion of the *Bombyx mori* genome sequence. This paper reviews the progress of the domesticated silkworm genome, particularly focusing on its genetic map, physical map and functional genome. This review also presents proteomics, the proteomic technique and its application in silkworm research.

Published On: Acta Biochimical et Biophysica Sinica, 2008, 40(7), 601-611.

College of Animal Sciences, Zhejiang University, Hangzhou 310029, China.
*Corresponding author E-mail: bxzhong@zju.edu.cn.

从基因组到蛋白质组：家蚕重大进展

周仲华　杨惠娟　钟伯雄*

摘要：家蚕作为唯一真正驯化的昆虫，不仅具有巨大的经济价值，同时也具有遗传学和分子生物学研究模型的价值。近年来，基因组学和蛋白质组学在家蚕研究上表现出了巨大的潜力，尤其是在家蚕基因组序列绘制完成以后。本文阐述了家蚕基因组研究进展，着重落脚于其遗传图谱、物理图谱和功能基因组。同时也展示了蛋白质组学、蛋白质组学技术及其在家蚕研究上的应用。

动物科学学院，浙江大学，杭州

Reference genes identified in the silkworm *Bombyx mori* during metamorphism based on oligonucleotide microarray and confirmed by qRT-PCR

Wang GH[1,2] Xia QY[1,2*] Cheng DJ[2] Duan J[1,2]

Zhao P[2] Chen J[2] Zhu L[2]

Abstract: Gene expression quantification at mRNA level is very important for postgenomic studies, as gene expression level is the reflection of the special biological function of the target gene. Methods used for gene expression quantification, such as microarray or quantitative real-time polymerase chain reaction (qRT-PCR), require stable expressed reference genes. Thus, finding suitable control genes is essential for gene quantification. In this study, a genome-wide survey of reference genes during metamorphism was performed on silkworm *Bombyx mori*. Twelve genes were chosen as putative reference genes based on a whole genome oligonucleotide microarray normalized by external controls. Then, qRT-PCR was employed for further validation and selection of potential reference gene candidates. The results were analyzed, and stable genes were selected using geNorm 3.4 and NormFinder software. Finally, considering factors from every aspect, translation initiation factor 4A, translation initiation factor 3 subunit 4, and translation initiation factor 3 subunit 5 (represented by *sw22934*, *sw14876*, and *sw13956*) were selected as reliable internal controls across the examined developmental stages, while cytoplasmic actin (*sw22671*), the commonly used reference gene in a previous study was shown to vary drastically throughout the examined developmental stages. For future research, we recommend the use of the geometric mean of those three stable reference genes as an accurate normalization factor for data normalization of different developmental stages during metamorphism.

Published On: Insect Science, 2008, 15(5), 405 - 413.

1 Institute of Agriculture and Life Sciences, Chongqing University, Chongqing 400030, China.

2 Key Sericultural Laboratory of Agricultural Ministry, School of Biotechnology, Southwest University, Chongqing 400716, China.

*Corresponding author E-mail: xiaqy@swu.edu.cn.

基于家蚕芯片及荧光定量PCR技术对家蚕表达发育时期看家基因的鉴定与筛选

王根洪[1,2] 夏庆友[1,2*] 程道军[2] 段 军[1,2]
赵 萍[2] 陈 杰[2] 祝 力[2]

摘要：基因表达水平的定量分析是基因功能研究的重要组成部分，而在大多数基因定量分析中，选择合适的看家基因决定了定量结果的可靠性。本文描述了基于家蚕全基因组DNA芯片及荧光定量PCR技术，筛选家蚕变态发育阶段表达水平相对稳定的看家基因的过程。我们首先对家蚕各个发育阶段的基因表达水平进行了全基因组DNA芯片扫描，选择了12个可能的候选看家基因；然后，通过荧光定量PCR技术，对这些基因的表达水平进行确认；最后，通过表达水平稳定性专业分析软件geNorm 3.4和NormFinder，确定了在家蚕变态发育阶段3个表达水平最稳定的基因translation initiation factor 4A，translation initiation factor 3 subunit 4和translation initiation factor 3 subunit 5（分别表示为*sw22934*，*sw14876*，*sw13956*）；而在此前的基因表达分析中，常被作为看家基因的cytoplasmic actin（*sw22671*），在变态发育过程中表达变化非常明显。据此，在以后的基因表达分析中，我们推荐采用此3个基因表达水平的几何平均数作为基因表达的归一化因子。

1 农学及生命科学研究院，重庆大学，重庆
2 农业部蚕桑学重点实验室，生物技术学院，西南大学，重庆

In silico identification of silkworm selenoproteomes

Chen P [1,2,3] Duan J [1] Jiang L [2] Liu Q [2]

Zhao P [1] Xia QY [1*] Xu HB [3]

Abstract: Selenium (Se) is an essential trace element *in vivo*. Its biological function is mainly exerted through selenoproteins. Selenocysteine (Sec), the active site of selenoproteins, is incorporated into the protein at an in-frame TGA codon under the guidance of Sec insertion sequence (SECIS) element in the 3′ untranslated region (UTR) of the gene. In this work, a method was developed and a series of programs were edited by PERL language to *in silico* identify selenoproteomes from the genome of domesticated silkworm (*Bombyx mori*). Out of 18 510 annotated genes, 6 348 was terminated with TGA codons, 249 containing both in-frame TGAs and SECIS elements in the 3′ UTRs. Alignments of those selenoprotein candidates with their cysteine (Cys)-containing homologs revealed that 52 genes had TGA/Cys pairs and similar flanking regions around the in-frame TGAs. Restricted by the patterns of SECIS elements only 5 genes were screened out to fully meet the requirements for selenoproteins. Among them glutathione S-transferase (GST) has been reported as a microbial selenoprotein, the other four are novel selenoproteins annotated as CG6024, CG5195, ATP-binding cassette transporter subfamily A (ABCA), and nuclear VCP-like protein. Derived from the general properties of GST, ABCA and VCP, silkworm selenoproteins may play important roles in redox regulation, Se storage and transportation, as well as cell apoptosis.

Published On: Chinese Science Bulletin, 2006, 51(23), 2860-2867.

1 Key Sericultural Laboratory of Agricultural Ministry, Southwest University, Chongqing 400716, China.

2 College of Life Science, Shenzhen University, Shenzhen 518060, China.

3 Department of Chemistry, Huazhong University of Science and Technology, Wuhan 430074, China.

*Corresponding author E-mail: xiaqy@swau.cq.cn.

家蚕硒蛋白组的计算机识别

陈 萍[1,2,3] 段 军[1] 蒋 亮[2] 刘 琼[2]
赵 萍[1] 夏庆友[1*] 徐辉碧[3]

摘要：硒是一种主要以硒蛋白形式发挥生物学功能的生物必需微量元素。硒代半胱氨酸(Selenocysteine，Sec)是硒蛋白的活性位点，它是在基因的3′末端非翻译区(UTR)，在Sec插入序列元件(SECIS)的指引下，被整合到硒蛋白阅读框内TGA的密码子处。TGA码既是终止码，又可翻译成硒代半胱氨酸，这使普通基因注释软件无法正确预测硒蛋白，导致现有数据库中许多物种的硒蛋白被错误注释或丢失。本研究基于已公布的家蚕基因组预测信息，采用PERL语言编程，对家蚕基因组中硒蛋白进行了计算机检索与分析，结果表明，在家蚕数据库18 510个已注释基因中，以TGA码终止的基因有6 348条，其中249条的3′UTR含有阅读框内TGAs和SECIS指引序列。对含半胱氨酸(Cys)的硒蛋白候选基因进行比对，发现了52基因具有TGA/Cys时，在阅读框内TGAs周围有相似的侧翼区域。受SECIS元件的限定，最终检索到完全具备硒蛋白特点的基因有5条，其中谷胱甘肽硫转移酶(GST)是一种已知微生物硒蛋白，而其他4种则是新硒蛋白，分别在已有基因表中被注释为CG6024蛋白、CG5195蛋白、ATP-结合盒转运蛋白A型(ABCA)和核VCP相似蛋白。通过对GST、ABCA和VCP主要性质的分析，推测家蚕硒蛋白在氧化调节、硒储存运输和细胞凋亡等过程中起重要作用。

1 农业部蚕桑学重点实验室，西南大学，重庆
2 生命科学学院，深圳大学，深圳
3 化学学院，华中科技大学，武汉

Aligning the proteome and genome of the silkworm, *Bombyx mori*

Zhang YZ [1#*] Xia QY [2#*] Xu J [1] Chen J [1] Nie ZM [1] Wang D [1] Zhang WP [1]
Chen JQ [1] Zheng QL [1] Chen Q [1] Kong LY [1] Ren XY [1] Wang J [1]
Lü ZB [1] Yu W [1] Jiang CY [1] Liu LL [1] Sheng Q [1] Jin YF [3] Wu XF [1]

Abstract: A technology of mass spectrometry (MS) was used in this study for the large-scale proteomic identification and verification of protein-encoding genes present in the silkworm (*Bombyx mori*) genome. Peptide sequences identified by MS were compared with those from an open reading frame (ORF) library of the *B. mori* genome and a cDNA library, to validate the coding attributes of ORFs. Two databases were created. The first was based on a 9× draft sequence of the silkworm genome and contained 14 632 putative proteins. The second was based on a *B. mori* pupal cDNA library containing 3 187 putative proteins of at least 30 amino acid residues in length. A total of 81 000 peptide sequences with a threshold score of 60% were generated by the MS/MS analysis, and 55 400 of these were chosen for a sequence alignment. By searching these two databases, 6 649 and 250 proteins were matched, which accounted for approximately 45.4% and 7.8% of the peptide sequences and putative proteins, respectively. Further analyses carried out by several bioinformatic tools suggested that the matches included proteins with predicted transmembrane domains (1 393) and preproteins with a signal peptide (976). These results provide a fundamental understanding of the expression and function of silkworm proteins.

Published On: Functional & Integrative Genomics, 2009, 9(4), 447-454.

1 The Key Laboratory of Bioreactor and Biopharmacy of Zhejiang Province, Institute of Biochemistry, Zhejiang Sci-Tech University, Hangzhou 310018, China.

2 Key Sericultural Laboratory of Agricultural Ministry, The Key Laboratory for Sericultural Sciences and Genomics of the Ministry of Education, College of Biotechnology, Southwest University, Chongqing 400716, China.

3 College of Life Sciences, Zhejiang University, Hangzhou 310029, China.

[#]These authors contributed equally.

[*]Corresponding author E-mail: yaozhou@chinagene.com; xiaqy@swu.edu.cn.

家蚕的蛋白质组和基因组比较

张耀洲[1#*] 夏庆友[2#*] 徐 杰[1] 陈 健[1] 聂作明[1] 王 丹[1] 张文平[1]
陈剑清[1] 郑青亮[1] 陈 琴[1] 孔令印[1] 任小元[1] 王 江[1] 吕正兵[1]
于 威[1] 蒋彩英[1] 刘立丽[1] 盛 清[1] 金勇丰[3] 吴祥甫[1]

摘要:本研究使用质谱技术(MS)对家蚕基因组中的蛋白质编码基因进行大规模蛋白质组鉴定和验证。将MS鉴定的肽段与来自家蚕基因组的开放阅读框(ORF)文库和cDNA文库编码的序列进行比较,以验证ORF的编码属性。创建了两个数据库。第一个是基于家蚕9倍基因组草图序列,包含14 632个预测的蛋白质。第二个是基于家蚕蛹cDNA文库,含有3 187个长度至少为30个氨基酸残基的预测蛋白质。通过MS/MS分析产生了总共81 000个阈值得分为60%的肽段序列,其中55 400个被选择用于序列比对。通过搜索这两个数据库,匹配了6 649个和250个蛋白质,分别占肽段序列和预测蛋白质的约45.4%和7.8%。进一步综合多个软件工具进行生物信息学分析,结果表明匹配的蛋白质包括具有预测跨膜结构域的蛋白质(1 393个)和具有信号肽的蛋白质前体(976个)。这些结果为了解家蚕蛋白质的表达和功能提供了基础。

1 浙江省家蚕生物反应器和生物医药重点实验室,生物化学研究所,浙江理工大学,杭州
2 农业部蚕桑学重点实验室,教育部家蚕基因组学重点实验室,生物技术学院,西南大学,重庆
3 生命科学学院,浙江大学,杭州

Transcriptome sequencing and positive selected genes analysis of *Bombyx mandarina*

Cheng TC[#] Fu BH[#] Wu YQ

Long RW Liu C Xia QY[*]

Abstract: The wild silkworm *Bombyx mandarina* is widely believed to be an ancestor of the domesticated silkworm, *Bombyx mori.* Silkworms are often used as a model for studying the mechanism of species domestication. Here, we performed transcriptome sequencing of the wild silkworm using an Illumina HiSeq2000 platform. We produced 100 004 078 high-quality reads and assembled them into 50 773 contigs with an N50 length of 1 764 bp and a mean length of 941.62 bp. A total of 33 759 unigenes were identified, with 12 805 annotated in the Nr database, 8 273 in the Pfam database, and 9 093 in the Swiss-Prot database. Expression profile analysis found significant differential expression of 1 308 unigenes between the middle silk gland (MSG) and posterior silk gland (PSG). Three sericin genes (*sericin 1*, *sericin 2*, and *sericin 3*) were expressed specifically in the MSG and three fibroin genes (*fibroin-H*, *fibroin-L*, and *fibroin / P25*) were expressed specifically in the PSG. In addition, 32 297 Single-nucleotide polymorphisms (SNPs) and 361 insertion-deletions (INDELs) were detected. Comparison with the domesticated silkworm *p50 / Dazao* identified 5 295 orthologous genes, among which 400 might have experienced or to be experiencing positive selection by Ka / Ks analysis. These data and analyses presented here provide insights into silkworm domestication and an invaluable resource for wild silkworm genomics research.

Published On: PLoS ONE, 2015, 10(3).

State Key Laboratory of Silkworm Genome Biology, Southwest University, Chongqing 400716, China.

[#]These authors contributed equally.

[*]Corresponding author E-mail: xiaqy@swu.edu.cn.

野蚕转录组测序和正向选择基因的分析

程廷才[#]　付博华[#]　吴玉乾
龙仁文　刘　春　夏庆友[*]

摘要：野蚕(*Bombyx mandarina*)被广泛认为是家蚕(*Bombyx mori*)的祖先。家蚕经常被用作研究物种驯化机制的模型。在这里,我们使用Illumina HiSeq2000平台对野蚕进行转录组测序。我们得到了100 004 078个高质量的读长,并将它们组装成50 773个重叠群且N50长度为1 764 bp,平均长度为941.62 bp。共鉴定了33 759个单基因簇,其中包括在Nr数据库中注释的12 805个,Pfam数据库中的8 273个和Swiss-Prot数据库中的9 093个。表达谱分析发现中部丝腺(MSG)和后部丝腺(PSG)之间有1 308个单基因簇存在显著的差异表达。3个丝胶蛋白基因(*sericin 1*,*sericin2*和*sericin 3*)在中部丝腺中特异性表达,3个丝素蛋白基因(*fibroin-H*,*fibroin-L*和*fibroin/P25*)在后部丝腺中特异表达。此外,检测到32 297个单核苷酸多态性和361个插入缺失。使用Ka/Ks分析,通过与家蚕*p50 / Dazao*鉴定到的5 295个同源基因比较,其中有400个基因可能经历或正在经历正向选择。这些数据和分析提供了对家蚕驯化的见解和野蚕基因组学研究的宝贵资源。

家蚕基因组生物学国家重点实验室,西南大学,重庆

The complete mitochondrial genome of the Chinese oak silkmoth, *Antheraea pernyi* (Lepidoptera: Saturniidae)

Liu YQ[1,2] Li YP[2] Pan MH[1] Dai FY[1]
Zhu XW[3] Lu C[1*] Xiang ZH[1]

Abstract: We determined the complete nucleotide sequence of the mitogenome from Chinese oak silkmoth, *Antheraea pernyi* (Lepidoptera: Saturniidae). The 15 566 bp circular genome contains a typical gene organization and order for lepidopteran mitogenomes. The mitogenome contains the lowest A+T content (80.16%) among the known lepidopteran mitogenome sequences. An unusual feature is the occurrence of more Ts than As, with a slightly negative AT skewness (-0.021), in the composition of the major genome strand. All protein-coding genes are initiated by ATN codons, except for cytochrome oxidase subunit I, which is proposed by the TTAG sequence as observed in other lepidopterans. All transfer RNAs (tRNAs) have a typical clover-leaf structure of mitochondrial tRNA, except for $tRNA^{Ser}$(AGN), the DHU arm of which could not form a stable stem-loop structure. Two aligned sequence blocks with a length of more than 50 bp and 90% of the sequence identity were identified in the A+T-rich region of the Saturniidae and Bombycoidae species.

Published On: Acta Biochimica et Biophysica Sinica (Shanghai), 2008, 40(8), 693-703.

1 The Key Sericultural Laboratory of Agricultural Ministry, Southwest University, Chongqing 400716, China.
2 Department of Sericulture, College of Bioscience and Biotechnology, Shenyang Agricultural University, Shenyang 110161, China.
3 Sericultural Experiment Station of Henan Province, Nanyang 474676, China.
*Corresponding author E-mail: lucheng@swu.edu.cn.

中国柞蚕完整线粒体基因组

刘彦群[1,2] 李玉萍[2] 潘敏慧[1] 代方银[1]
朱绪伟[3] 鲁 成[1*] 向仲怀[1]

摘要：我们鉴定了中国柞蚕完整的线粒体核苷酸序列(鳞翅目:天蚕蛾科)。柞蚕线粒体环形基因组长15 566 bp,其包含了鳞翅目线粒体基因组典型的基因组成和排列。在已知的鳞翅目昆虫线粒体基因组中A+T含量是最低的(80.16%)。一个不同寻常的特点是主要基因链的碱基T含量高于A,AT偏度为-0.021。除了细胞色素氧化酶亚基I,蛋白质编码基因都以ATN作为起始密码子,而其他鳞翅目中推测细胞色素氧化酶亚基I以TTAG作为假想起始密码子。除了tRNA^{Ser}(AGN),所有转移RNA(tRNA)有一个典型线粒体tRNA的三叶草结构,tRNA^{Ser}(AGN)的DHU臂不能形成一个稳定的茎环结构。在柞蚕和野桑蚕物种的富含A+T的区域中鉴定出两个比对序列,其长度超过50 bp,且具有90%的序列同一性。

1 农业部蚕桑学重点实验室,西南大学,重庆
2 蚕业研究所,生物科学技术学院,沈阳农业大学,沈阳
3 河南省蚕业科学研究院,南阳

RNA recognition motif (RRM)-containing proteins in *Bombyx mori*

Wang LL Zhou ZY*

Abstract: RNA recognition motif (RRM)-containing proteins play important roles in the processing of RNA and regulation of protein synthesis. *Bombyx mori* is a model species of Lepidoptera whose genome sequence is available; however, its RRM-containing proteins have not been thoroughly analyzed. In this study, 134 RRM-containing protein genes in *Drosophila melanogaster* were used to perform BLAST search against *B. mori* genome- and EST databases. Results showed that 123 genes have orthologues in the *B. mori* genome database and 89 in *B. mori* EST database, indicating some of the genes were not expressed in *B. mori*. 23 RRM-containing protein genes that have complete open reading frame were obtained through contig assembly in *B. mori* EST database and then registered in NCBI. Cluster analysis showed that these 23 new proteins formed two distinct clades. Especially, TAR DNA-binding protein-43 (TBPH) gene of *B. mori* is different from that of other species in that it does not have introns. Moreover, the length of *TBPH* of *B. mori* is much shorter than that of other species' *TBPH*, suggesting that it may play some roles in emergency responses.

Published On: African Journal of Biotechnology, 2009, 8(6), 1121-1126.

College of Life Sciences, Chongqing Normal University, Chongqing 400047, China.
*Corresponding author E-mail: zyzhou@cqnu.edu.cn

家蚕RNA识别结构域富含蛋白鉴定

王林玲　周泽扬*

摘要：含RNA识别结构域(RRM)的蛋白质在RNA处理和调节蛋白质合成中起到重要作用。家蚕作为基因组序列已知的鳞翅目昆虫模式生物，其RRM蛋白还没有被精细地分析过。在本研究中，利用134个果蝇的RRM富含蛋白基因在家蚕的基因组和EST数据库中进行比对分析。结果显示，其在家蚕基因组数据库中有123个同源基因，而在EST数据库中有89个同源基因，这也说明其中一些基因在家蚕中是不表达的。对EST数据库中片段重叠群进行组装，得到23个含有完整开放阅读框的RRM富含蛋白基因，并将序列提交至NCBI。聚类分析显示这23个蛋白属于两个进化分支。特别需要指明的是，家蚕TAR DNA结合蛋白43(TBPH)基因与其他物种有所不同，其不含有内含子。此外，相比其他物种，家蚕*TBPH*基因序列尤其短小，暗示其可能参与家蚕的应急反应。

生命科学学院，重庆师范大学，重庆

A genomewide survey of homeobox genes and identification of novel structure of the *Hox* cluster in the silkworm, *Bombyx mori*

Chai CL Zhang Z Huang FF Wang XY Yu QY
Liu BB Tian T Xia QY Lu C* Xiang ZH

Absract: Homeobox genes encode transcriptional factors that play crucial roles in a variety of developmental pathways from unicellular to multicellular eukaryotes. We have identified 102 homeobox genes in the typical insect of Lepidoptera, *Bombyx mori*, based on the newly assembled genome sequence with 9 × coverage. These identified homeobox genes were categorized into nine classes including at least 74 families. The available ESTs and microarray data at present confirmed that more than half of them were expressed during silkworm developmental processes. Orthologs of *pb*, *zen* and *ftz* were newly identified in the *Bombyx Hox* cluster on chromosome 6. Interestingly, a special group of 12 tandemly duplicated homeobox genes was found located between *Bmpb* and *Bmzen* in the *Bombyx Hox* cluster, suggesting that Hox cluster might have experienced a lineagespecific expansion in the silkworm. A detailed analysis on genome data reveals that a split exists between *Bmlab* and *Bmpb*. Our data provide valuable information for future research on the development and evolution of silkworm.

Published On: Insect Biochemistry and Molecular Biology, 2008, 38(12), 1111-1120.

Key Sericultural Laboratory of Agricultural Ministry, College of Biotechnology, Institute of Sericulture and Systems Biology, Southwest University, Chongqing 400716, China.
*Corresponding author E-mail: lucheng@swu.edu.cn.

家蚕同源异型盒基因的全基因组检测以及家蚕*Hox*基因新结构的鉴定

柴春利　张　泽　黄飞飞　王先燕　余泉友
刘彬斌　田　甜　夏庆友　鲁　成*　向仲怀

摘要:同源异形盒编码的转录因子可以在从单细胞到多细胞组成的真核生物的各种发育过程中发挥关键作用。以9倍覆盖度的新组装的基因组序列为基础,在经典的鳞翅目昆虫家蚕中我们鉴定了102种典型的同源异形盒基因。这些同源异形盒基因被分成9类至少包含74个家族。可获得的序列标签和数据证明了在家蚕的发育过程中超过一半的同源异形盒基因都有表达,最新鉴定出*pb*、*zen*和*ftz*基因的直系同源物定位于家蚕6号染色体上*Hox*基因簇中。有趣的是,目前发现有一组12个串联的同源异形盒基因位于家蚕*Hox*基因的*Bmpb*和*Bmzen*之间,这表明在家蚕体内的*Hox*基因簇存在一个特殊的线性扩增。这为以后对家蚕发育和进化的研究提供了线索。

农业部蚕桑学重点实验室,生物技术学院,蚕业与系统生物学研究所,西南大学,重庆

SilkDB v2.0: a platform for silkworm (*Bombyx mori*) genome biology

Duan J[1,2] Li RQ[3] Cheng DJ[1] Fan W[3] Zha XF[1] Cheng TC[1,2]
Wu YQ[2] Wang J[3] Kazuei Mita[4] Xiang ZH[1] Xia QY[1,2*]

Abstract: The SilkDB is an open-access database for genome biology of the silkworm (*Bombyx mori*). Since the draft sequence was completed and the SilkDB was first released 5 years ago, we have collaborated with other groups to make much remarkable progress on silkworm genome research, such as the completion of a new high-quality assembly of the silkworm genome sequence as well as the construction of a genome-wide microarray to survey gene expression profiles. To accommodate these new genomic data and house more comprehensive genomic information, we have reconstructed SilkDB database with new web interfaces. In the new version (v2.0) of SilkDB, we updated the genomic data, including genome assembly, gene annotation, chromosomal mapping, orthologous relationship and experiment data, such as microarray expression data, Expressed Sequence Tags (ESTs) and corresponding references. Several new tools, including SilkMap, Silkworm Chromosome Browser (SCB) and BmArray, are developed to access silkworm genomic data conveniently. SilkDB is publicly available at the new URL of http://www.silkdb.org.

Published On: Nucleic Acids Research, 2010, 38, 453-456.

1 Key Sericultural Laboratory of Agricultural Ministry, Southwest University, Chongqing 400716, China.
2 The Institute of Agriculture and Life Sciences, Chongqing University, Chongqing 400030, China.
3 Beijing Genomics Institute at Shenzhen, Shenzhen 518083, China.
4 National Institute of Agrobiological Sciences, Owashi 1-2, Tsukuba 305-8634, Japan.
*Corresponding author E-mail: xiaqy@swu.edu.cn.

SilkDB v2.0: 家蚕(*Bombyx mori*)基因组生物学平台

段　军[1,2]　李瑞强[3]　程道军[1]　樊　伟[3]　查幸福[1]　程廷才[1,2]
吴玉乾[2]　王　俊[3]　Kazuei Mita[4]　向仲怀[1]　夏庆友[1,2*]

摘要：SilkDB是一个开放式的家蚕基因组生物学数据库。自从完成家蚕基因组框架图和SilkDB首次发布以来，本研究团队与其他团队合作，在家蚕基因组研究上取得了重大的进展，例如家蚕基因组精细图的完成，以及全基因组基因表达谱芯片的构建。为了整合这些新的基因组数据并提供更全面的基因组信息，我们使用新的Web界面对SilkDB进行了重建。在SilkDB的新版本(v2.0)中，我们更新了基因组数据，包括基因组组装、基因注释、染色体定位、直系同源关系以及实验数据，如基因芯片数据、表达序列标签(EST)和相应的参考文献。一些新的工具也整合到数据库中，如SilkMap，家蚕染色体的浏览器(SCB)和BmArray等，为访问家蚕基因组数据提供了方便。SilkDB可以通过新的网址http://www.silkdb.org进行公开访问。

1　农业部蚕桑学重点实验室，西南大学，重庆
2　农学及生命科学研究院，重庆大学，重庆
3　北京基因组研究所(深圳)，深圳
4　国立农业科学研究所，筑波，日本

Genomic sequence, organization and characteristics of a new nucleopolyhedrovirus isolated from *Clanis bilineata* larva

Zhu SY[1] Yi JP[2] Shen WD[3] Wang LQ[1]
He HG[4] Wang Y[4] Li B[3] Wang WB[1*]

Abstract: Baculoviruses are well known for their potential as biological agents for controlling agricultural and forest pests. They are also widely used as expression vectors in molecular cloning studies. The genome sequences of 48 baculoviruses are currently available in NCBI databases. As the number of sequenced viral genomes increases, it is important for the authors to present sufficiently detailed analyses and annotations to advance understanding of them. In this study, the complete genome of *Clanis bilineata* nucleopolyhedrovirus (*Clbi*NPV) has been sequenced and analyzed in order to understand this virus better. Results: The genome of *Clbi*NPV contains 135 454 base pairs (bp) with a G+C content of 37%, and 139 putative open reading frames (ORFs) of at least 150 nucleotides. One hundred and twenty-six of these ORFs have homologues with other baculovirus genes while the other 13 are unique to *Clbi*NPV. The 30 baculovirus core genes are all present in *Clbi*NPV. Phylogenetic analysis based on the combined *pif-2* and *lef-8* sequences places *Clbi*NPV in the Group II Alphabaculoviruses. This result is consistent with the absence of *gp* 64 from the *Clbi*NPV genome and the presence instead of a fusion protein gene, characteristic of Group II. Blast searches revealed that *Clbi*NPV encodes a photolyase-like gene sequence, which has a 1-bp deletion when compared with photolyases of other baculoviruses. This deletion disrupts the sequence into two small photolyase ORFs, designated *Clbiphr-1* and *Clbiphr-2*, which correspond to the CPD-DNA photolyase and FAD-binding domains of photolyases, respectively. Conclusion: *Clbi*NPV belongs to the Group II Alphabaculoviruses and is most closely related to *Orle*NPV, *LdM*NPV, *TnS*NPV, *Ecob*NPV and *Chch*NPV. It contains a variant DNA photolyase gene, which only exists in *Chch*NPV, *TnS*NPV and *Splt*GV among the baculoviruses.

Published On: BMC Genomics, 2009, 10.

1 Institute of Life Sciences, Jiangsu University, Zhenjiang 212013, China.
2 Shanghai Entry-Exit Inspection and Quarantine Bureau, Shanghai 200135, China.
3 School of Life Sciences, Soochow University, Suzhou 215123, China.
4 School of Food and Biological Engineering, Jiangsu University, Zhenjiang 212013, China.
*Corresponding author E-mail: wenbingwang@ujs.edu.cn.

从豆天蛾幼虫分离出的新的核型多角体病毒的基因组序列、组成和特征

朱姗颖[1] 易建平[2] 沈卫德[3] 王利群[1]
何华纲[4] 王 勇[4] 李 兵[3] 王文兵[1*]

摘要:杆状病毒具有作为生物杀虫剂控制农业和森林害虫的作用。它们还被广泛用作分子克隆的表达载体。在最新的NCBI数据库中有48个杆状病毒的基因组序列得到解析,随着病毒基因组测序数量的增加,研究者必须提出足够的分析和注释来推进对它们的理解。在本研究中,为了更好地了解该病毒,我们对豆天蛾进行了全基因组测序和分析。豆天蛾杆状病毒(*Clbi*NPV)基因组全长135 454 bp,G+C 含量为37%,大于150个核苷酸的开放阅读框有139个。其中126个与其他病毒有同源性,13个是豆天蛾特有的基因。30个杆状病毒核心基因均存在于*Clbi*NPV中。以*pif-2*和*lef-8*基因为参照的进化分析表明其归为II类阿尔法杆状病毒。Blast比对发现,*Clbi*NPV编码一个类光解酶基因序列,与其他杆状病毒的光解酶相比,该基因序列有1 bp的缺失。这一缺失将该序列分裂为2个小的光解酶读码框,分别为*Clbiphr-1*和*Clbiphr-2*,分别对应CPD-DNA光解酶和光解酶的FAD结合域。结果表明,*ClbiNPV*属于II类阿尔法杆状病毒,与*Orle*NPV,*LdM*NPV,*Tn*SNPV,*Ecob*NPV和*Chch*NPV等病毒亲缘关系较近。*Chch*NPV,*Tn*SNPV和*Splt*GV具有变异的光解酶基因。

1 生命科学研究院,江苏大学,镇江
2 上海市出入境检验检疫局,上海
3 生命科学学院,苏州大学,苏州
4 食品与生物工程学院,江苏大学,镇江

Gene expression divergence and evolutionary analysis of the *drosomycin* gene family in *Drosophila melanogaster*

Deng XJ[1,2] Yang WY[2] Huang YD[3] Cao Y[2]
Wen SY[4*] Xia QY[5] Xu PL[1]

Abstract: *Drosomycin* (*Drs*) encoding an inducible 44-residue antifungal peptide is clustered with six additional genes, *Dro1, Dro2, Dro3, Dro4, Dro5,* and *Dro6*, forming a multigene family on the 3L chromosome arm in *Drosophila melanogaster.* To get further insight into the regulation of each member of the drosomycin gene family, here we investigated gene expression patterns of this family by either microbe-free injury or microbial challenges using real time RT-PCR. The results indicated that among the seven *drosomycin* genes, *Drs, Dro2, Dro3, Dro4,* and *Dro5* showed constitutive expressions. Three out of five, *Dro2*, *Dro3*, and *Dro5*, were able to be upregulated by simple injury. Interestingly, *Drs* is an only gene strongly upregulated when D*rosophila* was infected with microbes. In contrast to these five genes, *Dro1* and *Dro6* were not transcribed at all in either noninfected or infected flies. Furthermore, by 5' rapid amplification of cDNA ends, two transcription start sites were identified in *Drs* and *Dro2*, and one in *Dro3, Dro4*, and *Dro5*. In addition, NF kB binding sites were found in promoter regions of *Drs, Dro2, Dro3,* and *Dro5,* indicating the importance of NF kB binding sites for the inducibility of *drosomycin* genes. Based on the analyses of flanking sequences of each gene in *D. melanogaster* and phylogenetic relationship of *drosomycins* in *D. melanogaster* species-group, we concluded that gene duplications were involved in the formation of the *drosomycin* gene family. The possible evolutionary fates of *drosomycin* genes were discussed according to the combining analysis of gene expression pattern, gene structure, and functional divergence of these genes.

Published On: Journal of Biomedicine and Biotechnology, 2009.

1 The Key Laboratory of Gene Engineering of Ministry of Education, Sun Yat-sen University, Guangzhou 510275, China.
2 College of Animal Science, South China Agricultural University, Guangzhou 510642, China.
3 Biopharmaceutical Research and Development Center, Jinan University, Guangzhou 510632, China.
4 College of Natural Resource and Environment, South China Agricultural University, Guangzhou 510642, China.
5 Key Sericultural Laboratory of Agriculture Ministry, College of Sericulture and Biotechnology, Southwest University, Chongqing 400716, China.
*Corresponding author E-mail: shywen@scau.edu.cn.

黑腹果蝇*drosomycin*基因家族的基因表达差异与进化分析

邓小娟[1,2] 杨婉莹[2] 黄亚东[3] 曹 阳[2]
温硕洋[4*] 夏庆友[5] 徐培林[1]

摘要：编码诱导型44-残基抗真菌肽的*Drosomycin*（Drs）与另外6个基因*Dro1*，*Dro2*，*Dro3*，*Dro4*，*Dro5*和*Dro6*聚集在一起，在果蝇的3L染色体臂上形成一个多基因家族。为了进一步了解*Drosomycin*基因家族的每个成员的调控情况，我们通过使用实时RT-PCR的微生物免疫损伤或微生物挑战调查了该家族的基因表达模式。结果表明，在7个*Drosomycin*基因中，*Drs*，*Dro2*，*Dro3*，*Dro4*和*Dro5*呈组成型表达。其中3/5的基因，*Dro2*，*Dro3*和*Dro5*，都能被简单的伤害导致表达上调。有趣的是，*Drs*是唯一一个在果蝇被微生物感染时强烈上调的基因。与这5个基因相反，*Dro1*和*Dro6*在非感染或感染的果蝇中都完全没有转录。此外，通过5′cDNA末端的快速扩增，在*Drs*和*Dro2*中鉴定出两个转录起始位点，在*Dro3*，*Dro4*和*Dro5*中鉴定出一个。此外，在*Drs*，*Dro2*，*Dro3*和*Dro5*的启动子区域中发现NF-κB结合位点，表明NF-κB结合位点对于*Drosomycin*基因的诱导能力的重要性。根据黑腹果蝇各基因的侧翼序列分析和黑腹果蝇物种组中*Drosomycins*的系统发育关系，我们得出结论，基因重复参与了*Drosomycin*基因家族的形成。根据基因表达模式，基因结构和这些基因的功能差异的组合分析，讨论了*Drosomycin*基因的可能的进化命运。

1 教育部基因工程重点实验室，中山大学，广州
2 动物学院，华南农业大学，广州
3 生物制药研发中心，暨南大学，广州
4 自然资源与环境学院，华南农业大学，广州
5 农业部蚕桑学重点实验室，蚕桑与生物技术学院，西南大学，重庆

BmSE, a SINE family with 3′ ends of (ATTT) repeats in domesticated silkworm (*Bombyx mori*)

Xu JS[1,2] Liu T[3] Li D[3]
Zhang Z[3] Xia QY[3] Zhou ZY[1,2,3*]

Abstract: Short interspersed elements (SINEs), which are mainly composed of *Bm1*, are abundant in the domesticated silkworm. A 294 bp novel SINE family, designated as *BmSE*, was identified by mining the database of the complete *Bombyx mori* genome. A representational *BmSE* element is flanked by an 11 bp target site duplication sequence posterior poly (A) at the 3′ end and has the sequence motifs of an internal promoter of RNA polymerase III, which are similar to that of *Bm1*. The repetitive elements of *BmSE* are widely distributed in all 28 chromosomes of the genome and share the common (ATTT) repeats at the ends. GC-content distribution shows that *BmSE* tends to accumulate preferably in the region of higher AT content than that of *Bm1*. A high proportion of the *BmSEs* are mapped to the coding sequence introns, whereas several elements are also present in the UTR of some transcripts, indicating that *BmSEs* are indeed exonized with UTRs. Of the 615 identified structural variants (SVs) of *BmSE* among the 40 domesticated and wild silkworms, only 230 SVs were found in the domesticated silkworms, indicating that many recent SV events of *BmSE* occurred after domestication, which was probably due to its mobilization. Our analysis might assist in developing *BmSE* as a potential marker and in understanding the evolutionary roles of SINEs in the domesticated silkworm.

Published On: Journal of Genetics and Genomics, 2009, 37(2), 125-135.

1 Laboratory of Animal Biology, Chongqing Normal University, Chongqing 400047, China.
2 Engineering Research Center of Bioactive Substances, Chongqing Normal University, Chongqing 400047, China.
3 Institute of Sericulture and System Biology, Southwest University, Chongqing 400716, China.
*Corresponding author E-mail: zyzhou@cqnu.edu.cn.

BmSE，一个具有(ATTT)3′端重复序列的SINE家族

许金山[1,2] 刘 铁[3] 李 东[3]
张 泽[3] 夏庆友[3] 周泽扬[1,2,3*]

摘要：主要由*Bm1*组成的短散布元素(*SINEs*)在驯化家蚕中含量丰富。通过分析完整的蚕蛾基因组的数据库，鉴定到一个长度为294 bp的*SINE*家族，命名为*BmSE*。代表性*BmSE*元件在3′末端侧接11 bp靶位点重复序列，并具有RNA聚合酶Ⅲ内部启动子的序列基序，与*Bm1*类似。*BmSE*的重复元件广泛分布在基因组的28个染色体中，并在末端共享共同(ATTT)重复。GC含量分布表明，*BmSE*倾向于积累在AT含量高于*Bm1*的区域内。很大一部分*BmSE*被定位到编码序列内含子中，而某些转录本的UTR中也存在若干个元件，这表明*BmSE*实际上已被UTR外显子化，在40种家蚕和野生蚕中共鉴定到615个确定的*BmSE*结构变异体(SV)，家蚕中仅发现230个SV，表明*BmSE*最近发生的许多SV事件发生在驯化后，这可能是由于其被驯化。我们的分析可能有助于将*BmSE*开发为潜在的靶标，并且了解*SINE*在家蚕进化中的作用。

1 动物生物学实验室，重庆师范大学，重庆
2 生物活性物质工程研究中心，重庆师范大学，重庆
3 蚕学与系统生物学研究所，西南大学，重庆

Identification of novel members reveals the structural and functional divergence of lepidopteran-specific Lipoprotein_11 family

Zhang Y　Dong ZM　Liu SP
Yang Q　Zhao P*　Xia QY

Abstract: 30K proteins (30KPs) are classified into the lepidopteran-specific Lipoprotein_11 family. They are involved in various physiological processes such as energy storage, embryonic development, and immune response in the silkworm. To date, 30KPs were only found in *Bombyx mori* and *Manduca sexta*. Moreover, the C-termini of ENF peptide binding proteins (ENF-BPs) show similarity to 30 KPs. ENF peptides are multifunctional insect cytokines and involved in growth regulation and defense reaction, whereas ENF-BPs act as active regulators of ENF peptides. In order to get insights into this gene family in Lepidoptera, we performed an extensive survey of lepidopteran-derived genome and EST datasets. We identified 73 30KP homologous genes in 12 lepidopteran species, of which 56 are novel members. The structural and phylogenetic analyses revealed that these genes could be classified into three groups: *ENF-BP* genes, typical *30KP* genes, and s*erine/threonine-rich 30KP* (*S/T-rich 30KP*) genes. The C-terminal regions are common to all the three subfamilies, but the N-termini are highly variable. We found a novel subfamily of Lipoprotein_11 and named it S/T-rich 30KP according to its exclusive S/T-rich domain in the N terminus. ENF-BP was also found to contain a special domain in the N terminus, which is homologous to Pp-0912 of *Pseudomonas putida*. Microarray data and semi-quantitative RT-PCR showed that the three groups have their respective temporal - spatial expression patterns. *S/T-rich 30KP* genes have enriched expression in the mature testis and might be involved in spermiogenesis or fertilization. Typical *30KP* genes are expressed mainly in the fat body and integument at the larvae and pupae stages. *ENF-BP* genes are expressed predominantly in the hemocyte. The differential spatial - temporal expression profilesrevealed the functional divergence of three Lipoprotein_11 subfamilies.

Published On: Functional & Integrative Genomics, 2012, 12(4), 705-715.

State Key Laboratory of Silkworm Genome Biology, Southwest University, Chongqing 400716, China.
*Corresponding author E-mail: zhaop@swu.edu.cn.

Lipoprotein_11家族新成员的鉴定
揭示了鳞翅目特异性Lipoprotein_11的结构和功能差异

张　艳　董照明　刘仕平
杨　强　赵　萍* 夏庆友

摘要：30K蛋白（30KPs）被归为鳞翅目昆虫所特有的Lipoprotein_11家族，它们参与家蚕体内多种生理过程，比如能量存储、胚胎发育、免疫反应等。迄今为止，30K蛋白只在家蚕和烟草天蛾中被发现。并且，ENF肽结合蛋白（ENF-BPs）的C端与30K蛋白相似，ENF肽是一种多功能的昆虫细胞因子，参与了生长调控和防御反应，而ENF肽结合蛋白是ENF肽的活性调节剂。为了研究鳞翅目中这一家族的基因，我们对鳞翅目昆虫衍生的基因组及EST数据库进行了扩展调查。我们在12个鳞翅目物种中确定了73个30K蛋白同源基因，其中有56个是新发现的30K蛋白基因。结构和系统进化分析表明，这些基因可以被归入以下3个类群：ENF结合蛋白基因、典型的30K蛋白基因及富含丝氨酸/苏氨酸的30K蛋白基因，这3个亚族的C端区域相同，而N端却高度变化。我们发现了Lipoprotein_11家族的一个全新亚族，根据其N端特有富含丝氨酸/苏氨酸的结构域将其命名为富含丝氨酸/苏氨酸的30K蛋白（S/T-rich 30KP）。还发现ENF结合蛋白的N端也有一个特殊的结构域，它与恶臭假单胞菌（*Pseudomonas putida*）的Pp-0912同源。微矩阵芯片数据与半定量RT-PCR表明，上述3个亚族分别具有不同的瞬时空间表达模式，富含丝氨酸/苏氨酸的30K蛋白在成熟的精巢中表达量丰富，其可能参与了精子形成或受精的过程。典型的30K蛋白基因主要在幼虫及蛹的脂肪体和表皮中表达。ENF结合蛋白基因主要在血细胞中表达。差异化的时空表达谱展示了上述3种Lipoprotein_11亚族的功能差异。

家蚕基因组生物学国家重点实验室，西南大学，重庆

Genome-wide analysis of the WW domain-containing protein genes in silkworm and their expansion in eukaryotes

Meng G　Dai FY*　Tong XL　Li NN

Ding X　Song JB　Lu C*

Abstract: WW domains are protein modules that mediate protein-protein interactions through recognition of proline-rich peptide motifs and phosphorylated serine/threonine-proline sites. WW domains are found in many different structural and signaling proteins that are involved in a variety of cellular processes. WW domain-containing proteins (WWCPs) and complexes have been implicated in major human diseases including cancer as well as in major signaling cascades such as the Hippo tumor suppressor pathway, making them targets for new diagnostics and therapeutics. There are a number of reports about the WWCPs in different species, but systematic analysis of the *WWCP* genes and its ligands is still lacking in silkworm and the other organisms. In this study, *WWCP* genes and PY motif-containing proteins have been identified and analyzed in 56 species including silkworm. Whole-genome screening of *B. mori* identified thirty-three proteins with thirty-nine WW domains located on thirteen chromosomes. In the 39 silkworm WW domains, 15 domains belong to the Group I WW domain; 14 domains were in Group II/III, 9 domains derived from 8 silkworm WWCPs could not be classified into any group, and Group IV contains only one WW domain. Based on gene annotation, silkworm *WWCP* genes have functions in multi-biology processes. A detailed list of WWCPs from the other 55 species was sorted in this work. In 14 623 silkworm predicted proteins, nearly 18 % contained PY motif, nearly 30 % contained various motifs totally that could be recognized by WW domains. Gene Ontology and KEGG analysis revealed that dozens of WW domain-binding proteins are involved in Wnt, Hedgehog, Notch, mTOR, EGF and Jak-STAT signaling pathway. Tissue expression patterns of *WWCP* genes and potential WWCP-binding protein genes on the third day of the fifth instar (L5D3) were examined by microarray analysis. Tissue expression profile analysis found that several *WWCP* genes and poly-proline or PY motif-containing protein genes took tissue - or gender-dependent expression manner in silkworms. We further analyzed *WWCPs* and PY motif-containing proteins in representative organisms of invertebrates and vertebrates. The results showed that there are no less than 16 and up to 29 WWCPs in insects, the average is 22. The number of WW domains in insects is no less than 19, and up to 47, the average is 36. In vertebrates, excluding the *Hydrobiontes*, the number of WWCPs is no less than 34 and up to 49, the average is 43. The number of WW domains in vertebrates is no less than 56 and up to 85, the average is 73. Phylogenetic analysis revealed that most homologous genes of the WWCP subfamily in vertebrates were duplicated during evolution and functions diverged. Nearly 1 000 PY motif-containing protein genes were found in insect genomes and nearly 2 000 genes in vertebrates. The different distributions of *WWCP* genes and PY motif-containing protein genes in different species revealed a possible positive correlation with organism complexity. In conclusion, this comprehensive bio-information analysis of WWCPs and its binding ligands would provide rich fundamental knowledge and useful information for further exploration of the function of the WW domain-containing proteins not only in silkworm, but also in other species.

Published On: Molecular Genetics and Genomics, 2015, 290(3), 807-824.

State Key Laboratory of Silkworm Genome Biology, Key Laboratory for Sericulture Functional Genomics and Biotechnology of Agricultural Ministry, Southwest University, Chongqing, 400716, China.

*Corresponding author E-mail: fydai@swu.edu.cn; lucheng@swu.edu.cn.

全基因组分析家蚕中含有WW结构域的蛋白质基因及其在真核生物中的扩增

孟　刚　代方银[*]　童晓玲　李念念
丁　鑫　宋江波　鲁　成[*]

摘要：WW域是通过识别富含脯氨酸的肽基序和磷酸化丝氨酸/苏氨酸－脯氨酸位点介导蛋白质-蛋白质相互作用的蛋白质结构域。WW域被发现存在于许多不同的结构和信号蛋白中，这些蛋白质参与多种细胞过程。含WW结构域的蛋白质（WWCPs）和复合物已被牵涉到包括癌症在内的主要人类疾病以及Hippo肿瘤抑制通路等主要信号级联中，使其成为新的诊断和治疗的靶标。关于不同物种的WWCPs有很多报道，但在家蚕和其他生物体中仍然缺乏*WWCP*基因及其配体的系统分析。本研究对包括蚕在内的56种生物的*WWCP*基因和含PY基序的蛋白进行了鉴定和分析。家蚕全基因组筛选发现了位于13条染色体上的33个具有39个WW结构域的蛋白质，在39个家蚕WW结构域中，15个属于Ⅰ类WW结构域，14个是Ⅱ/Ⅲ类WW结构域，9个结构域来源于8个家蚕WWCPs但不能归入任何一类，Ⅳ类只含有一个WW结构域。基于基因注释，家蚕*WWCP*基因在多生物学过程中具有功能。本研究对其他55个种类的WWCP进行了详细的列表分类。在14 623个蚕预测蛋白质中，近18%含有PY基序，近30%含有各种可以被WW域识别的基序。Gene Ontology和KEGG分析显示，许多WW结构域结合蛋白参与Wnt，Hedgehog，Notch，mTOR，EGF和Jak-STAT信号通路。通过芯片分析检测五龄第3天（L5D3）的*WWCP*基因和潜在的WWCP结合蛋白基因的组织表达模式。组织表达谱分析发现，几种*WWCP*基因和含聚脯氨酸或含PY基序的蛋白质基因在家蚕中具有组织或性别依赖性表达的方式。我们进一步分析了无脊椎动物和脊椎动物的代表性生物中的*WWCPs*和含PY基序的蛋白质。结果表明，昆虫中WWCPs不低于16个，最多29个，平均为22个，昆虫WW结构域数不少于19个，最多47个，平均为36个。在脊椎动物中，不包括水生动物，WWCPs的数量不低于34个，最多49个，平均为43个。脊椎动物中WW结构域的数量不低于56个，最多85个，平均为73个。系统发育分析显示，大多数同源脊椎动物中WWCP亚科的基因在进化过程中被重复，功能发生了分化。在昆虫基因组中发现近1 000个含有PY基序的蛋白基因，脊椎动物中发现近2 000个基因。不同种类的*WWCP*基因和含有PY基序的蛋白质基因的不同分布显示出其与生物复杂性可能存在正相关性。总之，此次对WWCPs及其结合配体的全面生物信息分析，将为进一步探索含WW结构域的蛋白质的功能提供丰富的基础知识和有用信息，不仅在家蚕中，在其他物种中也是如此。

家蚕基因组生物学国家重点实验室，农业部蚕桑学重点实验室，西南大学，重庆

Nuclear receptors in *Bombyx mori*: Insights into genomic structure and developmental expression

Cheng DJ[1] Xia QY[1,2*] Duan J[1,2] Wei L[3]
Huang C[1] Li ZQ[1] Wang GH[1,2] Xiang ZH[1]

Abstract: Nuclear receptors (NRs) function as ligand-dependent transcription factors and are involved in diverse biological processes in different animals. The updated assembly of complete genome sequence of the *Bombyx mori* enabled a systematic analysis of the NRs in the five holometabolous insects including *B. mori*, *Drosophila melanogaster*, *Anopheles gambiae*, *Apismel lifera*, and *Tribolium castaneum*. As a result, nineteen NRs were identified in the *B. mori* genome, each of eighteen NRs has 1 : 1 : 1 : 1 ortholog in the other four insects. Interestingly, the average intron number of ligand-binding domain (LBD) of each NR gene in *B. mori* was 2.4, much higher than that in the other four insects; the genomic position of introns in LBDs of all orthologs for each NR presents more diversity. Phylogenetic trees of all NRs from the five insects were consistent or aberrant with classical phylogeny of these insect species. The characteristics in number, genomic structure and phylogeny of all NRs revealed their evolutionary conservation and divergence during insect evolution. The expression patterns of several NR genes displayed temporal specificity similar to that in *D. melanogaster* and may be associated with the key biological processes during silkworm metamorphosis. The RNAi of *BmβFTZ-F1* resulted in abnormality in larva-pupa transition, further suggesting it is also crucial for silkworm metamorphosis. In conclusion, the present study provided new insights into the structure, evolution, expression, and functions of silkworm NRs.

Published On: Insect Biochemistry and Molecular Biology, 2018, 38(12), 1130-1137.

1 Key Sericultural Laboratory of Agricultural Ministry, Southwest University, Chongqing 400716, China.
2 Institute of Agronomy and Life Sciences, Chongqing University, Chongqing 400030, China.
3 School of Life Science, Southwest University, Chongqing 400716, China.
*Corresponding author E-mail: xiaqy@swu.edu.cn.

家蚕核受体基因结构和表达分析

程道军[1]　夏庆友[1,2*]　段　军[1,2]　魏　玲[3]
黄　重[1]　李志清[1]　王根洪[1,2]　向仲怀[1]

摘要：核受体作为配体依赖的转录因子在不同的动物中参与了不同的生物过程。我们利用家蚕基因组的精细图鉴定了核受体家族成员，并与果蝇、冈比亚按蚊、蜜蜂和赤拟谷盗等完全变态昆虫进行了系统性的分析。我们在家蚕中一共鉴定到19个核受体蛋白，其中有18个与果蝇、按蚊、蜜蜂和赤拟谷盗的基因呈明显的直系同源关系。有趣的是，家蚕每个核受体基因的配体结合结构域总是聚为一类，且平均包含2.4个内含子，比其他4种昆虫中的高，而这些内含子所在的基因组的位置也具有极大的多样性。核受体的系统发育树分析结果表明，所有的核受体在5种昆虫中的保守性高。这些核受体在数量、基因结构和系统发育中的特征揭示了其在昆虫进化过程中的保守性和差异性。多个核受体的表达模式显示了其与果蝇在时空特征方面的相似性，同时可能与家蚕在变态发育中关键的生物学过程相关。干涉家蚕核受体*βFTZ-F1*基因导致了幼虫–蛹转变的异常，进一步表明其在家蚕变态过程中的重要性。总之，我们的研究结果以全新的角度在家蚕核受体的结构、进化、表达和功能方面进行了系统分析。

1　农业部蚕桑学重点实验室，西南大学，重庆
2　农学及生命科学研究院，重庆大学，重庆
3　生命科学学院，西南大学，重庆

The genomic underpinnings of apoptosis in the silkworm, *Bombyx mori*

Zhang JY[#] Pan MH[#] Sun ZY Huang SJ
Yu ZS Liu D Zhao DH Lu C[*]

Abstract:

Background: Apoptosis is regulated in an orderly fashion by a series of genes, and has a crucial role in important physiological processes such as growth development, immunological response and so on. Recently, substantial studies have been undertaken on apoptosis in model animals including humans, fruit flies, and the nematode. However, the lack of genomic data for silkworms limits their usefulness in apoptosis studies, despite the advantages of silkworm as a representative of Lepidoptera and an effective model system. Herein we have identified apoptosis-related genes in the silkworm *Bombyx mori* and compared them to those from insects, mammals, and nematodes.

Results: From the newly assembled genome databases, a genome-wide analysis of apoptosis-related genes in *Bombyx mori* was performed using both nucleotide and protein Blast searches. Fifty-two apoptosis-related candidate genes were identified, including five caspase family members, two tumor necrosis factor (TNF) superfamily members, one Bcl-2 family member, four baculovirus IAP (inhibitor of apoptosis) repeat (BIR) domain family members and 1 RHG (Reaper, Hid, Grim, and Sickle; *Drosophila* cell death activators) family member. Moreover, we identified a new caspase family member, *BmCaspase-New*, two splice variants of *BmDronc*, and *Bm3585*, a mammalian TNF superfamily member homolog. Twenty-three of these apoptosis-related genes were cloned and sequenced using cDNA templates isolated from *BmE-SWU1* cells. Sequence analyses revealed that these genes could have key roles in apoptosis.

Conclusions: *Bombyx mori* possesses potential apoptosis-related genes. We hypothesized that the classic intrinsic and extrinsic apoptotic pathways potentially are active in *Bombyx mori*. These results lay the foundation for further apoptosis-related study in *Bombyx mori*.

Published On: BMC Genomics, 2010, 11, 611.

Key Sericultural Laboratory of Agricultural Ministry, Institute of Sericulture and Systems Biology, Southwest University, Chongqing 400716, China.

[#]These authors contributed equally.

[*]Corresponding author E-mail: lucheng@swu.edu.cn.

家蚕细胞凋亡的基因组基础分析

张金叶[#] 潘敏慧[#] 孙志亚 黄淑静
于子舒 刘 迪 赵丹红 鲁 成[*]

摘要：背景——细胞凋亡受到一系列基因的有序调节，并且细胞凋亡在诸如生长发育、免疫应答等重要的生理过程中扮演重要的角色。最近，已经对包括人、果蝇和线虫在内的模式动物的细胞凋亡进行了大量研究。然而，尽管家蚕作为鳞翅目的代表和有效的模型系统具有优势，但是家蚕基因组数据的缺乏限制了其在细胞凋亡研究中的可用性。该研究中，我们鉴定了家蚕中与细胞凋亡相关的基因，并与其他的昆虫、哺乳动物和线虫中的相关基因进行了比较。结果——在新组装的家蚕基因组数据库中，通过核酸和蛋白质的比对，对家蚕细胞凋亡相关基因进行了全基因组分析。鉴定了52个细胞凋亡相关候选基因，其中包括5个半胱氨酸蛋白酶家族（caspase）成员、2个肿瘤坏死因子（TNF）超家族成员、1个细胞凋亡相关基因-2（Bc1-2）家族成员、4个杆状病毒细胞凋亡抑制子（IAP）重复结构域家族成员和1个细胞死亡激活家族RHG成员。此外，我们鉴定了一个新的caspase家族成员 *BmCaspase-New*、*BmDronc* 的两种剪接突变体，以及哺乳动物肿瘤坏死因子超家族成员的同系物 *Bm3585*。以来源于家蚕胚胎细胞（*BmE-SWU1*）的cDNA为模板，对其中23个细胞凋亡相关基因进行了克隆并测序。序列分析表明这些基因在细胞凋亡过程中具有重要的作用。结论——家蚕拥有潜在的细胞凋亡相关基因。我们猜测经典的内在和外在的细胞凋亡途径在家蚕中是活跃的。这些结果为进一步开展家蚕细胞凋亡相关的研究奠定了基础。

农业部蚕桑学重点实验室，蚕学与系统生物学研究所，西南大学，重庆

Transcriptome analysis of the silkworm (*Bombyx mori*) by high-throughput RNA sequencing

Li YN[1#] Wang GZ[1,2#] Tian J[1] Liu HF[1] Yang HP[1] Yi YZ[1]
Wang JH[1] Shi XF[1] Jiang F[1] Yao B[2*] Zhang ZF[1*]

Abstract: The domestic silkworm, *Bombyx mori*, is a model insect with important economic value for silk production that also acts as a bioreactor for biomaterial production. The functional complexity of the silkworm transcriptome has not yet been fully elucidated, although genomic sequencing and other tools have been widely used in its study. We explored the transcriptome of silkworm at different developmental stages using high-throughput paired-end RNA sequencing. A total of about 3.3 gigabases (Gb) of sequence was obtained, representing about a 7-fold coverage of the *B. mori* genome. From the reads that were mapped to the genome sequence; 23 461 transcripts were obtained, 5 428 of them were novel. Of the 14 623 predicted protein-coding genes in the silkworm genome database, 11 884 of them were found to be expressed in the silkworm transcriptome, giving a coverage of 81.3%. A total of 13 195 new exons were detected, of which, 5 911 were found in the annotated genes in the Silkworm Genome Database (SilkDB). An analysis of alternative splicing in the transcriptome revealed that 3 247 genes had undergone alternative splicing. To help with the data analysis, a transcriptome database that integrates our transcriptome data with the silkworm genome data was constructed and is publicly available at http://124.17.27.136/gbrowse2/. To our knowledge, this is the first study to elucidate the silkworm transcriptome using high-throughput RNA sequencing technology. Our data indicate that the transcriptome of silkworm is much more complex than previously anticipated. This work provides tools and resources for the identification of new functional elements and paves the way for future functional genomics studies.

Published On: PLoS ONE, 2012, 7(8).

1 Biotechnology Research Institute, Chinese Academy of Agricultural Sciences, Beijing 100081, China.
2 Feed Research Institute, Chinese Academy of Agricultural Sciences , Beijing 100081, China.
[#]These authors contributed equally.
[*]Corresponding author E-mail: yaobin@mail.caas.net.cn; zhifangzhang@yahoo.com.

通过高通量测序技术对家蚕的转录组进行分析

李铁女[1#] 王国增[1,2#] 田 健[1] 刘慧芬[1] 杨慧鹏[1] 易咏竹[1]
王金辉[1] 史晓峰[1] 江 峰[1] 姚 斌[2*] 张志芳[1*]

摘要：家蚕是一种模式昆虫，因为能够产丝而具有重要的经济价值，同时也可作为生产生物材料的生物反应器。在家蚕研究中，尽管基因组测序和其他的工具已经被广泛地使用，但是家蚕转录的复杂性仍未被完全阐明。我们利用高通量双端RNA测序探索了家蚕在不同发育时期的转录组情况。总共获得3.3 Gb的序列信息，约是家蚕基因组的7倍覆盖率。综合能比对的基因组上的Reads信息，获得了23 461个转录本，其中5 428个转录本是新获得的。在家蚕基因组数据库中14 623个预测的蛋白编码基因中有11 884个在家蚕转录组中表达，覆盖率为81.3%。共检测到13 195个新的外显子，其中有5 911个存在于家蚕基因组数据库（SilkDB）注释的基因中。对转录本进行选择性剪接分析，揭示3 247个基因进行选择性剪接。为了有助于数据的分析，我们构建了转录组数据库，该数据库将我们的转录组数据与家蚕基因组数据整合在一起，并公开于http://124.17.27.136/gbrowse2/。据我们所知，这是首次通过高通量RNA测序对家蚕进行转录组分析。我们的数据表明家蚕的转录比之前预测的要更复杂。这部分工作为新功能元件的鉴定提供了工具和资源，为今后的功能基因组学研究铺平了道路。

1 生物技术研究所，中国农业科学院，北京
2 饲料研究所，中国农业科学院，北京

Genome-wide comparison of genes involved in the biosynthesis, metabolism, and signaling of juvenile hormone between silkworm and other insects

Cheng DJ* Meng M Peng J
Qian WL Kang LX Xia QY*

Abstract: Juvenile hormone (JH) contributes to the regulation of larval molting and metamorphosis in insects. Herein, we comprehensively identified 55 genes involved in JH biosynthesis, metabolism and signaling in the silkworm (*Bombyx mori*) as well as 35 in *Drosophila melanogaster*, 35 in *Anopheles gambiae*, 36 in *Apis mellifera*, 47 in *Tribolium castaneum*, and 44 in *Danaus plexippus*. Comparative analysis showed that each gene involved in the early steps of the mevalonate (MVA) pathway, in the neuropeptide regulation of JH biosynthesis, or in JH signaling is a single copy in *B. mori* and other surveyed insects, indicating that these JH-related pathways or steps are likely conserved in all surveyed insects. However, each gene participating in the isoprenoid branch of JH biosynthesis and JH metabolism, together with the *FPPS* genes for catalyzing the final step of the MVA pathway of JH biosynthesis, exhibited an obvious duplication in Lepidoptera, including *B. mori* and *D. plexippus*. Microarray and real-time RT-PCR analysis revealed that different copies of several JH-related genes presented expression changes that correlated with the dynamics of JH titer during larval growth and metamorphosis. Taken together, the findings suggest that duplication-derived copy variation of JH-related genes might be evolutionarily associated with the variation of JH types between Lepidoptera and other insect orders. In conclusion, our results provide useful clues for further functional analysis of JH-related genes in *B. mori* and other insects.

Published On: Genetics and Molecular Biology, 2014, 37(2), 444-459.

State Key Laboratory of Silkworm Genome Biology, Southwest University, Chongqing 400716, China.
*Corresponding author E-mail: chengdj@swu.edu.cn; xiaqy@swu.edu.cn.

全基因组比较分析家蚕和其他昆虫中参与保幼激素生物合成、代谢和信号传导的基因

程道军* 孟 勐 彭 健
钱文良 康丽霞 夏庆友*

摘要：在昆虫中，保幼激素（JH）参与调控幼虫的蜕皮和变态发育。在这里，我们全面鉴定了参与JH的生物合成、代谢和信号传导的基因，其中家蚕（*Bombyx mori*）中有55个，黑腹果蝇中有35个，冈比亚按蚊中有35个，蚜虫中有36个，赤拟谷盗中有47个，以及黑脉金斑蝶中有44个。比较分析显示，每个基因均涉及甲羟戊酸（MVA）通路的早期过程、JH生物合成的神经肽调节过程，或在*B. mori*和已调查昆虫的JH信号通路中均属于单拷贝基因，这表明在所有调查的昆虫中这些JH相关的通路与过程都相对保守。然而，参与JH生物合成和代谢的类异戊二烯分支的每个基因以及用于催化JH生物合成的MVA途径的最终过程中的*FPPS*基因在鳞翅目中呈现多拷贝，包括在家蚕和黑脉金斑蝶中。芯片技术和RT-PCR分析显示，在幼虫生长和变态期间几种JH相关基因的不同拷贝的变化与JH滴度的动态变化具有一定的相关性。综上所述，本研究结果表明，在鳞翅目和其他昆虫中JH相关基因的拷贝数的变异在进化上可能与JH类型的多样性相关。总之，我们的研究结果为进一步分析家蚕和其他昆虫中的JH相关基因的功能提供了可靠的线索。

家蚕基因组生物学国家重点实验室，西南大学，重庆

In silico identification of *BESS-DC* genes and expression analysis in the silkworm, *Bombyx mori*

Rao ZC[1] Duan J[2] Xia QY[2] Feng QL[1*]

Abstract: BESS domain is a protein binding domain that can interact each other or with other domains. In this study, 323 BESS domain containing (BESS-DC) proteins were identified in 3 328 proteomes. These *BESS-DC* genes pertain to 41 species of five phyla, most of which are arthropod insects. A BESS domain contains two α-helixes linked by a coil or β-turn. Phylogenetic tree and architecture analysis show that the BESS domain seems to generate along with the DNA-binding MADF domain. Two hundred thirty three *BESS-DC* genes (71.1%) contain at least one MADF domain, while 59 genes (18.2%) had only the BESS domain. In addition to BESS and MADF domains, some of genes also contain other ligand binding domains, such as DAO, DUS and NAD_C. Nineteen genes (5.8%) are associated with other DNA binding domains, such as *Myb* and *BED*. The *BESS-DC* genes can be divided into 17 subfamilies, eight of which have more than one clade. In *Bombyx mori*, 12 *BESS-DC* genes that do not contain intron in the BESS domain region were localized to eight chromosomes. Real-time PCR results showed that most of the *B. mori BESS-DC* genes were highly expressed from late larval stage to adult stage. Sequence comparison and evolution analyses suggest a hypothesis that the *BESS-DC* genes may play a role in central nervous system development, long term memory and metamorphosis of insects of different phyla.

Published On: Gene, 2016, 575(2), 478-487.

1 Guangzhou Key Laboratory of Insect Development Regulation and Application Research, School of Life Sciences, South China Normal University, Guangzhou 510631, China.

2 State Key Laboratory of Silkworm Genome Biology, Southwest University, 400716, China.

*Corresponding author E-mail: qlfeng@scnu.edu.cn.

家蚕*BESS-DC*基因的生物信息学鉴定与表达分析

饶中臣[1]　段　军[2]　夏庆友[2]　冯启理[1*]

摘要：BESS结构域是蛋白中能够相互作用或与其他结构互作的一种结构域。本文从3 328个蛋白组数据中鉴定了323个含有BESS结构域的蛋白。这些BESS-DC蛋白来自5个门的41个物种，主要分布于节肢动物门的昆虫纲。BESS结构域由2段α-螺旋区域组成，通过无规则卷曲或β折叠连接。进化树结合结构分析的结果显示，BESS结构域还似乎形成了一个可与DNA结合的MADF结构域。233个*BESS-DC*基因（71.1%）包含至少一个MADF结构域，而59个基因（18.2%）仅含有BESS结构域。除了BESS和MADF结构域，一些基因同样含有其他的一些配体结合结构域，例如，DAO、DUS和NAD_C。19个基因（5.8%）与其他的DNA结合结构域相关，例如*Myb*和*BED*基因。*BESS-DC*基因家族可以分成17个亚族，其中8个含有1个以上的进化分支。家蚕有12个*BESS-DC*基因，它们的BESS结构域中不含有内含子，且分布于8条染色体上。定量PCR的结果显示，家蚕大部分的*BESS-DC*基因在幼虫末期到成虫期有高水平的表达。序列比对和进化分析的结果显示，*BESS-DC*基因可能在中枢神经系统的发育、长期记忆及昆虫变态发育过程中起着重要的作用。

1　广州市昆虫发育调控与应用研究重点实验室，生命科学学院，华南师范大学，广州

2　家蚕基因组生物学国家重点实验室，西南大学，重庆

Identification and characterization of the *cyclin-dependent kinases* gene family in silkworm, *Bombyx mori*

Li YN Jiang F Shi XF Liu XJ Yang HP Zhang ZF*

Abstract: Cyclin-dependent protein kinases (CDKs) play key roles at different checkpoint regulations of the eukaryotic cell cycle. However, only few studies of lepidoptera CDK family proteins have been reported so far. In this study, we performed the cDNA sequencing of 10 members of the CDK family in *Bombyx mori*. Gene structure analysis suggested that *CDK12* and *CDC2L1* owned two and three isoforms, respectively. Phylogenetic analysis showed that *CDK* genes in different species were highly conserved, implying that they evolved independently even before the split between vertebrates and invertebrates. We found that the expression levels of *BmCDKs* in 13 tissues of fifth-instar day 3 larvae were different: *CDK1*, *CDK7*, and *CDK9* had a high level of expression, whereas *CDK4* was low-level expressed and was detected only in the testes and fat body cells. Similar expression profiles of *BmCDKs* during embryo development were obtained. Among the variants of *CDK12*, *CDK12* transcript variant A had the highest expression, and the expression of *CDC2L1* transcript variant A was the highest among the variants of *CDC2L1*. It was shown from the RNAi experiments that the silencing of *CDK1*, *CDK10*, *CDK12*, and *CDC2L1* could influence the cells from G_0/G_1 to S phase transition.

Published On: DNA and Cell Biology, 2016, 35(1), 13-23.

Biotechnology Research Institute, Chinese Academy of Agricultural Sciences, Beijing 100081, China.
*Corresponding author E-mail: zhifangzhang@yahoo.com.

鉴定并分析家蚕细胞周期蛋白依赖性蛋白激酶基因家族

李轶女　江　峰　石小峰　刘兴健　杨慧鹏　张志芳*

摘要: 细胞周期蛋白依赖性蛋白激酶(CDKs)在调控真核生物细胞周期的不同节点中具有关键作用。然而,到目前为止只有很少的关于鳞翅目昆虫的CDK家族蛋白被报道。在本研究中,我们分析了家蚕CDK家族的10个成员的cDNA序列,基因结构分析表明*CDK12*和*CDC2L1*分别具有2个和3个特异的异构体,系统发育分析表明,CDK基因在不同物种中高度保守,这表明其在脊椎动物和无脊椎动物分离之前就已独立进化。我们同时发现家蚕*CDK*基因的表达水平在五龄第3天时期的13个组织中都有差异:*CDK1*、*CDK7*和*CDK9*有高表达,而*CDK4*低表达,只在精巢和脂肪体细胞中能检测到。有趣的是,在胚胎发育时期各*BmCDKs*的表达模式是相似的。此外,在*CDK12*的转录异构体中,*CDK12* A亚型转录异构体的表达量最高;在*CDC2L1*的转录异构体中,*CDC2L1*的A型转录异构体的表达量最高。RNAi实验表明,*CDK1*、*CDK10*、*CDK12*和*CDC2L1*的沉默会影响细胞从G_0/G_1到S期的转变。

生物技术研究所,中国农业科学院,北京

Identification and analysis of YELLOW protein family genes in the silkworm, *Bombyx mori*

Xia AH[3#] Zhou QX[1,3#] Yu LL[3] Li WG[1] Yi YZ[2]
Zhang YZ[2*] Zhang ZF[3*]

Abstract:

Background: The major royal jelly proteins / yellow (MRJP / YELLOW) family possesses several physiological and chemical functions in the development of *Apis mellifera* and *Drosophila melanogaster*. Each protein of the family has a conserved domain named MRJP. However, there is no report of MRJP/YELLOW family proteins in the Lepidoptera.

Results: Using the YELLOW protein sequence in *Drosophila melanogaster* to BLAST silkworm EST database, we found a gene family composed of seven members with a conserved MRJP domain each and named it YELLOW protein family of *Bombyx mori*. We completed the cDNA sequences with RACE method. The protein of each member possesses a MRJP domain and a putative cleavable signal peptide consisting of a hydrophobic sequence. In view of genetic evolution, the whole *Bm* YELLOW protein family composes a monophyletic group, which is distinctly separate from *Drosophila melanogaster* and *Apis mellifera*. We then showed the tissue expression profiles of *Bm* YELLOW protein family genes by RT-PCR.

Conclusion: A *Bombyx mori* YELLOW protein family is found to be composed of at least seven members. The low homogeneity and unique pattern of gene expression by each member among the family ensure us to prophesy that the members of *Bm* YELLOW protein family would play some important physiological functions in silkworm developments.

Published On: BMC Genomics, 2006, 7.

1 The Sericultural Research Institute, Chinese Academy of Agricultural Sciences, Zhenjiang 212018, China.

2 Institute of Biochemistry, Zhejiang Sci-Tech University, Hangzhou 310018, China.

3 The Biotechnology Research Institute, National Engineering of crop germplasm and genetic improvement, Chinese Academy of Agricultural Sciences, Beijing 100081, China.

#These authors contributed equally.

*Corresponding author E-mail:yaozhou@chinagene.com; zhifangzhang@yahoo.com.

家蚕 YELLOW 蛋白家族基因的鉴定与分析

夏爱华[3#]　周庆祥[1,3#]　于琳琳[3]　李卫国[1]
易咏竹[2]　张耀洲[2*]　张志芳[3*]

摘要：背景——主要蜂王浆蛋白(MRJP /YELLOW)家族在蜜蜂和黑腹果蝇的发育过程中具有多种生理生化功能。这类家族蛋白都含有一个MRJP保守结构域，然而，在鳞翅目昆虫中还未见MRJP / YELLOW蛋白家族报道。结果——利用果蝇的YELLOW蛋白序列在家蚕EST数据库中进行BLAST比对分析，我们发现了一个由7个含保守MRJP结构域的成员组成的基因家族，命名为家蚕YELLOW蛋白家族。我们通过RACE方法克隆并完成了cDNA测序，发现该家族每一个成员的蛋白质都具有MRJP结构域和一个推测可切割的由疏水性氨基酸组成的信号肽序列。从遗传进化的角度看，整个家蚕YELLOW蛋白家族组成一个单系群，明显有别于黑腹果蝇和蜜蜂的同源家族。然后，我们通过RT-PCR分析鉴定了家蚕YELLOW蛋白家族基因组织表达谱。结论——家蚕YELLOW蛋白家族由至少7个成员组成，根据家蚕这些成员的低同源性和独特的基因表达模式，我们推测家蚕YELLOW蛋白家族在家蚕发育过程中起着重要的生理作用。

1　蚕业研究所，中国农业科学院，镇江
2　生物化学研究所，浙江理工大学，杭州
3　生物技术研究所，作物基因组与遗传改良研究室，中国农业科学院，北京

Identification and expression pattern of the chemosensory protein gene family in the silkworm, *Bombyx mori*

Gong DP Zhang HJ Zhao P Lin Y Xia QY* Xiang ZH

Abstract: Insect chemosensory proteins (CSPs) as well as odorant-binding proteins (OBPs) have been supposed to transport hydrophobic chemicals to receptors on sensory neurons. Compared with OBPs, CSPs are expressed more broadly in various insect tissues. We performed a genome-wide analysis of the candidate *CSP* gene family in the silkworm. A total of 20 candidate *CSPs*, including 3 gene fragments and 2 pseudogenes, were characterized based on their conserved cysteine residues and their similarity to CSPs in other insects. Some of these genes were clustered in the silkworm genome. The gene expression pattern of these candidates was investigated using RT-PCR and microarray, and the results showed that these genes were expressed primarily in mature larvae and the adult moth, suggesting silkworm CSPs may be involved in development. The majority of silkworm *CSP* genes are expressed broadly in tissues including the antennae, head, thorax, legs, wings, epithelium, testes, ovaries, pheromone glands, wing disks, and compound eyes.

Published On: Insect Biochemistry and Molecular Biology, 2007, 37(3), 266-277.

The Key Sericultural Laboratory of Agricultural Ministry, Southwest University, Chongqing 400716, China.
*Corresponding author E-mail: xiaqy@swu.edu.cn.

家蚕化学感受蛋白基因家族的鉴定和表达特征分析

龚达平　张辉洁　赵　萍　林　英　夏庆友*　向仲怀

摘要：昆虫化学感受蛋白（CSPs）与气味结合蛋白（OBPs）被认为具有将不溶性的化学物质运送到感觉神经受体上的功能。与OBPs相比，CSPs在各种昆虫组织中表达更广泛。我们对家蚕*CSP*基因家族进行了全基因组分析，基于其保守的半胱氨酸残基及其与其他昆虫中CSPs的相似性，共鉴定到20个候选*CSPs*基因，包括3个基因片段和2个假基因，部分基因在基因组中成簇分布。利用RT-PCR和芯片数据研究了这些基因的表达模式，结果表明，这些基因主要在成熟幼虫和成虫中表达，暗示家蚕CSPs可能参与个体发育过程。大多数家蚕*CSP*基因在触角、头、胸、足、翅、表皮、精巢、卵巢、信息素腺、翅盘以及复眼等组织中都有表达。

农业部蚕桑学重点实验室，西南大学，重庆

Analysis of the structure and expression of the 30K protein genes in silkworm, *Bombyx mori*

Sun Q Zhao P* Lin Y Hou Y Xia QY Xiang ZH

Abstract: A group of lipoproteins with molecular sizes of approximately 30 kDa, referred to as 30K proteins, are synthesized in fat body cells in the fifth instar larvae of silkworm, *Bombyx mori*. Analyzing the silkworm genome and its expressed sequence tags (ESTs), we found 10 genes encoding 30K proteins, which are mainly distributed in three subfamilies. Of these, seven coding proteins were found to harbor the degrading sites of 30kP protease A, although the number of degrading sites may be different. As some potential core promoters and regulatory elements were supposed to be essential for gene transcription, the expression profiles of these genes were examined by semi-quantitative reverse transcription polymerase chain reaction. Eight 30K protein genes were detected to express luxuriantly in the fat body, while two were hardly expressed. Such results suggest that these 30K proteins may have different functions, and their adjacent regulatory elements play a crucial role in regulating their transcription.

Published On: Insect Science, 2007,14(1),5-14.

The Key Laboratory of Sericulture of Agriculture Ministry, College of Sericulture and Biotechnology, Southwest University, Chongqing 400716, China.

*Corresponding author E-mail: zhaopingxqy@163.com.

家蚕30K蛋白基因的结构和表达分析

孙 全 赵 萍* 林 英 侯 勇 夏庆友 向仲怀

摘要：一群分子质量约为30 kDa的脂蛋白被命名为30K蛋白，这些蛋白质在家蚕五龄幼虫的脂肪体细胞中合成分泌。本研究通过对家蚕基因组和EST序列的信息学分析，发现了10个编码30K蛋白的基因，这些基因主要分布于3个亚家族中，其中有7个30K蛋白基因的编码蛋白有30 kP蛋白酶A的降解位点，并且降解位点数量在基因间存在差异。对30K蛋白基因结构分析，发现都具有对基因转录调控有重要作用的启动子位点和调节元件。通过半定量PT-PCR分析30K蛋白基因在脂肪体中的表达特点，发现除了2个基因没有表达外，其他8个基因都有高量表达。这些结果暗示30K蛋白在家蚕中发挥了不同的生物学功能，而且30K蛋白基因的邻近调控元件可能在其转录调控中扮演着重要的角色。

农业部蚕桑学重点实验室，蚕学与生物技术学院，西南大学，重庆

Species-specific expansion of C2H2 zinc-finger genes and their expression profiles in silkworm, *Bombyx mori*

Duan J[1,2] Xia QY[1,2*] Cheng DJ[1] Zha XF[1] Zhao P[1] Xiang ZH[1]

Abstract: Most C2H2 zinc-finger proteins (ZFPs) function as sequence-specific DNA-binding transcription factors, and play important roles in a variety of biology processes, such as development, differentiation, and tumor suppression. By searching the silkworm genome with a HMM model of C2H2 zinc-fingers, we have identified a total of 338 C2H2 ZFPs. Most of the *ZFP* genes were clustered on chromosomes and showed uneven distribution in the genome. Over one third of genes were concentrated on chromosome 11, 15 and 24. Phylogenetic analysis classified all silkworm *C2H2 ZFPs* into 75 families; 63 of which belong to evolutionarily conserved families. In addition, 188 *C2H2 ZFP* genes (55.6%) are species-specific to the silkworm. A species-specific expansion of a family with 39 members in a tandem array on chromosome 24 may explain the higher number of species-specific *ZFPs* in silkworm compared to other organisms. The expression patterns of *C2H2 ZFP* genes were also examined by microarray analysis. Most of these genes were actively expressed among different tissues on day 3 of the fifth instar. The results provide insight into the biological functions of the silkworm *C2H2 ZFP* genes in metamorphism and development.

Published On: Insect Biochemistry and Molecular Biology, 2008, 38(12), 1121-1129.

1 The Key Sericultural Laboratory of Agricultural Ministry, Southwest University, Chongqing 400716, China.

2 The Institute of Agricultural and Life Sciences, Chongqing University, Chongqing 400030, China.

*Corresponding author E-mail: xiaqy@swu.edu.cn.

家蚕C2H2型锌指蛋白基因物种特异性扩增及表达谱分析

段　军[1,2]　夏庆友[1,2*]　程道军[1]　查幸福[1]　赵　萍[1]　向仲怀[1]

摘要：C2H2型锌指蛋白是一类重要的转录调控因子，具有DNA序列结合特异性，在发育、分化、肿瘤抑制等过程中发挥关键作用。基于C2H2锌指结构域的HMM模型，在家蚕基因组中共发现了338个C2H2型锌指蛋白基因，这些基因在染色体上分布不均衡，绝大部分呈串联重复分布，其中超过1/3的基因集中分布在家蚕第11、15和24号染色体上。遗传聚类分析将家蚕C2H2型锌指蛋白基因分类为75个家族，其中63个属于进化上保守的家族。另外，188个(55.6%)C2H2型锌指蛋白基因属于家蚕特异基因。在家蚕第24号染色体上有一个特异基因家族，该家族的39个成员串联分布在一起，可能正是因为这个基因家族发生了扩增，导致家蚕具有比其他物种更多的特异C2H2型锌指蛋白基因。另外，通过基因芯片对家蚕C2H2型锌指蛋白基因的表达情况进行了研究，发现大多数基因在家蚕五龄第3天的不同组织中具有转录活性。这些结果为进一步研究家蚕C2H2型锌指蛋白基因在变态和发育中的作用提供了线索。

1　农业部蚕桑学重点实验室，西南大学，重庆
2　农学及生命科学研究院，重庆大学，重庆

Characterization of multiple *CYP9A* genes in the silkworm, *Bombyx mori*

Ai JW[1,2,3] Yu QY[1,2] Cheng TC[1,2] Dai FY[1,2]
Zhang XS[1,2] Zhu Y[1,2*] Xiang ZH[1*]

Abstract: Based on the advances in the silkworm genome project, a new genome-wide analysis of *cytochrome P450* genes was performed, focusing mainly on gene duplication. All four *CYP9A* subfamily members from the silkworm, *Bombyx mori*, were cloned by RT-PCR and designated *CYP9A19-CYP9A22* by the P450 Nomenclature Committee. They each contain an open reading frame of 1 593 bp in length and encode a putative polypeptide of 531 amino acids. Both nucleic acid and amino acid sequences share very high identities with one another. The typical motifs of insect cytochrome P450, including the heme-binding region, helix-C, helix-I, helix-K, and PERF, show high sequence conservation among the multiple proteins. Alignment with their cDNA sequences revealed that these paralogues share identical gene structures, each comprising ten exons and nine introns of variable sizes. The locations of their introns (all nine introns follow the GT-AG rule) are absolutely conserved. *CYP9A19*, *CYP9A20*, and *CYP9A21* form a tandem cluster on chromosome 17, whereas *CYP9A22* is separated from the cluster by four tandem alcohol-dehydrogenase-like genes. Their phylogenetic relationships and structural comparisons indicated that these paralogues arose as the results of gene duplication events. RT-PCR detected their mRNAs in different "first line of defense" tissues, as well as in several other organs, suggesting diverse functions. Tissue selective expression also indicates their functional divergence. The identified *CYP9A* genes have not yet been found outside the Lepidoptera, and are probably unique to the Lepidoptera. They show high sequence and structural similarities to each other, indicating that the Lepidoptera-specific P450s may be of functional importance. This analysis constitutes the first report of the clustering, spatial organization, and functional divergence of P450 in the silkworm.

Published On: Molecular Biology Reports, 2010, 37(3), 1657-1664.

1 The Key Sericultural Laboratory of the Agricultural Ministry, Southwest University, Chongqing 400716, China.
2 College of Biotechnology, Southwest University, Chongqing 400716, China.
3 The Sericultural Research Institute of Hunan, Changsha 410127, China.
*Corresponding author E-mail: zhu@swu.edu.cn; xbxzh@swu.edu.cn.

家蚕多个*CYP9A*基因的特征分析

艾均文[1,2,3]　余泉友[1,2]　程廷才[1,2]　代方银[1,2]
张学松[1,2]　朱　勇[1,2*]　向仲怀[1*]

摘要：基于家蚕基因组计划的进展，重新开展了对细胞色素P450基因的重复序列的全基因组分析。我们通过RT-PCR方法克隆了家蚕4个*CYP9A*亚组成员，并由P450命名委员会命名为*CYP9A19*–*CYP9A22*。它们各自含有长度为1 593 bp的开放阅读框，并编码531个氨基酸的多肽。它们的核苷酸和氨基酸序列都具有非常高的相似性。昆虫细胞色素P450的典型基序包括血红素结合区、螺旋C、螺旋I、螺旋K和PERF，在多种蛋白质中表现出高度的序列保守性。与其cDNA序列的比对显示，这些旁系同源物具有相同的基因结构，每个包含10个外显子和9个大小不等的内含子。其内含子(所有9个内含子遵循GT-AG规则)的位置是绝对保守的。*CYP9A19*、*CYP9A20*和*CYP9A21*在17号染色体上形成串联簇，而*CYP9A22*被4个串联的醇–脱氢酶样基因隔开。它们的系统发生关系和结构比较表明，这些旁系同源物是基因重复事件的结果。RT-PCR在不同的"第一道防线"组织以及其他几个器官中检测到它们的mRNA，表明它们具有多种功能。组织选择性表达也表明了其功能差异。鉴定的*CYP9A*基因尚未在鳞翅目外发现，可能是鳞翅目特有的。它们显示出高度序列和结构相似性，表明鳞翅目特异性P450可能具有重要的功能。我们的研究首次报道了家蚕中P450的聚类、空间组织和功能分化。

1　农业部蚕桑学重点实验室，西南大学，重庆
2　生物技术学院，西南大学，重庆
3　湖南省蚕桑科学研究所，长沙

Identification and analysis of Toll-related genes in the domesticated silkworm, *Bombyx mori*

Cheng TC Zhang YL Liu C Xu PZ Gao ZH Xia QY* Xiang ZH

Abstract: Silkworm (*Bombyx mori*), a model system for Lepidoptera, has contributed enormously to the study of insect immunology especially in humoral immunity. But little is known about the molecular mechanism of immune response in the silkworm. Toll receptors are a group of evolutionarily ancient proteins, which play a crucial role in the innate immunity of both insects and vertebrates. In human, Toll-like receptors (TLRs) are the typical pattern recognition receptors for different kinds of pathogen molecules. Toll-related receptors in *Drosophila*, however, were thought to function as cytokine receptors in immune response and embryogenesis. We have identified 11 putative Toll-related receptors and two Toll analogs in the silkworm genome. Phylogenetic analysis of insect Toll family and human TLRs showed that BmTolls is grouped with *Drosophila* Tolls and *Anopheles*Tolls. These putative proteins are typical transmembrane receptors flanked by the extracellular leucine-rich repeat (LRR) domain and the cytoplasmic TIR domain. Structural prediction of the TIR domain alignment found five stranded sheets and five helices, which are alternatingly joined. Microarray data indicated that *BmToll* and *BmToll-2* were expressed with remarkable enrichment in the ovary, suggesting that they might play a role in the embryogenesis. However, the enriched expression of *BmToll-2* and *-4* in the midgut suggested that the proteins they encode may be involved in immune defense. Testis-specific expression of *BmToll-10* and *-11* and *BmToLK-2* implies that these may be involved in sex-specific biological functions. The RT-PCR results indicated that 10 genes were induced or suppressed with different degrees after their immune system was challenged by different invaders. Expression profiles of *BmTolls* and *BmToLKs* reported here provide insight into their role in innate immunity and development.

Published On: Developmental & Comparative Immunology, 2008, 32(5), 464-475.

Key Sericultural Laboratory of Agricultural Ministry, College of Life Science, Southwest University, Chongqing 400716, China.
*Corresponding author E-mail: xiaqy@swu.edu.cn.

家蚕Toll相关基因的鉴定和分析

程廷才　张雨丽　刘　春　许平震　高志宏　夏庆友*　向仲怀

摘要：家蚕作为鳞翅目的模式昆虫，对于昆虫免疫学特别是体液免疫相关领域的研究贡献巨大。然而，关于家蚕免疫应答分子机制的研究报道却很少。Toll受体是一类进化保守的蛋白质，在昆虫和脊椎动物的先天免疫系统中有重要作用。在人类中，Toll样受体是典型的模式识别受体，能识别不同类型的病原模式分子。在果蝇中，Toll相关受体被认为作为细胞因子受体参与免疫应答和胚胎发育等生物学过程。我们在家蚕基因组中鉴定了11个可能的Toll相关受体和2个Toll受体的类似物。系统进化分析显示家蚕的Toll受体与果蝇和按蚊的Toll受体聚为一群。这些假定的蛋白是典型的跨膜受体，位于细胞外富含亮氨酸重复结构域和胞质结构域的两侧。结构预测分析显示，TIR结构包含了交替排列的5个折叠结构和5个螺旋结构。基于基因芯片的表达谱分析结果显示*BmToll*和*BmToll-2*在卵巢显著富集表达，暗示了它们可能在胚胎发生过程中起作用。*BmToll-2*和*BmToll-4*在中肠组织富集表达，推测它们所编码的蛋白可能参与免疫防御。有趣的是，3个基因*BmToll-10*、*BmToll-11*和*BmToLK-2*具有精巢特异表达特征，暗示了它们可能具有性别特异的生物学功能。RT-PCR分析结果显示10个基因的表达在不同微生物感染后发生了不同程度上调或抑制。本研究开展的*BmTolls*和*BmToLKs*的表达分析将为探讨家蚕Toll相关受体基因在家蚕先天免疫和发育过程中发挥的功能提供重要参考。

农业部蚕桑学重点实验室，生命科学学院，西南大学，重庆

Identification of *MBF2* family genes in *Bombyx mori* and their expression in different tissues and stages and in response to *Bacillus bombysepticus* infection and starvation

Zhou CY[#] Zha XF[#] Liu C Han MJ Zhang LY
Shi PP Wang H Zheng RW Xia QY[*]

Abstract: The *Multiprotein bridge factor 2* (*MBF2*) gene was first identified as a co-activator involved in *Bm*FTZ-F1-mediated activation of the *Fushi tarazu* gene. Herein, nine homologous genes of *MBF*2 gene are identified. Evolutionary analysis showed that this gene family is insect-specific and that the family members are closely related to *response to pathogens* (*REPAT*) genes. Tissue distribution analysis revealed that these genes could be expressed in a tissue-specific manner. Developmental profiles analysis showed that the *MBF2* gene family members were highly expressed in the different stages. Analysis of the expression patterns of nine *MBF2* family genes showed that *Bacillus bombysepticus* treatment induced the up-regulation of several *MBF2* family genes, including *MBF2-4, -7, -9, -8.* Furthermore, we found the *MBF2* family genes were modulated by starvation and the expression of these genes recovered upon re-feeding, except for *MBF2-5, -9.* These findings suggested roles for these proteins in insect defense against pathogens and nutrient metabolism, which has an important guiding significance for designing pest control strategies.

Published On: Insect Science, 2016, 23(4), 502-512.

State Key Laboratory of Silkworm Genome Biology, Southwest University, Chongqing 400716, China.
[#]These authors contributed equally.
[*]Corresponding author E-mail: xiaqy@swu.edu.cn.

家蚕*MBF2*家族基因的鉴定及其在不同组织和时期的表达以及对家蚕饥饿和黑胸败血芽孢杆菌的应答反应

周春燕[#]　查幸福[#]　刘　春　韩民锦　张李颖
石盼盼　王　鹤　郑仁文　夏庆友[*]

摘要：多蛋白桥梁因子2（MBF2）最初在丝腺中被分离，并被鉴定为家蚕核受体因子FTZ-F1调控果蝇*ftz*基因的共转录激活因子。本研究中，我们鉴定了9个*MBF2*的同源基因。进化分析结果显示*MBF2*家族是昆虫特有的，并且与病原体的应答相关。组织表达谱表明该家族成员基因的表达具有组织特异性。时期表达谱分析结果显示该家族基因表达呈现明显的时期特异性。我们发现*MBF2-4*，*-7*，*-9*及*MBF2-8*在黑胸败血芽孢杆菌*Bb*诱导后均上调表达，表明它们参与了家蚕免疫反应。此外，除了*MBF2-5*和*MBF2-9*，其余基因在饥饿后表达量会发生变化，并且在重新喂食后表达水平可以恢复。这些结果表明*MBF2*家族基因在家蚕防御反应及营养代谢过程中发挥重要作用，这对害虫防治具有重要的指导意义。

家蚕基因组生物学国家重点实验室，西南大学，重庆

Identification and characterization of *piggy*Bac-like elements in the genome of domesticated silkworm, *Bombyx mori*

Xu HF Xia QY* Liu C Cheng TC
Zhao P Duan J Zha XF Liu SP

Abstract: *piggy*Bac is a short inverted terminal repeat (ITR) transposable element originally discovered in *Trichoplusia ni*. It is currently the preferred vector of choice for enhancer trapping, gene discovery and identifying gene function in insects and mammals. Many *piggy*Bac-like sequences have been found in the genomes of phylogenetically species from fungi to mammals. We have identified 98 *piggy*Bac-like sequences (*BmPBLE1 -98*) from the genome data of domesticated silkworm (*Bombyx mori*) and 17 fragments from expressed sequence tags (ESTs). Most of the *BmPBLE1-98* probably exist as fossils. A total of 21 *BmPBLEs* are flanked by ITRs and TTAA host dinucleotides, of which 5 contain a single ORF, implying that they may still be active. Interestingly, 16 *BmPBLEs* have CAC/GTG not CCC/GGG as the characteristic residues of ITRs, which is a surprising phenomenon first observed in the *piggy*Bac families. Phylogenetic analysis indicates that many *BmPBLEs* have a close relation to mammals, especially to *Homo sapiens*, only a few being grouped with the *T. ni piggy*Bac element. In addition, horizontal transfer was probably involved in the evolution of the *piggy*Bac-like elements between *B. mori* and *Daphnia pulicaria*. The analysis of the *BmPBLEs* will contribute to our understanding of the characteristic of the *piggy*Bac family and application of *piggy*Bac in a wide range of insect species.

Published On: Molecular Genetics and Genomics, 2006, 276(1), 31-40.

The Key Sericultural Laboratory of Agricultural Ministry of China, Southwest University, Chongqing 400716, China.
*Corresponding author E-mail: xiaqy@swu.edu.cn.

家蚕基因组中*piggy*Bac类似转座元件的鉴定和描述

徐汉福　夏庆友*　刘　春　程廷才
赵　萍　段　军　查幸福　刘仕平

摘要：*piggy*Bac是最初在粉纹夜蛾中发现的短的反向重复(ITR)转座元件。目前,它是昆虫和哺乳动物增强子捕获、基因捕获和鉴定基因功能的首选载体。从真菌到哺乳动物的基因组中已经发现了许多*piggy*Bac类似序列。我们从家蚕(*Bombyx mori*)的基因组数据和来自表达序列标签(ESTs)的17个片段中共鉴定出98个*piggy*Bac类似序列(*BmPBLE1-98*)。大多数*BmPBLE1-98*可能没有生物活性。共有21个*BmPBLE*含有ITR序列和TTAA宿主二核苷酸,其中5个含有单个ORF,这意味着它们仍然是活跃的。有趣的是,16个*BmPBLE*以CAC/GTG而不是CCC/GGG作为ITR的特征残基,这是在*piggy*Bac家族首次观察到的令人惊讶的现象。系统发育分析表明,许多*BmPBLE*与哺乳动物,特别是人的*piggy*Bac密切相关,只有少数与粉纹夜蛾*piggy*Bac原件相似。此外,本研究发现*piggy*Bac在*B. mori*和*Daphnia pulicaria*之间存在水平转移。对*BmPBLEs*的分析将有助于我们了解*piggy*Bac家族的特征,从而促进*piggy*Bac在昆虫中的应用。

农业部蚕桑学重点实验室,西南大学,重庆

The UDP-glucosyltransferase multigene family in *Bombyx mori*

Huang FF[1] Chai CL[1] Zhang Z[1,2] Liu ZH[1]
Dai FY[1] Lu C[1*] Xiang ZH[1]

Abstract:

Background: Glucosidation plays a major role in the inactivation and excretion of a great variety of both endogenous and exogenous compounds. A class of UDP-glycosyltransferases (UGTs) is involved in this process. Insect UGTs play important roles in several processes, including detoxication of substrates such as plant allelochemicals, cuticle formation, pigmentation, and olfaction. Identification and characterization of *Bombyx mori UGT* genes could provide valuable basic information for this important family and explain the detoxication mechanism and other processes in insects.

Results: Taking advantage of the newly assembled genome sequence, we performed a genome-wide analysis of the candidate *UGT* family in the silkworm, *B. mori*. Based on UGT signature and their similarity to UGT homologs from other organisms, we identified 42 putative silkworm *UGT* genes. Most of them are clustered on the silkworm chromosomes, with two major clusters on chromosomes 7 and 28, respectively. The phylogenetic analysis of these identified 42 UGT protein sequences revealed five major groups. A comparison of the silkworm *UGTs* with homologs from other sequenced insect genomes indicated that some *UGTs* are silkworm-specific genes. The expression patterns of these candidate genes were investigated with known expressed sequence tags (ESTs), microarray data, and RT-PCR method. In total, 36 genes were expressed in tissues examined and showed different patterns of expression profile, indicating that these *UGT* genes might have different functions.

Conclusion: *B. mori* possesses a largest insect *UGT* gene family characterized to date, including 42 genes. Phylogenetic analysis, genomic organization and expression profiles provide an overview for the silkworm UGTs and facilitate their functional studies in future.

Published On: BMC Genomics, 2008, 9, 563.

1 Key Sericultural Laboratory of Agricultural Ministry, Institute of Sericulture and Systems Biology, Southwest University, Chongqing 400716, China.

2 Institute of Agricultural and Life Sciences, Chongqing University, Chongqing 400044, China.

*Corresponding author E-mail: lucheng@swu.edu.cn.

家蚕尿苷二磷酸-糖基转移酶多基因家族的分析

黄飞飞[1]　柴春利[1]　张　泽[1,2]　刘增虎[1]
代方银[1]　鲁　成[1*]　向仲怀[1]

摘要：背景——糖基化作用在抑制和排除一系列内生和外源化合物的毒性方面起重要作用。一类尿苷二磷酸-糖基转移酶（UGTs）参与催化糖基化反应。昆虫的UGTs在多种生理学过程中起重要作用，包括底物的解毒，例如一些植物中的异种化感物，以及角质层形成，色素沉着和嗅觉作用等过程。对家蚕*UGT*基因进行鉴别和分析可以为解释这一重要家族在昆虫的解毒机制及其他过程中的作用提供有价值的信息。结果——利用新组装的基因组数据，我们对家蚕的*UGT*基因进行了分析。根据糖基转移酶的特征序列以及与其他物种的同源性，我们在家蚕基因组中共鉴别出了42个*UGT*基因。大部分基因成簇分布于染色体上，主要有两大基因群分别位于7号和28号染色体上。系统发生分析表明，这42个基因主要分为5大类群。对家蚕的*UGT*基因与其他已进行基因组测序的昆虫的同源基因进行比较发现，其中一些基因是家蚕特有的*UGT*基因。利用表达序列标签（ESTs）数据，基因芯片数据和RT-PCR对家蚕的*UGT*基因进行了表达模式的研究，总共有36个基因在组织中有表达证据，很多基因表现不同的表达模式，表明这些*UGT*基因可能具有不同的功能。家蚕基因组中存在42个*UGT*基因，这是迄今为止报道的拥有这一家族基因数目最多的昆虫。对家蚕*UGT*的系统发生分析，基因组结构和表达模式研究为以后研究这一基因家族的功能提供了重要基础。

1　农业部蚕桑学重点实验室，蚕学与系统生物学研究所，西南大学，重庆
2　农学与生命科学研究院，重庆大学，重庆

Annotation and expression of carboxylesterases in the silkworm, *Bombyx mori*

Yu QY[1,2] Lu C[2*] Li WL[2] Xiang ZH[2] Zhang Z[1,2*]

Abstract:

Background: Carboxylesterase is a multifunctional superfamily and ubiquitous in all living organisms, including animals, plants, insects, and microbes. It plays important roles in xenobiotic detoxification, and pheromone degradation, neurogenesis and regulating development. Previous studies mainly used Dipteran *Drosophila* and mosquitoes as model organisms to investigate the roles of the insect COEs in insecticide resistance. However, genome-wide characterization of COEs in phytophagous insects and comparative analysis remain to be performed.

Results: Based on the newly assembled genome sequence, 76 putative *COEs* were identified in *Bombyx* mori. Relative to other Dipteran and Hymenopteran insects, alpha-esterases were significantly expanded in the silkworm. Genomics analysis suggested that *BmCOEs* showed chromosome preferable distribution and 55% of which were tandem arranged. Sixty-one *BmCOEs* were transcribed based on cDNA/ESTs and microarray data. Generally, most of the *COEs* showed tissue specific expressions and expression level between male and female did not display obvious differences. Three main patterns could be classified, i.e. midgut-, head and integument-, and silk gland-specific expressions. Midgut is the first barrier of xenobiotics peroral toxicity, in which COEs may be involved in eliminating secondary metabolites of mulberry leaves and contaminants of insecticides in diet. For head and integument-class, most of the members were homologous to odorant-degrading enzyme (ODE) and antennal esterase. RT-PCR verified that the ODE-like esterases were also highly expressed in larvae antenna and maxilla, and thus they may play important roles in degradation of plant volatiles or other xenobiotics.

Conclusion: *B. mori* has the largest number of insect *COE* genes characterized to date. Comparative genomic analysis suggested that the gene expansion mainly occurred in silkworm alpha-esterases. Expression evidence indicated that the expanded genes were specifically expressed in midgut, integument and head, implying that these genes may have important roles in detoxifying secondary metabolites of mulberry leaves, contaminants in diet, and odorants. Our results provide some new insights into functions and evolutionary characteristics of COEs in phytophagous insects.

Published On: BMC Genomics, 2009, 10.

1 Institute of Agricultural and Life Sciences, Chongqing University, Chongqing 400044, China.

2 Key Sericultural Laboratory of the Agricultural Ministry of China, Southwest University, Chongqing 400716, China.

*Corresponding author E-mail: lucheng@swu.edu.cn; zezhang@swu.edu.cn.

家蚕羧酸酯酶的注释和表达

余泉友[1,2] 鲁 成[2*] 李文乐[2] 向仲怀[2] 张 泽[1,2*]

摘要：背景——羧酸酯酶(COEs)是一种多功能超家族酶，遍布于包括动物、植物和微生物在内的所有生物体中。它在脱毒、信息素降解、神经发生和调节发育中发挥着重要作用。先前的研究主要使用双翅目果蝇和蚊子作为模式生物来研究昆虫COEs在杀虫剂抗性中的作用。然而，植食性昆虫的COEs在全基因组的表征和比较分析仍然需要进一步研究。结果——基于新组装的基因组序列，在蚕蛾中鉴定预测出76个*COEs*。相对于其他双翅目和膜翅目昆虫，α-酯酶在家蚕中出现明显的基因扩增。基因组学分析表明，*BmCOEs*呈现染色体优先分布，其中55%是串联排列的。基于cDNA / EST和芯片数据共鉴定61个*BmCOEs*转录本。一般来说，大多数*COEs*具有组织特异性表达，雌性和雄性之间的表达水平没有明显差异。可分为三种主要的表达模式，即中肠、头和表皮以及丝腺特异性表达。中肠是食用外源毒性异物后的第一个屏障，其中COEs可能参与降解桑叶的次生代谢产物和饲料中的杀虫剂等。而头部与表皮中的大多数COEs与气味降解酶(ODE)和触角酯酶同源。RT-PCR证实ODE样酯酶也在幼虫触角和上颌骨中高度表达，因此它们可能在植物挥发物或其他异生物降解中起重要作用。结论——家蚕是迄今为止发现的具有*COE*基因数量最多的昆虫。比较基因组分析表明，家蚕α-酯酶存在基因扩增。表达证据表明扩增的*COE*基因在中肠、表皮和头部特异性表达，这意味着这些基因可能在降解桑叶的次生代谢物、食物中的异物和有气味的物质等过程中起重要作用。我们的研究结果为植食性昆虫中COEs的功能和进化特征提供了一些新的见解。

1 农学及生命科学研究院，重庆大学，重庆

2 农业部蚕桑重点实验室，西南大学，重庆

Genome-wide identification and characterization of ATP-binding cassette transporters in the silkworm, *Bombyx mori*

Liu SM　Zhou S　Tian L　Guo EE
Luan YX　Zhang JZ　Li S*

Abstract: BACKGROUND: The ATP-binding cassette (ABC) transporter superfamily is the largest transporter gene family responsible for transporting specific molecules across lipid membranes in all living organisms. In insects, ABC transporters not only have important functions in molecule transport, but also play roles in insecticide resistance, metabolism and development.

RESULTS: From the genome of the silkworm, *Bombyx mori*, we have identified 51 putative *ABC* genes which are classified into eight subfamilies (A-H) by phylogenetic analysis. Gene duplication is very evident in the *ABCC* and *ABCG* subfamilies, whereas gene numbers and structures are well conserved in the *ABCD*, *ABCE*, *ABCF*, and *ABCH* subfamilies. Microarray analysis revealed that expression of 32 silkworm *ABC* genes can be detected in at least one tissue during different developmental stages, and the expression patterns of some of them were confirmed by quantitative real-time PCR. A large number of *ABC* genes were highly expressed in the testis compared to other tissues. One of the *ABCG* genes, *BmABC002712*, was exclusively and abundantly expressed in the Malpighian tubule implying that *BmABC002712* plays a tissue-specific role. At least 5 *ABCG* genes, including *BmABC005226*, *BmABC005203*, *BmABC005202*, *BmABC010555*, and *BmABC010557*, were preferentially expressed in the midgut, showing similar developmental expression profiles to those of 20-hydroxyecdysone (20E)-response genes. 20E treatment induced the expression of these *ABCG* genes in the midgut and RNA interference-mediated knockdown of *USP*, a component of the 20E receptor, decreased their expression, indicating that these midgut-specific *ABCG* genes are 20E-responsive.

CONCLUSION　In this study, a genome-wide analysis of the silkworm ABC transporters has been conducted. A comparison of ABC transporters from 5 insect species provides an overview of this vital gene superfamily in insects. Moreover, tissue- and stage-specific expression data of the silkworm *ABCG* genes lay a foundation for future analysis of their physiological function and hormonal regulation.

Published On: BMC Genomics, 2011, 12, 491.

Key Laboratory of Developmental and Evolutionary Biology, Institute of Plant Physiology and Ecology, Shanghai Institutes for Biological Sciences, Chinese Academy of Sciences, Shanghai 200032, China.
*Corresponding author E-mail: shengli@sippe.ac.cn.

家蚕基因组中ATP结合转运蛋白的鉴定与描述

刘淑敏　周　顺　田　铃　郭恩恩

栾云霞　张建珍　李　胜*

摘要：背景：ATP结合转运蛋白（ATP-binding cassette，ABC）超家族蛋白是最大的转运家族，在所有生物中均负责转运特殊分子跨越脂质膜。昆虫中ABC结合转运蛋白不仅在分子运输中具有重要功能，而且在昆虫对杀虫剂的抗药性、代谢和发育中都发挥重要功能。结果：在家蚕的基因组，我们鉴定到51个可能的*ABC*基因，系统进化分析将它们分为A–H八个亚家族，其中*ABCC*和*ABCG*亚家族中基因重复非常明显，而基因数量和结构在*ABCD*、*ABCE*、*ABCF*和*ABCH*亚家族中则非常保守。基因芯片结果显示，有32个家蚕*ABC*基因能够在不同发育时期的、至少一种组织中被检测到，其中有些基因的表达模式通过实时荧光定量PCR进行了确认。与其他组织相比，大量*ABC*基因在精巢中高量表达。*ABCG*亚家族成员*BmABC002712*在马氏管中特异、高量表达，暗示其在马氏管中发挥组织特异性的功能。至少5个*ABCG*基因，包括*BmABC005226*、*BmABC005203*、*BmABC005202*、*BmABC010555*和*BmABC010557*主要在中肠中表达，与20E应答基因具有相似的表达模式。20E处理诱导中肠中*ABCG*基因的表达，而20E受体复合物中的*USP* RNAi后，它们的表达水平下调，显示这些中肠特异的*ABCG*基因受20E调控。结论：在本研究中，对家蚕ABC转运蛋白基因进行了全基因组分析。通过对包括家蚕在内的5种昆虫ABC转运蛋白的比较，对这一重要的基因超家族进行了综合论述。此外，家蚕*ABCG*亚家族基因的组织和时空特异表达数据为将来研究它们的生理功能和激素调控过程奠定了基础。

昆虫发育与进化生物学重点实验室，上海生命科学研究院植物生理生态研究所，中国科学院，上海

Genome-wide analysis of the ATP-binding cassette (ABC) transporter gene family in the silkworm, *Bombyx mori*

Xie XD[1] Cheng TC[1*] Wang GH[1]
Duan J[1] Niu WH[1] Xia QY[2*]

Abstract: The ATP-binding cassette (ABC) superfamily is a larger protein family with diverse physiological functions in all kingdoms of life. We identified 53 ABC transporters in the silkworm genome, and classified them into eight subfamilies (A-H). Comparative genome analysis revealed that the silkworm has an expanded *ABCC* subfamily with more members than *Drosophila melanogaster, Caenorhabditis elegans*, or *Homo sapiens*. Phylogenetic analysis showed that the *ABCE* and *ABCF* genes were highly conserved in the silkworm, indicating possible involvement in fundamental biological processes. Five multidrug resistance-related genes in the *ABCB* subfamily and two multidrug resistance-associated-related genes in the *ABCC* subfamily indicated involvement in biochemical defense. Genetic variation analysis revealed four *ABC* genes that might be evolving under positive selection. Moreover, the silkworm *ABCC4* gene might be important for silkworm domestication. Microarray analysis showed that the silkworm *ABC* genes had distinct expression patterns in different tissues on day 3 of the fifth instar. These results might provide new insights for further functional studies on the *ABC* genes in the silkworm genome.

Published On: Molecular Biology Reports, 2012, 39(7), 7281-7291.

1 Institute of Agricultural and Life Science, Chongqing University, Chongqing 400044, China.
2 State Key Laboratory of Silkworm Genome Biology, Southwest University, Chongqing 400716, China.
*Corresponding author E-mail: chengtc@cqu.edu.cn; xiaqy@swu.edu.cn.

家蚕ATP结合盒(ABC)转运蛋白基因家族的全基因组分析

谢小东[1] 程廷才[1*] 王根洪[1]
段 军[1] 牛维环[1] 夏庆友[2*]

摘要:ATP结合盒(ABC)超家族是一个在所有生命界都具有不同生理功能的较大蛋白家族。在家蚕基因组中,我们共鉴定到53种ABC转运蛋白,并将其分为8个亚家族(A-H)。对比基因组分析发现家蚕的*ABCC*亚家族比黑腹果蝇、秀丽隐杆线虫或人的更大,成员更多。进化分析显示家蚕的*ABCE*和*ABCF*基因高度保守,暗示着它们可能参与基本的生物学过程。*ABCB*亚家族的5个多药耐药相关基因和*ABCC*亚家族的2个多药耐药相关基因参与生化防御。遗传变异分析发现4个可能在正选择下发生进化的*ABC*基因。此外,家蚕*ABCC4*基因或许对家蚕的驯化具有重要的作用。芯片分析显示家蚕*ABC*基因在五龄第3天的不同组织中具有不同的表达模式。这些结果可能为家蚕基因组中*ABC*基因的进一步功能研究提供新的见解。

1 农学及生命科学研究院,重庆大学,重庆
2 家蚕基因组生物学国家重点实验室,西南大学,重庆

Genome-wide identification and expression profiling of the fatty acid desaturase gene family in the silkworm, *Bombyx mori*

Chen QM Cheng DJ Liu SP
Ma ZG Tan X Zhao P*

Abstract: Fatty acid desaturases exist in all living organisms and play important roles in many different biologic processes, such as fatty acid metabolism, lipid biosynthetic processes, and pheromone biosynthetic processes. Using the available silkworm genome sequence, we identified 14 candidate fatty acid desaturase genes. Eleven genes contain 3 conserved histidine cluster motifs and 4 transmembrane domains, but their N-terminal residues exhibit obvious diversity. Phylogenetic analysis revealed that there are 6 groups; *Bmdesat*1 and *Bmdesat5-8* were clustered into group 2, which is involved in Δ11 desaturation activity, and *Bmdesat3-4* were grouped in group 1, which is involved in Δ9 desaturation activity. Twelve of the 14 genes have expressed sequence tag evidence. Microarray data and reverse transcription polymerase chain reaction analysis demonstrated that *Bmdesat3-4* and *Bmdesat10* were expressed from the larval to moth stages and in multiple tissues on day 3 of 5th instar larvae. *Bmdesat9*, *Bmdesat11*, and *Bmdesat14* were expressed during the pupal and late-embryonic stage, suggesting that they may take part in fatty acid metabolism to provide energy. These results provide some insights into the functions of individual fatty acid desaturases in silkworm.

Published On: Genetics and Molecular Research, 2014, 13 (2), 3747-3760.

State Key Laboratory of Silkworm Genome Biology, Southwest University, Chongqing 400716, China.
*Corresponding author E-mail: zhaop@swu.edu.cn.

家蚕脂肪酸脱氢酶基因家族的全基因组鉴定和表达谱分析

陈全梅　程道军　刘仕平
马振刚　谭　祥　赵　萍*

摘要：脂肪酸脱氢酶存在于所有生物体内，并在不同的生命过程中发挥重要作用，如脂肪酸的代谢、脂质的生物合成及信息素的生物合成等。利用已知的家蚕基因序列，我们鉴定了14个备选的脂肪酸脱氢酶基因。11个基因含有3个保守的组氨酸簇基序和4个跨膜结构域，但它们的N末端残基存在明显差异。系统发育分析表明，这些基因分为6组，*Bmdesat1*和*Bmdesat5–8*为第2组，参与Δ11脱氢活性；*Bmdesat3–4*为第1组，参与Δ9脱氢活性；其中12个基因有表达序列标签。芯片数据和RT-PCR分析显示*Bmdesat3–4*和*Bmdesat10*从幼虫到成虫期都有表达，在五龄第3天的幼虫的多种组织中也有表达。*Bmdesat9*、*Bmdesat11*和*Bmdesat14*在蛹期及晚期胚胎期表达，这表明它们可能参与脂肪酸代谢，提供能量。这些结果为不同的家蚕脂肪酸脱氢酶功能研究提供了新的视野。

家蚕基因组生物学国家重点实验室，西南大学，重庆

Characters and expression of the gene encoding DH, PBAN and other FXPRLamide family neuropeptides in *Antheraea pernyi*

Wei ZJ[1,2*] Hong GY[1] Jiang ST[1] Tong ZX[3] Lu C[2]

Abstract: The full-length cDNA encoding diapause hormone (DH) and pheromone biosynthesis activating neuropeptide (PBAN) in *Antheraea pernyi* (Anp-DH-PBAN) were cloned and sequenced by rapid amplification of cDNA ends methods. The *Anp-DH-PBAN* cDNA encodes a 196-amino acid (aa) prehormone that contains a 24-aa DH-like peptide, a 33-aa PBAN and three other neuropeptides, all of which share a common C-terminal pentapeptide motif FXPR / KL (X = G, T, S). The Anp-DH-PBAN shows highest homology (82%) to that of *Samia cynthia ricini* at amino acid level. Northern blots demonstrate the presence of a 0.8-kb transcript in the brain and suboesophageal ganglion (SG) complex. During the early pupal stages, the *Anp-DH-PBAN* mRNA contents increase consistently in both diapause- and non-diapause-destined pupae. From day 7, mRNA in diapause pupae dropped promptly and maintained a lower level than that in the non-diapause type. The *Anp-DH-PBAN* mRNA was expressed mainly in SG, at much lower levels in brain and thoracic ganglia (TG), but not in non-neural tissues. FXPRLamide (Phe-X-Pro-Arg-Leu) peptide immunoreactivity was detected in the SG, TG and terminal abdominal ganglion of *A. pernyi* by whole-mount immunocytochemistry. The titres of FXPRLamide peptides in the haemolymph in diapause type are consistently lower than in non-diapause insects. In non-diapause individuals, there are two peaks of FXPRLamide titres, day 2 in the wandering stage and day-5 in the pupal stage respectively.

Published On: Journal of Applied Entomology, 2008, 132 (1), 59-67.

1 Department of Biotechnology, Hefei University of Technology, Hefei 230002, China.

2 Sericultural Laboratory of Agriculture Ministry, Southwest University, Chongqing 400716, China.

3 Wild Silkmoth Research Center, Sericultural Research Institute of Liaoning Province, Fengcheng 118100, China.

*Corresponding author E-mail: zjwei@hfut.edu.cn.

柞蚕中编码DH、PBAN和其他FXPRLamide家族神经肽的基因的特征和表达

魏兆军[1,2*] 洪桂云[1] 姜绍通[1] 仝振祥[3] 鲁 成[2]

摘要：通过cDNA末端快速扩增法克隆并测序了柞蚕中编码滞育激素(DH)和信息素生物合成活化神经肽(PBAN)的全长cDNA。*Anp-DH-PBAN* cDNA编码196个氨基酸(aa)激素前体，其含有24个氨基酸的类DH肽、33个氨基酸PBAN和另外3个神经肽，所有这些神经肽共享一个共同的C末端五肽基序FXPR/KL(X=G,T,S)。Anp-DH-PBAN在氨基酸水平上与蓖麻蚕的同源性最高(82%)。Northern印迹显示在脑和咽下神经节(SG)复合物中存在0.8 kb转录体。在蛹期早期，*Anp-DH-PBAN* mRNA含量在滞育和非滞育的蛹中均持续增加。从第7天起，滞育蛹中的mRNA迅速下降，维持在非滞育蛹的水平。*Anp-DH-PBAN* mRNA主要在SG中表达，在脑和胸下神经节(TG)中低得多，但在非神经组织中不表达。用免疫组织化学的方法检测FXPRLamide(Phe-X-Pro-Arg-Leu)多肽在柞蚕SG、TG和腹部末端神经节中的免疫活性，滞育型柞蛋白淋巴中FXPRLamide肽的滴度一直低于非滞育个体。在非滞育个体中，FXPRLamide滴度存在两个峰值，分别是上蔟第2天和蛹第5天。

1 生物技术学院，合肥工业大学，合肥
2 农业部蚕桑学重点实验室，西南大学，重庆
3 蚕业科学研究所，辽宁省农业科学院，凤城

Transcriptome analysis of the brain of the silkworm *Bombyx mori* infected with *Bombyx mori* nucleopolyhedrovirus: a new insight into the molecular mechanism of enhanced locomotor activity induced by viral infection

Wang GB[1] Zhang JJ[1] Shen YW[1] Zheng Q[1]
Feng M[1] Xiang XW[2] Wu XF[1*]

Abstract: Baculoviruses have been known to induce hyperactive behavior in their lepidopteran hosts for over a century. As a typical lepidopteran insect, the silkworm *Bombyx mori* displays enhanced locomotor activity (ELA) following infection with *B. mori* nucleopolyhedrovirus (*Bm*NPV). Some investigations have focused on the molecular mechanisms underlying this abnormal hyperactive wandering behavior due to the virus; however, there are currently no reports about *B. mori.* Based on previous studies that have revealed that behavior is controlled by the central nervous system, the transcriptome profiles of the brains of *Bm*NPV-infected and non-infected silkworm larvae were analyzed with the RNA-Seq technique to reveal the changes in the *Bm*NPV-infected brain on the transcriptional level and to provide new clues regarding the molecular mechanisms that underlies *Bm*NPV-induced ELA. Compared with the controls, a total of 742 differentially expressed genes (*DEGs*), including 218 up-regulated and 524 down-regulated candidates, were identified, of which 499, 117 and 144 *DEGs* could be classified into GO categories, KEGG pathways and COG annotations by GO, KEGG and COG analyses, respectively. We focused our attention on the *DEGs* that are involved in circadian rhythms, synaptic transmission and the serotonin receptor signaling pathway of *B. mori.* Our analyses suggested that these genes were related to the locomotor activity of *B. mori* via their essential roles in the regulations of a variety of behaviors and the down-regulation of their expressions following *Bm*NPV infection. These results provide new insight into the molecular mechanisms of *Bm*NPV-induced ELA.

Published On: Journal of Invertebrate Pathology, 2015, 128, 37-43.

1 College of Animal Science, Zhejiang University, Hangzhou 310029, China.
2 Zhejiang Marine Development Research Institute, Zhoushan 316000, China.
*Corresponding author E-mail: wuxiaofeng@zju.edu.cn.

感染*Bm*NPV后家蚕脑的转录组分析：病毒感染诱导宿主运动增强行为的分子机制

王国宝[1] 张健家[1] 沈运旺[1] 郑 秦[1]
冯 敏[1] 相兴伟[2] 吴小锋[1*]

摘要：在鳞翅目昆虫中，发现杆状病毒可以诱导宿主运动行为增强这一现象已经有一个多世纪。家蚕是一种典型的鳞翅目昆虫，在感染家蚕核型多角体病毒（*Bm*NPV）后表现出运动行为增强的现象（ELA）。目前，有一些研究在关注这种由病毒感染引起的异常多动行为的分子机制。但是，在家蚕中却未见报道。以前的研究表明这种行为受中枢神经系统调控，因此，我们利用RNA测序技术分析*Bm*NPV感染和未感染的家蚕幼虫脑的转录组，旨在揭示转录组水平上*Bm*NPV感染后家蚕脑的变化，为*Bm*NPV诱发的ELA的分子机制提供新的依据。与对照组相比，共鉴定到742个差异基因（*DEGs*），包括218个上调候选基因和524个下调候选基因，其中499、117、144个*DEGs*可分别进行GO分类、KEGG通路和COG注释分析。我们主要关注涉及家蚕的周期性节律、突触传导、5-羟色胺受体信号通路的*DEGs*。我们的分析表明，这些基因调控着昆虫的各种行为并且在*Bm*NPV感染后会下调表达，从而影响家蚕的运动行为。这些结果为研究*Bm*NPV诱发的ELA的分子机制提供了新思路。

1 动物科学学院，浙江大学，杭州
2 浙江省海洋开发研究院，舟山

Analysis of oral infection and helicase gene of the nucleopolyhedroviruses isolated from *Philosamia cynthia ricini* and *Antheraea pernyi*

Chen Y[1] Li B[2] Shen WD[2]

Zhu SY[1] Wu Y[1] Wang WB[1*]

Abstract: *Philosamia cynthia ricini* is an important commercial silkworm in Asia. In this report, a nucleopolyhedrovirus isolated from *P. cynthia ricini* (*Phcy*NPV) larva was purified and compared with *Antheraea pernyi* nucleopolyhedrovirus (*Anpe*NPV), a pathogen of *A. pernyi*, another commercial silkworm in China. The two viruses had similar polyhedral morphology and shared high sequence homologue of viral fragments including the *p143* gene. However, the restriction fragments, digested with *Sal* I, *Xho* I, *Hin*d III and *Pst* I, respectively, were different. The cross-infectivity of the two viruses was also tested. *Anpe*NPV caused 57% mortality in larvae of *P. cynthia ricini*, whereas *Phcy*NPV did not kill larvae of *A. pernyi*. Results indicated that *Phcy*NPV and *Anpe*NPV had closed relatedness, and that *Phcy*NPV might be a variant of *Anpe*NPV.

Published On: Biocontrol Science and Technology, 2008, 18(9), 967-973.

1 Institute of Life Sciences, Jiangsu University, Zhenjiang 212013, China.

2 School of Life Sciences, Soochow University, Suzhou 215123, China.

*Corresponding author E-mail: wenbingwang@ujs.edu.cn.

比较分析蓖麻蚕和柞蚕核型多角体病毒的经口感染途径及解旋酶基因

陈　言[1]　李　兵[2]　沈卫德[2]
朱姗颖[1]　吴　岩[1]　王文兵[1*]

摘要：亚洲蓖麻蚕是重要的产丝昆虫。我们从蓖麻蚕中分离到杆状病毒，与感染柞蚕（另一种中国经济家蚕）的柞蚕杆状病毒进行比较。两者在多角体外形上很相似，并同时含有高度同源性病毒片段（包括*p143*基因在内），但分别用限制性内切酶*Sal*I，*Xho*I，*Hin*dIII 和 *Pst*I 消化后的片段在条带上有差异。同时对这两种病毒进行了交叉感染能力检测，柞蚕杆状病毒可以造成57%的蓖麻蚕个体发生死亡，但蓖麻蚕杆状病毒不能导致柞蚕死亡。结果表明蓖麻蚕杆状病毒和柞蚕杆状病毒有较近的亲缘关系，蓖麻蚕杆状病毒可能是柞蚕杆状病毒的一个突变种。

1　生命科学研究院，江苏大学，镇江
2　生命科学学院，苏州大学，苏州

The odorant binding protein gene family from the genome of silkworm, *Bombyx mori*

Gong DP[1,2] Zhang HJ[1] Zhao P[1]
Xia QY[1,3*] Xiang ZH[1]

Abstract:

Background: Chemosensory systems play key roles in the survival and reproductive success of insects. Insect chemoreception is mediated by two large and diverse gene superfamilies, chemoreceptors and odorant binding proteins (OBPs). OBPs are believed to transport hydrophobic odorants from the environment to the olfactory receptors.

Results: We identified a family of *OBP*-like genes in the silkworm genome and characterized their expression using oligonucleotide microarrays. A total of forty-four *OBP* genes were annotated, a number comparable to the 57 *OBPs* known from *Anopheles gambiae* and 51 from *Drosophila melanogaster*. As seen in other fully sequenced insect genomes, most silkworm *OBP* genes are present in large clusters. We defined six subfamilies of *OBPs*, each of which shows lineage-specific expansion and diversification. EST data and *OBP* expression profiles from multiple larvae tissues of day three fifth instars demonstrated that many *OBPs* are expressed in chemosensory-specific tissues although some *OBPs* are expressed ubiquitously and others exclusively in non-chemosensory tissues. Some atypical *OBPs* are expressed throughout development. These results reveal that, although many *OBPs* are chemosensory-specific, others may have more general physiological roles.

Conclusion: Silkworms possess a number of *OBPs* genes similar to other insects. Their expression profiles suggest that many *OBPs* may be involved in olfaction and gustation as well as general carriers of hydrophobic molecules. The expansion of *OBP* gene subfamilies and sequence divergence indicate that the silkworm *OBP* family acquired functional diversity concurrently with functional constraints. Further investigation of the *OBPs* of the silkworm could give insights in the roles of *OBPs* in chemoreception.

Published On: BMC Genomics, 2009, 10(332).

1 The Key Sericultural Laboratory of Agricultural Ministry, Southwest University, Chongqing 400716, China.

2 Key Laboratory for Tobacco Quality Control, Ministry of Agriculture, Tobacco Research Institute, Chinese Academy of Agricultural Science, Qingdao 266101, China.

3 Institute of Agronomy and Life Sciences, Chongqing University, Chongqing 400030, China.

*Corresponding author E-mail: xiaqy@swu.edu.cn.

家蚕基因组中的气味结合蛋白基因家族的分析

龚达平[1,2]　张辉洁[1]　赵　萍[1]
夏庆友[1,3*]　向仲怀[1]

摘要：背景——化学感受系统在昆虫的生存和繁殖过程中扮演着非常重要的角色。昆虫的化学感受主要是由两大类分化的基因超家族介导的，即气味结合蛋白和化学感受器基因超家族。气味结合蛋白从外部运输疏水的气味分子到嗅觉感受器。结果——我们在家蚕基因组中鉴定到了一个类气味结合蛋白基因家族，并利用芯片分析了它们的表达特征。家蚕基因组中共有44个注释为气味结合蛋白基因，和果蝇（51个）、按蚊（57个）中的气味结合蛋白基因数量相当。与其他昆虫基因组中的气味结合蛋白基因一样，家蚕的气味结合蛋白基因在基因组中成簇分布。这些基因可分成6个亚家族，每个亚家族表现出明显的世系扩张和分化。EST芯片数据和五龄第3天大量的组织表达谱分析表明，许多气味结合蛋白基因在化学感受特异的组织器官中表达，但也有部分气味结合蛋白成员在多个组织中，或在非化学感受组织中表达。一些非典型味结合蛋白在整个发育过程中都表达，这表明部分气味结合蛋白基因是化学特异的，还有些基因可能具有更多的普遍生理作用。结论——家蚕具有与其他昆虫相似的*OBPs*基因。它们的表达模式表明许多*OBPs*与味觉和嗅觉有关，也可以作为疏水分子的载体，*OBP*基因亚家族的扩张和序列的差异，表明家蚕*OBP*家族在获得功能多样性的同时，其功能也受到了限制。对家蚕*OBPs*基因的深入研究有助于阐明其在化学感受中的作用。

1　农业部蚕桑学重点实验室，西南大学，重庆
2　农业部烟草质量控制重点实验室，中国农业科学院烟草研究所，青岛
3　农学及生命科学研究院，重庆大学，重庆

The basic helix-loop-helix transcription factor family in *Bombyx mori*

Wang Y[1] Chen KP[2*] Yao Q[2]
Wang WB[2] Zhu Z[1]

Abstract: The basic helix-loop-helix (bHLH) proteins are a superfamily of transcription factors that play important roles in a wide range of developmental processes in higher organisms. bHLH family members have been identified in a dozen of organisms including fruit fly, mouse and human. We identified 52 bHLH members in silico in the silkworm genome. Phylogenetic analyses revealed that they belong to 39 bHLH families with 21, 10, 12, 1, 7 and 1 members in groups A, B, C, D, E and F, respectively. Genes that encode ASCb, NeuroD, Oligo, MyoRb, Figα and Mad were not found in the silkworm genome. The present study provides important background information for future studies using the silkworm as a model system for insect development. Besides, the in-group phylogenetic analysis was demonstrated to be effective in classifying identified bHLH sequences into corresponding families, which can be helpful in the classification of bHLH members of other organisms.

Published On: Development, Genes and Evolution, 2007, 217(10), 715-723.

1 Department of Biotechnology, Faculty of Food and Biological Engineering, Jiangsu University, Zhenjiang 212013, China.
2 Institute of Life Science, Jiangsu University, Zhenjiang 212013, China.
*Corresponding author E-mail: kpchen@ujs.edu.cn.

家蚕碱性螺旋-环-螺旋(bHLH)转录因子家族分析

王 勇[1] 陈克平[2*] 姚 勤[2]
王文兵[2] 朱 帜[1]

摘要:碱性螺旋-环-螺旋(bHLH)蛋白是高等生物发育过程中起重要作用的转录因子超家族。bHLH家族成员已经在包括果蝇、小鼠和人类等的十几种生物体中被鉴定到。我们在家蚕基因组中鉴定到了52个bHLH成员。系统发育分析显示,它们属于39个bHLH家族,分别在A,B,C,D,E和F组中含有21,10,12,1,7和1个成员。在家蚕基因组中没有发现编码ASCb,NeuroD,Oligo,MyoRb,Figα和Mad的基因。本研究为今后以家蚕作为昆虫发育模型系统的研究提供了重要的背景信息。此外,组内系统发育分析被证明可以有效地将鉴定到的bHLH序列分类到相应的家族中,这有助于其他生物体中bHLH成员的分类。

1 生物技术系,食品与生物工程学院,江苏大学,镇江
2 生命科学研究院,江苏大学,镇江

Identification and characterization of 30K protein genes found in *Bombyx mori* (Lepidoptera: Bombycidae) transcriptome

Shi XF[1,2] Li YN[1] Yi YZ[3] Xiao XG[2] Zhang ZF[1*]

Abstract: The 30K proteins, the major group of hemolymph proteins in the silkworm, *Bombyx mori* (Lepidoptera: Bombycidae), are structurally related with molecular masses of 30 kDa and are involved in various physiological processes, e.g., energy storage, embryonic development, and immune responses. For this report, known 30K protein gene sequences were used as Blastn queries against sequences in the *B. mori* transcriptome (SilkTransDB). Twenty-nine cDNAs (*Bm30K-1-29*) were retrieved, including four being previously unidentified in the Lipoprotein_11 family. The genomic structures of the 29 genes were analyzed and they were mapped to their corresponding chromosomes. Furthermore, phylogenetic analysis revealed that the 29 genes encode three types of 30K proteins. The members increased in each type is mainly a result of gene duplication with the appearance of each type preceding the differentiation of each species included in the tree. Real-Time Quantitative Polymerase Chain Reaction (Q-PCR) confirmed that the genes could be expressed, and that the three types have different temporal expression patterns. Proteins from the hemolymph was separated by SDS-PAGE, and those with molecular mass of 30 kDa were isolated and identified by mass spectrometry sequencing in combination with searches of various databases containing *B. mori* 30K protein sequences. Of the 34 proteins identified, 13 are members of the 30K protein family, with one that had not been found in the SilkTransDB, although it had been found in the *B. mori* genome. Taken together, our results indicate that the 30K protein family contains many members with various functions. Other methods will be required to find more members of the family.

Published On: Journal of Insect Science, 2015, 15(1).

1 The Biotechnology Research Institute, Chinese Academy of Agricultural Sciences, Beijing 100081, China.

2 The College of Biological Sciences, China Agricultural University, Beijing 100094, China.

3 The Sericultural Research Institute, Chinese Academy of Agricultural Sciences, Zhenjiang 212018, China.

*Corresponding author E-mail: zhifangzhang@yahoo.com.

家蚕转录组中30K蛋白基因的鉴定与表征

石小峰[1,2] 李轶女[1] 易咏竹[3] 肖兴国[2] 张志芳[1*]

摘要：家蚕(鳞翅目昆虫)血淋巴蛋白中的主要成分是30K蛋白,其分子质量大小为30 kDa左右,涉及多种生理学过程,如能量储存、胚胎发育、免疫应答等。本研究利用已知的30K蛋白基因序列在家蚕转录组数据中进行同源性比对。29个cDNAs(*Bm30K-1-29*)被比对,包括4个之前在脂蛋白11家族中被确认的。对这29个基因的结构进行分析,并将其分别定位到相应的染色体上。此外系统进化发育分析表明这29个基因编码了3种类型的30K蛋白,从进化树中可得知,每种类型的增加主要是因为基因在物种分化之前发生了复制。实时定量PCR结果表明这些基因均有表达,且3种类型的时间表达模式不同。通过SDS-PAGE分离血淋巴中的蛋白质,并将分子质量为30 kDa的蛋白质分离,并结合多个数据库中含30K蛋白序列结果的质谱测序分析来鉴定这些蛋白质。34个蛋白质被鉴定到,其中13个是30K蛋白家族的成员,其中一种在SilkTransDB中没有发现,尽管它已经在家蚕基因组中发现。总之,本研究结果表明30K蛋白家族包含的成员行使着多种功能,需要通过其他的方法来寻找这个家族的其他成员。

1 生物技术研究所,中国农业科学院,北京

2 生物科学学院,中国农业大学,北京

3 蚕业研究所,中国农业科学院,镇江

A genome-wide survey for host response of silkworm, *Bombyx mori* during pathogen *Bacillus bombysepticus* infection

Huang LL[1,3] Cheng TC[2] Xu PZ[1]
Cheng DJ [1] Fang T[1] Xia QY[1,2*]

Abstract: Host-pathogen interactions are complex relationships, and a central challenge is to reveal the interactions between pathogens and their hosts. *Bacillus bombysepticus* (*Bb*) which can produces spores and parasporal crystals was firstly separated from the corpses of the infected silkworms (*Bombyx mori*). *Bb* naturally infects the silkworm can cause an acutefuliginosa septicaemia and kill the silkworm larvae generally within one day in the hot and humid season. *Bb* pathogen of the silkworm can be used for investigating the host responses after the infection. Gene expression profiling during four time-points of silkworm whole larvae after *Bb* infection was performed to gain insight into the mechanism of *Bb*-associated host whole body effect. Genome-wide survey of the host genes demonstrated many genes and pathways modulated after the infection. GO analysis of the induced genes indicated that their functions could be divided into 14 categories. KEGG pathway analysis identified that six types of basal metabolic pathway were regulated, including genetic information processing and transcription, carbohydrate metabolism, amino acid and nitrogen metabolism, nucleotide metabolism, metabolism of cofactors and vitamins, and xenobiotic biodegradation and metabolism. Similar to *Bacillus thuringiensis* (*Bt*), *Bb* can also induce a silkworm poisoning-related response. In this process, genes encoding midgut peritrophic membrane proteins, aminopeptidase N receptors and sodium / calcium exchange protein showed modulation. For the first time, we found that *Bb* induced a lot of genes involved in juvenile hormone synthesis and metabolism pathway upregulated. *Bb* also triggered the host immune responses, including cellular immune response and serine protease cascade melanization response. Real time PCR analysis showed that *Bb* can induce the silkworm systemic immune response, mainly by the Toll pathway. Anti-microorganism peptides (AMPs), including of *Attacin*, *Lebocin*, *Enbocin*, *Gloverin* and *Moricin* families, were upregulated at 24 hours post the infection.

Published On: PLoS ONE, 2009, 4(12).

1 Institute of Sericulture and Systems Biology, Southwest University, Chongqing 400716, China.
2 Institute of Agronomy and Life Science, Chongqing University, Chongqing 400030, China.
3 Institute of Economic Crops Breeding and Cultivation, Sichuan Academy of Agricultural Sciences, Chengdu 610066, China.
*Corresponding author E-mail: xiaqy@swu.edu.cn.

家蚕对病原细菌黑胸败血菌的全基因组水平的应答调查

黄璐琳[1,3]　程廷才[2]　许平震[1]
程道军[1]　方　婷[1]　夏庆友[1,2*]

摘要：宿主–病原相互作用是一种复杂的关系，揭示病原及其宿主间的相互作用机制是当前研究的核心问题。黑胸芽孢杆菌（*Bacillus bombysepticus*，*Bb*）能产生芽孢和伴孢晶体，是从病蚕的尸体上首次分离鉴定获得的。在温度较高且潮湿的环境下，家蚕感染*Bb*后能在一天之内发生黑胸败血病而死亡。家蚕*Bb*病原能用于研究感染后的宿主应答。通过对*Bb*感染后4个时间点的家蚕幼虫基因表达谱分析，用以深入探究*Bb*与宿主相互作用的分子机制。全基因组调查发现大量宿主基因和相关信号途径在*Bb*感染家蚕后受到调控。GO分类显示，被诱导表达的基因大致可被分为14个类别。基于KEGG通路分析，发现6种基础的代谢途径受到调节，包括遗传信息的加工和转录、糖代谢、氨基酸和氮代谢、核苷酸代谢、辅因子和维生素的代谢以及体内异物的生物降解和代谢。和苏云金芽孢杆菌（*Bacillus thuringiensis*，*Bt*）相似的是，*Bb*也能够诱导与家蚕中毒相关的反应。此外，*Bb*能够诱导保幼激素合成和代谢途径中许多基因的上调表达。*Bb*也能触发家蚕的免疫应答，包括细胞免疫和丝氨酸蛋白酶黑色素级联反应。RT-PCR分析显示，*Bb*主要通过Toll途径诱导家蚕系统性免疫应答，*Attacin*、*Lebocin*、*Enbocin*、*Gloverin*和*Moricin*等抗菌肽基因成员在病原诱导24 h后上调表达。

1　蚕学与系统生物学研究所，西南大学，重庆
2　农学及生命科学研究院，重庆大学，重庆
3　经济作物育种及栽培研究所，四川省农业科学研究院，成都

Immunoglobulin superfamily is conserved but evolved rapidly and is active in the silkworm, *Bombyx mori*

Huang LL[1,3#] Cheng TC[2#] Xu PZ[1]

Duan J[2] Fang T[1] Xia QY[1,2*]

Abstract: Immunoglobulin superfamily (IgSF) proteins are known for their abilities to specifically recognize and adhere to cells. In this paper, we predicted the presence of 133 IgSF proteins in the silkworm (*Bombyx mori*) genome. Comparison with similar proteins in other model organisms (*Caenorhabditis elegans*, *Drosophila melanogaster*, *Anopheles gambiae*, *Apis mellifera* and *Homosapiens*) indicated that IgSF proteins are conserved but have rapidly evolved from worms to human beings. However, these proteins are well conserved amongst insects. Silkworm microarray-based expression data showed tissue expression of 57 *IgSF* genes and microbe-induced differential expression of 37 genes. Based on the expression data, we can conclude that the silkworm IgSF is active.

Published On: Insect Molecular Biology, 2009, 18(4), 517-530.

1 Institute of Sericulture and Systems Biology, Southwest University, Chongqing 400716, China.

2 Institute of Agronomy and Life Science, Chongqing University, Chongqing 400030, China.

3 Sichuan Academy of Agricultural Sciences, Chengdu 610066,China

#These authors contributed equally.

*Corresponding author E-mail: xiaqy@swu.edu.cn.

家蚕免疫球蛋白超家族基因的保守性、快速进化及活化特征分析

黄璐琳[1,3#]　程廷才[2#]　许平震[1]
段　军[2]　方　婷[1]　夏庆友[1,2*]

摘要：免疫球蛋白超家族（IgSF）是一类能识别并连接细胞的蛋白质。本研究通过对家蚕基因组数据库分析，在家蚕基因组中鉴定出了133个IgSF蛋白。比较家蚕同线虫、果蝇、按蚊、蜜蜂和人等模式生物中的IgSF蛋白发现，该类蛋白从线虫到人进行了快速的进化并且相对保守，但在昆虫中却非常保守。家蚕幼虫五龄第3天各组织的表达芯片数据分析表明，57个家蚕*IgSF*基因在这一时期检测到了组织表达。分析家蚕脂肪体在微生物诱导后的芯片表达谱发现，37个*IgSF*基因具有诱导差异表达。这些表达数据提示家蚕IgSF是具有活性的。

1　蚕学与系统生物学研究所，西南大学，重庆
2　农学及生命科学研究院，重庆大学，重庆
3　四川省农业科学研究院，成都

Genome-wide identification and expression analysis of serine proteases and homologs in the silkworm *Bombyx mori*

Zhao P[1] Wang GH[1,2] Dong ZM[1] Duan J[1,2]
Xu PZ[1] Cheng TC[1,2] Xiang ZH[1] Xia QY[1,2*]

Abstract:

Background: Serine proteases (SPs) and serine proteases homologs (SPHs) are a large group of proteolytic enzymes, with important roles in a variety of physiological processes, such as cell signalling, defense and development. Genome-wide identification and expression analysis of serine proteases and their homologs in the silkworm might provide valuable information about their biological functions.

Results: In this study, 51 *SP* genes and 92 *SPH* genes were systematically identified in the genome of the silkworm *Bombyx mori*. Phylogenetic analysis indicated that six gene families have been amplified species-specifically in the silkworm, and the members of them showed chromosomal distribution of tandem repeats. Microarray analysis suggests that many silkworm-specific genes, such as members of *SP_fam12*, *13*, *14* and *15*, show expression patterns that are specific to tissues or developmental stages. The roles of SPs and SPHs in resisting pathogens were investigated in silkworms when they were infected by *Escherichia coli*, *Bacillus bombysepticus, Batrytis bassiana* and *B. mori* nucleopolyhedrovirus, respectively. Microarray experiment and real-time quantitative RT-PCR showed that 18 *SP* or *SPH* genes were significantly up-regulated after pathogen induction, suggesting that *SP* and *SPH* genes might participate in pathogenic microorganism resistance in *B. mori.*

Conclusion: Silkworm *SP* and *SPH* genes were identified. Comparative genomics showed that SP and SPH genes belong to a large family, whose members are generated mainly by tandem repeat evolution. We found that silkworm has species-specific *SP* and *SPH* genes. Phylogenetic and microarray analyses provide an overview of the silkworm SPs and SPHs, and facilitate future functional studies on these enzymes.

Published On: BMC Genomics, 2010, 11.

1 Key Sericultural Laboratory of Agricultural Ministry, Southwest University, Chongqing 400716, China.
2 Institute of Agricultural and Life Sciences, Chongqing University, Chongqing 400044, China.
*Corresponding author E-mail: xiaqy@swu.edu.cn.

家蚕丝氨酸蛋白酶和同系物的全基因组鉴定和表达分析

赵　萍[1]　王根洪[1,2]　董照明[1]　段　军[1,2]
许平震[1]　程廷才[1,2]　向仲怀[1]　夏庆友[1,2*]

摘要：背景——丝氨酸蛋白酶(SPs)和丝氨酸蛋白酶同系物(SPHs)属于蛋白水解酶,在各种生理过程如细胞信号传导、防御和发育中具有重要作用。丝氨酸蛋白酶及其同系物在家蚕基因组中的识别和表达分析可能提供有关其生物学功能的有价值的信息。结果——本研究,在家蚕基因组中共鉴定到家蚕51个*SP*基因和92个*SPH*基因。系统进化分析表明,6个基因家族在家蚕中特异性基因加倍,其成员在染色体上串联重复排列。芯片分析表明,许多家蚕特异性基因,如*SP_fam12*,*13*,*14*和*15*的成员,显示出组织特异性或者时期特异性的表达模式。当家蚕分别感染大肠杆菌、芽孢杆菌、蚕白僵菌和家蚕核型多角体病毒时,研究了SPs和SPHs在抵抗病原体中的作用。芯片与定量PCR结果显示18个*SP*或*SPH*基因在病原诱导后显著上调表达,表明*SP*和*SPH*基因可能参与家蚕中的致病微生物防御。讨论——本文鉴定了家蚕*SP*和*SPH*基因。比较基因组学显示,*SP*和*SPH*基因属于一个大家族,其成员主要是通过串联重复进化产生的。我们发现家蚕具有特异性的*SP*和*SPH*基因。利用系统进化和芯片数据分析对家蚕SPs和SPHs进行了概述,将有效促进这些酶在未来的功能研究。

1　家蚕基因组生物学国家重点实验室,西南大学,重庆
2　农学及生命科学研究院,重庆大学,重庆

Genome-wide identification and immune response analysis of serine protease inhibitor genes in the silkworm, *Bombyx mori*

Zhao P[1#] Dong ZM[1#] Duan J[1,2] Wang GH[1,2]
Wang LY[1] Li YS[1] Xiang ZH[1] Xia QY[1,2*]

Abstract: In most insect species, a variety of serine protease inhibitors (SPIs) have been found in multiple tissues, including integument, gonad, salivary gland, and hemolymph, and are required for preventing unwanted proteolysis. These SPIs belong to different families and have distinct inhibitory mechanisms. Herein, we predicted and characterized potential *SPI* genes based on the genome sequences of silkworm, *Bombyx mori*. As a result, a total of eighty *SPI* genes were identified in *B. mori*. These *SPI* genes contain 10 kinds of SPI domains, including serpin, Kunitz_BPTI, Kazal, TIL, amfpi, Bowman-Birk, Antistasin, WAP, Pacifastin, and alpha-macroglobulin. Sixty-three SPIs contain single SPI domain while the others have at least two inhibitor units. Some SPIs also contain non-inhibitor domains for protein-protein interactions, including EGF, ADAM_ spacer, spondin_N, reeler, TSP_1 and other modules. Microarray analysis showed that fourteen *SPI* genes from lineage-specific TIL family and Group F of serpin family had enriched expression in the silk gland. The roles of SPIs in resisting pathogens were investigated in silkworms when they were infected by four pathogens. Microarray and qRT-PCR experiments revealed obvious up-regulation of 8, 4, 3 and 3 *SPI* genes after infection with *Escherichia coli*, *Bacillus bombysepticus*, *Beauveria bassiana* or *B. mori* nuclear polyhedrosis virus (*Bm*NPV), respectively. On the contrary, 4, 11, 7 and 9 *SPI* genes were down-regulated after infection with *E. coli,B. bombysepticus*, *B. bassiana* or *Bm*NPV, respectively.These results suggested that these *SPI* genes may be involved in resistance to pathogenic microorganisms. These findings may provide valuable information for further clarifying the roles of SPIs in the development, immune defence, and efficient synthesis of silk gland protein.

Published On: PLoS ONE, 2012, 7(2).

1 State Key Laboratory of Silkworm Genome Biology, Southwest University, Chongqing 400716, China.
2 Institute of Agricultural and Life Sciences, Chongqing University, Chongqing 400030, China.
#These authors contributed equally.
*Corresponding author E-mail: xiaqy@swu.edu.cn.

家蚕体内丝氨酸蛋白酶抑制剂的全基因组鉴定与免疫反应分析

赵　萍[1#]　董照明[1#]　段　军[1,2]　王根洪[1,2]
王凌燕[1]　李游山[1]　向仲怀[1]　夏庆友[1,2*]

摘要：许多昆虫的不同组织（包括表皮、生殖腺、唾液腺、血淋巴）中发现多种丝氨酸蛋白酶抑制剂（SPIs）。在昆虫体内，丝氨酸蛋白酶抑制剂可防止不必要的蛋白水解。这些不同家族的丝氨酸蛋白酶抑制剂具有不同的抑制机理。在这篇文章中，我们基于家蚕基因组序列，成功地预测出并描述了家蚕体内可能存在的*SPI*基因的特征。我们在家蚕体内鉴定出80个*SPI*基因，它们含有10种SPI结构域，包括serpin、Kunitz_BPTI、Kazal、TIL、amfpi、Bowman-Birk、Antistasin、WAP、Pacifastin和alpha-macroglobulin。其中有63个丝氨酸蛋白酶抑制剂只包含一个SPI结构域而其他的则至少含有2个抑制单元。一些丝氨酸蛋白酶抑制剂也含有蛋白互作的非抑制剂结构域，包括EGF、ADAM_spacer、spondin_N、reeler、TSP_1和其他结构。芯片分析表明，来自世系特异TIL家族和serpin家族F组的14个*SPI*基因在丝腺中高量表达。我们也调查了被4种病原感染时，丝氨酸蛋白酶抑制剂抵御病原的作用。芯片分析和qRT-PCR实验表明，被大肠杆菌、黑胸败血菌、球孢白僵菌和家蚕核型多角体病毒感染后，8、4、3和3个*SPI*基因表达分别显著上调，相反地，4、11、7和9个*SPI*基因表达分别被抑制。结果表明这些*SPI*基因可能参与了蚕体对病原微生物的抵御。上述研究结果可能为进一步阐释丝氨酸蛋白酶抑制剂在家蚕生长发育、免疫防御及丝腺高效地合成蛋白质中的作用提供了有效信息。

1　家蚕基因组生物学国家重点实验室，西南大学，重庆，
2　农学及生命科学研究院，重庆大学，重庆

Bombyx mori transcription factors: genome-wide identification, expression profiles and response to pathogens by microarray analysis

Huang LL Cheng TC Xu PZ Fang T Xia QY*

Abstract: Transcription factors are present in all living organisms, and play vital roles in a wide range of biological processes. Studies of transcription factors will help reveal the complex regulation mechanism of organisms. So far, hundreds of domains have been identified that show transcription factor activity. Here, 281 reported transcription factor domains were used as seeds to search the transcription factors in genomes of *Bombyx mori* L. (Lepidoptera: Bombycidae) and four other model insects. Overall, 666 transcription factors including 36 basal factors and 630 other factors were identified in *B. mori* genome, which accounted for 4.56% of its genome. The silkworm transcription factors expression profiles were investigated in relation to multiple tissues, developmental stages, sexual dimorphism, and responses to oral infection by pathogens and direct bacterial injection. These all provided rich clues for revealing the transcriptional regulation mechanism of silkworm organ differentiation, growth and development, sexual dimorphism, and response to pathogen infection.

Published On: Journal of Insect Science, 2012, 12.

State Key Laboratory of Silkworm Genome Biology, Southwest University, Chongqing 400716, China.
*Corresponding author E-mail: xiaqy@swu.edu.cn.

通过基因芯片数据对家蚕转录因子进行全基因组鉴定、表达谱分析和响应病原菌感染分析

黄璐琳　程廷才　许平震　方　婷　夏庆友*

摘要：转录因子存在于所有生物体内，并在多数生物学过程中发挥至关重要的作用。转录因子的研究有助于揭示生物体复杂的调节机制。到目前为止，人们已经鉴定到数百个具有转录因子活性的结构域。本文利用已报道的281个转录因子结构域去搜索家蚕（鳞翅目：蚕蛾科）以及其他4种模式昆虫的转录因子。在家蚕基因组中共鉴定到666个转录因子，包括36个基本转录因子和630个其他转录因子，占家蚕基因组的4.56%。并且对雌雄家蚕的不同组织、不同发育阶段、经口感染病原菌和直接感染细菌后转录因子的组织表达谱进行了研究。这些结果可以为家蚕器官分化、生长发育、雌雄异形和响应病原菌感染的转录调节机制研究提供线索。

家蚕基因组生物学国家重点实验室，西南大学，重庆

Genome-wide transcriptional response of silkworm (*Bombyx mori*) to infection by the microsporidian *Nosema bombycis*

Ma ZG[1,3#] Li CF[1,3#] Pan GQ[1,3] Li ZH[1,3] Han B[1,3] Xu JS[2] Lan XQ[1] Chen J[1,3]
Yang DL[1,3] Chen QM[1,3] Sang Q[1,3] Ji XC[1,3] Li T[1,3] Long MX[1,3] Zhou ZY[1,2,3*]

Abstract: Microsporidia have attracted much attention because they infect a variety of species ranging from protists to mammals, including immunocompromised patients with AIDS or cancer. Aside from the study on *Nosema ceranae*, few works have focused on elucidating the mechanism in host response to microsporidia infection. *Nosema bombycis* is a pathogen of silkworm pébrine that causes great economic losses to the silkworm industry. Detailed understanding of the host (*Bombyx mori*) response to infection by *N. bombycis* is helpful for prevention of this disease. A genome-wide survey of the gene expression profile at 2, 4, 6 and 8 days post-infection by *N. bombycis* was performed and results showed that 64, 244, 1 328, 1 887 genes were induced, respectively. Up to 124 genes, which are involved in basal metabolism pathways, were modulated. Notably, *B. mori* genes that play a role in juvenile hormone synthesis and metabolism pathways were induced, suggesting that the host may accumulate JH as a response to infection. Interestingly, *N. bombycis* can inhibit the silkworm serine protease cascade melanization pathway in hemolymph, which may be due to the secretion of serpins in the microsporidia. *N. bombycis* also induced up-regulation of several cellular immune factors, in which CTL11 has been suggested to be involved in both spore recognition and immune signal transduction. Microarray and real-time PCR analysis indicated the activation of silkworm Toll and JAK/STAT pathways. The notable up-regulation of antimicrobial peptides, including *gloverins*, *lebocins* and *moricins*, strongly indicated that antimicrobial peptide defense mechanisms were triggered to resist the invasive microsporidia. An analysis of *N. bombycis*-specific response factors suggested their important roles in anti-microsporidia defense. Overall, this study primarily provides insight into the potential molecular mechanisms for the host-parasite interaction between *B. mori* and *N. bombycis* and may provide a foundation for further work on host-parasite interaction between insects and microsporidia.

Published On: PLoS ONE, 2013, 8(12).

1 State Key Laboratory of Silkworm Genome Biology, Southwest University, Chongqing 400716, China.
2 College of Life Sciences, Chongqing Normal University, Chongqing 401331, China.
3 Key Laboratory for Sericulture Functional Genomics and Biotechnology of Agricultural Ministry, Southwest University, Chongqing 400716, China.
#These authors contributed equally.
*Corresponding author E-mail: zyzhou@swu.edu.cn.

家蚕微孢子虫诱导家蚕全基因组转录应答

马振刚[1,3#] 李春峰[1,3#] 潘国庆[1,3] 李致宏[1,3] 韩 冰[1,3] 许金山[2] 蓝希钳[1] 陈 洁[1,3]
杨东林[1,3] 陈全梅[1,3] 桑 颀[1,3] 季小存[1,3] 李 田[1,3] 龙梦娴[1,3] 周泽扬[1,2,3*]

摘要：微孢子虫寄主范围十分广泛，可以感染从原生动物到哺乳动物的诸多物种，其中包括艾滋病患者或者癌症患者等免疫缺陷的病人，因此微孢子虫受到了研究者的广泛关注。目前，除了东方蜜蜂微孢子虫外，很少有研究关注微孢子虫感染引起的宿主反应机制。家蚕微孢子虫是家蚕微粒子病的病原体，这种传染病在养蚕区的广泛传播给蚕业生产带来了巨大的经济损失。详细了解家蚕对家蚕微孢子虫的应答机制将有助于对该病的防治。本研究在基因组水平上调查了微孢子虫感染家蚕后第2、4、6和8天的基因的诱导情况，结果表明，分别有64、244、1 328和1 887个基因被诱导表达。其中，多达124个与代谢通路相关的基因被诱导。值得注意的是，家蚕中在保幼激素合成和代谢中起到重要作用的相关基因被诱导表达，暗示宿主可能通过积累保幼激素作为对病原入侵的应答。有趣的是，家蚕微孢子虫能够抑制家蚕血淋巴的丝氨酸蛋白酶级联的黑化反应通路，这可能是由家蚕微孢子虫分泌丝氨酸蛋白酶抑制剂引起的。家蚕微孢子虫还诱导家蚕的一些细胞免疫因子发生了上调表达，其中包括在孢子识别和免疫信号传递等过程中行使功能的C型凝集素11。基因芯片分析和荧光定量PCR分析的结果表明，家蚕受到感染后其Toll通路和JAK/STAT通路被活化。而*gloverins*、*lebocins*和*moricins*等抗菌肽分子的显著上调表达暗示了抗菌肽防御被触发来抵抗微孢子虫的入侵。对家蚕微孢子特异反应免疫因子的分析结果暗示了它们在微孢子虫防御过程中的重要功能。总的来说，这项研究分析了家蚕与家蚕微孢子的宿主—病原互作的分子机制，其可以为昆虫—微孢子虫互作机制的深入研究奠定坚实的基础。

1 家蚕基因组生物学国家重点实验室，西南大学，重庆
2 生命科学学院，重庆师范大学，重庆
3 农业部蚕桑功能基因组与生物技术重点实验室，西南大学，重庆

Genome-wide identification and comprehensive analyses of the kinomes in four pathogenic microsporidia species

Li Z[1] Hao YJ[1] Wang LL[1] Xiang H[3] Zhou ZY[1,2*]

Abstract: Microsporidia have attracted considerable attention because they infect a wide range of hosts, from invertebrates to vertebrates, and cause serious human diseases and major economic losses in the livestock industry. There are no prospective drugs to counteract this pathogen. Eukaryotic protein kinases (ePKs) play a central role in regulating many essential cellular processes and are therefore potential drug targets. In this study, a comprehensive summary and comparative analysis of the protein kinases in four microsporidia-*Enterocytozoon bieneusi*, *Encephalitozoon cuniculi*, *Nosema bombycis* and *Nosema ceranae*-was performed. The results show that there are 34 ePKs and 4 atypical protein kinases (aPKs) in *E. bieneusi*, 29 ePKs and 6 aPKs in *E. cuniculi*, 41 ePKs and 5 aPKs in *N. bombycis*, and 27 ePKs and 4 aPKs in *N. ceranae*. These data support the previous conclusion that the microsporidian kinome is the smallest eukaryotic kinome. Microsporidian kinomes contain only serine-threonine kinases and do not contain receptor-like and tyrosine kinases. Many of the kinases related to nutrient and energy signaling and the stress response have been lost in microsporidian kinomes. However, cell cycle-, development- and growth-related kinases, which are important to parasites, are well conserved. This reduction of the microsporidian kinome is in good agreement with genome compaction, but kinome density is negatively correlated with proteome size. Furthermore, the protein kinases in each microsporidian genome are under strong purifying selection pressure. No remarkable differences in kinase family classification, domain features, gain and/or loss, and selective pressure were observed in these four species. Although microsporidia adapt to different host types, the coevolution of microsporidia and their hosts was not clearly reflected in the protein kinases. Overall, this study enriches and updates the microsporidian protein kinase database and may provide valuable information and candidate targets for the design of treatments for pathogenic diseases.

Published On: PLoS ONE, 2014, 9(12).

1 College of Life Sciences, Chongqing Normal University, Chongqing, 400013, China.
2 State Key Laboratory of Silkworm Genome Biology, Southwest University, Chongqing 400716, China.
3 College of Animal Science and Technology, Southwest University, Chongqing 400716, China.
*Corresponding author E-mail: zyzhou@swu.edu.cn.

全基因组鉴定和综合分析四种致病性微孢子虫的蛋白激酶组

李　治[1]　郝友进[1]　王林玲[1]　向　恒[3]　周泽扬[1,2*]

摘要：微孢子虫因其具有广泛的宿主，包括脊椎动物和无脊椎动物，并可导致严重的人类疾病和畜牧业经济损失，从而受到广泛的重视。对该病原虫的防治没有理想的药物。真核生物的蛋白激酶因其在许多基础细胞的调控进程中扮演着重要的角色，而被作为潜在的药物靶标。本研究详细地总结和比较分析了4种微孢子虫的蛋白激酶，包括比氏肠道微孢子虫、兔脑炎微孢子虫、家蚕微孢子虫、蜜蜂微孢子虫。分析结果表明，比氏肠道微孢子虫具有34个典型蛋白激酶和4个非典型蛋白激酶，兔脑炎微孢子虫具有29个典型蛋白激酶和6个非典型蛋白激酶，家蚕微孢子虫具有41个典型蛋白激酶和5个非典型蛋白激酶，蜜蜂微孢子虫具有27个典型蛋白激酶和4个非典型蛋白激酶，该数据再次印证了微孢子虫的蛋白激酶组是迄今发现的最小真核生物激酶组的结论。微孢子虫的蛋白激酶组仅由丝氨酸-苏氨酸类激酶组成，不具有受体类和酪氨酸类激酶。微孢子虫的激酶组中，许多与营养和能量信号，以及与胁迫应答相关的激酶已丢失。但是，对寄生虫很重要的细胞周期、发育、生长相关的激酶在微孢子虫基因组中被很好地保留下来。微孢子虫蛋白激酶组的这一缩减特征与其基因组的压缩具有很好的同趋性，但激酶组的密度却与蛋白质组的大小呈负相关关系。此外，本研究中4种微孢子虫的蛋白激酶均受到强烈的正向选择压力作用。蛋白激酶的家族分类、结构域特征、获得与丢失、所受选择压力在4种微孢子虫之间并没有显著的差异。虽然这4种微孢子虫各自的宿主类型并不相同，但它们与宿主的系统进化特性却并未在蛋白激酶的组成方面清晰体现。总体而言，本研究丰富和更新了微孢子虫的蛋白激酶数据库信息，为针对该病原虫的防控提供了候选药物设计靶标和有价值的信息。

1　生命科学学院，重庆师范大学，重庆
2　家蚕基因组生物学国家重点实验室，西南大学，重庆
3　动物科技学院，西南大学，重庆

第十二章

素材创制与品种培育

Chapter 12

我国的养蚕业一直面临严重的病害威胁，每年因蚕病爆发造成的损失占蚕业生产总收入的20%左右，其中病毒病的威胁最为严重。传统消毒方法可以在一定程度上预防蚕病，但具有很大的局限性。项目组在研究家蚕病毒与宿主互作机制的基础上，利用转基因等分子育种手段创制了一批家蚕抗病毒素材，为后续抗性品种培育奠定了基础。

抗性候选基因的筛选及鉴定：家蚕质型多角体病毒（*Bm*CPV）只侵染家蚕中肠，利用抑制消减杂交法和转录组测序对感染*Bm*CPV后的中肠差异表达基因进行了分析；家蚕核型多角体病毒（*Bm*NPV）是对蚕业生产危害最为严重的一类病原，*Bm*NPV可以感染家蚕的几乎全部组织器官，利用*Bm*NPV抗性和易感家蚕品系，通过组织的mRNA荧光差异显示PCR和转录组测序、组织和消化液的蛋白质组学分析等方法，对差异表达基因进行了筛选。对部分差异表达基因进行了定量PCR检测，进而筛选出抗性候选基因进行克隆，然后对其时空表达模式、亚细胞定位、病毒诱导表达模式进行了检测。进一步研究发现，在细胞水平增量表达V-ATPase c亚基和*BmSpry*可以抑制*Bm*NPV的增殖。鉴定BmNPV的主抗基因和受体基因还需要后续更多的研究。另外，项目组研究发现*Bm*SPI39有效地抑制了白僵菌毒性蛋白酶CDEP-1，提高了家蚕的存活率。*TIF-4A*可能是对病毒感染或高温胁迫下的家蚕进行基因表达分析的最佳内参基因。

抗病毒策略及素材创制。首先，建立了家蚕抗性检测平台，制定了多个操作标准，提高了抗性检测结果的真实性、可靠性和可重复性。接着，根据病毒的侵染过程，选择不同靶标基因，分别在侵染早期、mRNA转录水平、蛋白合成水平和免疫信号通路等四个关键环节抑制病毒增殖。然后，对抗病毒策略进行优化，对影响转基因干涉宿主抗性的因素进行了系统分析和比较；利用诱导型启动子和增强子介导抗病基因的表达，使其表达量随着病原量的增加而显著增加。最后，改进和整合不同抗性策略，利用家蚕当家品种“芙蓉”，制备了同时增量表达抗性基因和干涉多个病毒基因的转基因家蚕SW-H。在感染高剂量的*Bm*NPV以后，SW-H的死亡率比非转基因对照降低78%，同时其经济性状未受到影响。SW-H成为第一个能在感染的多个阶段抑制病毒的转基因高抗病毒动物。项目组因此受邀在国际昆虫学经典杂志IBMB发表家蚕抗病毒研究的综述文章。

总的来说，项目组建立了家蚕分子育种平台技术和理论体系，已进入动物分子育种研究领域前沿，对其他动物的分子育种具有重要的参考价值与借鉴意义。家蚕转基因抗性品系在经过安全性检测后，有望应用于蚕业生产，减少家蚕死亡并增加蚕农经济收入。

蒋亮

The progress and future of enhancing antiviral capacity by transgenic technology in the silkworm *Bombyx mori*

Jiang L Xia QY*

Abstract: *Bombyx mori* is a common lepidopteran model and an important economic insect for silk production. *B. mori* nucleopolyhedrovirus (*Bm*NPV) is a typical pathogenic baculovirus that causes serious economic losses in sericulture. *B. mori* and *Bm*NPV are a model of insect host and pathogen interaction including invasion of the host by the pathogen, host response, and enhancement of host resistance. The antiviral capacity of silkworms can be improved by transgenic technology such as overexpression of an endogenous or exogenous antiviral gene, RNA interference of the *Bm*NPV gene, or regulation of the immune pathway to inhibit *Bm*NPV at different stages of infection. Antiviral capacity could be further increased by combining different methods. We discuss the future of an antiviral strategy in silkworm, including possible improvement of anti-*Bm*NPV, the feasibility of constructing transgenic silkworms with resistance to multiple viruses, and the safety of transgenic silkworms. The silkworm model could provide a reference for disease control in other organisms.

Published On: Insect Biochemistry and Molecular Biology, 2014, 48, 1-7.

State Key Laboratory of Silkworm Genome Biology, Southwest University, Chongqing 400716, China.
*Corresponding author E-mail: xiaqy@swu.edu.cn.

家蚕转基因抗病毒研究进展与展望

蒋　亮　夏庆友*

摘要：家蚕是具有重要经济价值的鳞翅目模式昆虫。家蚕核型多角体病毒（*Bm*NPV）是一个典型的致病杆状病毒，给蚕丝产业造成了严重的经济损失。家蚕和*Bm*NPV是昆虫宿主和病原体相互作用的一个模型，包括病原体侵染、宿主响应和增强宿主抗性等。利用转基因技术可以提高家蚕抗病毒能力，增量表达内源或外源抗病毒基因、RNA干扰*Bm*NPV基因和调控免疫通路能够在不同感染阶段抑制*Bm*NPV，整合不同抗病毒策略可以进一步提高家蚕抗性。我们讨论了家蚕抗病毒策略的未来，包括抗*Bm*NPV的可能的改进策略，构建对多种病毒具有抗性的转基因家蚕的可行性以及转基因蚕的安全性。家蚕抗病毒研究模型为其他物种的抗病研究提供了参考。

家蚕基因组生物学国家重点实验室，西南大学，重庆

Selection of reference genes for analysis of stress-responsive genes after challenge with viruses and temperature changes in the silkworm *Bombyx mori*

Guo HZ Jiang L Xia QY*

Abstract: Viruses and high temperature (HT) are the primary threats to silkworms. Changes in the expression of stress-response genes can be measured using quantitative polymerase chain reaction (qPCR) after exposure to viruses or HT. However, appropriate reference genes (RGs) for qPCR data normalization have not been established in this organism. In this study, we summarized the RGs used in the previous silkworm studies after infection with *Bombyx mori* nucleopolyhedrovirus (*Bm*NPV), *B. mori* cytoplasmic polyhedrosis virus (*Bm*CPV), or *B. mori* densovirus (*Bm*DNV) or after HT treatment. The expression levels of these RGs were extracted from silkworm transcriptome data to screen for candidate RGs that were unaffected by the experimental conditions. *Actin-1* (*A1*), *actin-3* (*A3*), *glyceraldehyde-3-phosphate dehydrogenase* (*GAPDH*), and *translation initiation factor 4a* (*TIF-4A*) were selected for further qPCR verification. The results of RNA-seq and qPCR showed that *GAPDH* and *TIF-4A* were suitable RGs after *Bm*NPV challenge or HT stress, whereas *TIF-4A* was an appropriate RG for *Bm*CPV or *Bm*DNV-Z challenge in silkworms. These results suggested that *TIF-4A* may be the most appropriate RG for gene expression analysis after challenge with viruses or HT in silkworms.

Published On: Molecular Genetics and Genomics, 2016, 291(2), 999-1004.

State Key Laboratory of Silkworm Genome Biology, Southwest University, Chongqing 400716, China.
*Corresponding author E-mail: xiaqy@swu.edu.cn.

家蚕受病毒和高温胁迫后对应激反应基因进行分析的家蚕内参基因的筛选

郭慧珍　蒋　亮　夏庆友*

摘要：病毒和高温(HT)是对家蚕的主要威胁。在病毒或高温胁迫下，应激反应基因表达的变化可以使用定量聚合酶链式反应(qPCR)进行测定。然而，在家蚕中尚未建立适用于qPCR数据归一化的合适内参基因(RGs)。本研究总结了适用于先前家蚕研究中的家蚕核型多角体病毒(*Bm*NPV)、家蚕质型多角体病毒(*Bm*CPV)、家蚕浓核病毒(*Bm*DNV)或高温处理后的内参基因。这些内参基因的表达量来源于家蚕转录组数据，用来筛选不受实验条件影响的候选内参基因。选择*Actin-1*(*A1*)，*Actin-3*(*A3*)，甘油醛-3-磷酸脱氢酶(*GAPDH*)和*translation initiation factor 4a*(*TIF-4A*)用于进一步qPCR验证。RNA-seq和qPCR结果显示，*GAPDH*和*TIF-4A*是适用于*Bm*NPV或高温处理的内参基因，而*TIF-4A*为适合于*Bm*CPV或*Bm*DNV-Z处理的内参基因。综上结果表明，*TIF-4A*可能是家蚕病毒或高温胁迫下进行基因表达分析的最佳内参基因。

家蚕基因组生物学国家重点实验室，西南大学，重庆

Suppression of intestinal immunity through silencing of *TCTP* by RNAi intransgenic silkworm, *Bombyx mori*

Hu CM Wang F Ma SY Li XY

Song L Hua XT Xia QY*

Abstract: Intestinal immune response is a front line of host defense. The host factors that participate in intestinal immunity response remain largely unknown. We recently reported that Translationally Controlled Tumor Protein(*Bm*TCTP) was obtained by constructing a phage display cDNA library of the silkworm midgut and carrying out high throughput screening of pathogen binding molecules. To further address the function of *Bm*TCTP in silkworm intestinal immunity, transgenic RNAi silkworms were constructed by microinjection *piggy*Bac plasmid to *Dazao* embryos. The antimicrobial capacity of transgenic silkworm decreased since the expression of gut antimicrobial peptide from transgenic silkworm was not sufficiently induced during oral microbial challenge. Moreover, dynamic ERK phosphorylation from transgenic silkworm midgut was disrupted. Taken together, the innateimmunity of intestinal was suppressed through disruption of dynamic ERK phosphorylation after oral microbialinfection as a result of RNAi-mediated knockdown of midgut *TCTP* in transgenic silkworm.

Published On: Gene, 2015, 574 (1), 82-87.

State Key Laboratory of Silkworm Genome Biology, Southwest University, Chongqing 400716, China.

*Corresponding author E-mail: xiaqy@swu.edu.cn.

通过转基因干涉家蚕肠道 *TCTP* 抑制肠道免疫

胡翠美　王　峰　马三垣　李显扬
宋　亮　化晓婷　夏庆友*

摘要：肠道免疫应答是宿主防御的前线，参与肠道免疫应答的宿主因子大部分是未知的。通过构建家蚕肠道cDNA噬菌体展示文库并利用病原识别分子对文库进行高通量筛选的方法，获得了翻译控制的肿瘤蛋白（*Bm*TCTP）。为了进一步了解*Bm*TCTP在家蚕肠道免疫中的功能，通过显微注射*piggy*Bac质粒到*Dazao*胚胎中，获得了转基因干涉家蚕。由于食下感染微生物不能充分诱导转基因家蚕肠道抗菌肽的表达，因此转基因家蚕的抗微生物能力下降。而且，转基因家蚕肠道的动态ERK磷酸化紊乱。综上所述，转基因干涉家蚕肠道*TCTP*将导致微生物感染后肠道ERK磷酸化紊乱进而肠道免疫受抑制。

家蚕基因组生物学国家重点实验室，西南大学，重庆

TIL-type protease inhibitors may be used as targeted resistance factors to enhance silkworm defenses against invasive fungi

Li YS[1,2] Zhao P[1] Liu HW[1] Guo XM[1]
He HW[1] Zhu R[3] Xiang ZH[1] Xia QY[1*]

Abatract: Entomopathogenic fungi penetrate the insect cuticle using their abundant hydrolases. These hydrolases, which include cuticle-degrading proteases and chitinases, are important virulence factors. Our recent findings suggest that many serine protease inhibitors, especially TIL-type protease inhibitors, are involved in insect resistance to pathogenic microorganisms. To clarify the molecular mechanism underlying this resistance to entomopathogenic fungi and identify novel genes to improve the silkworm antifungal capacity, we conducted an in-depth study of serine protease inhibitors. Here, we cloned and expressed a novel silkworm TIL-type protease inhibitor, *Bm*SPI39. In activity assays, *Bm*SPI39 potently inhibited the virulence protease CDEP-1 of *Beauveria bassiana*, suggesting that it might suppress the fungal penetration of the silkworm integument by inhibiting the cuticle-degrading proteases secreted by the fungus. Phenol oxidase activation studies showed that melanization is involved in the insect immune response to fungal invasion, and that fungus-induced excessive melanization is suppressed by *Bm*SPI39 by inhibiting the fungal cuticle-degrading proteases. To better understand the mechanism involved in the inhibition of fungal virulence by protease inhibitors, their effects on the germination of *B. bassiana conidia* was examined. *Bm*SPI38 and *Bm*SPI39 significantly inhibited the germination of *B. bassiana conidia.* Survival assays showed that *Bm*SPI38 and *Bm*SPI39 markedly improved the survival rates of silkworms, and can therefore be used as targeted resistance proteins in the silkworm. These results provided new insight into the molecular mechanisms whereby insect protease inhibitors confer resistance against entomopathogenic fungi, suggesting their potential application in medicinal or agricultural fields.

Published On: Insect Biochemistry and Molecular Biology, 2015, 57, 11-19.

1 State Key Laboratory of Silkworm Genome Biology, Southwest University, Chongqing 400716, China.
2 Vitamin D Research Institute, Shannxi University of Technology, Hanzhong 723001, China.
3 School of Management, Shannxi University of Technology, Hanzhong 723001, China.
*Corresponding author E-mail: xiaqy@swu.edu.cn.

TIL型蛋白酶抑制剂可作为靶标抗性因子增强家蚕对入侵真菌的防御

李游山[1,2]　赵　萍[1]　刘华伟[1]　郭晓朦[1]
何华伟[1]　朱　瑞[3]　向仲怀[1]　夏庆友[1*]

摘要:昆虫病原真菌通过大量的水解酶侵入昆虫表皮。这些水解酶是重要的毒力因子,包括表皮降解蛋白酶和几丁质酶。我们的研究表明,许多丝氨酸蛋白酶抑制剂参与昆虫对病原微生物的防御,尤其是TIL型蛋白酶抑制剂。为了探究昆虫防御病原真菌的分子机制,鉴定新基因以提高家蚕抵御真菌的能力,我们对丝氨酸蛋白酶抑制剂进行了深入的研究。在这项研究中,我们克隆并表达了一个新型TIL型蛋白酶抑制剂——*Bm*SPI39。在酶活性测定实验中,*Bm*SPI39有效地抑制了白僵菌毒性蛋白酶CDEP-1,表明其可能通过抑制真菌分泌的表皮降解蛋白酶活性来抵御真菌的入侵。酚氧化酶活性研究表明,黑化反应参与家蚕对真菌入侵的免疫应答,*Bm*SPI39通过抑制家蚕表皮降解酶活性来抑制由真菌诱导的过量黑化反应。为了更好地理解蛋白酶抑制剂抵御真菌的机制,我们检测了蛋白酶抑制剂对白僵菌孢子萌发的影响,结果表明*Bm*SPI38和*Bm*SPI39显著地抑制了白僵菌孢子的萌发。存活率实验表明*Bm*SPI38和*Bm*SPI39有效地提高了家蚕的存活率,因此可以作为提高家蚕真菌抗性的靶标蛋白。这些结果为研究昆虫蛋白酶抑制剂抵御致病真菌的分子机制提供了新视野,表明了蛋白酶抑制剂在医药和农业领域具有潜在应用价值。

1　家蚕基因组生物学国家重点实验室,西南大学,重庆
2　维生素D研究所,陕西理工大学,汉中
3　管理学院,陕西理工大学,汉中

Overexpression of host plant urease in transgenic silkworms

Jiang L[#] Huang CL[#] Sun Q Guo HZ Peng ZW Dang YH
Liu WQ Xing DX Xu GW Zhao P Xia QY[*]

Abstract: *Bombyx mori* and mulberry constitute a model of insect-host plant interactions. Urease hydrolyzes urea to ammonia and is important for the nitrogen metabolism of silkworms because ammonia is assimilated into silk protein. Silkworms do not synthesize urease and acquire it from mulberry leaves. We synthesized the artificial DNA sequence *ure-as* using the codon bias of *B. mori* to encode the signal peptide and mulberry urease protein. A transgenic vector that overexpresses *ure-as* under control of the silkworm midgut-specific *P2* promoter was constructed. Transgenic silkworms were created via embryo microinjection. RT-PCR results showed that *urease* was expressed during the larval stage and qPCR revealed the expression only in the midgut of transgenic lines. Urea concentration in the midgut and hemolymph of transgenic silkworms was significantly lower than in a nontransgenic line when silkworms were fed an artificial diet. Analysis of the daily body weight and food conversion efficiency of the fourth and fifth instar larvae and economic characteristics indicated no differences between transgenic silkworms and the nontransgenic line. These results suggested that overexpression of host plant *urease* promoted nitrogen metabolism in silkworms.

Published On: Molecular Genetics and Genomics, 2015, 290(3), 1117-1123.

State Key Laboratory of Silkworm Genome Biology, Southwest University, Chongqing 400716, China.
[#]These authors contributed equally.
[*]Corresponding author E-mail: xiaqy@swu.edu.cn.

家蚕中转基因过表达宿主植物脲酶的研究

蒋 亮[#] 黄春林[#] 孙 强 郭慧珍 彭正文 党颖慧
刘纬强 邢东旭 徐国文 赵 萍 夏庆友[*]

摘要:家蚕和桑树是昆虫和宿主植物相互作用的一个模型。脲酶(urease)对家蚕的氮代谢很重要,它将尿素水解为氨,氨又被同化成丝蛋白。家蚕自身不能合成脲酶,只能从桑叶中摄取。我们根据家蚕密码子偏好性合成一条人工DNA序列*ure-as*,编码信号肽和桑树脲酶蛋白。构建家蚕中肠特异启动子*P2*介导*ure-as*增量表达的转基因载体,通过显微注射制备转基因家蚕。RT-PCR结果显示*urease*在转基因家蚕的幼虫期表达,qPCR显示*urease*在中肠特异表达。在人工饲料饲养条件下,转基因家蚕中肠和血淋巴中的尿素浓度显著低于非转基因对照。分析四、五龄幼虫的逐日体重和食物转化效率以及经济性状,显示转基因家蚕和非转基因家蚕之间没有显著差异。这些结果表明增量表达宿主植物*urease*促进了家蚕的氮代谢。

家蚕基因组生物学国家重点实验室,西南大学,重庆

Differentially expressed genes in the midgut of silkworm infected with cytoplasmic polyhedrosis virus

Wu P[1] Li MW[1] Wang X[2] Zhao P[2]
Wang XY[1] Liu T[1] Qin GX[1] Guo XJ[1,2*]

Abstract: Understanding of the responsive and interactive mechanism between the host cells and *Bombyx mori* cytoplasmic polyhedrosis virus (*Bm*CPV) is crucial to the diagnosis of CPV-caused disease and the development of new control measures. In this report, we employed suppression subtractive hybridization to compare differentially expressed genes in the midguts of CPV-infected and normal silkworm larvae. 36 genes and 20 novel ESTs were obtained from 2 reciprocal subtractive libraries. Three up-regulated genes (*ferritin, rpL11* and *alkaline nuclease*) and 3 down-regulated genes (*serine protease, trypsin-like protease* and *inhibitor of apoptosis protein*) were identified by quantitative real-time PCR. The transcript differences of these 6 genes at 6, 12, 24, 48 and 72 h post-inoculation both in CPV-infected and normal midguts were compared. Our results indicated that ferritin and rpL11 were increased during the early stage (6-12 h p.i.) of CPV infection, whereas alkaline nuclease was increased during the late stage (24-72 h p.i.) of CPV infection. The expression of serine protease and trypsin-like protease is decreased at 24-72 h after CPV infection, while the expression of inhibitor of apoptosis protein is decreased throughout the infective stage. Our results provide new clues for investigating the molecular mechanism of *Bm*CPV infection.

Published On: African Journal of Biotechnology, 2009, 8 (16), 3711-3720.

1 Key Laboratory of Genetic Improvement of Silkworm and Mulberry, Sericultural Research Institute, Chinese Academy of Agricultural Sciences, Zhenjiang 212018, China.
2 School of Biotechnology and Environmental Engineering, Jiangsu University of Science and Technology, Zhenjiang 212018, China.
*Corresponding author E-mail: guoxijie@126.com.

家蚕感染质型多角体病毒后中肠基因的差异表达分析

吴　萍[1]　李木旺[1]　王　秀[2]　赵　盼[2]
王修业[1]　刘　挺[1]　覃光星[1]　郭锡杰[1,2*]

摘要：理解宿主细胞和家蚕质型多角体病毒（*Bm*CPV）之间的反应和相互作用机制对于CPV引起的疾病的诊断和开发新控制措施至关重要。在本研究中，我们采用抑制消减杂交法比较CPV感染和正常蚕幼虫中肠的差异表达基因。从2个交互消减文库数据中获得36个基因和20个新的EST序列。通过实时定量PCR鉴定证实了3个上调基因（铁蛋白基因、*rpL11*和碱性核酸酶基因）和3个下调基因（丝氨酸蛋白酶基因、胰蛋白酶样蛋白酶基因和凋亡蛋白抑制剂基因）。进一步比较了在正常中肠和CPV接种感染后6 h，12 h，24 h，48 h和72 h这6个基因的转录本差异。结果表明，铁蛋白基因和*rpL11*在CPV感染的早期（6—12 h p.i.）显著增加，而碱性核酸酶基因在感染的晚期（24—72 h p.i.）明显增加。丝氨酸蛋白酶基因和胰蛋白酶样蛋白酶基因在CPV感染后24—72 h呈表达降低趋势，而细胞凋亡蛋白抑制剂基因在整个感染阶段表达水平降低。我们的研究结果将为调查*Bm*CPV感染的分子机制提供了新的线索。

1　农业部蚕桑遗传改良重点实验室，中国农业科学院蚕业研究所，镇江
2　生物技术与环境工程学院，江苏科技大学，镇江

Cytoplasmic polyhedrosis virus-induced differential gene expression in two silkworm strains of different susceptibility

Gao K[1,2,3] Deng XY[1] Qian HY[2,3]
Qin GX[2,3] Hou CX[2,3] Guo XJ[2,3*]

Abstract: Digital gene expression (DGE) was performed to investigate the gene expression profiles of *4008* and *p50* silkworm strains at 48 h after oral infection with *Bm*CPV. 3 668 437 clean tags were identified in the *Bm*CPV infected *p50* silkworms and 3 540 790 clean tags in the control *p50*. By contrast, 4 498 263 clean tags were identified in the *Bm*CPV-infected 4008 silkworms and 4 164 250 clean tags in the control *4008*. A total of 691 differentially expressed genes were detected in the infected *4008* DGE library and 185 were detected in the infected *p50* DGE library, respectively. The expression profiles identified some important differentially expressed genes involved in signal transduction, enzyme activity and apoptotic changes, some of which were verified using quantitative real-time PCR (qRT-PCR). These results provide important clues on the molecular mechanism of *Bm*CPV invasion and resistance mechanism of silkworms against *Bm*CPV infection.

Published On: Gene, 2014, 539(2), 230-237.

1 College of Biotechnology and Chemical Engineering, Jiangsu University of Science and Technology, Zhenjiang 212003, China.
2 Sericultural Research Institute, Jiangsu University of Science and Technology, Zhenjiang 212018, China.
3 Sericultural Research Institute, Chinese Academy of Agricultural Sciences, Zhenjiang 212018, China.
*Corresponding author E-mail: guoxijie@126.com.

两个不同抗性家蚕品种感染*Bm*CPV的差异表达基因分析

高　坤[1,2,3]　邓祥元[1]　钱荷英[2,3]
覃光星[2,3]　侯成香[2,3]　郭锡杰[2,3*]

摘要：采用数字基因表达谱（DGE）技术对*4008*和*p50*两个不同抗性家蚕品种在经口感染*Bm*CPV 48 h后的差异表达基因进行了分析。在对照组和感染*Bm*CPV组的*p50*家蚕中分别得到了3 668 437个和3 540 790个标签，而在对照组和感染*Bm*CPV组的*4008*家蚕中分别得到了4 498 263个和4 164 250个标签；在*4008*家蚕的DGE中得到了691个差异表达基因，而在*p50*家蚕的DGE中得到了185个差异表达基因。在差异基因表达谱中鉴定获得一些与信号转导、酶活性和细胞凋亡相关基因的表达变化，并通过荧光实时定量PCR技术对其中的一些差异基因进行了验证。相关研究结果为从分子水平上阐明*Bm*CPV侵染家蚕的机理及家蚕对*Bm*CPV的抗性机理提供了重要依据。

1　生物与化学工程学院，江苏科技大学，镇江
2　蚕业研究所，江苏科技大学，镇江
3　蚕业研究所，中国农业科学院，镇江

Digital gene expression analysis in the midgut of *4008* silkworm strain infected with cytoplasmic polyhedrosis virus

Gao K[1,2,3] Deng XY[1] Qian HY[2,3]
Qin GX[2,3] Guo XJ[2,3*]

Abstract: Digital Gene Expression was performed to investigate the midgut transcriptome profile of *4008* silkworm strain orally infected with *Bm*CPV. A total of 4 498 263 and 4 258 240 clean tags were obtained from the control and *Bm*CPV-infected larvae. A total of 752 differentially expressed genes were detected, of which 649 were upregulated and 103 were downregulated. Analysis results of the Kyoto Encyclopedia of Genes and Genomes pathway showed that 334 genes were involved in the ribosome and RNA transport pathways. Moreover, 408 of the 752 differentially expressed genes have a GO category and can be categorized into 41 functional groups according to molecular function, cellular component and biological process. Differentially expressed genes involved in signaling, gene expression, metabolic process, cell death, binding, and catalytic activity changes were detected in the expression profiles. Quantitative real-time PCR was performed to verify the expression of these genes. The upregulated expression levels of Calreticulin, FK506-binding protein, and protein kinase c inhibitor gene probably led to a calcium-dependent apoptosis in the *Bm*CPV-infected cells. The results of this study may serve as a basis for future research not only on the molecular mechanism of *Bm*CPV invasion but also on the anti-*Bm*CPV mechanism of silkworm.

Published On: Journal of Invertebrate Pathology, 2014, 115, 8-13.

1 College of Biotechnology and Chemical Engineering, Jiangsu University of Science and Technology, Zhenjiang 212003, China.
2 Sericultural Research Institute, Chinese Academy of Agricultural Sciences, Zhenjiang 212018, China.
3 Sericultural Research Institute, Jiangsu University of Science and Technology, Zhenjiang 212018, China.
*Corresponding author E-mail: guoxijie@126.com.

家蚕品种*4008*感染*Bm*CPV的数字基因表达谱分析

高　坤[1,2,3]　邓祥元[1]　钱荷英[2,3]
覃光星[2,3]　郭锡杰[2,3*]

摘要：本研究构建了家蚕*4008*品种经口感染质型多角体病毒（*Bm*CPV）后中肠组织的数字基因表达谱。从对照组和感染*Bm*CPV组幼虫中肠中分别获得4 498 263个标签和4 258 240个标签。共发现了752个差异表达基因，其中649个上调表达，103个下调表达。KEGG分析结果显示，334个基因参与了核糖体和RNA的运输，并且752个差异表达基因中有408个基因可以归属到GO分类中，根据这些差异基因参与的分子功能、细胞成分和生物过程可将其分为41个功能团。在表达谱中检测到涉及信号传递、基因表达、代谢过程、细胞死亡、结合和催化活性变化的差异表达基因。通过实时荧光定量PCR技术验证了其中一些基因的差异表达。推测其中的CRT、FK506结合蛋白和蛋白激酶C抑制剂基因表达水平的上调可能诱导了某个钙依赖的细胞凋亡途径。该研究成果不仅可以为今后研究*Bm*CPV侵染家蚕的分子机制提供理论依据，还可以为研究家蚕抵御*Bm*CPV感染的机制奠定基础。

1　生物与化学工程学院，江苏科技大学，镇江
2　蚕业研究所，中国农业科学院，镇江
3　蚕业研究所，江苏科技大学，镇江

Transcriptome analysis of interactions between silkworm and cytoplasmic polyhedrosis virus

Jiang L[#] Peng ZW[#] Guo YB[#] Cheng TC Guo HZ
Sun Q Huang CL Zhao P Xia QY[*]

Abstract: *Bombyx mori* cytoplasmic polyhedrosis virus (*Bm*CPV) specifically infects silkworm midgut (MG) and multiplication occurs mainly in posterior midgut (PM). In this study, MG and fat body (FB) were extracted at 0, 3, 24, and 72 h after *Bm*CPV infection. The total sequence reads of each sample were more than 1 510 000, and the mapping ratio exceeded 95.3%. Upregulated transcripts increased in MG during the infection process. Gene ontology (GO) categories showed that antioxidants were all upregulated in FB but not in MG. *BGI001299*, *BGI014434*, *BGI012068*, and *BGI009201* were MG-specific genes with transmembrane transport function, the expression of which were induced by *Bm*CPV. *BGI001299*, *BGI014434*, and *BGI012068* expressed in entire MG and may be involved in *Bm*CPV invasion. *BGI009201* expressed only in PM and may be necessary for *Bm*CPV proliferation. *BmPGRP-S2* and *BGI012452* (a putative serine protease) were induced by *Bm*CPV and may be involved in immune defense against *Bm*CPV. The expression level of *Bm*CPV *S1*, *S2*, *S3*, *S6*, and *S7* was high and there was no expression of S9 in MG 72 h, implying that the expression time of structural protein coding genes is earlier. These results provide insights into the mechanism of *Bm*CPV infection and host defense.

Published On: Scientific Reports, 2016, 6, 24894.

State Key Laboratory of Silkworm Genome Biology, Southwest University, Chongqing 400716, China.
[#]These authors contributed equally.
[*]Corresponding author E-mail: xiaqy@swu.edu.cn.

转录组分析家蚕和质型多角体病毒的相互作用

蒋　亮[#]　彭正文[#]　郭由兵[#]　程廷才　郭慧珍
孙　强　黄春林　赵　萍　夏庆友[*]

摘要：家蚕质型多角体病毒（*Bm*CPV）专性感染家蚕中肠（MG），主要在中肠后部（PM）增殖。在*Bm*CPV感染后的0 h、3 h、24 h和72 h，提取家蚕中肠和脂肪体（FB）RNA用于转录组测序。每个样本的总序列超过1 510 000个，覆盖比例超过95.3%。中肠上调表达的转录本数量随着感染进程不断增加。GO分析发现脂肪体中的抗氧化相关基因全部上调表达，但中肠中的未上调。具有跨膜转运功能的中肠特异表达基因*BGI001299*、*BGI014434*、*BGI012068*和*BGI009201*被*Bm*CPV诱导上调表达。*BGI001299*、*BGI014434*和*BGI012068*在整个中肠中表达，可能参与了*Bm*CPV的入侵；*BGI009201*只在中肠后部表达，可能是*Bm*CPV增殖的必需基因。*BmPGRP-S2*和*BGI012452*（假定的丝氨酸蛋白酶）被*Bm*CPV诱导上调表达，可能参与对*Bm*CPV的免疫防御。在感染后72 h的中肠，*Bm*CPV的*S1*、*S2*、*S3*、*S6*、*S7*高表达而*S9*不表达，暗示结构蛋白编码基因的表达时间较早。这些结果有助于我们理解*Bm*CPV的侵染机制和宿主防御机制。

家蚕基因组生物学国家重点实验室，西南大学，重庆

Cloning and expression analysis of a peptidoglycan recognition protein in silkworm related to virus infection

Gao K[1,2,3] Deng XY[3] Qian HY[1,2]
Qin GX[1,2] Hou CX[1,2] Guo XJ[1,2*]

Abstract: In this study, the full-length cDNA of a peptidoglycan recognition protein named *Bm*PGRP-S3 was identified from the silkworm, *Bombyx mori* by rapid amplification of cDNA ends. It is 807 bp and comprises the following: a 5'- untranslated region (UTR) with a length of 112 bp, a 3′-UTR with a length of 92 bp including a polyadenylation signal sequence (AATAAA) and a poly(A) tail. The longest open reading frame (ORF) of *BmPGRP-S3* is 603 bp and encodes a polypeptide of 200 amino acids with a predicted molecularweight of 22.3 kDa including a PGRP domain. Sequence similarity and phylogenic analysis results indicated that *Bm*PGRP-S3 belongs to the group of insect PGRPs and is closer to *Bm*PGRP-S4 with the highest identity of 68%. Fluorescent quantitative realtime PCR results revealed that the mRNA transcripts of *BmPGRP-S3* were presented in all of the tissues, but were highest in the midgut. In the silkworm larvae infected with *B. mori* cytoplasmic polyhedrosis virus (*Bm*CPV), the relative expression level of *BmPGRP-S3* was upregulated. The DNA segment of amature *Bm*PGRP-S3 peptide was inserted into the expression plasmid pET-28a(+) to construct a recombinant expression plasmid.Western blot results revealed that mature *Bm*PGRP-S3 could be detected in the hemolymph and midgut which were the most important immune tissues in silkworm. All the results suggested that *Bm*PGRP-S3 may play an important role in the immune response of silkworm to *Bm*CPV infection and provided helpful information for further studying the function of *Bm*PGRP-S3 in silkworm.

Published On: Gene, 2014, 552(1), 24-31.

1 Sericultural Research Institute, Jiangsu University of Science and Technology, Zhenjiang 212018, China.
2 Sericultural Research Institute, Chinese Academy of Agricultural Sciences, Zhenjiang 212018, China.
3 College of Biotechnology, Jiangsu University of Science and Technology, Zhenjiang 212003, China.
*Corresponding author E-mail: guoxijie@126.com.

家蚕感染病毒相关的肽聚糖识别蛋白的克隆与表达分析

高　坤[1,2,3]　邓祥元[3]　钱荷英[1,2]
覃光星[1,2]　侯成香[1,2]　郭锡杰[1,2*]

摘要：本研究采用RACE技术克隆获得家蚕一个肽聚糖识别蛋白（*Bm*PGRP-S3）的cDNA全长序列。该cDNA全长为807 bp，包含一个112 bp 5′非翻译区和一个92 bp的3′非翻译区，3′非翻译区包含有终止密码子、加尾信号AATAAA和poly（A）尾；其最大开放阅读框（ORF）为603 bp，编码一个200个氨基酸组成的蛋白质，其中包含一个识别肽聚糖的PGRP结构域，预测蛋白质分子质量为22.3 kDa。多序列比对和系统进化树显示，家蚕*Bm*PGRP-S3与家蚕的*Bm*PGRP-S4的进化关系最近，同源相似性为68%。实时荧光定量PCR（qRT-PCR）分析表明*BmPGRP-S3*在家蚕各个组织中均有表达，其中在中肠中的表达量最高，且家蚕中肠*BmPGRP-S3*在家蚕感染质型多角体病毒（*Bombyx mori* cytoplasmic polyhedrosis virus，*Bm*CPV）后的表达量明显上调。通过构建重组表达质粒pET-28a-BmPGRP-S3对*Bm*PGRP-S3蛋白的成熟肽进行重组表达，Western blot分析表明*Bm*PGRP-S3在家蚕的血淋巴和中肠等重要的免疫组织中可以检测到。综上所述，*Bm*PGRP-S3在家蚕感染*Bm*CPV过程中可能起着重要的免疫防御作用，蛋白质的成功表达也为进一步研究其作用机理奠定了基础。

1　蚕业研究所，江苏科技大学，镇江
2　蚕业研究所，中国农业科学院，镇江
3　生物技术学院，江苏科技大学，镇江

dsRNA interference on expression of a RNA-dependent RNA polymerase gene of *Bombyx mori* cytoplasmic polyhedrosis virus

Pan ZH[1,2,3] Gao K[1,2] Hou CX[1,2] Wu P[1,2]
Qin GX[1,2] Geng T[1,2] Guo XJ[1,2*]

Abstract: *Bombyx mori* cytoplasmic polyhedrosis virus (*Bm*CPV) is one of the major viral pathogens in silkworm. Its infection often results in significant losses to sericulture. Studies have demonstrated that RNAi is one of the important anti-viral mechanisms in organisms. In this study, three dsRNAs targeting the RNA-dependent RNA polymerase (*RDRP*) gene of *Bm*CPV were designed and synthesized with 2′-F modification to explore their interference effects on *Bm*CPV replication in silkworm larvae. The results showed that injecting dsRNA in the dosage of 4-6 ng per mg body weight into the 5th instar larvae can interfere with the *Bm*CPV-*RDRP* expression by 93% after virus infection and by 99.9% before virus infection. In addition, the expression of two viral structural protein genes (*genome RNA segments 1* and *5*) was also decreased with the decrease of *RDRP* expression, suggesting that RNAi interference of *Bm*CPV-*RDRP* expression could affect viral replication. The study provides an effective method for investigating virus replication as well as the virus-host interactions in the silkworm larvae using dsRNA.

Published On: Gene, 2015, 565(1), 56-61.

1 Sericultural Research Institute, Jiangsu University of Science and Technology, Zhenjiang 212018, China.
2 Sericultural Research Institute, Chinese Academy of Agricultural Sciences, Zhenjiang 212018, China.
3 School of Biology and Basic Medical Science, Soochow University, Suzhou 215123, China.
*Corresponding author E-mail: guoxijie@126.com.

dsRNA 干扰家蚕质型多角体病毒RNA依赖型的RNA聚合酶基因的表达

潘中华[1,2,3] 高 坤[1,2] 侯成香[1,2] 吴 萍[1,2]
覃光星[1,2] 耿 涛[1,2] 郭锡杰[1,2*]

摘要：家蚕质型多角体病毒（*Bm*CPV）是家蚕主要的病毒病原之一，其感染引起的病害常给养蚕业造成重大损失。研究表明RNAi是生物体重要的抗病毒机制之一。本研究以*Bm*CPV的依赖于RNA的RNA聚合酶（RNA-dependent RNA polymerase，*RDRP*）基因为靶标，设计并合成了3条2′-F修饰的dsRNA，探讨它们在家蚕体内对*Bm*CPV复制的干涉作用。结果表明，按4—6 ng/mg的剂量对家蚕五龄幼虫进行体腔注射dsRNA能有效抑制*Bm*CPV-*RDRP*基因的表达，感染病毒后注射dsRNA其抑制效率为93%，感染病毒前注射其抑制效率达99.9%。此外，*RDRP*基因的表达被抑制后，*Bm*CPV的2个结构蛋白基因（病毒基因组RNA第1和第5片段）的表达水平也被下调，表明干涉*Bm*CPV-*RDRP*基因能够影响病毒的复制。本研究为利用dsRNA研究家蚕幼虫病毒复制和病毒与宿主的互作提供了一个有效的方法。

1 蚕业研究所，江苏科技大学，镇江
2 蚕业研究所，中国农业科学院，镇江
3 基础医学与生物科学学院，苏州大学，苏州

Transgenic breeding of anti-*Bombyx mori* L. nuclear polyhedrosis virus silkworm *Bombyx mori*

Yang HJ[1] Fan W[1] Wei H[1] Zhang JW[2] Zhou ZH[1]
Li JY[1] Lin JR[3] Ding N[2] Zhong BX[1*]

Abstract: Silkworm strains resistant to *Bombyx mori* L. nuclear polyhedrosis virus were obtained through transgenic experiments. *piggy*Bac transposon with an A3 promoter were randomly inserted into the silkworm, driving the enhanced green fluorescent protein (*EGFP*) reporter gene into the silkworm genome. Polymerase chain reaction results verified the insertion of the extraneous *EGFP* gene, and fluorescence microscopy showed that the *EGFP* was expressed in the midgut tissue. The morbidity ratio of the nuclear polyhedrosis decreased from 90% in the original silkworm strain to 66.7% in the transgenic silkworm strain. Compared with the resistance to the *Bombyx mori* L. nuclear polyhedrosis virus in the *Qiufeng* strain, which is commonly used in the production, there was an increase of 33 centesimal points in the transgenic silkworms. The antivirotic character in the *Chunhua*×*Qiuyue* strain, which was bred from a different transgenic family, was about 10 centesimal points higher than that in the *Qiufeng*×*Baiyu*, another crossbreed used in production. Our results indicated a good application value of the transposon-inserted mutation in the breeding of anti-*Bm*NPV silkworm strain.

Published On: Acta Biochimica et Biophysica Sinica, 2008, 40(10), 873-876.

1 College of Animal Sciences, Zhejiang University, Hangzhou 310029, China.
2 Huzhou Academy of Agriculture Sciences, Huzhou 313000, China.
3 College of Animal Sciences, South China Agricultural University, Guangzhou 510642, China.
*Corresponding author E-mail: bxzhong@zju.edu.cn.

通过转基因技术培育抗核型多角体病毒的家蚕

杨慧娟[1] 范 伟[1] 危 浩[1] 张金卫[1] 周仲华[1]
李建营[1] 林健荣[3] 丁 农[2] 钟伯雄[1*]

摘要:本研究通过转基因技术获得了具有抗核型多角体病毒的家蚕品种。携带A3启动子启动*EGFP*报告基因的*piggy*Bac转座子元件被随机地插入到家蚕基因组内。聚合酶链式反应结果证实外源*EGFP*片段插入到家蚕染色体上。通过荧光显微镜观察,检测到*EGFP*在中肠组织中表达。相同攻毒条件下,核型多角体病毒病的发病率从非转基因家蚕品种90%减少到转基因家蚕品种66.7%。与生产上常用的抗家蚕核型多角体病毒品种*Qiufeng*相比较,转基因家蚕的抗性也比*Qiufeng*提高了33%,比另一个来自不同转基因家族的杂交种*Qiufeng*×*Baiyu*抗病性提高10%。结果表明,通过转座插入产生突变在抗*Bm*NPV家蚕的品系的选育中具有良好的后用价值。

1 动物科学学院,浙江大学,杭州
2 湖州农业科学研究院,湖州
3 动物科学学院,华南农业大学,广州

V-ATPase is involved in silkworm defense response against *Bombyx mori* nucleopolyhedrovirus

Lü P[1,2#] Xia HC[2#] Gao L[3] Pan Y[4] Wang Y[1] Cheng X[1,5] Lü HG[2]
Lin F[6] Chen L[1] Yao Q[2] Liu XY[2] Tang Q[1] Chen KP[1,2*]

Abstract: Silkworms are usually susceptible to the infection of *Bombyx mori* (*B. mori*) nucleopolyhedrovirus (*Bm*NPV), which can cause significant economic loss. However, some silkworm strains are identified to be highly resistant to *Bm*NPV. To explore the silkworm genes involved in this resistance in the present study, we performed comparative real-time PCR, ATPase assay, over-expression and sub-cellular localization experiments. We found that when inoculated with *Bm*NPV both the expression and activity of V-ATPase were significantly up-regulated in the midgut column cells (not the goblet cells) of *Bm*NPV-resistant strains (*NB* and *BC8*), the main sites for the first step of *Bm*NPV invasion, but not in those of a *Bm*NPV-susceptible strain *306*. Furthermore, this up-regulation mainly took place during the first 24 hours post inoculation (hpi), the essential period required for establishment of virus infection, and then was down-regulated to normal levels. Amazingly, transient overexpression of V-ATPase c subunit in *Bm*NPV-infected silkworm cells could significantly inhibit *Bm*NPV proliferation. To our knowledge this is the first report demonstrating clearly that V-ATPase is indeed involved in the defense response against *Bm*NPV. Our data further suggests that prompt and potent regulation of V-ATPase may be essential for execution of this response, which may enable fast acidification of endosomes and/or lysosomes to render them competent for degradation of invading viruses.

Published On: PLoS ONE, 2013, 8(6).

1 School of Food and Biological Engineering, Jiangsu University , Zhenjiang 212013, China.
2 Institute of Life Sciences, Jiangsu University, Zhenjiang 212013, China.
3 School of Medical Science and Laboratory Medicine, Jiangsu University, Zhenjiang 212013, China.
4 The Laboratory Animal Research Center, Jiangsu University, Zhenjiang 212013, China.
5 Nanchang Key Laboratory of Applied Fermentation Technology, Jiangxi Agricultural University, Nanchang 330045, China.
6 Zhejiang Institute of Freshwater Fisheries, Huzhou 313001, China.
[#]These authors contributed equally.
[*]Corresponding author E-mail: kpchen@ujs.edu.cn.

V-ATP酶涉及家蚕对家蚕核型多角体病毒的防御反应

吕　鹏[1,2#]　夏恒传[2#]　高　路[3]　潘　晔[4]　王　勇[1]　程　新[1,5]　吕洪刚[2]
林　峰[6]　陈　亮[1]　姚　勤[2]　刘晓勇[2]　唐　琦[1]　陈克平[1,2*]

摘要：家蚕通常易感染家蚕核型多角体病毒(*Bm*NPV)，导致重大的经济损失。然而，一些家蚕品系对*Bm*NPV具有高度抗性。为了探索这种抗性涉及的基因，我们比较了实时PCR、ATP酶测定、过表达和亚细胞定位实验的结果。我们发现当接种*Bm*NPV时，V-ATP酶的表达和活性在*Bm*NPV抗性品系(*NB*和*BC8*)的中肠柱细胞(不是杯状细胞)中均显著上调，该细胞是*Bm*NPV侵袭第一步的主要位点，但在*Bm*NPV易感品系*306*中则没有上调。此外，这种上调可能主要发生在接种后24 h，是病毒感染所需的基本时间，然后V-ATP酶的表达被调节至正常水平。令人惊讶的是，*Bm*NPV感染的家蚕细胞中V-ATPase c亚基的瞬时过表达可显著抑制*Bm*NPV增殖。据我们所知，这是首次报告V-ATPase确实涉及对*Bm*NPV的防御反应。我们的数据进一步表明V-ATP酶的迅速和有效的调节对执行该反应可能是必不可少的，它可能会使分泌体和/或溶酶体快速酸化，使它们能够降解入侵的病毒。

1　食品与生物工程学院，江苏大学，镇江
2　生命科学研究院，江苏大学，镇江
3　医学科学与实验医学院，江苏大学，镇江
4　实验动物中心，江苏大学，镇江
5　南昌市发酵应用技术重点实验室，江西农业大学，南昌
6　浙江省淡水水产研究所，湖州

Identification of a new sprouty protein responsible for the inhibition of the *Bombyx mori* nucleopolyhedrovirus reproduction

Jin SK[1] Cheng TC[1] Jiang L[1] Lin P[1]

Yang Q[2] Xiao Y[2] Takahiro Kusakabe[3] Xia QY[1*]

Abstract: The rat sarcoma-extracellular signal regulated kinase mitogen-activated protein kinases pathway, one of the most ancient signaling pathways, is crucial for the defense against *Bombyx mori* nucleopolyhedrovirus (*Bm*NPV) infection. Sprouty (Spry) proteins can inhibit the activity of this pathway by receptor tyrosine kinases. We cloned and identified a new *B. mori* gene with a Spry domain similar to the Spry proteins of other organisms, such as fruitfly, mouse, human, chicken, Xenopus and zebrafish, and named it *BmSpry*. The gene expression analysis showed that *BmSpry* was transcribed in all of the examined tissues and in all developmental stages from embryo to adult. *BmSpry* also induced expression of *Bm*NPV in the cells. Our results indicated: (1) the knock-down of *BmSpry* led to increased *Bm*NPV replication and silkworm larvae mortality; (2) overexpression of *BmSpry* led to reduced *Bm*NPV replication; and (3) *BmSpry* regulated the activation of ERK and inhibited *Bm*NPV replication. These results showed that *BmSpry* plays a crucial role in the antiviral defense of the silkworm both *in vitro* and *in vivo*.

Published On: PLoS ONE, 2014, 9(6).

1 State Key Laboratory of Silkworm Genome Biology, Southwest University, Chongqing 400716, China.

2 Sericulture and Farm Product Processing Research Institute, Guangdong Academy of Agricultural Sciences, Guangzhou 510640, China.

3 Laboratory of Silkworm Science, Kyushu University Graduate School of Bioresource and Bioenvironmental Sciences, Fukuoka 819-0395, Japan.

*Corresponding author E-mail: xiaqy@swu.edu.cn.

一种抑制家蚕核型多角体病毒复制的新蛋白Sprouty的鉴定

金盛凯[1] 程廷才[1] 蒋 亮[1] 林 平[1]
杨 琼[2] 肖 阳[2] Takahiro Kusakabe[3] 夏庆友[1*]

摘要:大鼠肉瘤-细胞外信号调节丝裂原活化蛋白激酶途径,是最古老的信号通路之一,对于防御家蚕核型多角体病毒(*Bm*NPV)感染至关重要。Sprouty(Spry)蛋白可通过受体酪氨酸激酶抑制该途径的活性。我们克隆并鉴定了一种新的家蚕基因,其Spry结构域与其他生物体的Spry蛋白质相似,如果蝇、小鼠、人、鸡、非洲爪蟾和斑马鱼,并命名为*BmSpry*。基因表达分析表明,*BmSpry*在所有检查的组织和从胚胎到成虫的所有发育阶段都有转录。*BmSpry*还诱导细胞中*Bm*NPV的表达。我们的研究结果表明:(1)*BmSpry*基因的敲低导致*Bm*NPV复制和蚕幼虫死亡率增加;(2)*BmSpry*的过表达导致*Bm*NPV复制减少;(3)*BmSpry*调节ERK的激活并抑制*Bm*NPV复制。这些结果表明,*BmSpry*在体外和体内对蚕的抗病毒防御发挥着至关重要的作用。

1 家蚕基因组生物学国家重点实验室,西南大学,重庆
2 蚕业与农产品加工研究所,广东省农业科学院,广州
3 蚕业科学研究室,九州大学生物资源和生物环境科学研究生院,福冈,日本

Resistance to *Bombyx mori* nucleopolyhedrovirus via overexpression of an endogenous antiviral gene in transgenic silkworms

Jiang L[1] Wang GH[1] Cheng TC[1] Yang Q[2] Jin SK[1]
Lu G[1] Wu FQ[2] Xiao Y[2] Xu HF[1] Xia QY[1*]

Abstract: Transgenic technology is a powerful tool for improving disease-resistant species. *Bm*lipase-1, purified from the midgut juice of *Bombyx mori*, showed strong antiviral activity against *B. mori* nucleopolyhedrovirus (*Bm*NPV). In an attempt to create an antiviral silkworm strain for sericulture, a transgenic vector overexpressing the *Bmlipase-1* gene was constructed under the control of a baculoviral immediate early-1 (IE1) promoter. Transgenic lines were generated via embryo microinjection. The mRNA level of *Bmlipase-1* in the midguts of the transgenic line was 27.3 % higher than that of the non-transgenic line. After feeding the silkworm with different amounts of *Bm*NPV, the mortality of the transgenic line decreased to approximately 33% compared with the non-transgenic line when the virus dose was 10^6 OB/larva. These results imply that overexpressing endogenous antiviral genes can enhance the antiviral resistance of silkworms.

Published On:Archives of Virology, 2012, 157(7), 1323-1328.

1 State Key Laboratory of Silkworm Genome Biology, Southwest University, Chongqing 400716, China.

2 Sericulture and Farm Product Processing Research Institute, Guangdong Academy of Agricultural Sciences, Guangzhou 510610, China.

*Corresponding author E-mail: xiaqy@swu.edu.cn.

增量表达内源抗病毒基因提高转基因家蚕对核型多角体病毒的抗性

蒋　亮[1]　王根洪[1]　程廷才[1]　杨　琼[2]　金盛凯[1]
陆　改[1]　吴福泉[2]　肖　阳[2]　徐汉福[1]　夏庆友[1*]

摘要：转基因技术是改良抗病品种的有力工具。从家蚕肠液中纯化出的*Bm*lipase-1蛋白具有较强的抗病毒活性。为了创制抗病毒家蚕品种，我们构建了IE1启动子介导*Bmlipase-1*增量表达的转基因载体，通过胚胎显微注射制备了转基因家蚕品系。转基因家蚕中肠*Bmlipase-1*的mRNA含量比非转基因的高27.3%。取不同剂量的*Bm*NPV进行添食，当感染剂量为10^6 OB/头时，转基因家蚕品系的死亡率比非转基因品系降低了33%。这些结果表明增量表达内源抗病毒基因能够增强家蚕的抗病毒能力。

1　家蚕基因组生物学国家重点实验室，西南大学，重庆
2　蚕业与农产品加工研究所，广东省农业科学院，广州

Resistance to *Bm*NPV via overexpression of an exogenous gene controlled by an inducible promoter and enhancer in transgenic silkworm, *Bombyx mori*

Jiang L[1] Cheng TC[1] Zhao P[1] Yang Q[2] Wang GH[1]
Jin SK[1] Lin P[1] Xiao Y[2] Xia QY[1*]

Abstract: The *hycu-ep32* gene of *Hyphantria cunea* NPV can inhibit *Bombyx mori* nucleopolyhedrovirus (*Bm*NPV) multiplication in co-infected cells, but it is not known whether the overexpression of the *hycu-ep32* gene has an antiviral effect in the silkworm, *Bombyx mori*. Thus, we constructed four transgenic vectors, which were under the control of the 39 K promoter of *Bm*NPV (*39 KP*), *Bombyx mori A4* promoter (*A4P*), *hr3* enhancer of *Bm*NPV combined with *39 KP*, and *hr3* combined with *A4P*. Transgenic lines were created via embryo microinjection using practical diapause silkworm. qPCR revealed that the expression level of *hycu-ep32* could be induced effectively after *Bm*NPV infection in transgenic lines where *hycu-ep32* was controlled by *hr3* combined with *39 KP* (i.e., HEKG). After oral inoculation of *Bm*NPV with 3×10^5 occlusion bodies per third instar, the mortality with HEKG-B was approximately 30% lower compared with the non-transgenic line. The economic characteristics of the transgenic lines remained unchanged. These results suggest that overexpression of an exogenous antiviral gene controlled by an inducible promoter and enhancer is a feasible method for breeding silkworms with a high antiviral capacity.

Published On: PLoS ONE, 2012, 7(8).

1 State Key Laboratory of Silkworm Genome Biology, Southwest University, Chongqing 400716, China.
2 Sericulture and Farm Product Processing Research Institute, Guangdong Academy of Agricultural Sciences, Guangzhou 510610, China.
*Corresponding author E-mail: xiaqy@swu.edu.cn.

诱导型启动子和增强子介导外源基因的增量表达提高转基因家蚕对*Bm*NPV的抗性

蒋　亮[1]　程廷才[1]　赵　萍[1]　杨　琼[2]　王根洪[1]
金盛凯[1]　林　平[1]　肖　阳[2]　夏庆友[1*]

摘要:在共感染细胞中,美国白蛾核型多角体病毒的*hycu-ep32*可以抑制家蚕核型多角体病毒(*Bm*NPV)的增殖。目前还不清楚*hycu-ep32*是否在家蚕中具有抗病毒活性,因此我们构建了4个转基因载体,分别由*Bm*NPV的39 K启动子(*39 KP*)、家蚕的*A4*启动子(*A4P*)、*hr3*增强子*+39 KP*和*hr3+A4P*驱动,通过胚胎显微注射家蚕实用滞育蚕种制备转基因家蚕品系。qPCR检测结果表明,在*hr3+39 KP*介导*hycu-ep32*表达的转基因品系*HEKG*中在感染*Bm*NPV后可有效诱导*hycu−ep32*的表达水平。在三龄起蚕时以3×10^5 OB/头经口添食*Bm*NPV后,HEKG-B的死亡率比非转基因的降低了约30%。转基因品系的经济性状没有发生变化。这些结果表明,利用诱导型启动子和增强子介导外源抗病毒基因的增量表达是培育抗病毒家蚕的一种有效方法。

1　家蚕基因组生物学国家重点实验室,西南大学,重庆
2　蚕业与农产品加工研究所,广东省农业科学院,广州

Comparison of factors that may affect the inhibitory efficacy of transgenic RNAi targeting of baculoviral genes in silkworm, *Bombyx mori*

Jiang L[1,2#] Zhao P[1,2#] Wang GH[1,2] Cheng TC[1,2] Yang Q[3]
Jin SK[1,2] Lin P[1,2] Xiao Y[3] Sun Q[2] Xia QY[1,2*]

Abstract: *Bombyx mori* nucleopolyhedrovirus (*Bm*NPV) is the primary pathogen affecting *B. mori*. This virus could be combated via RNAi of *Bm*NPV genes in transgenic silkworm. However, several factors may affect the resistance of transgenic RNAi silkworm, such as the connection pattern of gene fragments and spacers ("head to head" or "tail to tail"), and the selection of promoters and target genes. In this study, we constructed several transgenic RNAi vectors using different phase genes (*ie-1*, *helicase*, *gp64*, and *vp39*) and promoters (*Bm*NPV IE1 promoter (*IE1P*), *IE1P* combined with *hr3* enhancer of *Bm*NPV, and *B. mori A4* promoter (*A4P*). Transgenic lines were generated via embryo microinjection using a practical silkworm strain. We analyzed the anti-*Bm*NPV ability, virus gene mRNA level, and *Bm*NPV content of these transgenic larvae. The results showed that "head to head" was better than "tail to tail," *IE1P* combined with *hr3* was better than *IE1P* and *A4P*, and an immediate early gene was the best target for RNAi.

Published On: Antiviral Research, 2013, 97(3), 255-263.

1 State Key Laboratory of Silkworm Genome Biology, Southwest University, Chongqing 400716, China.
2 College of Biotechnology, Southwest University, Chongqing 400716, China.
3 Sericulture & Farm Product Processing Research Institute, Guangdong Academy of Agricultural Sciences, Guangzhou 510610, China.
#These authors contributed equally.
*Corresponding author E-mail: xiaqy@swu.edu.cn.

家蚕杆状病毒转基因干涉抑制效果影响因素的比较

蒋　亮[1,2#]　赵　萍[1,2#]　王根洪[1,2]　程廷才[1,2]　杨　琼[3]
金盛凯[1,2]　林　平[1,2]　肖　阳[3]　孙　强[2]　夏庆友[1,2*]

摘要：家蚕核型多角体病毒（*Bm*NPV）是家蚕的主要病原，干涉病毒基因可以提高转基因家蚕对*Bm*NPV的抗性。但是，某些因素会影响转基因干涉家蚕的抗病毒能力，比如，基因片段和间隔序列的连接模式（“头对头”或“尾对尾”）以及启动子和靶标基因的选择。我们利用不同时期的病毒基因（*ie-1*、*helicase*、*gp64*和*vp39*）和不同启动子（*IE1P*、*IE1P+hr3*和*A4P*）构建了多个转基因干涉载体，通过胚胎显微注射家蚕实用品种制备了转基因品系。对这些转基因品系的抗*Bm*NPV能力、病毒基因mRNA含量和病毒DNA含量进行了检测，发现“头对头”连接模式优于“尾对尾”，*IE1P+hr3*优于*IE1P*和*A4P*，病毒极早期基因是最好的干涉靶标。

1　家蚕基因组生物学国家重点实验室，西南大学，重庆
2　生物技术学院，西南大学，重庆
3　蚕业与农产品加工研究所，广东省农业科学院，广州

Short hairpin RNA expression for enhancing the resistance of *Bombyx mori* (*Bm*) to nucleopolyhedrovirus *in vitro* and *in vivo*

Roy Bhaskar[1] Zhou F[1] Liang S[1]
Yue WF[2] Niu YS[1] Miao YG[1*]

Abstract: A new paradigm of RNAi technology has been studied for enhancing the resistance to virus in plants and animals. Previous studies have shown that the *Bombyx mori* (*Bm*) *U6* promoter based shRNA is an effective tool for inducing RNAi in *Bombyx mori* cell line. However, widespread knockdown and induction of phenotypes in *Bm* larvae have not been fully demonstrated. In this study, we examined *Bm U6* promoter based shRNA expression for suppressing *Bm* nucleopolyhedrovirus (NPV) in the *Bm* cell line and silkworm larvae. We measured the relative expression level of replication genes of *Bm*NPV in hemolymph of silkworm larvae and *BmN* cells transfected with recombinant targeting shRNA by quantitative real time polymerase chain reaction (PCR). These results indicated that the recombinant shRNA expression system was a useful tool for resistance to *Bm*NPV *in vivo* and *in vitro*. The approach opens the door of RNAi technology as a wide range of strategies that offer a technically simpler, cheaper, and quicker gene-knockdown by recombinant shRNA for future genetics in silkworm *Bm* and other related species.

Published On: African Journal of Biotechnology, 2013, 12(49), 6801-6808.

1 Key Laboratory of Animal Virology of Ministry of Agriculture, College of Animal Sciences, Zhejiang University, Hangzhou 310058, China.

2 College of Animal Sciences, Zhejiang A & F University, Lin'an 311300, China.

*Corresponding author E-mail: miaoyg@zju.edu.cn.

过表达短发夹RNA增强体外和体内家蚕对*Bm*NPV抗性

Roy Bhaskar[1] 周 芳[1] 梁 爽[1]
岳万福[2] 牛艳山[1] 缪云根[1*]

摘要：新型RNAi技术被用于增强动植物的抗病毒能力。先前研究表明基于短发夹RNA（shRNA）的家蚕*U6*启动子可以在家蚕细胞系中有效地诱导RNAi发生。然而，家蚕幼虫体内大范围地敲低或诱导表型产生至今尚未完全证明。在本文中，我们检测了基于shRNA的*U6*启动子在家蚕*Bm*N细胞和幼虫体内对家蚕核型多角体病毒（*Bm*NPV）的抑制作用。利用定量PCR，我们在转染了目标基因shRNA的家蚕血淋巴和*BmN*细胞中测定了*Bm*NPV的复制基因表达水平。结果显示，重组shRNA表达系统在体内外都能有效抑制*Bm*NPV。该方法拓展了RNAi技术的应用范围，并且提供了一个更简便、廉价和快捷的基因敲低技术，为未来家蚕遗传学研究和其他类似物种研究提供了新的视野。

1 农业部动物病毒重点实验室，动物科学学院，浙江大学，杭州
2 动物科学学院，浙江农林大学，临安

Resistance of transgenic silkworm to *Bm*NPV could be improved by silencing *ie-1* and *lef-1* genes

Zhang PJ[1#] Wang J[1#] Lu Y[1] Hu YY[1]

Xue RY[1,2] Cao GL[1,2#] Gong CL[1,2*]

Abstract: RNA interference (RNAi) - mediated viral inhibition has been used in several organisms for improving viral resistance. In the present study, we reported the use of transgenic RNAi in preventing *Bombyx mori* nucleopolyhedrovirus (*Bm*NPV) multiplication in the transgenic silkworm *B. mori.* We targeted the *Bm*NPV *immediate-early-1* (*ie-1*) and *late expression factor-1* (*lef*-1) genes in the transiently transfected *BmN* cells, in the stable transformed *BmN* cell line and in the transgenic silkworms. We generated four *piggy*Bac-based vectors containing short double-stranded *ie-1* RNA (*sdsie-1*), short double-stranded *lef*-1 RNA (*sdslef-1*), *long double-stranded ie*-1 RNA (*ldsie-1*) and both *sdsie-1 and sdslef-1* (*sds-ie1-lef1*) expression cassettes. Strong viral repression was observed in the transiently transfected cells and in the stable transformed *BmN* cells transfected with *sds-ie-1*, *sdslef-1*, *ldsie-1* or *sds-ie-lef.* The decrease of *ie-1* mRNA level in the sds-ie1-lef1 transiently transfected cells was most obvious among the cells transfected with different vectors. The inhibitory effect of viral multiplication was decreased in a viral dose-dependent manner; the infection ratio of transfected cells for *sds-ie-1, sdslef-1, ldsie-1 and sds-ie-lef* decreased by 18.83%, 13.73%, 6.93% and 30.63%, respectively, compared with control cells 5 days after infection. We generated transgenic silkworms using transgenic vector *piggyantiIE-lef1-neo* with *sds-ie1-lef1* expression cassette; the fourth instar larvae of transgenic silkworms of generation G5 exhibited stronger resistance to *Bm*NPV, the mortalities for the transgenic silkworms and control silkworms were 60% and 100%, respectively, at 11 days after inoculation with *Bm*NPV (10^6 occlusion bodies per mL). These results suggest that double-stranded RNA expression of essential genes of *Bm*NPV is a feasible method for breeding silkworms with a high antiviral capacity.

Published On: Gene Therapy, 2014, 21(1), 81-88.

1 School of Biology and Basic Medical Sciences , Soochow University, Suzhou 215123, China.

2 National Engineering Laboratory for Modern Silk, Soochow University, Suzhou 215123, China.

#These authors contributed equally.

*Corresponding author E-mail: gongcl@suda.edu.cn.

通过沉默*ie-1*和*lef-1*基因可以提高转基因家蚕对*Bm*NPV的抗性

张鹏杰[1#]　王　健[1#]　陆　叶[1]　胡莹莹[1]
薛仁宇[1,2]　曹广力[1,2#]　贡成良[1,2*]

摘要：RNA 干扰（RNAi）介导的病毒抑制已被用于提高对病毒的抵抗力。在本研究中，我们报道了 RNAi 技术抑制 *Bm*NPV 的增殖。利用 *Bm*NPV 复制必需基因 *ie-1*、*lef-1* 分别构建了含有短双链 *ie-1* RNA（*sdsie-1*）、短双链 *lef-1* RNA（*sdslef-1*）、长双链 *ie-1* RNA（*ldsie-1*）和既含有短双链 *ie-1* RNA 也含有短双链 *lef-1* RNA（*sds-ie1-lef1*）表达盒的 4 种 *piggy*Bac 载体，在 *BmN* 中转染 4 种载体均有强烈的抑制病毒作用，其中在转染 *sds-ie1-lef1* 载体的细胞中病毒 *ie-1* mRNA 水平下降最明显。对病毒增殖的抑制作用具有剂量依赖性；转染 sds-*ie-1*，*sdslef-1*，*ldsie-1* 和 *sds-ie-lef* 后，与对照相比，病毒感染 5 d 时的感染率分别降低了 18.83%，13.73%，6.93% 和 30.63%。用 *piggyantiIE-lef1-neo* 质粒和 *sds-ie1-lef1* 表达盒获得的转基因家蚕的第五代四龄家蚕对 *Bm*NPV 的抗性增强。*Bm*NPV（10^6 OB/mL）感染家蚕 11 d 后，转基因家蚕和对照蚕的死亡率分别为 60% 和 100%。这些结果表明，*Bm*NPV 复制必需基因双链 RNA 的表达是培育 *Bm*NPV 高抗家蚕品种的有效方法。

1　基础医学与生物科学学院，苏州大学，苏州
2　现代丝绸国家工程实验室，苏州大学，苏州

*piggy*Bac transposon-derived targeting shRNA interference against the *Bombyx mori* nucleopolyhedrovirus (*Bm*NPV)

Zhou F　Chen RT　Lu Y

Liang S　Wang MX　Miao YG*

Abstract: The *Bombyx mori* nucleopolyhedrovirus (*Bm*NPV) is one of the most destructive diseases in silkworm, which has caused the main damage to sericulture industry. In this study, we developed a system of RNAi to prevent the *Bm*NPV infection using the *piggy*Bac transposon-derived targeting short hairpin RNA (shRNA) interference. The shRNAs targeting the genes of *i. e. - 1*, *lef-1*, *lef-2* and *lef-3* of *Bm*NPV were designed and used to inhibit the intracellular replication or multiplication of *Bm*NPV in *Bm* cells. The highest activity was presented in the shRNA targeting the *i. e. - 1c* of *Bm*NPV, of which the inhibition rate reached 94.5 % *in vitro*. Further a stable *Bm* cell line of *piggy*Bac transposon-derived targeting shRNA interference against *Bm*NPV was established, which has a highly efficacious suppression on virus proliferation. These results indicated that the recombinant shRNA expression system was a useful tool for resistance to *Bm*NPV *in vitro*. The approach by recombinant shRNAs opens a door of RNAi technology as a strategy that offering technically simpler, cheaper, and quicker gene knockdown for promising research and biotechnology application on silkworm lethal diseases.

Published On: Molecular Biology Reports, 2014, 41(12), 8247-8254.

Key Laboratory of Animal Virology of Ministry of Agriculture, College of Animal Sciences, Zhejiang University, Hangzhou 310058, China.

*Correspongding author E-mail: miaoyg@zju.edu.cn.

基于*piggy*Bac转座子系统靶向shRNA干涉抵抗家蚕核型多角体病毒(*Bm*NPV)

周　芳　陈瑞婷　陆　岳
梁　爽　王梅仙　缪云根*

摘要：家蚕核型多角体病毒(*Bm*NPV)是家蚕中最具破坏性的病之一，对蚕业生产造成了巨大的影响。在本研究中，我们通过*piggy*Bac转座子衍生的靶向短发夹RNA(shRNA)的RNAi系统来预防*Bm*NPV感染。我们设计了靶*Bm*NPV *ie-1*，*lef-1*，*lef-2*和*lef-3*基因的shRNA，从而来抑制*Bm*NPV在家蚕细胞内的复制和增殖。在这些shRNA中，抑制活性最高的为靶向*ie-1c*的shRNA，其抑制效率可达94.5%。进一步还构建了*piggy*Bac转座子衍生的稳定表达靶抗*Bm*NPV的靶向shRNA细胞系，该细胞系可以高度抑制*Bm*NPV的增殖。这些结果说明，重组shRNA系统可以作为一个有效的体外抗病毒工具。该方法拓展了RNAi技术的应用范围，并且提供了一个更简便、廉价和快捷的基因敲低技术，对未来家蚕疾病研究和生物技术应用开阔了新的视野。

农业部动物病毒学重点实验室，动物科学学院，浙江大学，杭州

Identification of a midgut-specific promoter in the silkworm *Bombyx mori*

Jiang L[#] Cheng TC[#] Dang YH Peng ZW Zhao P

Liu SP Jin SK Lin P Sun Q Xia QY[*]

Abstract: The midgut is an important organ for digestion and absorption of nutrients and immune defense in the silkworm *Bombyx mori*. In an attempt to create a tool for midgut research, we cloned the 1 080 bp P2 promoter sequence (*P2P*) of a highly expressed midgut-specific gene in the silkworm. The transgenic line (*P2*) was generated via embryo microinjection, in which the expression of *EGFP* was driven by *P2P*. There was strong green fluorescence only in the midgut of *P2*. RT-PCR and Western blot showed that *P2P* was a midgut-specific promoter with activity throughout the larval stage. A transgenic truncation experiment suggested that regions -305 to -214 and +107 to +181 were very important for *P2P* activity. The results of this study revealed that we have identified a midgut-specific promoter with a high level of activity in the silkworm that will aid future research and application of silkworm genes.

Published On: Biochemical Biophysical Research Communications. 2013, 433(4), 542-546.

State Key Laboratory of Silkworm Genome Biology, Southwest University, Chongqing 400716, China.

[#]These authors contributed equally.

[*]Corresponding author E-mail: xiaqy@swu.edu.cn.

家蚕中肠特异启动子的鉴定

蒋　亮[#]　程廷才[#]　党颖慧　彭正文　赵　萍
刘仕平　金盛凯　林　平　孙　强　夏庆友[*]

摘要：中肠是家蚕的重要器官，具有消化、吸收营养物质和免疫防御等功能。为了给中肠研究创制一个工具，我们克隆了一个在家蚕中肠特异高量表达基因的启动子序列，全长1 080 bp，命名为*P2P*。通过胚胎显微注射我们成功制备了*P2P*驱动*EGFP*表达的转基因家蚕品系*P2*。荧光观察发现*P2*的中肠发出强烈的绿色荧光，RT-PCR和Western blot检测结果表明，*P2P*是一个中肠特异启动子并在整个幼虫期都具有活性。转基因截短实验表明，-305至-214和+107至+181是*P2P*的活性关键区域。这些结果表明我们鉴定了一个高活性的家蚕中肠特异启动子，这将有助于家蚕基因的功能研究和应用研究。

家蚕基因组生物学国家重点实验室，西南大学，重庆

A transgenic animal with antiviral properties that might inhibit multiple stages of infection

Jiang L[1#] Zhao P[1#] Cheng TC[1] Sun Q[1,2] Peng ZW[1] Dang YH[1]
Wu XW[1] Wang GH[1] Jin SK[1] Lin P[1] Xia QY[1,2*]

Abstract: *Bombyx mori* nucleopolyhedrovirus (*Bm*NPV) is the primary pathogen of silkworms, causing severe economic losses in sericulture. To create antiviral silkworm strains, we constructed a transgenic vector in which the dsRNA for five tandem *Bm*NPV genes was controlled by the *Bm*NPV *hr3* enhancer and *IE1* promoter. The antivirus gene *Bmlipase-1* was driven by *B. mori* midgut-specific promoter *P2*. Transgenic strains (SW-H) were generated via embryo microinjection using the practical silkworm strain *SW*. After infection with a high dose of *Bm*NPV, the survival rates of *SW-H* and non-transgenic *SW* were 64% and 13%, respectively. *SW-H* could be the first transgenic animal that is highly antiviral and that might inhibit the virus at multiple stages of infection.

Published On: Antiviral Research, 2013, 98(2), 171-173.

1 State Key Laboratory of Silkworm Genome Biology, Southwest University, Chongqing 400716, China.
2 College of Biotechnology, Southwest University, Chongqing 400716, China.
#These authors contributed equally.
*Corresponding author E-mail: xiaqy@swu.edu.cn.

在感染的多个阶段抑制病毒的转基因动物

蒋 亮[1#] 赵 萍[1#] 程廷才[1] 孙 强[1,2] 彭正文[1] 党颖慧[1]
吴向伟[1] 王根洪[1] 金盛凯[1] 林 平[1] 夏庆友[1,2*]

文摘：家蚕核型多角体病毒（*Bm*NPV）是家蚕的主要病原，给蚕业生产造成了严重的经济损失。为了创制抗病毒家蚕品种，我们构建了一个转基因载体，该载体中*hr3*增强子和*IE1*启动子介导5个病毒基因的串联干涉，家蚕中肠特异启动子*P2*介导抗病毒基因*Bmlipase-1*的增量表达。通过胚胎显微注射家蚕实用种*SW*，制备了转基因品系*SW-H*。感染高剂量的*Bm*NPV后，*SW-H*和非转基因*SW*的存活率分别为64%和13%。*SW-H*是第一个能在感染的多个阶段抑制病毒的转基因高抗病毒动物品种。

1 家蚕基因组生物学国家重点实验室，西南大学，重庆
2 生物技术学院，西南大学，重庆

Molecular cloning and characterization of a putative cDNA encoding endoglucanase IV from *Trichoderma viride* and its expression in *Bombyx mori*

Li XH Zhang P Liang S Zhou F Wang MX
Roy Bhaskar Firdose Ahmad Malik Niu YS Miao YG*

Abstract: The development of cellulase production technology has greatly contributed to the successful use of cellulosic materials as renewable carbon sources. In this study, a putative endoglucanase IV (EG IV) complementary DNA was cloned from the mycelium of a strain of the filamentous fungus *Trichoderma viride* using a PCR-based exon-splicing method and expressed in both a silkworm *BmN* cell line and in silkworm larvae. Western blot analysis detected a band of 42 kDa in *BmN* cells after infection with a recombinant mBacmid/*Bm*NPV/EG IV baculovirus. Sequence alignment analysis of the *T. viride* EG IV gene showed two domains that were highly conserved with glycosyl hydrolases and a funga-type cellulose-binding domain. Analysis of variance showed that silkworms infected with recombinant baculoviruses exhibited significantly higher enzyme activity that was 48.84% higher than silkworms infected with blank baculoviruses and 46.61% higher than normal silkworms. The expressed bioactive EG IV was also stable at the pH range from 5.0 to 10.0. The availability of large quantities of bioactive EG IV in silkworm provided a possibility to produce cellulase transgenic silkworm, which express bioactive cellulose specially in its digestive tract and improve its metabolism efficiency of mulberry leaves. Its application in the sericulture industry may be very promising.

Published On: Applied Biochemistry and Biotechnology, 2012, 166(2), 309-320.

Key Laboratory of Animal Virology of Ministry of Agriculture, College of Animal Sciences, Zhejiang University, Hangzhou 310029, China.

*Corresponding author E-mail: miaoyg@zju.edu.cn.

编码绿色木霉内切葡聚糖酶IV的假定cDNA的分子克隆和鉴定及其在家蚕中的表达

李兴华 张 鹏 梁 爽 周 芳 王梅仙
Roy Bhaskar Firdose Ahmad Malik 牛艳山 缪云根*

摘要：纤维素酶生产技术的发展使得利用纤维素材料作为可再生碳源成为可能。在本研究中，使用基于PCR的外显子剪接方法从丝状真菌绿色木霉菌株的菌丝体中克隆了一种假定的内切葡聚糖酶IV（EG IV）基因，并在家蚕*BmN*细胞系和家蚕幼虫中表达。Western blot结果表明，在用重组mBacmid/*Bm*NPV/EG IV杆状病毒感染后，在*BmN*细胞中检测到42 kDa的条带。绿色木霉EG IV基因的序列比对分析显示了两个与糖基水解酶和真菌型纤维素结合结构域高度保守的结构域。方差分析表明，重组杆状病毒感染的蚕具有较高的酶活性，比空白杆状病毒感染蚕高48.84%，比正常蚕高46.61%。表达的EG IV在pH 5.0—10.0生物活性也是稳定的。利用家蚕大量生产生物活性EG IV的可行性为生产纤维素酶转基因家蚕提供了可能，该家蚕在其消化道中特异表达生物活性纤维素，可以提高其桑叶的代谢效率。这在蚕桑业的应用中可能非常具有前景。

农业部动物病毒学重点实验室，动物科技学院，浙江大学，杭州

第十三章

其他研究

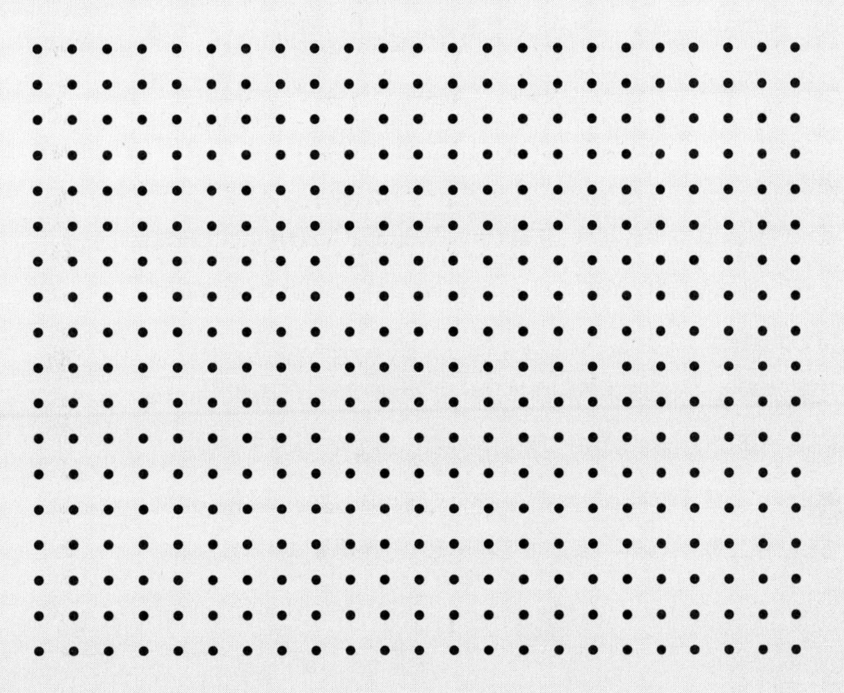

Chapter 13

基因组计划完成以后，家蚕成为了众多鳞翅目昆虫研究的典型案例。除了丝蛋白合成、变态发育机制、免疫机制等重要科学问题得到良好阐释以外，许多先进的手段也在家蚕研究中得到良好发展，为昆虫学研究提供了借鉴。

结构生物学研究手段是解析生物学机制的重要工具，在家蚕基因组及功能基因组研究突飞猛进的情况下，利用结构生物学手段解析家蚕重要蛋白结构，阐明其生物学机制显得尤为重要。在"973"项目的大力支持下，中国科技大学、中国科学院生物物理研究所、新加坡国立大学生物系等结构生物学团队陆续加入到了家蚕重要蛋白质结构研究中，成立了结构生物学联合研发平台，并成功解析了家蚕丝素重链蛋白Fib-H N端、家蚕贮藏蛋白arylphorin复合物、家蚕30K蛋白、家蚕*Bm*NPV-ORF75等重要蛋白质结构。丝蛋白的结构一直以来都是科学研究的难点，Fib-H N端结构的解析为蚕丝性能与结构关系揭示奠定了理论基础。

基因组计划的完成为家蚕分子生物学研究提供了极为便利的条件，在基因组与数据库网站的不断发展下，人们可以方便地鉴定并克隆与其他昆虫甚至脊椎动物同源的靶标基因，虽然它们在家蚕中的具体作用尚不十分清楚，但是了解它们分布模式、表达特征，获得其序列的序列信息，将为揭示这些基因在家蚕或者昆虫中的作用提供参考信息。

侯勇

N-terminal domain of *Bombyx mori* fibroin mediates the assembly of silk in response to pH decrease

He YX[#] Zhang NN[#] Li WF Jia N Chen BY
Zhou K Zhang JH Chen YX Zhou CZ[*]

Abstract: Fibroins serve as the major building blocks of silk fiber. As the major component of fibroin, the fibroin heavy chain is a considerably large protein comprising N-terminal and C-terminal hydrophilic domains and 12 highly repetitive Gly-Ala-rich regions flanked by internal hydrophilic blocks. Here, we show the crystal structure of the fibroin N-terminal domain (FibNT) at pH 4.7, revealing a remarkable double-layered anti-parallel β-sheet with each layer comprising two FibNT molecules entangled together. We also show that FibNT undergoes a pH-responsive conformational transition from random coil to β-sheets at around pH 6.0. Dynamic light scattering demonstrates that FibNT tends to oligomerize as pH decreases to 6.0, and electron microscopy reveals micelle-like oligomers. Our results are consistent with the micelle assembly model of silk fibroin and, more importantly, show that the N-terminal domain in itself has the capacity to form micelle-like structures in response to pH decrease. Structural and mutagenesis analyses further reveal the important role of conserved acidic residues clustered in FibNT, such as Glu56 and Asp100, in preventing premature β-sheet formation at neutral pH. Collectively, we suggest that FibNT functions as a pH-responsive self-assembly module that could prevent premature β-sheet formation at neutral pH yet could initiate fibroin assembly as pH decreases along the lumen of the posterior silk gland to the anterior silk gland.

Published On: Journal of Molecular Biology, 2012, 418(3-4), 197-207.

Hefei National Laboratory for Physical Sciences at the Microscale, University of Science and Technology of China, Hefei 230027, China.

[#]These authors contributed equally.

[*]Corresponding author E-mail: zcz@ustc.edu.cn.

家蚕丝素的N端结构域介导了蚕丝的组装以响应pH值的降低

何永兴[#] 张楠楠[#] 李卫芳 贾 宁 陈宝玉
周 康 张家海 陈宇星 周丛照[*]

摘要：丝素是丝纤维的主要构建成分。作为丝素的主要组成部分，丝素重链是一个相当大的蛋白质，由N端和C端的亲水性结构域和12个富含Gly-Ala的疏水性区域所组成。我们解析了在pH4.7时丝素N端结构域（FibNT）的晶体结构，揭示了一个显著的双层反向平行的β片层，每一层由两个缠绕在一起的FibNT分子组成。研究还表明，FibNT在pH为6.0左右发生了构象转变，从随机卷曲到β-折叠。动态光散射显示pH下降至6.0时，FibNT倾向于寡聚化，同时我们用电镜亦观察到了FibNT的胶束状寡聚化。我们的研究结果与丝素的胶束组装模型相一致，更重要的是发现了N端结构域本身有能力形成胶束状结构以应对pH的降低。结构和突变分析进一步揭示了聚集在FibNT（如Glu56和Asp100）中的保守酸性残基在防止中性pH值下过早形成β-片层的作用。综上所述，我们得出FibNT作为一个pH响应的自组装元件，可以防止在中性pH条件下过早形成β-片层，当pH沿着后部丝腺到前部丝腺发生下降时，可以启动蚕丝蛋白的自组装。

合肥微尺度物质科学国家研究中心，中国科学技术大学，合肥

Structural insights into the unique inhibitory mechanism of the silkworm protease inhibitor serpin18

Guo PC[#] Dong ZM[#] Zhao P Zhang Y
He HW Tan X Zhang WW Xia QY[*]

Abstract: Serpins generally serve as inhibitors that utilize a mobile reactive center loop (RCL) as bait to trap protease targets. Here, we present the crystal structure of serpin18 from *Bombyxmori* at 1.65 Å resolution, which has a very short and stable RCL. Activity analysis showed that the inhibitory targetof serpin18 is a cysteine protease rather than a serine protease. Notably, this inhibitiory reaction results from the formation of an intermediate complex, which then follows for the digestion of protease and inhibitor into small fragments. This activity differs from previously reported modes of inhibition for serpins. Our findings have thus provided novel structural insights into the unique inhibitory mechanism of serpin18. Furthermore, one physiological target of serpin18, fibroinase, was identified, which enables us to better define the potential role for serpin18 in regulating fibroinase activity during *B. mori* development.

Published On: Scientific Reports, 2015, 5.

State Key Laboratory of Silkworm Genome Biology, Southwest University, Chongqing 400716, China.
[#]These authors contributed equally.
[*]Corresponding author E-mail: xiaqy@swu.edu.cn.

从结构上分析家蚕蛋白酶抑制剂 serpin18 的独特抑制机制

郭鹏超[#] 董照明[#] 赵 萍 张 艳
何华伟 谭 祥 张薇薇 夏庆友[*]

摘要:丝氨酸蛋白酶抑制剂(serpins)通常利用一个可移动的反应中心环作为诱饵诱捕蛋白酶靶点。在这篇文章中,本研究以1.65Å的分辨率展示了家蚕serpin18的晶体结构,发现其具有一个很短且稳定的反应中心环。活性检测表明serpin18的抑制靶标为半胱氨酸蛋白酶而不是丝氨酸蛋白酶。值得注意的是,抑制活性来源于中间反应复合物的形成,随后蛋白酶和抑制剂被消化成小片段。该活性与先前报道的serpin抑制模式不同。本研究提供了一种从结构上研究serpin18独特抑制机制的新型方法。此外,本研究鉴定了一种serpin18的生理学靶标——丝素酶,这有助于更好地理解serpin18调控丝蛋白酶活性在家蚕发育过程中的潜在功能。

家蚕基因组生物学国家重点实验室,西南大学,重庆

Crystal structure of *Bombyx mori* arylphorins reveals a 3:3 heterohexamer with multiple papain cleavage sites

Hou Y [1,2,3] Li JW [1] Li Y [1] Dong ZM [1] Xia QY[1,3] Y. Adam Yuan [2,3,4*]

Abstract: In holometabolous insects, the accumulation and utilization of storage proteins (SPs), including arylphorins and methionine-rich proteins, are critical for the insect metamorphosis. SPs function as amino acids reserves, which are synthesized in fat body, secreted into the larval hemolymph and taken up by fat body shortly before pupation. However, the detailed molecular mechanisms of digestion and utilization of SPs during development are largely unknown. Here, we report the crystal structure of *Bombyx mori* arylphorins at 2.8 Å, which displays a heterohexameric structural arrangement formed by trimerization of dimers comprising two structural similar arylphorins. Our limited proteolysis assay and microarray data strongly suggest that papain-like proteases are the major players for *B. mori* arylphorins digestion *in vitro* and *in vivo*. Consistent with the biochemical data, dozens of papain cleavage sites are mapped on the surface of the heterohexameric structure of *B. mori* arylphorins. Hence, our results provide the insightful information to understand the metamorphosis of holometabolous insects at molecular level.

Published On: Protein Science, 2014, 23(6), 735-746.

1 State Key Laboratory of Silkworm Genome Biology, College of Biotechnology, Southwest University, Chongqing 400716, China.

2 Department of Biological Sciences and Center for Bioimaging Sciences, National University of Singapore, Singapore 117543, Singapore.

3 SWU-NUS Joint Laboratory in Structural Genomics, Southwest University, Chongqing 400716, China.

4 National University of Singapore (Suzhou) Research Institute, Suzhou Jiangsu, 215123, China.

*Corresponding author E-mail: dbsyya@nus.edu.sg.

家蚕贮藏蛋白晶体结构显示其为含有多个木瓜蛋白酶酶切位点的3:3六聚体

侯　勇[1,2,3]　李建伟[1]　李　懿[1]　董照明[1]　夏庆友[1,3]　Y. Adam Yuan[2,3,4*]

摘要：在完全变态昆虫中，贮藏蛋白(SPs，包括芳香族贮藏蛋白和富含蛋氨酸贮藏蛋白)的积累和利用对昆虫的变态发育是十分重要的。贮藏蛋白作为氨基酸的供体，由脂肪体合成并分泌到血淋巴中，在化蛹前的短暂时间内又被脂肪体所吸收。然而，贮藏蛋白在昆虫发育中的消化和利用机制仍然未知。本文中，我们报道了家蚕贮藏蛋白的晶体结构，其分辨率达到了2.8 Å。该蛋白是由两个结构相似的芳香族贮藏蛋白形成二聚物，再次3倍化后形成的异源六聚体。有限的蛋白水解实验和芯片数据强烈表明，类木瓜蛋白酶是芳香族贮藏蛋白在体内外消化的主要参与者。与生物化学数据一致的是，大量的木瓜蛋白酶水解位点出现在家蚕芳香族贮藏蛋白的异六聚体结构表面。因此，我们的结果为从分子水平上理解完全变态昆虫的变态发育过程提供了有价值的信息。

1　家蚕基因组生物学国家重点实验室，生物技术学院，西南大学，重庆

2　生物科学与生物成像科学中心，新加坡国立大学，新加坡

3　西南大学-新加坡国立大学结构基因组学联合实验室，西南大学，重庆

4　新加坡国立大学(苏州)研究所，苏州

Crystal structure of juvenile hormone epoxide hydrolase from the silkworm *Bombyx mori*

Zhou K[1] Jia N[1] Hu C[1] Jiang YL[1] Yang JP[1]
Chen YX[1] Li S[2*] Li WF[1*] Zhou CZ[1*]

Abstract: The juvenile hormone (JH) is a kind of epoxide containing sesquiterpene ester secreted by a pair of corpora allatum behind the brain of insects. It controls the metamorphosis development of insects together with the ecdysone. Thus the synthesis and degradation of JH are tightly regulated in different development stages. The degradation of JH is catalyzed by two hydrolases, juvenile hormone epoxide hydrolase (JHEH) and juvenile hormone esterase(JHE). JHEH is responsible for opening the epoxide ring of JH to produce JH diol, whereas JHE catalyzes the removal of the methyl ester moiety of JH to form JH acid. JHEH belongs to the microsomal epoxide hydrolase (mEH) (EC 3.3.2.9) family, which is one of the most widely distributed families of epoxide hydrolases (EHs). EHs can transform epoxides to compounds with decreased chemical reactivity, increased water solubility, and altered biological activity. In addition to participating in the catabolism of JH in insects, mEHs also play important roles in cytoprotection, steroid metabolism, bile acid transport, and xenobiotic metabolism. To date, the only structure of the mEH from the fungus *Aspergillus niger* (termed AnEH, PDB 1QO7) revealed a typical α/β-hydrolase core composed of a twisted eightstranded β-sheet packing on both sides with several α-helices. Structural analyses suggested a bimolecular nucleophilic substitution (S_N2) reaction mechanism involving a standard nucleophile-histidine-acid catalytic triad of Asp-His-Glu/Asp. However, the mechanism of substrate recognition and catalysis of mEHs remains unclear. Here we report the crystal structure of *Bombyx mori* JHEH (*Bm*JHEH) at 2.30 Å resolution. Structural analyses together with molecular simulation reveal insights into the specific binding of JH in the active-site pocket. These findings increase our understanding of the substrate recognition and catalysis of mEHs and might help the design of JH-derived pesticides.

Published On: Proteins-Structure Function and Bioinformatics, 2014, 82(11), 3224-3229.

1 Hefei National Laboratory for Physical Sciences at the Microscale, School of Life Sciences, University of Science and Technology of China, Hefei 230027, China.

2 Key Laboratory of Developmental and Evolutionary Biology, Shanghai Institute of Plant Physiology and Ecology, Shanghai Institutes for Biological Sciences, Chinese Academy of Sciences, Shanghai 200032, China.

*Corresponding author E-mail: shengli@sippe.ac.cn; liwf@ustc.edu; zcz@ustc.edu.cn.

家蚕保幼激素环氧水解酶的晶体结构

周　康[1]　贾　宁[1]　胡　晨[1]　江永亮[1]　杨捷频[1]
陈宇星[1]　李　胜[2*]　李卫芳[1*]　周丛照[1*]

摘要：保幼激素(JH)是由昆虫脑后的一对咽侧体分泌的一种倍半萜烯类化合物。它与蜕皮激素共同调控昆虫的变态发育。因此，JH的合成和降解在不同的发育阶段受到严格的调控。JH的降解是由两种水解酶[保幼激素环氧水解酶(JHEH)和保幼激素酯酶(JHE)]催化。JHEH负责打开JH的环氧环生成保幼激素二醇，而JHE催化去除保幼激素的甲酯部分形成保幼激素酸。JHEH属于微粒体环氧化物水解酶(mEH)(EC 3.3.2.9)家族，mEH是环氧化物水解酶(EHs)中分布最广泛的家族之一。EHs可以将环氧化合物转化为化学活性降低、水溶性增加和生物活性转变的化合物。mEHs除了参与昆虫JH的分解代谢外，还在细胞保护、类固醇代谢、胆酸转运、外源物质代谢等方面发挥重要作用。迄今为止，唯一的mEH的结构报道来自黑曲霉(被称为AnEH，PDB 1QO7)，结构显示一个经典的α/β-水解酶核心由八链扭曲成的β-折叠组成，该β-折叠的两边包装了几个α-螺旋。结构分析表明，双分子亲核取代(S_N2)反应机制涉及标准的亲核物-组氨酸-核酸(Asp-His-Glu/Asp)催化三元体系。然而，mEHs的底物识别和催化机制尚不清楚。在这里，我们报道了分辨率2.30 Å的家蚕JHEH(*Bm*JHEH)的晶体结构。结构分析和分子模拟揭示了JH在活性位点口袋中的具体结合方式。这些发现增加了我们对mEHs底物识别和催化的理解，可能有助于JH衍生农药的开发。

1　合肥微尺度物质科学国家研究中心，生命科学学院，中国科学技术大学，合肥

2　昆虫发育与进化生物学重点实验室，上海植物生理生态研究所，上海生命科学研究院，中国科学院，上海

Crystal structure of the 30K protein from the silkworm *Bombyx mori* reveals a new member of the beta-trefoil superfamily

Yang JP Ma XX He YX Li WF
Kang Y Bao R Chen YX* Zhou CZ*

Abstract: The hemolymph of the fifth instar larvae of the silkworm *Bombyx mori* contains a group of homologous proteins with a molecular weight of approximately 30 kDa, termed *B. mori* low molecular weight lipoproteins (*Bm*lps), which account for about 5% of the total plasma proteins. These so-called "30K proteins" have been reported to be involved in the innate immune response and transportation of lipid and/or sugar. To elucidate their molecular functions, we determined the crystal structure of a 30K protein, *Bm*lp7, at 1.91 angstrom. It has two distinct domains: an all-α N-terminal domain (NTD) and an all-β C-terminal domain (CTD) of the β-trefoil fold. Comparative structural analysis indicates that *Bm*lp7 represents a new family, adding to the 14 families currently identified, of the β-trefoil superfamily. Structural comparison and simulation suggest that the NTD has a putative lipid-binding cavity, whereas the CTD has a potential sugar-binding site. However, we were unable to detect the binding of either lipid or sugar. Therefore, further investigations are needed to characterize the molecular function of this protein.

Published On: Journal of Structural Biology, 2011, 175(1), 97-103.

Hefei National Laboratory for Physical Sciences at the Microscale, University of Science and Technology of China, Hefei, 230027, China.
*Corresponding author E-mail: cyxing@ustc.edu.cn; zcz@ustc.edu.cn.

家蚕30K蛋白是β三叶草家族中的一个新成员

杨捷频　马萧萧　何永兴　李卫芳
康　炎　包　锐　陈宇星*　周丛照*

摘要:家蚕五龄幼虫血淋巴中存在着一组分子质量大约30 kDa的同源蛋白质,称为家蚕低分子质量脂蛋白(*Bm*lps),大约占总血淋巴蛋白的5%。这些"30K蛋白"已被报道参与先天免疫反应和脂质/糖运输。为了阐明其分子功能,我们解析了30K蛋白*Bm*lp7的晶体结构(1.91 Å)。*Bm*lp7有两个不同的结构域: all-α的N端结构域(NTD)和all-β的C端结构域(CTD)。结构比较分析表明,*Bm*lp7的CTD为β-三叶草超家族结构域,与已知结构的14个β-三叶草家族成员相比较,30K蛋白家族为一个新的三叶草家族。根据结构比较推测*Bm*lp7的NTD可能具有结合脂的能力,而CTD存在结合糖的潜在位点,因而可能间接地参与了昆虫天然免疫反应。然而,我们的实验未能检测到它与糖或者脂分子结合,可能该蛋白的天然底物不是我们实验中用的糖或者脂分子,因此对这一蛋白质的功能还需要进一步研究。

合肥微尺度物质科学国家实验室,中国科学技术大学,合肥

Crystal structures of holo and Cu-deficient Cu/Zn-SOD from the silkworm *Bombyx mori* and the implications in amyotrophic lateral sclerosis

Zhang NN He YX Li WF Teng YB
Yu J Chen YX Zhou CZ*

Abstract: Cu / Zn superoxide dismutases (Sod1) are a large family of cytosolic antioxidant proteins involved in responses to oxidative stress. They catalyze the disproportionation of superoxide radicals into less toxic hydrogen peroxide and dioxygen. Mutations in human *Cu/Zn-SOD* (*hSod1*) are associated with about 20% of the familial cases of amyotrophic lateral sclerosis (fALS), a fatal neurodegenerative disorder in heterozygotes. Although the mechanism of this toxicity is still unknown, the aberrant aggregation of hSod1 mutants is strongly suggested to play an important role in the etiology of fALS. The Cu/Zn-SOD from the silkworm *Bombyx mori* (*Bm*Sod1) shares a sequence identity of 63% with hSod1, with 56 different residues out of the total 153 residues and residue T76 which could be mapped to the corresponding ALS-linked hSOD1 mutation sites. Here, we determined the crystal structure of *Bm*Sod1 at high resolution. The overall structures of the holo as well as the Zn-*Bm*Sod1 preserve the typical Sod1 tertiary structure. Crystal packing of both Cu/Zn-*Bm*Sod1 and Zn-*Bm*Sod1 showed helical fibril arrangement in the crystalline environment via newly gained nonnative interfaces. The nonnative interface I is also formed by the zinc loop (involving residues P73, S74, S75, and A76) from one dimer packing onto the exposed edge of β6 (involving S98, I99, Q100, and S102) from the neighboring subunit, leading to helical fibrils. The buried interface is up to 650Å^2. The helical fibrils further stick together to form a water-filled nanotube with an outer diameter of 120 Å and a waterfilled inner cavity of 40 Å in diameter. The nonnative interface II between helical fibrils buried up to 450 Å^2 on average, involving interactions between the electrostatic loop residues E132-L133 and zinc loop residues E66-K67. The fibril arrangement of Cu/Zn-*Bm*Sod1 and Zn-*Bm*Sod1 in the crystalline environment suggests an ALS-linked mutant-like nature of the wild-type *Bm*Sod1. The nonnative interface involved in fibril assembly is located in the electrostatic loop, zinc loop, and strand β6, all of which are suggested as major candidates for forming aggregation-prone interfaces. We speculated that Sod1 might be an intrinsically aggregation-prone protein during earlier stages of phylogenesis, but underwent substantial site mutations opposing self-aggregation during evolution because endogenous protein aggregates are highly toxic and even lethal to mammalian/human neurons.

Published On: Proteins, 2010, 78(8), 1999-2004.

Hefei National Laboratory for Physical Sciences at the Microscale, University of Science and Technology of China, Hefei 230026, China.
*Corresponding author E-mail: zcz@ustc.edu.cn.

家蚕Cu/Zn-SOD全酶及其无铜晶体结构及其对肌萎性脊髓侧索硬化症的提示

张楠楠　何永兴　李卫芳　滕衍斌
俞　江　陈宇星　周丛照*

摘要：铜锌超氧化物歧化酶（Sod1）是参与抗氧化应激反应的一大类抗氧化蛋白。Sod是一种含有金属元素的活性蛋白酶，是一种能够催化超氧化物通过歧化反应转化为氧气和过氧化氢的酶。人类铜锌超氧化物歧化酶（hSod1）突变体与20%家族肌萎性脊髓侧索硬化症（fALS）这一致命的神经系统退行性疾病有密切的关系。虽然对毒性机理尚不清楚，Sod1的异常聚集被认为是致病原因之一。家蚕中铜锌超氧化物歧化酶（*Bm*Sod1）与hSod1有63%的同源性，全序列的153个氨基酸中的56个氨基酸残基和76位Thr可以与ALS-linked hSOD1的突变位点相对应。我们得到了Cu/Zn-*Bm*Sod1和Zn-*Bm*Sod1的晶体结构，两者的整体结构呈现出典型的Sod1四级结构。在晶体堆积中，Cu/Zn-*Bm*Sod1和Zn-*Bm*Sod1通过一个新的相互作用面形成纤维结构。两个晶体结构中都存在两种新的非天然相互作用面并可以形成直径约为120 Å和内径约为40 Å的中空纳米管状结构。非天然相互作用界面I的面积约为650 $Å^2$，主要由二体锌环上的Pro73、Ser74、Ser75和Ala76与相邻二体β6折叠片上Ser98、Ile99、Gln100和Ser102。非天然相互作用界面II是螺旋纤维之间的相互作用面，面积约为450 $Å^2$，主要是静电环上的Gln132和Leu133与锌环上的Gln66与Lys67之间的氢键作用。在结晶环境下，Cu/Zn-*Bm*Sod1和Zn-*Bm*Sod1形成纤维结构说明野生型*Bm*Sod1具有类似ALS相关突变体的性质。非天然作用界面主要存在于静电环、锌环和β6，这些位置常常被认为是ALS相关hSod1突变体聚集倾向的主要潜在位点。我们推测，Sod1可能在早期的进化中是一种具有聚集倾向的蛋白质，但在进化过程中氨基酸的替换突变阻止了蛋白的聚集，而这种聚集对于哺乳动物或人类的神经元来说是有毒性的，甚至是致命的。

合肥微尺度物质科学国家实验室，中国科学技术大学，合肥

Crystal structure of *Bombyx mori* nucleopolyhedrovirus ORF75 reveals a pseudo-dimer of thiol oxidase domains with a putative substrate-binding pocket

Hou Y[1,2,3] Xia QY[2,3] Adam Yuan [1,3,4*]

Abstract: *Bombyx mori* nucleopolyhedrovirus (*Bm*NPV) triggers the global shutdown of host silkworm gene expression and protein synthesis approximately 12-18 h post-infection. Genome sequence analysis suggests that *Bm*NPV ORF75 could be a flavin adenine dinucleotide (FAD)-linked thiol oxidase essential for virion assembly and virus propagation. Here, we report the crystal structure of *Bm*NPV ORF75 at 2.1 Å (0.21 nm). The structure of *Bm*NPV ORF75 resembles that of the thiol oxidase domain of human quiescin thiol oxidase (QSOX), displaying a pseudo-dimer of canonical and non-canonical thiol oxidase domains. However, *Bm*NPV ORF75 is further dimerized by its Cterminal canonical thiol oxidase domain. Within the unique quaternary structural arrangement, the FAD-binding pocket and the characteristic CXXC motif from each monomer is 35 Å (3.5 nm) away from that of its corresponding molecule, which suggests that *Bm*NPV ORF75 might adopt a deviant mechanism from that of QSOX to catalyse disulfide bond formation. Our thiol oxidase activity assay on the point mutations of the conserved residues participating in FAD recognition reveals an aromatic cage next to the FAD isoalloxazine moiety for substrate binding. These data suggest that the thiol oxidase activity of *Bm*NPV ORF75 could be critical to catalyse the formation of the disulfide bonds of certain *Bm*NPV proteins essential for *Bm*NPV virion assembly.

Published On: Journal of General Virology, 2012, 93, 2142-2151.

1 Department of Biological Sciences and Center for Bioimaging Sciences, National University of Singapore, Singapore 117543, Singapore.

2 State Key Laboratory of Silkworm Genome Biology, College of Biotechnology, Southwest University, Chongqing 400716, China.

3 SWU-NUS Joint Laboratory in Structural Genomics, Southwest University, Chongqing 400716, China.

4 National University of Singapore (Suzhou) Research Institute, Suzhou Jiangsu, 215123, China.

*Corresponding author E-mail: dbsyya@nus.edu.sg.

家蚕核型多角体病毒ORF75蛋白的晶体结构揭示该蛋白是一个拥有硫基氧化酶底物结合口袋结构域的假二聚体

侯　勇[1,2,3]　夏庆友[2,3]　Adam Yuan[1,3,4*]

摘要：家蚕核型多角体病毒(*Bm*NPV)在感染家蚕12—18 h后，会关闭大量宿主基因的表达与蛋白质合成。基因组序列分析显示，*Bm*NPV ORF75可能是一个黄素腺嘌呤二核苷酸(FAD)结合硫氧化酶，其对于病毒粒子的组装和病毒的增殖起着必需作用。本文对*Bm*NPV ORF75蛋白的晶体结构进行了报道，分辨率达到2.1 Å。*Bm*NPV ORF75的结构包含一个人类静息素硫基氧化酶(QSOX)结构域，并且是一个包括典型和非典型硫基氧化酶结构域的假二聚体。不同的是，*Bm*NPV ORF75是通过其C端典型硫基氧化酶结构域进行进一步的二聚化的。在独特的四级结构组装中，发现FAD结合口袋和来自各单体的特征CXXC基序与其相应的分子距离35 Å，这表明*Bm*NPV ORF75可能采取偏差机理催化QSOX中二硫键的形成。硫基氧化酶活性测定实验表明，将FAD识别的保守残基进行突变，显示ORF75是通过FAD异环嗪部分的芳香族口袋与底物进行结合。这些数据表明*Bm*NPV ORF75的硫基氧化酶活性对催化*Bm*NPV某些关键蛋白质的二硫键形成至关重要，对*Bm*NPV病毒粒子的组装是必需的。

1　生物科学与生物成像科学中心，新加坡国立大学，新加坡
2　家蚕基因组生物学国家重点实验室，生物技术学院，西南大学，重庆
3　西南大学-新加坡国立大学结构基因组学联合实验室，西南大学，重庆
4　新加坡国立大学(苏州)研究所，苏州

Cloning and homologic analysis of *Tpn I* gene in silkworm *Bombyx mori*

Zhao Y[1,2] Yao Q[1] Tang XD[1] Wang QH[1]
Yin HJ[1] Hu ZG[1] Lu J[1] Chen KP[1*]

Abstract: The troponin complex is composed of three subunits, Troponin C (the calcium sensor component) and Troponin T and I (structural proteins). Tpn C is encoded by multiple genes in insects, while the Tpn T and Tpn I proteins are encoded by single genes. Tpn I binds to actin and Tpn T binds to tropomyosin. We cloned and sequenced the *Tpn I* (AY873787) gene from *Bombyx mori* that encodes 225 amino acids and contains a conserved motif seen in *Drosophila virilis* and *Anopheles gambiae*. Bioinformatic analysis suggests that its deduced amino sequence shares 81.3% and 78.7% homology with the *Tpn I* genes of *A.gambiae* and *D.virili*, respectively.

Published On: African Journal of Biotechnology, 2007, 6(6), 672-676.

1 Institute of Life Sciences, Jiangsu University, Zhenjiang 212013,China.

2 Sericultural Research Institute, Chinese Academy of Agricultural Sciences, Zhenjiang 212018, China.

*Corresponding author E-mail: kpchen@ujs.edu.cn.

家蚕 *Tpn I* 基因的克隆和同源性分析

周　阳[1,2]　姚　勤[1]　唐旭东[1]　王庆华[1]
尹慧娟[1]　胡志刚[1]　陆　健[1]　陈克平[1*]

摘要：肌钙蛋白复合物由肌钙蛋白C(钙传感器组分)、肌钙蛋白T和I(结构蛋白)3个亚基组成。Tpn C由昆虫中的多个基因编码，而Tpn T和Tpn I蛋白则由单个基因编码。Tpn I结合肌动蛋白，Tpn T则与原肌球蛋白结合。我们克隆并测序了家蚕 *Tpn I* 基因（AY873787)，该基因编码225个氨基酸并含有在果蝇和冈比亚按蚊中发现的保守基序。生物信息学分析表明，其编码的氨基酸序列分别与冈比亚按蚊和果蝇的 *Tpn I* 具有81.3%和78.7%的同源性。

1　生命科学研究院，江苏大学，镇江
2　蚕业科学研究所，中国农业科学院，镇江

Cloning and characterization of hydroxypyruvate isomerase (EC 5.3.1.22) gene in silkworm *Bombyx mori*

Lü HG[1] Chen KP[1*] Yao Q[1] Wang L[2]

Abstract: The sequence of hydroxypyruvate isomerase gene was obtained in NCBI. In this study, the hydroxypyruvate isomerase gene of *Bombyx mori* was identified and annotated with bioinformatics tools. The result was confirmed by RT-PCR, prokaryotic expression, mass spectrographic analysis and sub-cellular localization. The hydroxypyruvate isomerase cDNA comtains a 783 bp ORF, and has 4 exons. The deduced protein has 260 amino acid residues with the predicted molecular weight of 29 169.30 Da, isoelectric point of 6.10, and contains conserved PRK09997 and Hfi domains. The hydroxypyruvate isomerases of *Nasonia vitripennis* and *Bombyx mori* have a high homology. Through RT-PCR analysis, we found that this transcript was present in testis, ovary, blood-lymph, fat body, midgut, silk gland and tuba Malpighii. This protein was located in cytoplasm through immunohistochemistry. We submitted the cloned gene under the accession number EU344910. The enzyme has been classified under accession number EC 5.3.1.22.

Published On: International Journal of Industrial Entomology, 2008, 17(2), 189-195.

1 Institute of Life Sciences, Jiangsu University, Zhenjiang 212013, China.

2 Beijing Entry-Exit Inspection and Quarantine Bureau, Beijing 100026, China.

*Corresponding author E-mail: kpchen@ujs.edu.cn.

家蚕中羟基丙酮酸异构酶(EC 5.3.1.22)基因的克隆和表征

吕洪刚[1] 陈克平[1*] 姚 勤[1] 汪 琳[2]

摘要：从NCBI中获得了羟基丙酮酸异构酶的基因序列。在本研究中，我们用生物信息学方法鉴定并注释了家蚕的羟基丙酮酸异构酶基因。进一步通过RT-PCR、原核表达、质谱分析和亚细胞定位证实了这一结果。羟基丙酮酸异构酶cDNA含有一个783 bp的ORF，具有4个外显子。推测其蛋白质由260个氨基酸残基组成，理论分子质量为29 169.30 Da，等电点为6.10，并含有保守的PRK09997和Hfi结构域。丽蝇蛹集金小蜂和家蚕的羟基丙酮酸异构酶具有高的同源性。通过RT-PCR分析，我们发现该转录本存在于精巢、卵巢、血淋巴、脂肪体、中肠、丝腺和马氏管中。该蛋白质通过免疫组织化学定位于细胞质中。我们以保藏号EU344910提交了该克隆基因。该酶已被分类，登录号为EC 5.3.1.22。

1 生命科学研究院，江苏大学，镇江

2 北京出入境检验检疫局，北京

Molecular characters and expression analysis of the gene encoding eclosion hormone from the Asian corn borer, *Ostrinia furnacalis*.

Wei ZJ[1,2*] Hong GY[1] Wei HY[3] Jiang ST[1] Lu C[2]

Abstract: Using rapid amplification of cDNA ends (RACE), the cDNA encoding eclosion hormone (EH) was cloned from the brain of *Ostrinia furnacalis*. The full *Osf-EH* cDNA is 986 bp and contains a 267 bp open reading frame encoding an 88 amino acid preprohormone, which including a hydrophobic 26 amino acid signal peptide and a 62 amino acid mature peptide. The mature Osf-EH shows high identity with *Manduca sexta* (95.2%), *Helicoverpa armigera* (91.9%) and *Bombyx mori* (85.5%), but low identify with *Tribolium castaneum* (63.6%), *Drosophila melanogaster* (56.5%) and *Apis mellifera* (54.8%). Using the HMMSTR Prediction Server, the 3D structure of Osf-EH was modeled. There are four beta-turns and three alpha-helixes predicted in Osf-EH, with the pattern of beta-beta-alpha-alpha-beta-beta-alpha. Northern blotting analysis indicated a 1.0 kb transcript present only in the brain. The *Osf-EH* mRNA can not be detected in other neural tissues, such as the suboesophageal ganglion, thoracic ganglion, abdominal ganglion and other non-neural tissues, such as the midgut, fat body and epidermis. The *Osf-EH* mRNA content in the brain was measured using the combined method of quantitative RT-PCR and Southern blotting, which reached its highest level the day before the molt.

Published On: DNA Sequence, 2008, 19(3), 301-307.

1 Department of Biotechnology, Hefei University of Technology, Hefei 230009, China.

2 Sericultural Laboratory of Agriculture Ministry, Southwest University, Chongqing 400716, China.

3 College of Agronomy, Jiangxi Agricultural University, Nanchang 330045, China.

*Corresponding author E-mail: zjwei@hfut.edu.cn.

亚洲玉米螟编码羽化激素基因的分子特征和表达分析

魏兆军[1,2*] 洪桂云[1] 魏洪义[3] 姜绍通[1] 鲁 成[2]

摘要：采用快速扩增cDNA末端(RACE),从亚洲玉米螟脑中克隆编码羽化激素(EH)的cDNA。完整的*Osf-EH* cDNA长度为986 bp,含一个长267 bp的开放阅读框,并编码包含88个氨基酸的前基素原。其中包含一个由26个疏水性氨基酸组成的信号肽和一个含62个氨基酸的成熟肽。成熟的Osf-EH与烟草天蛾(95.2%)、棉铃虫(91.9%)和家蚕(85.5%)具有很高的相似性,但与赤拟谷盗(63.6%)、果蝇(56.5%)和欧洲蜜蜂(54.8%)的相似性较低。我们使用HMMSTR预测服务器,对Osf-EH的3D结构进行了建模。在Osf-EH中预测到4个β-转角和3个α-螺旋,其模式为β-β-α-α-β-β-α。Northern blot分析显示其仅存在于脑中的1.0 kb转录本中。*Osf-EH* mRNA在其他神经组织(如食管神经节、胸神经节、腹部神经节)和其他非神经组织(如中肠、脂肪体和表皮)中没有检测到。使用定量RT-PCR和Southern blot相结合的方法测量脑中的*Osf-EH* mRNA含量,其在蜕皮前一天达到最高水平。

1 生物技术系,合肥工业大学,合肥

2 农业部蚕桑学重点实验室,西南大学,重庆

3 农学院,江西农业大学,南昌

Transcription level of messenger RNA per gene copy determined with dual-spike-in strategy

Zhang Y[1#] Wei ZG[2#] Li YY[3,4#]
Chen YH[2] Shen WD[2] Lu CD[1*]

Abstract: To quantify the transcription level of a gene, we have conceived a novel concept, transcription level of messenger RNA (mRNA) per gene copy, which was determined with a dual-spike-in strategy. In this strategy, an exogenous DNA was added as the spike reference for target DNA in addition to the exogenous RNA as the reference for target RNA. After the mRNA-to-DNA ratio of a target gene was estimated by real-time polymerase chain reaction (PCR), it was first normalized with the mRNA-to-DNA ratio of the exogenous reference. The normalized ratio was multiplied by the ratio of exogenous RNA to exogenous DNA to obtain the transcription level of mRNA per gene copy. This quantified transcription value allows one to compare the expression of a target gene in different tissues or the expression in a specified tissue under different conditions.

Published On: Analytical Biochemistry, 2009, 394(2), 202-208.

1 State Key Laboratory of Molecular Biology, Institute of Biochemistry and Cell Biology, Shanghai Institutes for Biological Sciences, Chinese Academy of Sciences, Shanghai 200031, China.
2 School of Basic Medicine and Life Sciences, Soochow University, Suzhou 215123, China.
3 Shanghai Center for Bioinformation Technology, Shanghai 200235, China.
4 Bioinformatics Center, Key Laboratory of Systems Biology, Shanghai Institutes for Biological Sciences, Chinese Academy of Sciences, Shanghai 200031, China.
#These authors contributed equally.
*Corresponding author E-mail: cdlu@sibs.ac.cn.

双跟踪策略确定每个基因拷贝的信使RNA转录水平

张　翌[1#]　卫正国[2#]　李媛媛[3,4#]
陈玉华[2]　沈卫德[2]　陆长德[1*]

摘要：为了测定基因的转录水平，我们提出了一个新的概念，即利用双跟踪（dual-spike-in）策略来测定每个基因拷贝的mRNA转录水平。在这个策略中，我们除了将外源RNA作为目标RNA的参考外，还加入外源DNA来跟踪目标DNA。通过实时定量PCR计算目标基因的mRNA与DNA的比率，首次实现了利用外源参照基因的mRNA与DNA的比率进行归一化处理。通过归一化的比率乘以外源RNA和DNA的比率，可以获得每个基因拷贝的mRNA的转录水平。这一数值可以比较目的基因在不同的组织或在不同条件下特定组织的表达水平。

1　分子生物学重点实验室，生物化学和细胞生物学研究所，上海生命科学研究院，中国科学院，上海
2　基础医学与生物科学学院，苏州大学，苏州
3　上海生物信息技术中心，上海
4　生物信息中心，系统生物学重点实验室，上海生命科学研究院，中国科学院，上海

Identification and characterization of an arginine kinase as a major allergen from silkworm (*Bombyx mori*) larvae

Liu ZG[1*] Xia LX[1] Wu YL[1]
Xia QY[2] Chen JJ[1] Kenneth H. Roux[3]

Abstract: The silkworm, *Bombyx mori*, is an important insect in the textile industry and its pupa are used in Chinese cuisine and traditional Chinese medicine. The silk, urine and dander of silkworms is often the cause of allergies in sericulture workers and the pupa has been found to be a food allergen in China. Recent studies have focused on reporting cases of silkworm allergies, but only a few studies have addressed the specific allergens present in the *B. mori* silkworm. We collected sera from 10 patients with a positive skin prick test to silkworm crude extract (SCE) and analyzed these samples by Western blot and ELISA. The cDNA of arginine kinase from the *B. mori* silkworm was also cloned and expressed in high yield in *Escherichia coli*. Allergenicity and cross-allergenicity of the recombinant *B. mori* arginine kinase (r*Bm*AK) were investigated by ELISA inhibition assay. Collected sera all reacted to a 42 kDa protein in a Western blot with SCE as the antigen. Preincubation of sera with r*Bm*AK eliminated the reactivity of the patients' sera to this 42 kDa band. All patient sera also exhibited positive reactivity to SCE in an ELISA assay. *Bm*AK also demonstrated cross-reactivity with a recombinant AK from cockroach. Arginine kinase from the *B. mori* silkworm is a major allergen and crossreacts with cockroach AK.

Published On: International Archives of Allergy and Immunology, 2009, 150(1), 8-14.

1 Institute of Allergy and Immunology, School of Medicine, Shenzhen University, Shenzhen 518060, China.
2 Institute of Sericulture and Systems Biology, Southwest University, Chongqing 400716, China.
3 Department of Biological Science, Florida State University, Tallahassee 32306, USA.
*Corresponding author E-mail: lzg195910@126.com.

家蚕主要过敏原精氨酸激酶的鉴定和表征

刘志刚[1*] 夏立新[1] 邬玉兰[1]
夏庆友[2] 陈佳杰[1] Kenneth H. Roux[3]

摘要：背景——家蚕是纺织工业中的重要经济昆虫，其蛹被用于中国菜和中药。家蚕丝、尿液和鳞毛往往是养蚕工人过敏的原因，而蚕蛹在中国被证明是食物过敏原。最近的研究已集中报道了家蚕过敏病例，但只有少数研究是针对目前家蚕中存在的特定过敏原。方法——利用皮肤点刺实验，我们收集了10名对家蚕粗提物(SCE)呈阳性的患者血清，通过Western blot和ELISA对该血清进行分析。我们还克隆了家蚕精氨酸激酶的cDNA并在大肠杆菌中高量表达，再通过ELISA抑制实验对重组家蚕精氨酸激酶(r*Bm*AK)的过敏性和交叉过敏性进行研究。结果——以SCE作为抗原进行Western blot，结果发现，所收集血清都与其中的42 kDa蛋白质发生了反应。而提前与r*Bm*AK孵育的病人血清不能与这个42 kDa蛋白条带发生反应。ELISA实验中，所有病人的血清都能与SCE发生反应。家蚕AK和重组蟑螂AK具有交叉活性。结论——总而言之，家蚕精氨酸激酶是主要过敏原，且和蟑螂AK发生交叉反应。

1 过敏反应和免疫学研究所，医学院，深圳大学，深圳
2 蚕学与系统生物学研究所，西南大学，重庆
3 生命科学学院，佛罗里达州立大学，塔拉哈西，美国

Genotyping of acetylcholinesterase in insects

Li B[1,2] Wang YH[2] Liu HT[2] Xu YX[1,2]

Wei ZG[1,2] Chen YH[1,2] Shen WD[1,2*]

Abstract: To investigate the genotyping criteria for the insect acetylcholinesterase gene (*ace*), we cloned two types of *ace* genes in domestic (*Bombyx mori*) and wild silkworm (*Bombyx mandarina*) through RT-PCR. The cloned genes were named *Bm-ace 1, Bm-ace 2, Bmm-ace 1* and *Bmm-ace 2*, respectively. The ORFs of *Bm-ace 1* and *Bmm-ace 1* contained 2 025 base pairs, encoding 683 amino acid residues (AA' s). The predicted protein has a molecular weight (MW) of 76.955 kDa and an isoelectric point (pI) of 6.36. The *Bm-ace 2* and *Bm-ace 2* genes contained 1 917 bp nucleotides, encoding 638 AA' s. The predicted protein has a MW of 71.675 kDa and a pI of 5.49. Both *ace*1 and *ace*2 contain signature domains of acetylcholinesterases. Homology analysis of 18 NCBI downloaded insect AChEs peptide sequences and the 4 AChEs deducted in this study revealed that type 1 and type 2 insect AChEs had significant differences. Type 2 sequence is more conserved than type 1. Near the active centers of both types of AChEs, 48 strictly conserved AA' s (336-384) are present, and homology of these two peptide fragments was only 54.16%. Meanwhile, at AA positions 280-297, type 1 and type 2 AChEs both have conserved sequences with the similarity of the two being 52.94%. In type 2 AChEs, a uniquely conserved peptide sequence is found at positions 226-239 (QHLRVRHHQDKPL). We propose to use the above mentioned three conserved regions as criteria for insect acetylcholinesterases gene genotyping. This will benefit the genotyping of other acetylcholinesterase genes and the study of their functions.

Published On: Pesticide Biochemistry and Physiology, 2010, 98(1), 19-25.

1 National Engineering Laboratory for Modern Silk, Soochow University, Suzhou 215123, China.

2 School of Basic Medicine and Biological Sciences, Soochow University, Suzhou 215123, China.

*Corresponding author E-mail: sdlibing@hotmail.com.

昆虫乙酰胆碱酯酶的基因分型

李　兵[1,2]　王燕红[2]　刘海涛[2]　许雅香[1,2]
卫正国[1,2]　陈玉华[1,2]　沈卫德[1,2*]

摘要：为了研究昆虫乙酰胆碱酯酶(*ace*)的基因分型标准，我们采用RT-PCR方法分析克隆了两种类型的家蚕和野桑蚕的*ace*基因，分别命名为*Bm-ace1*，*Bm-ace2*，*Bmm-ace1*和*Bmm-ace2*。*Bm-ace1*和*Bmmace1*的ORF是2 025 bp，编码683个氨基酸，编码蛋白的分子质量是76.955 kDa，等电点为6.36；*Bm-ace2*和*Bmm-ace2*的ORF是1 917 bp，编码638个氨基酸，编码蛋白的分子质量为71.675 kDa，等电点pI是5.49。*ace1*和*ace2*都包含乙酰胆碱酯酶的特征域。对来源于NCBI上的18个AChEs和本研究克隆的4个AChEs的氨基酸序列进行同源进化分析，结果表明1型和2型AChEs之间存在显著的差异，2型较1型更加保守。在两种AChEs的活性中心附近区域，均包含48个非常保守的氨基酸序列(336—384)，两种AChEs片段的同源性只有54.16%。同时，在氨基酸的280—297位点，1型和2型AChEs均存在相对保守的序列，同源性只有52.94%。在2型AChEs中还存在其特有的保守序列226—239(QHLRVRHHQDKPL)。我们建议采用上述3个特征区域作为昆虫乙酰胆碱酯酶基因分型的依据。这将有利于对其他乙酰胆碱酯酶基因进行分型并研究其功能。

1　现代丝绸国家工程实验室，苏州大学，苏州
2　基础医学与生物科学学院，苏州大学，苏州

Molecular cloning, expression and characterization of *BmIDGF* gene from *Bombyx mori*

Pan Y[1] Chen KP[1*] Xia HC[1] Yao Q[1] Gao L[1,2]
Lü P[1] Huo J[1] He YQ[1] Wang L[3]

Abstract: Imaginal disc growth factors (IDGF) play a key role in insect development, but their mechanism remains unclear. In this study, we cloned a novel *IDGF* gene in *Bombyx mori* and designated it as *BmIDGF*. We found that the *BmIDGF* gene contains eight exons and seven introns, encoding a peptide of 434 amino-acid residues. The protein was predicted to contain one conserved motif of the glycosyl hydrolases family 18 and fall into group V chitinases. Sequence alignment showed that *BmIDGF* shares extensive homology with other invertebrate IDGF. RT-PCR analysis showed that *BmIDGF* is expressed in all developmental stages of silkworm larvae and various larvae tissues, which was further confi rmed by Western blot analysis. Subcellular localization analysis indicated that *Bm*IDGF is located in the extracellular space. We also successfully expressed it in *E. coli* and further characterized it by SDS-PAGE and mass spectrometry. Taken together, our data suggests that *Bm*IDGF is a chitinase-like extracellular protein, and provides an excellent platform for subsequent studies on its enzyme activity and role in *B. mori* development.

Published On: Zeitschrift Fur Naturforschung Section C—A Journal of Biosciences, 2010, 65(3-4), 277-283.

1 Institute of Life Sciences, Jiangsu University, Zhenjiang 212013, China.
2 School of Medical Science and Laboratory Medicine, Jiangsu University, Zhenjiang 212013, China.
3 Beijing Entry-Exit Inspection and Quarantine Bureau, Beijing 100026, China.
*Corresponding author E-mail: kpchen@ujs.edu.cn.

家蚕*BmIDGF*基因的分子克隆、表达和表征

潘　晔[1]　陈克平[1*]　夏恒传[1]　姚　勤[1]　高　路[1,2]
吕　鹏[1]　霍　娟[1]　何远清[1]　汪　琳[3]

摘要：成虫原基生长因子（IDGF）在昆虫发育中起关键作用，但其机制尚不清楚。在本研究中，我们在家蚕中克隆了一种新的*IDGF*基因，并将其命名为*BmIDGF*。我们发现*BmIDGF*基因含有8个外显子和7个内含子，编码含有434个氨基酸残基的肽链。蛋白质被预测含有糖基水解酶家族18的一个保守基序并归于几丁质酶V组。序列比对显示*BmIDGF*与其他无脊椎动物*IDGF*具有广泛的同源性。RT-PCR分析显示，*BmIDGF*在家蚕幼虫和各种幼虫组织的所有发育阶段都表达，并进一步通过蛋白质印迹分析进行了确定。亚细胞定位分析表明*Bm*IDGF位于细胞外隙。我们还在大肠杆菌中成功表达了*Bm*IDGF，并通过SDS-PAGE和质谱进行了进一步表征。总之，实验的数据表明，*Bm*IDGF是一种几丁质酶类型的胞外蛋白，本研究为后续研究其酶活性和在家蚕发育中的功能提供了一个极好的平台。

1　生命科学研究所，江苏大学，镇江
2　江苏大学，镇江
3　北京出入境检验检疫局，北京

Structure-guided activity restoration of the silkworm glutathione transferase Omega GSTO3-3

Chen BY[1#] Ma XX[1#] Guo PC[1] Tan X[2] Li WF[1] Yang JP[1]

Zhang NN[1] Chen YX[1] Xia QY[2] Zhou CZ[1*]

Abstract: Glutathione transferases (GSTs) are ubiquitous detoxification enzymes that conjugate hydrophobic xenobiotics with reduced glutathione. The silkworm *Bombyx mori* encodes four isoforms of GST Omega (GSTO), featured with a catalytic cysteine, except that *Bm*GSTO3-3 has an asparagine substitution of this catalytic residue. Here, we determined the 2.20-angstrom crystal structure of *Bm*GSTO3-3, which shares a typical GST overall structure. However, the extended C-terminal segment that exists in all the four *Bm*GSTOs occupies the G-site of *Bm*GSTO3-3 and makes it unworkable, as shown by the activity assays. Upon mutation of Asn29 to Cys and truncation of the C-terminal segment, the *in vitro* GST activity of *Bm*GSTO3-3 could be restored. These findings provided structural insights into the activity regulation of GSTOs.

Published On: Journal of Molecular Biology, 2011, 412(2), 204-211.

1 Hefei National Laboratory for Physical Sciences at the Microscale,School of Life Sciences, University of Science and Technology of China, Hefei 230027, China.

2 Key Sericultural Laboratory of Agricultural Ministry, Institute of Sericulture and Systems Biology,Southwest University, Chongqing 400716, China.

[#]These authors contributed equally.

[*]Corresponding author E-mail: zcz@ustc.edu.cn.

家蚕谷胱甘肽转移酶Omega GSTO3-3的结构诱导的活性恢复

陈宝玉[1#] 马萧萧[1#] 郭鹏超[1] 谭 祥[2] 李卫芳[1]
杨捷频[1] 张楠楠[1] 陈宇星[1] 夏庆友[2] 周丛照[1*]

摘要：谷胱甘肽S转移酶(GSTs)是一类普遍存在于生物体内的解毒酶。它可以识别GSH从而结合有毒的底物。家蚕中存在4个Omega家族的谷胱甘肽S转移酶,能催化半胱氨酸,但*Bm*GSTO3-3的该催化残基被天冬酰胺残基取代。本研究,我们解析了家蚕中Omega家族的其中一个蛋白（Omega3）的晶体结构(2.20 Å),发现*Bm*GSTO3-3具有典型的GST折叠方式。对*Bm*GSTO3-3酶活的测定实验发现由于存在于所有4个BmGSTOs中的扩展C端占用了*Bm*GSTO3-3的G位点而失去活性。当第29位的Asn突变为Cys,并将C末端截短后,*Bm*GSTO3-3的体外活性得到恢复。这些发现为GSTO3的活性调节提供了结构上的见解。

1 合肥微尺度物质科学国家实验室,中国科学技术大学,合肥
2 农业部蚕桑学重点实验室,蚕学与系统生物学研究所,西南大学 ,重庆

A syntenic coding region for vitelline membrane proteins in four lepidopteran insects

Xu YM Zou ZL Zha XF Xiang ZH He NJ*

Abstract: The vitelline membrane is the inner layer of the eggshell, but the genomic information available for vitelline membrane proteins (VMPs) in Lepidoptera is limited. In the present study, we identified a syntenic coding region for *VMPs* in four lepidopteran genomes (*Bombyx mori*, *Manduca sexta*, *Danaus plexippus* and *Heliconius melpomene*) and four putative *VMP* coding genes located within it. RT-PCR results showed *Bombyx VMP* coding genes expressed prior to the early choriogenesis stage in follicles. Alignment analyses revealed that the vitelline membrane domain was shared between Lepidoptera and Diptera. However, the third cysteine residue conserved in dipteran VMPs was absent in those of Lepidoptera. In addition, another conserved region was identified in lepidopteran VMPs.

Published On: Journal of Molecular Biochemistry, 2012, 1, 155-160.

State Key Laboratory of Silkworm Genome Biology, Southwest University, Chongqing 400716, China.
*Corresponding author E-mail: hejia@swu.edu.cn.

卵黄膜蛋白在四个鳞翅目昆虫中的同义编码区域

徐云敏　邹自良　查幸福　向仲怀　何宁佳*

摘要：卵黄膜是卵壳的内层，但鳞翅目中可用于卵黄膜蛋白（VMP）的基因组信息是有限的。在本研究中，我们在4个鳞翅目昆虫（*Bombyx mori*，*Manduca sexta*，*Danaus plexippus* 和 *Heliconius melpomene*）以及位于其中的4个推定的*VMP*编码基因中鉴定了*VMP*的同义编码区。RT-PCR结果显示在卵泡早期绒毛膜发生之前表达*Bombyx VMP*编码基因。比对分析显示，鳞翅目和双翅目共享了卵黄膜结构域。然而，鳞翅目中没有保留VMPs的第三个半胱氨酸残基。此外，在鳞翅目昆虫VMPs中也发现了另一个保守区域。

家蚕基因组生物学国家重点实验室，西南大学，重庆

Molecular characterization of a peritrophic membrane protein from the silkworm, *Bombyx mori*

Hu XL Chen L Yang R Xiang XW Wu XF*

Abstract: The peritrophic membrane lines the gut of most insects at one or more stages of their life cycles. It facilitates the digestive processes in the guts and protects from invasion by pathogens or food particles. In the current study, a novel PM protein, designated as *Bm*Mtch, was identified from the silkworm, *Bombyx mori*. The open reading frame of *BmMtch* is 888 bp in length, encoding 295 amino acid residues consisting of two domains (Mito_carr domains) and three transmembrane regions. They are localized on the 11th chromosome as single copy with one exon only. Quantitative real time PCR analysis (qRT-PCR) revealed that *BmMtch* was mainly expressed in larval fat bodies, Malpighian tubules, testis and ovaries, and could be detected through all stages of the life cycle of silkworm. Immuno-fluorescence analysis indicated that *Bm*Mtch was localized within the goblet cell of larval midgut. Western blot analysis showed that *Bm*Mtch were detected in total proteins of PM and larval midgut. The characteristics of *Bm*Mtch indicated that *Bm*Mtch represents a novel member of insect PM proteins, without chitin-binding domains.

Published On: Molecular Biology Reports, 2013, 40(2), 1087-1095.

College of Animal Sciences, Zhejiang University, Hangzhou 310058, China.
*Corresponding author E-mail: wuxiaofeng@zju.edu.cn.

一种家蚕围食膜蛋白的分子表征

胡小龙　陈　琳　杨　锐　相兴伟　吴小锋*

摘要：围食膜存在于大多数昆虫中肠，并在其生命周期中的一个或多个阶段中存在。围食膜具有促进食物消化、使肠道上皮细胞免受外来食物的机械性损伤、阻止病原微生物入侵等功能。本文鉴定到了一种新的家蚕围食膜蛋白，命名为*Bm*Mtch。*BmMtch*基因开放阅读框长888 bp，编码295个氨基酸，含有2个Mito_carr结构域和3个跨膜区域，定位于第11号染色体，属于单外显子基因。qRT-PCR结果显示*BmMtch*基因在家蚕不同组织和不同发育时期均有表达，主要分布于幼虫脂肪体、马氏管、精巢和卵巢中。免疫荧光定位结果表明，该蛋白质存在于中肠杯状细胞中。Western blot分析表明在中肠及围食膜中均能检测到*Bm*Mtch的存在。由此可知，*Bm*Mtch是一种新的围食膜蛋白，但是该蛋白质无几丁质结合结构域。

动物科学学院，浙江大学，杭州

A novel method of silkworm embryo preparation for immunohistochemistry

You ZZ　Sun CF　Chen L　Yao Q　Chen KP*

Abstract: It is difficult to obtain intact embryos, especially intact early embryos, from insect eggs because of their small sizes. Based on the means traditionally used to get silkworm embryos and the previous approaches used for getting *Drosophila* embryos, we established a novel method of silkworm embryo preparation. The new method is straightforward and easy to operate. Silkworm embryos could be prepared without severe damage in large quantities by this new protocol. In addition, the novel method of silkworm embryo preparation is quite suitable for immunohistochemistry.

Published On: Biotechnology Letters, 2013, 35(8), 1209-1214.

Institute of Life Sciences, Jiangsu University, Zhenjiang 212013, China.
*Corresponding author E-mail: kpchen@ujs.edu.cn.

一种制备家蚕胚胎用于免疫组织化学实验的新方法

尤在芝　孙春凤　陈　亮　姚　勤　陈克平*

摘要：由于昆虫卵尺寸较小，所以难以从中获得完整的胚胎，特别是完整的早期胚胎。基于传统获得家蚕胚胎的方法和以前用于获得果蝇胚胎的方法，我们建立了蚕胚胎制备的新方法。这种新方法简单易操作。可以通过这种新方案制备大量没有受到严重损害的家蚕胚胎。另外，家蚕胚胎制备的新方法非常适合免疫组织化学测定。

生命科学研究院，江苏大学，镇江

Expression analysis of chlorophyllid α binding protein, a secretory, red fluorescence protein in the midgut of silkworm, *Bombyx mori*

Chen L Yang R Hu XL Xiang XW Wu XF*

Abstract: Chlorophyllid α binding protein (CHBP) was recently characterized by its ability to bind the prosthetic group of chlorophylls and little information is known regarding its expression. In the present study, we found that *CHBP* was expressed highly and exclusively in the midgut of silkworm, *Bombyx mori*. The expression level of *CHBP* was very high in the newly molted fifth instar larvae followed by gradual decline in the same instar. Our results demonstrated that CHBP was a secretory protein and located mainly in the apical of midgut epithelial cells. Real-time polymerase chain reaction analysis results showed that *CHBP* highly expressed in the anterior midgut, threefold and sixfold higher compared with that of the middle midgut and posterior midgut, respectively, and *CHBP* expression declined in darkness. In addition, the expression of *CHBP*was affected by high-dose virus or bacterium infection.

Published On: Insect Science, 2014, 21(1), 20-30.

Institute of Sericulture and Apiculture, College of Animal Sciences, Zhejiang University, Hangzhou, China.
*Corresponding author E-mail: wuxiaofeng@zju.edu.cn.

家蚕中肠红色荧光分泌蛋白叶绿酸α结合蛋白CHBP的表达分析

陈　琳　杨　锐　胡小龙　相兴伟　吴小锋*

摘要：最近研究发现叶绿酸α结合蛋白(CHBP)能够结合叶绿素辅基，但对其表达情况知之甚少。本研究发现*CHBP*在家蚕中肠高量特异表达。*CHBP*表达水平在五龄起蚕中达到最高，随后在同一龄期内逐渐降低。本文结果阐明了CHBP是一种分泌性蛋白，主要定位于中肠上皮细胞顶端。RT-PCR结果表明*CHBP*主要在中肠的前部大量表达，表达量比中肠中部和中肠后部分别高出3倍和6倍之多，并在黑暗中表达量降低。此外，*CHBP*表达会受到高浓度病毒和细菌感染的影响。

蚕蜂研究所，动物科学学院，浙江大学，杭州

Molecular cloning and characterization of a *Bombyx mori* gene encoding the transcription factor Atonal

Hu P[1#] Feng F[2#] Xia HC[1]
Chen L[1] Yao Q[1] Chen KP[1,2*]

Abstract: The *atonal* genes are an evolutionarily conserved group of genes encoding regulatory basic helix-loop-helix (bHLH) transcription factors. These transcription factors have a critical antioncogenic function in the retina, and are necessary for cell fate determination through the regulation of the cell signal pathway. In this study, the *atonal* gene was cloned from *Bombyx mori*, and the transcription factor was named *Bm*Atonal. Sequence analysis showed that the *Bm*Atonal protein shares extensive homology with other invertebrate Atonal proteins with the bHLH motif. Reverse transcription-polymerase chain reaction (RT-PCR) and Western blot analyses revealed that *Bm*Atonal was expressed in all developmental stages of *B. mori* and various larval tissues. The *Bm*Atonal protein was expressed in *Escherichia coli*, and polyclonal antibodies were raised against the purified protein. By immunofluorescence, the *Bm*Atonal protein was localized to both the nucleus and cytoplasm of *BmN* cells. After knocking out nuclear localization signals (NLS), the *Bm*Atonal protein was only detected in the cytoplasm. In addition, using the *B. mori* nuclear polyhedrosis virus (*Bm*NPV) baculovirus expression system, the recombinant *Bm*Atonal protein was successfully expressed in the *B. mori* cell line *BmN*. This work lays the foundation for exploring the biological functions of the *Bm*Atonal protein, such as identifying its potential binding partners and understanding the molecular control of the formation of sensory organs.

Published On: Zeitschrift Fur Naturforschung Section C-a Journal of Biosciences, 2014, 69(3-4), 155-164.

1 Institute of Life Sciences, Jiangsu University, Zhenjiang 212013, China.

2 School of Food and Biological Engineering, Jiangsu University, Zhenjiang 212013, China.

#These authors contributed equally.

*Corresponding author E-mail: kpchen@ujs.edu.cn.

家蚕基因编码转录因子Atonal的分子克隆和特征分析

胡　平[1#]　冯　凡[2#]　夏恒传[1]
陈　亮[1]　姚　勤[1]　陈克平[1,2*]

摘要：无性基因是编码调节性基因螺旋–环–螺旋(bHLH)转录因子的进化保守基因。这些转录因子在视网膜中具有关键的抗癌功能，并且是通过细胞信号通路调节细胞命运所必需的。在本研究中，从家蚕中克隆了无性基因，其转录因子被命名为*Bm*Atonal。序列分析表明，*Bm*Atonal蛋白与其他具有bHLH基序的无脊椎动物Atonal蛋白质具有广泛的同源性。RT-PCR和Western blot分析显示，*Bm*Atonal在家蚕幼虫全时期、全组织中均有表达。将*Bm*Atonal蛋白在大肠杆菌中进行表达，并且针对纯化的蛋白质制备多克隆抗体。通过免疫荧光方法，确定*Bm*Atonal蛋白定位于*BmN*细胞的细胞核和细胞质中。敲除核定位信号(NLS)后，仅在细胞质中检测到*Bm*Atonal蛋白。另外，使用家蚕核型多角体病毒(*Bm*NPV)表达系统，重组*Bm*Atonal蛋白在家蚕细胞系*BmN*中成功表达。这项工作为探索*Bm*Atonal蛋白的生物学功能奠定了基础，如识别其潜在的结合配体，了解感觉器官形成的分子机制。

1　生命科学研究院，江苏大学，镇江
2　食品与生物工程学院，江苏大学，镇江

*Bm*HSP20.8 is localized in the mitochondria and has a molecular chaperone function *in vitro*

Wu CC[1#] Wang C[1#] Li D[1] Liu Y[2]
Sheng Q[1] Lü ZB[1] Yu W[1] Nie ZM[1*]

Abstract: Heat shock proteins (HSPs) are abundant and ubiquitous in almost all organisms from bacteria to mammals. *BmHSP20.8* is a small HSP in *Bombyx mori* that contains a 561 bp open reading frame that encodes a protein of 186 amino acid residues with a predicted molecular mass of 20.8 kDa. The subcellular localization prediction indicated that *Bm*HSP20.8 is likely distributed in the mitochondria with a 51% probability. To identify the subcellular localization of *Bm*HSP20.8, three recombinant vectors were constructed and used to transfect *BmN* cells. The cytoplasmic and mitochondrial proteins were extracted 72 h after transfection. The Western blot showed that recombinant *Bm*HSP20.8 exists only in the mitochondria. To locate the mitochondrial localization signal domain of *Bm*HSP20.8 more accurately, we cloned four truncated recombinant vectors. The Western blot analysis of the cytoplasmic and mitochondrial proteins showed that the mitochondrial localization signal domain of *Bm*HSP20.8 is located between amino acids 143 to 186. We constructed the pETduet-HIS-SUMO-*Bm*HSP20.8 vector and a soluble *Bm*HSP20.8 was expressed. In a citrate synthase (CS) thermal aggregation experiment, we found that the recombinant *Bm*HSP20.8 protein can protect CS from aggregating at 43 and 48 ℃ and thus exhibited molecular chaperone activity. Taken together, the results showed that *Bm*HSP20.8 could be a mitochondrial protein and has a molecular chaperone activity, suggesting an important role in mitochondria.

Published On: Journal of Insect, 2015, 15.

1 College of Life Science, Zhejiang Sci-Tech University, Hangzhou, 310018, China.
2 Zhejiang Economic and Trade Polytechnic, Hangzhou, 310018, China.
#These authors contributed equally.
*Corresponding author E-mail: wuxinzm@126.com.

*Bm*HSP20.8蛋白定位于线粒体并具有分子伴侣功能

吴程程[1#] 王 婵[1#] 李 丹[1] 刘 悦[2]
盛 清[1] 吕正兵[1] 于 威[1] 聂作明[1*]

摘要：热激蛋白(Heat shock proteins，HSPs)是一类广泛存在于从细菌到哺乳动物的几乎所有生物中的蛋白质。*BmHSP20.8*是一种家蚕小热休克蛋白基因，其ORF框大小为561 bp，编码186个氨基酸残基，预测分子质量为20.8 kD。亚细胞定位预测发现*Bm*HSP20.8有51%可能性定位于线粒体。为了鉴定*Bm*HSP20.8蛋白的亚细胞定位情况，我们构建了3种真核表达载体并转染家蚕*BmN*细胞。72 h后提取细胞的胞质和线粒体蛋白，Western blot分析该蛋白的定位情况。结果表明*Bm*HSP20.8蛋白只存在于线粒体中。为了进一步确定*Bm*HSP20.8蛋白的线粒体定位信号区域，我们构建了4个截短型*Bm*HSP20.8真核表达载体，胞质和线粒体蛋白Western blot分析表明，*Bm*HSP20.8蛋白的线粒体定位信号区域位于该蛋白C端的143—186 aa区域。我们构建了pETduet-HIS-SUMO-*Bm*HSP20.8表达载体并表达了可溶的*Bm*HSP20.8蛋白。柠檬酸合酶(CS)的热聚集实验证实重组*Bm*HSP20.8蛋白可以阻止CS在43 ℃和48 ℃时的蛋白聚集，显示其具有分子伴侣活性。总的来说，*Bm*HSP20.8可能是一种线粒体蛋白，同时具有分子伴侣活性，暗示该蛋白可能在线粒体中担任重要的角色。

1 生命科学学院，浙江理工大学，杭州
2 浙江经贸职业技术学院，杭州

Expression analysis and tissue distribution of two 14-3-3 proteins in silkworm (*Bombyx mori*)

Kong LY[1] Lü ZB[1] Chen J[1] Nie ZM[1] Wang D[1]
Shen HD[1] Wang XD[1] Wu XF[1,2] Zhang YZ[1*]

Abstract: 14-3-3 proteins, which have been identified in a wide variety of eukaryotes, are highly conserved acidic proteins. In this study, we identified two genes in silkworm that encode 14-3-3 proteins (*Bm14-3 - 3ζ* and *Bm14-3-3ε*). Category of two 14-3-3 proteins was identified according to phylogenetic analysis. *Bm*14-3-3ζ shared 90% identity with that in *Drosophila*, while *Bm*14-3-3ε shared 86% identity with that in *Drosophila*. According to Western blot and real-time PCR analysis, the *Bm*14-3-3ζ expression levels are higher than *Bm*14-3-3ε in seven tissues and in four silkworm developmental stages examined. *Bm*14-3-3ζ was expressed during every stage of silkworm and in every tissue of the fifth instar larvae that was examined, but *Bm*14-3-3ε expression was not detected in eggs or heads of the fifth instar larvae. Both 14-3-3 proteins were highly expressed in silk glands. These results suggest that *Bm*14-3-3ζ expression is universal and continuous, while *Bm*14-3-3ε expression is tissue and stage-specific. Based on tissue expression patterns and the known functions of 14-3-3 proteins, it may be that both 14-3-3 proteins are involved in the regulation of gene expression in silkworm silk glands.

Published On: Biochimica et Biophysica Acta-General Subjects, 2007, 1770(12), 1598-1604.

1 College of Life Sciences, Zhejiang Sci-Tech University, Hangzhou 310018, China.
2 Institute of Biochemistry and Cell Biology, Chinese Academy of Sciences, Shanghai 200031, China.
*Corresponding author E-mail: yaozhou@chinagene.com.

家蚕中两种14-3-3蛋白的表达分析和组织分布

孔令印[1]　吕正兵[1]　陈　健[1]　聂作明[1]　王　丹[1]
沈红丹[1]　王雪东[1]　吴祥甫[1,2]　张耀洲[1*]

摘要：14-3-3蛋白已经在多种真核生物中鉴定出，是高度保守的酸性蛋白质。我们的研究鉴定了家蚕中编码14-3-3蛋白的两个基因（*Bm14-3-3ζ* 和 *Bm14-3-3ε*）。根据系统发育分析确定了两种14-3-3蛋白的系统分类。*Bm*14-3-3ζ与果蝇同源蛋白具有90%的相似性，而*Bm*14-3-3ε与果蝇同源蛋白具有86%的相似性。根据Western blot和实时定量PCR分析，发现在家蚕7个组织中和4个发育阶段，*Bm*14-3-3ζ的表达水平均高于*Bm*14-3-3ε。*Bm*14-3-3ζ在家蚕的每个阶段和五龄幼虫被分析的每个组织中都有表达，但在五龄幼虫的卵或头中未检测到*Bm*14-3-3ε表达。两种14-3-3蛋白在丝腺中均高丰度表达。这些结果表明*Bm*14-3-3ζ的表达是普遍和连续的，而*Bm*14-3-3ε的表达具有组织和发育阶段特异性。基于组织表达模式和14-3-3蛋白的已知功能，我们推测14-3-3蛋白可能参与家蚕丝腺中基因表达的调控。

1　生命科学学院，浙江理工大学，杭州
2　生物化学与细胞生物学研究所，中国科学院，上海

Expression analysis and characteristics of *profilin* gene from silkworm, *Bombyx mori*

Nie ZM Xu JT Chen J Lü ZB Wang D Sheng Q
Wu Y Wang XD Wu XF Zhang YZ *

Abstract: A recombinant *Bombyx mori* profilin protein (r*Bm*PFN) was overexpressed in *Escherichia coli* BL21. Purified r*Bm*PFN was used to generate anti-*Bm*PFN polyclonal antibody, which were used to determine the subcellular localization of *Bm*PFN. Immunostaining indicated that profilin can be found in both the nucleus and cytoplasm but is primarily located in the cytoplasm. Real-time RT-PCR and Western blot analyses indicated that, during the larvae stage, profilin expression levels are highest in the silk gland, followed by the gonad, and are lowest in the fat body. Additionally, *Bm*PFN expression begins during the egg stage, increases during the larvae stage, reaches a peak during the pupa stage, and decreases significantly in the moth. Therefore, we propose that *Bm*PFN may play an important role during larva stage development, especially in the silk gland.

Published On: Applied Biochemistry and Biotechnology, 2009, 158(1), 59-71.

Institute of Biochemistry, College of Life Sciences, Zhejiang Sci-Tech University, Hangzhou 310018, China.
*Corresponding author E-mail: yaozhou@chinagene.com.

家蚕*profilin*基因的表达和特征分析

聂作明　徐江涛　陈　健　吕正兵　王　丹
盛　清　吴　怡　王雪冬　吴祥甫　张耀洲*

摘 要：在大肠杆菌BL21中过表达家蚕重组profilin蛋白(r*Bm*PFN)。用纯化的r*Bm*PFN制备了*Bm*PFN多克隆抗体，并用于确定其亚细胞定位。免疫组化染色表明在细胞核和细胞质中均存在profilin，但其主要分布在细胞质中。实时定量RT-PCR和Western blot分析表明，*Bm*PFN蛋白在幼虫期的丝腺中表达水平最高，其次是在性腺中，在脂肪体中表达水平最低。此外，*Bm*PFN在卵期开始表达，在幼虫阶段表达量增加，在蛹期达到峰值，在蛾期显著降低。因此，我们推测*Bm*PFN可能在幼虫的发育过程，特别是在丝腺发育中起着重要的作用。

生物化学研究所，生命科学学院，浙江理工大学，杭州

Expression analysis and characteristics of *Bm-LOC778477* gene, a hypothetical protein from silkworm pupae, *Bombyx mori*

Yu W Tan AW Zhang YZ *

Abstract: This is the first identification of a gene encoding *Bm*-LOC778477 protein (a "hypothetical protein") from silkworm pupal cDNA library. The full-length *Bm-LOC778477* cDNA contained a 546-bp ORF encoding 187 amino acids. Using bioinformatics, the *Bm-LOC778477* gene was predicted on the W chromosome and 4 exons and 3 introns were determined when compared to the silkworm genome. The consensus splice sites of exon/intron junctions were consistent with Chambon's rule. In addition, TATA box and initiator (Inr) were also predicated in the promoter region of the gene. To elucidate whether *Bm-LOC778477* gene encoded a functional protein in silkworm, pGEX-4T-1-*Bm*-LOC778477 expression plasmid was constructed to express GST-LOC778477 fusion protein in *Escherichia coli* Rosetta. r*Bm*-LOC778477 was used to generate anti-*Bm*-LOC778477 polyclonal antibody, which were used to determine the subcellular localization of *Bm*-LOC778477. Immunostaining indicated that *Bm*-LOC778477 protein could be found in both the cytoplasm and nucleus. Western blot analysis indicated that *Bm*-LOC778477 was specific expression during the pupa stage and there was seldom expression during the larvae stage. Therefore, we propose that *Bm*-LOC778477 may play an important role during pupa stage development. These results may lay an important foundation of further function studies of *Bm*-LOC778477 protein.

Published On: African Journal of Biotechnology, 2009, 8(14), 3192-3198.

Institute of Biochemistry, College of Life Science, Zhejiang Sci-Tech University, Hangzhou 310018, China.

*Corresponding author E-mail: yaozhou@zstu.edu.cn.

家蚕蚕蛹中假定蛋白基因*Bm-LOC778477*的表达分析和特征

于　威　谭爱武　张耀洲*

摘要：本研究是对家蚕蚕蛹cDNA文库中编码*Bm*-LOC778477蛋白(一个“假定蛋白”)的基因的首次鉴定。*Bm-LOC778477*的全长cDNA包含一个546 bp的开放阅读框，编码187个氨基酸。生物信息学分析显示，*Bm-LOC778477*基因位于W染色体上，包含了4个外显子和3个内含子，外显子和内含子的衔接位点与Chambon法则一致。此外，在基因的启动子区域预测到TATA盒和起始子(Inr)的存在。为了阐释*Bm-LOC778477*基因是否在家蚕中编码有功能的蛋白，我们构建了pGEX-4T-1-*Bm*-LOC778477表达质粒，并在大肠杆菌Rosetta中表达了GST-LOC778477融合蛋白。用r*Bm*-LOC778477蛋白制备抗*Bm*-LOC778477多克隆抗体，抗体随后用于检测*Bm*-LOC778477的亚细胞定位。免疫染色显示在细胞质和细胞核中可检测到*Bm*-LOC778477蛋白存在。Western blot分析显示*Bm*-LOC778477在蛹期特异性表达，在幼虫期极少表达。因此，我们推测*Bm*-LOC778477可能在家蚕的蛹期发育中起重要作用。本研究结果为*Bm*-LOC778477蛋白功能的进一步研究奠定了重要基础。

生物化学研究所，生命科学学院，浙江理工大学，杭州

Cloning and partial characterization of a gene in *Bombyx mori* homologous to a human adiponectin receptor

Zhu MF Chen KP* Wang Y Guo ZJ Yin HJ Yao Q Chen HQ

Abstract: In this study, we report the cloning and characteristics of an adiponectin-like receptor gene from *Bombyx mori* (*BmAdipoR*) with highly conserved deduced amino-acid sequences and similar structure to the human adiponectin receptor (AdipoR). Structural analysis of the translated cDNA suggested it encoded a membrane protein with seven transmembrane domains. *BmAdipoR* was found to be expressed in multiple tissues and highly expressed in Malpighian tubules, fat body and testis. *Bm*NPV (*Bombyx mori* nucleopolyhedrovirus) bacmid system combined with confocal microscopy revealed that *Bm*AdipoR was targeted to the cell membrane. We also found that infection with *Bm*NPV did not have an effect on *Bm*AdipoR mRNA quantity in the midgut of susceptible *Bombyx mori* strain (*306*) at 48 h, but *BmAdipoR* mRNA quantity was increased significantly at 72 h. We concluded that *Bm*AdipoR was a membrane protein ubiquitously expressed in *Bombyx mori* tissues and that its expression was altered by treating with *Bm*NPV.

Published On: Acta Biochimica Polonica, 2008, 55(2), 241-249.

Institute of Life Sciences, Jiangsu University, Zhenjiang 212013, China.
*Corresponding author E-mail:Kpchen@ujs.edu.cn.

家蚕脂联素受体同源基因的克隆和功能初探

朱敏锋　陈克平*　王　勇　郭忠建　尹慧娟　姚　勤　陈慧卿

摘要:本研究在家蚕中克隆和分析了一个脂联素受体(*AdipoR*)基因类似物,其与人脂联素受体具有高度保守的氨基酸序列和结构特征。蛋白结构预测分析显示,家蚕脂联素受体基因类似物是一个具有7个跨膜结构域的膜蛋白。*BmAdipoR*在多个组织中表达,并在马氏管、脂肪体和精巢中高表达。结合*Bm*NPV (*Bombyx mori* nucleopolyhedrovirus)杆状病毒表达系统与共聚焦显微镜技术,发现*Bm*AdipoR定位于细胞膜。我们还发现易感蚕品系(*306*)感染*Bm*NPV 48 h后,*BmAdipoR* mRNA的表达量没有明显变化,但在72 h时*BmAdipoR* mRNA的量显著增加。综上,*Bm*AdipoR是一个在家蚕各组织中普遍表达的膜蛋白,而*Bm*NPV感染可以改变其表达水平。

生命科学研究院,江苏大学,镇江

Identification and characterization of a novel 1-Cys peroxiredoxin from silkworm, *Bombyx mori*

Wang Q[1] Chen KP[1*] Yao Q[1] Zhao Y[2]
Li YJ[1] Shen HX[1] Mu RH[1]

Abstract: Peroxiredoxins (Prxs) are believed to play an important role in insects for protection against the toxicity of reactive oxygen species (ROS). A gene encoding a novel *1-Cys Prx* was firstly identified and characterized from an expressed sequence tag database (EST) in the lepidopteran insect, *Bombyx mori*. The 1-Cys Prx of *B. mori* (*BmPrx*) cDNA contained an open reading frame of 672 bp encoding a protein of 223 amino acid residues with calculated molecular mass of 25 kDa and included conserved cysteine residues signature motifs (PVCT) in the N-terminus amino acid. Sequence comparison showed that *Bm*Prx shared 53% to 64% identity with other species 1-Cys proteins. RT-PCR revealed that the *BmPrx* transcripts were present in all tissues and developmental stages. The coding sequence was cloned and expressed as a 30 kDa protein in *Escherichia coli*. The purified enzyme acted as a catalyst in ferrithiocyanate system and protected supercoiled form of plasmid DNA from damage in metal-catalyzed oxidation (MCO) system *in vitro*. In addition, real-time PCR analysis indicated that significant rise of the transcripts level of *BmPrx* was induced by temperature stress including low and high temperature stimuli.

Published On: Comparative Biochemistry and Physiology-Part B:Biochemistry and Molecular Biology. 2008, 149(1), 176-182.

1 Institute of Life Sciences, Jiangsu University, Zhenjiang 212013, China.
2 Sericultural Research Institute, Chinese Academy of Agricultural Sciences, Zhenjiang 212018, China.
*Corresponding author E-mail:kpchen@ujs.edu.cn.

一个新的家蚕1-Cys过氧化物还原蛋白的鉴定和功能分析

王　强[1]　陈克平[1*]　姚　勤[1]　赵　远[2]
李怡佳[1]　申红星[1]　母润红[1]

摘要：抗氧化蛋白(Prxs)在昆虫中起重要作用，可以防止活性氧(ROS)的毒性。从家蚕表达序列标签数据库(EST)中，首次鉴定了一个新的 *1-Cys Prx* 基因，并对其功能进行了初步分析。家蚕 *1-Cys Prx* (*BmPrx*)的cDNA包含一个长度为672 bp的开放阅读框，编码223个氨基酸残基，预测其分子质量为25 kDa，并在N末端含有保守的半胱氨酸残基特征基序(PVCT)。序列比对分析表明，*Bm*Prx与其他物种的1-Cys蛋白质具有53%—64%的同源性。RT-PCR分析显示 *BmPrx* 转录本存在于所有组织和发育阶段。我们对其编码序列进行了克隆并在大肠杆菌表达系统中表达得到了30 kDa的蛋白质。使用纯化的酶进行酶活检测，发现其在铁硫铁氰酸盐体系中起催化剂的作用，并在体外金属催化氧化(MCO)体系中对超螺旋型质粒DNA起保护作用。此外，RT-PCR分析表明，低温和高温刺激的温度胁迫可诱导 *BmPrx* 的转录水平显著升高。

1　生命科学研究院，江苏大学，镇江
2　蚕业研究所，中国农业科学院，镇江

Cloning and characterization of the gene encoding an ubiquitin-activating enzyme E1 domain-containing protein of silkworm, *Bombyx mori*

Wu P[1] Li MW[1] Jiang YF[2] Wang ZS[2] Guo XJ[1,2*]

Abstract: *Bombyx mori* cytoplasmic polyhedrosis virus (*Bm*CPV) is one of the major viral pathogens for the silkworm. To date, the molecular mechanism of *Bm*CPV invasion is unclear. We cloned the full length complementary (c)DNA which encodes the ubiquitin-activating enzyme E1-domain containing protein1 (UbE1DC1) of *Bombyx mori* by using suppression subtractive hybridization (SSH) and rapid amplification of complementary(c)DNA ends (RACE). The full-length cDNA of *UbE1DC1* gene is 1 919 bp, consisting of a 100 bp 5′untranslated region, a 637 bp 3′untranslated region and an 1 182 bp open reading frame (ORF), encoding a 393 amino acid protein. The protein contained the THiF-MoeB-hesA-family domain, an adenosine triphosphate binding site, which belongs to the family of ubiquitin-activating enzyme E1. Reverse transcription-polymerase chain reaction analysis from the silkworm tissues, namely silk gland, hemocyte, fat body, gonad and midgut revealed that *UbE1DC1* was expressed in all the five tissues. The real-time quantitative polymerase chain reaction analysis indicated that the relative expression of *UbE1DC1* in the normal midgut was approximately 9.78-fold of that in the *Bm*CPV-infected midgut. It is implicated that *UbE1DC1* may play an important role in the interaction between the host and *Bm*CPV invasion.

Published On: Insect Science, 2010, 17, 75-83.

1 Sericultural Research Institute, Chinese Academy of Agricultural Sciences, Zhenjiang 212018, China.

2 College of Biotechnology and Environmental Engineering, Jiangsu University of Science and Technology, Zhenjiang 212018, China.

*Corresponding author E-mail: guoxijie@126.com.

编码家蚕含E1蛋白结构域的泛素活化酶基因的克隆与鉴定

吴　萍[1]　李木旺[1]　蒋云峰[2]　王资生[2]　郭锡杰[1,2*]

摘要：家蚕质型多角体病毒（*Bm*CPV）是主要的家蚕病毒病原体之一。到目前为止，*Bm*CPV入侵的分子机制还不清楚。我们采用抑制消减杂交（SSH）和cDNA末端快速扩增（RACE）技术克隆了编码家蚕含E1蛋白结构域的泛素活化酶UbE1DC1的全长互补DNA（cDNA）序列。*UbE1DC1*基因的cDNA全长1 919 bp，由100 bp的5′非翻译区，637 bp的3′非翻译区和1 182 bp的开放阅读框（ORF）组成，编码含393个氨基酸的蛋白质。UbE1DC1蛋白包含THiF-MoeB-hesA结构域，该结构域是三磷酸腺苷的结合位点，属于泛素活化酶E1家族。RT-PCR分析发现*UbE1DC1*在家蚕的5种组织，即丝腺、血细胞、脂肪体、性腺和中肠都有表达。qRT-PCR分析表明正常中肠的*UbE1DC1*相对表达量大约是*Bm*CPV感染中肠的9.78倍。这些结果暗示了*UbE1DC1*基因在宿主与*Bm*CPV入侵的互作中具有重要的功能。

1　蚕业研究所，中国农业科学院，镇江

2　生物与环境工程学院，江苏科技大学，镇江

Cloning, expression, and cell localization of a novel small heat shock protein gene: *BmHSP25.4*

Sheng Q Xia JY Nie ZM Zhang YZ*

Abstract: Using molecular approaches, a new member of the *Bombyx mori* small heat shock protein family was cloned and characterized. The isolated gene contains an open reading frame of 672 bp, encodes a polypeptide of 223 amino acid residues with a predicted molecular mass of 25.4 kDa, and is therefore named *Bm*HSP25.4. The gene codes for a protein homologous to the previously characterized HSP20.4 and HSP19.9. Western blot analysis revealed that *Bm*HSP25.4 existed in the fifth-instar larva's fatty body and blood tissues. Immunohistochemistry assay also showed that *Bm*HSP25.4 was located in the fifth-instar larva's fatty body. The results of above studies have indicated constitutive expression of *BmHSP25.4* in fatty body, blood tissues, and *Bm5* cells. Finally, we examined the effect of heat stress on localization of *Bm*HSP25.4 using anti-BmHSP25.4 polyclonal antibody by immunofluorescence. Under normal conditions, *Bm*HSP25.4 was mostly found in the cytoplasm. However, after heat treatment, most of *Bm*HSP25.4 distributed in the cell membrane. After 3 h of recovery following the heat shock treatment, the localization of *Bm*HSP25.4 was the same as that under normal conditions.

Published On: Applied Biochemistry and Biotechnology, 2010, 162(5), 1297-1305.

Institute of Biochemistry, Zhejiang Sci-Tech University, Hangzhou 310018, China.
*Corresponding author E-mail: yaozhou@chinagene.com.

一个新的小热激蛋白基因*BmHSP25.4*的克隆、表达和细胞定位

盛　清　夏佳音　聂作明　张耀洲*

摘要:采用分子生物学方法克隆并鉴定了一个家蚕小热激蛋白家族的一个新成员。分离的基因包含672 bp的开放阅读框,编码长度为223个氨基酸残基的蛋白质,预测其分子质量为25.4 kDa,因此命名为*Bm*HSP25.4。该基因编码蛋白与先前鉴定的HSP20.4和HSP19.9蛋白同源。Western blot分析显示,*Bm*HSP25.4存在于五龄幼虫脂肪体和血淋巴中。免疫组织化学测定也显示*Bm*HSP25.4位于五龄幼虫脂肪体。上述研究结果表明*BmHSP25.4*在脂肪体、血淋巴和*Bm5*细胞中组成型表达。最后,利用抗*Bm*HSP25.4多克隆抗体,采用免疫荧光法检查了热应激对*Bm*HSP25.4亚细胞定位的影响。在正常条件下,*Bm*HSP25.4主要存在于细胞质中。然而,热处理后,大部分*Bm*HSP25.4分布在细胞膜上。热激处理3 h后,*Bm*HSP25.4的定位又恢复到正常。

生物化学研究所,浙江理工大学,杭州

Molecular cloning and expression analysis of *Bmrbp1*, the *Bombyx mori* homologue of the *Drosophila* gene *rbp1*

Wang ZL Zha XF He NJ Xiang ZH Xia QY*

Abstract: RBP1 is an important splicing factor involved in alternative splicing of the pre-mRNA of *Drosophila* sex-determining gene *dsx*. In this work, the *Bombyx mori* homologue of the *rbp1* gene, *Bmrbp1*, was cloned. The pre-mRNA of *Bmrbp1* gene is alternatively spliced to produce four mature mRNAs, named *Bmrbp1-PA*, *Bmrbp1-PB*, *Bmrbp1-PC* and *Bmrbp1-PD*, with nucleotide lengths of 799 nt, 1 316 nt, 894 nt and 724 nt, coding for 142 aa, 159 aa, 91 aa and 117 aa, respectively. *Bm*RBP1-PA and *Bm*RBP1-PD contain a N terminal RNA recognization motif (RRM) and a C terminal arginine/serine-rich domain, while *Bm*RBP1-PB and *Bm*RBP1-PC only share a RRM. Amino acid sequence alignments showed that *Bm*RBP1 is conserved with its homologues in other insects and with other SR family proteins. The RT-PCR showed that *Bmrbp1-PA* was strongly expressed in all examined tissues and development stages, but *Bmrbp1-PB* was weakly expressed in these tissues and stages. The expression of both *Bmrbp1-PA* and *Bmrbp1-PB* showed no obvious sex difference. While the *Bmrbp1-PC* and *Bmrbp1-PD* were beyond detection by RT-PCR very likely due to their tissue/stage specificity. These results suggested that *Bmrbp1* should be a member of SR family splicing factors, whether it is involved in the sex-specific splicing of *Bmdsx* pre-mRNA needs further research.

Published On: Molecular Biology Reports, 2010, 37(5), 2525-2531.

Key Sericultural Laboratory of Agricultural Ministry, College of Sericulture and Biotechnology, Southwest University, Chongqing 400716, China.

*Corresponding author E-mail: xiaqy@swau.cq.cn.

家蚕中果蝇的同源基因*Bmrbp1*的分子克隆和表达分析

王子龙　查幸福　何宁佳　向仲怀　夏庆友*

摘要：RBP1是参与果蝇性别决定基因*dsx*前体mRNA选择性剪接的重要剪接因子。在本研究中，我们克隆了家蚕中*rbp1*基因的同源基因*Bmrbp1*。*Bmrbp1*基因的前体mRNA可以剪接产生4个成熟mRNA，命名为*Bmrbp1-PA*，*Bmrbp1-PB*，*Bmrbp1-PC*和*Bmrbp1-PD*，其核苷酸长度分别为799 nt，1 316 nt，894 nt和724 nt，分别编码142 aa，159 aa，91 aa和117 aa。*Bm*RBP1-PA和*Bm*RBP1-PD含有一个N末端RNA识别基序（RRM）和一个C末端富含精氨酸/丝氨酸的结构域，而*Bm*RBP1-PB和*Bm*RBP1-PC仅含RRM结构域。氨基酸序列比对显示*Bm*RBP1与其他昆虫和其他SR家族蛋白的同源性较高。RT-PCR显示*Bmrbp1-PA*在所有检测组织和发育阶段都高表达，但*Bmrbp1-PB*的表达量却比较低。*Bmrbp1-PA*和*Bmrbp1-PB*的表达无明显性别差异。*Bmrbp1-PC*和*Bmrbp1-PD*未检测到，可能是由于其组织或时期特异性所致。这些结果表明*Bmrbp1*是SR家族拼接因子的成员，但其是否参与*Bmdsx*前mRNA的性别特异性剪接还需要进一步研究。

农业部蚕桑学重点实验室，生物技术学院，西南大学，重庆

Structure characteristics and expression profiles of *Bombyx mori* α1 (Ⅳ) collagen gene, a temperature-sensitive lethality-related gene

Ji MM   Lu YJ   Gan LP   Niu YS   Sima YH   Xu SQ*

Abstract: The type IV collagen is a heterotrimer of two α1 (IV) and one α2 (IV) chains, which are encoded *by emb-9* and *let-2* genes in *Caenorhabditis elegans,* respectively, and the amino acid mutations in Gly-X-Y repeat region can cause temperature-sensitive lethality during late embryogenesis. Here, we introduced a way to quickly and effectively do research on *Bmlet-2* (*BmColIVα1*) gene from *Bombyx mori*, a central model for Lepidoptera. *BmColIV1* gene is 6.583 kbp in full length, containing four exons, and its cDNA is 3.847 kbp, which contains an ORF for a protein of 819 amino acids with a molecular weight of 82.1 kDa and an isoelectric point of 6.43. The predicted protein sequence has closer relationship with α1 (IV) collagen chain of other species, and contains two conserved regions of collagen and nine PFAM domains of collagen triple helix repeat regions. Moreover, transmembrane region analysis and subcellular localization analysis reveal that this protein belongs to extracellular secreted protein. The experimental expression profile demonstrated that the expression level of *BmColIVα1* was obviously higher in head and fat body on day 3 of the fifth instar. The mutations of *BmColIVα1* gene located in exon 2 and exon 3 were identified in temperature-sensitive lethal strain *sch* and wild strain *Dazao*. Seven mutations—namely, 242(T/A), 279 (T/A), 556(T/C), 675(C/T), 1 343(T/C), 1 463(A/G), and 1 574(T/C)—were novel, and five of them would modify the encoded amino acids, namely, 81(L/Q), 186(S/P), 448(L/S), 488(Q/R), and 525(V/A). The results of molecular modelling showed that the five amino acids mutations were located in four regions of 3-helixs repetitive element sequence of ColIVα1 protein. *Bm*ColIVα1 protein is a member of the collagen family, and its gene might be a temperature-sensitive lethality-related gene.

Published On: Journal of Applied Entomology, 2010, 134(9-10), 727-736.

National Engineering Laboratory for Modern Silk, Department of Applied Biology, Medical College, Soochow University, Suzhou 215123, China.

*Corresponding author E-mail: szsqxu@suda.edu.cn.

家蚕Ⅳ型胶原蛋白质(ColⅣ)基因的结构特征与表达谱分析

季明明　鲁延军　甘丽萍　牛艳山　司马杨虎　徐世清*

摘要：线虫*emb-9*基因和*let-2*基因分别编码Ⅳ型胶原蛋白质α1链和α2链，两者的Gly-X-Y重复区域氨基酸残基的突变能引起胚胎期温敏性致死。在这里，我们介绍了一种快速有效地研究鳞翅目动物中枢模型家蚕中*Bmlet-2*(*BmColIVα1*)基因的方法。*BmColIVα1*基因全长6 583 bp，含4个外显子，cDNA全长3 847 bp，其包含了一个编码819个氨基酸蛋白质的ORF，该蛋白质分子质量82.1 kDa，等电点6.43。预测的蛋白序列与其他物种的胶原α1(IV)链更近，有2个胶原蛋白保守区和9个PFAM结构域，跨膜区域分析和亚细胞定位分析显示为细胞外分泌型蛋白质。基因表达谱分析结果表明，*BmColIVα1*基因在家蚕五龄第3天的头和脂肪体中表达量较高。在温敏性致死品系*sch*和大造品系中鉴定了位于外显子2和外显子3的*BmColIVα1*突变基因。7个突变位点242(T/A)、279(T/A)、556(T/C)、675(C/ T)、1 343(T/C)、1 463 (A/ G)、1 574(T/ C)是异常的，其中5个位点导致编码氨基酸的改变，即81(L/Q) 、186(S/P)、448(L/S)、488(Q/P)、525(V /A)。分子建模分析发现，这5个突变位于*Bm*ColⅣα1 的4个三倍螺旋重复区。*Bm*ColⅣα1是胶原蛋白质家族的一员，是温敏性致死相关基因。

现代丝绸国家工程实验室，应用生物学系，医学部，苏州大学，苏州

Molecular characterization and tissue localization of an F-Box only protein from silkworm, *Bombyx mori*

Shen YF[1,2] Zhang TC[1,2] Chen JQ[1,2] Lü ZB[1,2] Chen J[1,2] Wang D[1,2] Nie ZM[1,2]
He PA[1,2] Wang J[1,2] Zheng QL[1,2] Sheng Q[1,2] Wu XF[1,3] Zhang YZ [1,2*]

Abstract: The eukaryotic F-box protein family is characterized by an F-box motif that has been shown to be critical for the controlled degradation of regulatory proteins. We identified a gene encoding an F-box protein from a cDNA library of silkworm pupae, which has an ORF of 1 821 bp, encoding a predicted 606 amino acids. Bioinformatic analysis on the amino acid sequence shows that *Bm*FBXO21 has a low degree of similarity to proteins from other species, and may be related to the regulation of cell-cycle progression. We have detected the expression pattern of *Bm*FBXO21 mRNA and protein and performed immunohistochemistry at three different levels. Expression was highest in the spinning stage, and in the tissues of head, epidermis, and genital organs.

Published On: Comparative and Functional Genomics, 2009.

1 Institute of Biochemistry, Zhejiang Sci-Tech University, Hangzhou 310018, China.
2 Zhejiang Provincial Key Laboratory of Silkworm Bioreactor and Biomedicine, Hangzhou 310018, China.
3 Shanghai Institute of Biochemistry and Cell Biology, Chinese Academy of Sciences, Shanghai 200031, China.
*Corresponding author E-mail: yaozhou@chinagene.com.

家蚕F-box蛋白的分子鉴定与组织定位

沈亚芳[1,2] 张天成[1,2] 陈剑清[1,2] 吕正兵[1,2] 陈 健[1,2] 王 丹[1,2] 聂作明[1,2]
贺平安[1,2] 王江[1,2] 郑青亮[1,2] 盛 清[1,2] 吴祥甫[1,3] 张耀洲[1,2*]

摘要: 真核F-box蛋白家族的特点是拥有F-box基序,而该基序已被证明是调节蛋白可控性降解的关键。我们从蚕蛹的cDNA文库找到一个能够编码F-box蛋白的基因,该基因的ORF长度为1 821 bp,预测编码含606个氨基酸残基的多肽。通过对氨基酸序列的生物信息学分析发现,*Bm*FBXO21与其他物种中同源蛋白的相似度较低,其可能参与了细胞周期的调控。我们从mRNA、蛋白质和免疫组化3个方面对*Bm*FBXO21的表达模式进行检测。结果发现,从时期表达水平上看*Bm*FBXO21在吐丝期的表达量最高;从组织表达水平上看,在头部、表皮和生殖器官等组织中表达量最高。

1 生物化学研究所,浙江理工大学,杭州
2 浙江省家蚕生物反应器和生物医药重点实验室,杭州
3 上海生物化学与细胞生物学研究所,中国科学院,上海

Characterization of the gene *BmEm4*, a homologue of *Drosophila E(spl)m4*, from the silkworm, *Bombyx mori*

Zeng FH[1,2] Xie HX[3] Nie ZM[1,2] Chen J[1,2] Lü ZB[1,2] Chen JQ[1,2]
Wang D[1,2] Liu LL[1,2] Yu W[1,2] Sheng Q[1,2] Wu XF[1,2] Zhang YZ [1,2*]

Abstract: The *Drosophila E(spl)m4* gene contains some highly conserved motifs (such as the Brd box, GY box, K box, and CAAC motif) in its 3' untranslated region (3' UTR). It was shown to be a microRNA target gene in *Drosophila* and to play an important role in the regulation of neurogenesis. We identified a homologue of the *E(spl)m4* gene from *Bombyx mori* called *BmEm4* and examined the expression patterns of *BmEm4* mRNA and protein. There was a lack of correlation in the expression of the mRNA and protein between the different developmental stages, which raises the possibility of posttranscriptional regulation of the *BmEm4* mRNA. Consistent with this idea is the finding that the 3' UTR contains two putative binding sites for microRNAs. Moreover, given that the expression is the highest in the larval head, as confirmed by immunohistochemistry, we propose that *Bm*Em4 may also be involved in the regulation of neurogenesis. Immunostaining indicated that *Bm*Em4 is located primarily in the cytoplasm.

Published On: Comparative and Functional Genomics, 2009.

1 Institute of Biochemistry, Zhejiang Sci-Tech University, Hangzhou 310018, China.

2 Institute of Biochemistry, College of life Sciences, Zhejiang Provincial Key Laboratory of Silkworm Bioreactor and Biomedicine, Hangzhou 310018, China.

3 Xin Yuan Institute of Medicine and Biotechnology, Zhejiang Sci-Tech University, Hangzhou 310018, China.

*Corresponding author E-mail: yaozhou@chinagene.com.

家蚕中果蝇*E*(*spl*)*m4*同源基因*BmEm4*的特征

曾凤辉[1,2]　谢宏霞[3]　聂作明[1,2]　陈　健[1,2]　吕正兵[1,2]　陈剑清[1,2]
王　丹[1,2]　刘立丽[1,2]　于　威[1,2]　盛　清[1,2]　吴祥甫[1,2]　张耀洲[1,2*]

摘要：果蝇*E*(*spl*)*m4*基因的3′非翻译区(3′UTR)包含一些高度保守的基序(如Brd box,GY box,K box和CAAC基序)。它被证明是果蝇中的一个microRNA靶基因,并在神经发生的调节中发挥重要的作用。我们从家蚕中鉴定了*E*(*spl*)*m4*的同源基因*BmEm4*,并检测了*BmEm4* mRNA和蛋白质的表达模式。在家蚕不同的发育阶段该基因的mRNA水平和蛋白表达水平之间缺少相关性,从而提高了*BmEm4*的基因表达存在转录后调控的可能性。与这个想法相一致的是,我们发现该基因3′ UTR区包含两个microRNA的预测结合位点。此外,通过免疫组化证实,在家蚕幼虫的头部*Bm*Em4蛋白表达量最高,推测*Bm*Em4也可能参与家蚕神经发生调节。细胞免疫染色实验表明,*Bm*Em4主要位于细胞质中。

1　生物化学研究所,浙江理工大学,杭州
2　生物化学研究所,生命科学学院,浙江省家蚕生物反应器与生物医药重点实验室,杭州
3　新元医药生物技术研究所,浙江理工大学,杭州

Identification and functional characterization of the *Rad23* gene of the silkworm, *Bombyx mori*

Xu HP[1] Xu YS[1*] Wang HB[1,2] He D[1] Hideki Kawasaki[2]

Abstract: Rad23 is an NER (nucleotide excision repair) protein and it plays an important role in the UPP (ubiquitin-proteasome pathway). In the present study, *BmRad23* (a homologous gene of *Rad23* from *Bombyx mori*) was cloned and designated as *BmRad23*. The ORF (open reading frame) of the *BmRad23* cDNA encoded deduced 324 amino acids with a calculated molecular mass of 36.13 kDa and an estimated pI of 4.50. The deduced amino acid sequence of the *BmRad23* cDNA revealed several indispensable domains for the function of the Rad23 protein family, such as one UbL (ubiquitin-like) region domain and two UBA (ubiquitin-associated) domains. UV irradiation and treatment with chemical DNA-damaging reagent increased the expression of *BmRad23*. The *BmRad23* gene was expressed in all the examined organs, and elevated expression was observed in testis and ovary. Northern blot and immunoblot analyses showed enhanced expression of *BmRad23* after day 3 of the wandering stage in the silk gland. From the present results it is suggested that *Bm*Rad23 functions in the UPP during the silkworm metamorphosis as well as participating in the NER when the genetic material is damaged by UV irradiation and other genotoxic stresses.

Published On: Bioscience Reports, 2010, 30(1), 19-26.

1 Institute of Sericulture and Apiculture, College of Animal Sciences, Zhejiang University, Hangzhou 310029, China.

2 Faculty of Agriculture, Utsunomiya University, 350 Mine, Utsunomiya, Tochigi 321-8505, Japan.

*Corresponding author E-mail: xuyusong@zju.edu.cn.

家蚕*Rad23*基因的鉴定和功能特征

徐和平[1] 徐豫松[1*] 王华兵[1,2] 何 达[1] Hideki Kawasaki[2]

摘要：Rad23蛋白是一种核酸切除修复(NER)蛋白，并且在蛋白泛素水解代谢通路(UPP)中扮演着重要角色。在本研究中，家蚕中与*Rad23*同源的基因被克隆并命名为*BmRad23*。*BmRad23* cDNA的开放阅读框(ORF)编码324个氨基酸，预测分子质量为36.13 kDa，预测等电点为4.50。*BmRad23*基因编码的氨基酸序列包含了Rad23蛋白家族的一些不可或缺的结构域，如一个类泛素(ubiquitin-like)结构域和两个泛素相关(ubiquitin-associated)结构域。紫外线照射和DNA损伤化学试剂处理能增加*BmRad23*的表达。*BmRad23*基因在所检测的组织中均有表达，并且在精巢和卵巢中表达较高。Northern印迹和免疫印迹分析显示，上蔟第3天后的丝腺中*BmRad23*的表达量逐渐增多。从目前的结果可以看出，*Bm*Rad23在家蚕变态期间的UPP通路中起作用，当遗传物质被紫外线照射或被其他遗传毒性物质应激损伤时，将参与核酸切除修复通路。

1 蚕蜂研究所，动物科学学院，浙江大学，杭州

2 农学部，宇都宫大学，栃木，日本

All-trans retinoic acid affects subcellular localization of a novel *Bm*NIF3l protein: functional deduce and tissue distribution of *NIF3l* gene from silkworm (*Bombyx mori*)

Chen JQ[1#] Gai QJ[1#] Lü ZB[1] Chen J[1]
Nie ZM[1] Wu XF[1,2] Zhang YZ[1*]

Abstract: A novel cDNA sequence encoding a predicted protein of 271 amino acids containing a conserved NIF3 domain was found from a pupal cDNA library of silkworm. The corresponding gene was named *BmNIF3l* (*Bombyx mori* NGG1p interacting factor 3-like). It was found by bioinformatics that *BmNIF3l* gene consisted of five exons and four introns and *Bm*NIF3l had a high degree of homology to other NIF3-like proteins, especially in the N-terminal and C-terminal regions. A His-tagged *Bm*NIF3l fusion protein with a molecular weight of approximately 33.6 kDa was expressed and purified to homogeneity. We have used the purified fusion protein to produce polyclonal antibodies against *Bm*NIF3l for histochemical analysis. Subcellular localization revealed that *Bm*NIF3l is a cytoplasmic protein that responds to all-trans retinoic acid (ATRA). Western blotting and real-time reverse transcription polymerase chain reaction showed that the expression level of *BmNIF3l* is higher in tissues undergoing differentiation. Taken together, the results suggest that *Bm*NIF3l functions in transcription.

Published On: Archives of Insect Biochemistry and Physiology, 2010, 74(4), 217-231.

1 Institute of Biochemistry, Zhejiang Sci-Tech University, Hangzhou 31008, China.
2 China and Shanghai Institute of Biochemistry and Cell Biology, Chinese Academy of Sciences, Shanghai 200031, China.
#These authors contributed equally.
*Corresponding author E-mail: yaozhou@zstu.edu.cn.

全反式维甲酸对家蚕*Bm*NIF3l蛋白亚细胞定位的影响：*BmNIF3l*基因的功能演绎以及组织分布研究

陈剑清[1#]　盖其静[1#]　吕正兵[1]　陈　健[1]
聂作明[1]　吴祥甫[1,2]　张耀洲[1*]

摘要：我们在家蚕蛹cDNA文库中发现了一条编码含有NIF3结构域蛋白的cDNA序列，该序列编码271个氨基酸。该基因被命名为*BmNIF3l*（*Bombyx mori* NGG1p interacting factor 3-like）。生物信息学分析发现，该基因由5个外显子和4个内含子组成，*Bm*NIF3l蛋白质的氨基酸序列与其他NIF3类蛋白质序列具有高度的同源性，尤其是在N末端和C末端区域。我们将*Bm*NIF3l与His标签进行融合表达，融合蛋白的分子质量约为33.6 kDa。将纯化的融合蛋白免疫兔子产生*Bm*NIF3l 的多克隆抗体用于组织化学分析。亚细胞定位实验结果发现，*Bm*NIF3l 是一种胞质蛋白并对全反式维甲酸（ATRA）有反应。通过Western blot和荧光定量PCR实验，我们发现*BmNIF3l*在正在分化的组织中表达水平较高。总之，以上结果均表明，*Bm*NIF3l 在转录过程中发挥作用。

1　生物化学研究所，浙江理工大学，杭州
2　生物化学与细胞生物学研究所，中国科学院，上海

Purification, characterization and cloning of a chymotrypsin inhibitor (CI-9) from the hemolymph of the silkworm, *Bombyx mori*

Zhao P[1,2] Xia QY[1*] Li J[1]

Hiroshi Fujii[2] Yutaka Banno[2] Xiang ZH[1]

Abstract: Hemolymph chymotrypsin inhibitor 9 (CI-9) from the hemolymph of the silkworm, *Bombyxmori*, was purified by ammonium sulfate precipitation, Butyl Toyopearl hydrophobic chromatography, gel filtration through Sephadex C-50 and chymotrypsin-sepharose 4B affinity chromatography. Checked by Native PAGE and SDS-PAGE in combination with silver staining, the final preparation appeared homogeneous. In tricine SDS-PAGE, CI-9 displayed a molecular weight of 7.5 kDa, which was determined to be 7 167 Da with the Voyager TOFMass analyser. The pI value for CI-9, revealed by 2D-PAGE (two-dimensional polyacrylamide gel electrophoresis), was 4.3. CI-9 exhibited inhibitory activity at a temperature as high as 100 ℃ and a stability against a wide range of pH (1-12). In N-terminal amino-acid analysis of CI-9, 40 amino acid residues were obtained. The C-terminal 22 amino acid residues were deduced by subsequently cloned cDNA and genomic fragments. MW and pI of CI-9 were predicted to be 7 170.98 Da and 4.61, respectively, on the website. Its low molecular weight, high stability, conserved active site and Kunitz domain showed that CI-9 is a Kunitz-type CI. The difference of sequence and pI between CI-9 and other Kunitz type Cis indicated that it is a novel chymotrypsin inhibitor.

Published On:The Protein Journal, 2007, 5(26), 349-357.

1 Key Sericultural Laboratory of Agricultural Ministry, Southwest University, Chongqing, 400716, China.

2 Institute of Genetic Resource, Kyushu University, Fukuoka, 812-8581, Japan.

*Corresponding author E-mail: xiaqy@swu.edu.cn.

家蚕血淋巴糜蛋白酶抑制剂(CI-9)的纯化、表征和克隆

赵 萍[1,2] 夏庆友[1*] 李 娟[1]
Hiroshi Fujii[2] Yutaka Banno[2] 向仲怀[1]

摘要：本实验通过硫酸铵溶液沉淀，Butyl-Toyopearl 疏水层析，Sephadex C-50 凝胶过滤和 chymotrypsin-sepharose 4B 亲和层析的方法，从家蚕血淋巴中提取、纯化血淋巴胰凝乳蛋白酶抑制剂 9（CI-9）。经 Native PAGE 和 SDS-PAGE 结合银染检测，确定 CI-9 蛋白样品为单一蛋白质。在 Tricine SDS-PAGE 结果中，CI-9 显示 7.5 kDa 的分子质量，用 Voyager TOFMass 分析仪确定分子质量为 7 167 Da。通过 2D-PAGE 显示的 CI-9 的 pI 为 4.3。CI-9 在高达 100 ℃的温度下仍然表现出抑制活性，并且在 pH 1—12 时表现出稳定性。对 CI-9 的 N–末端氨基酸序列进行分析后，获得 40 个氨基酸残基。随后以 cDNA 和基因组片段为模板克隆并推导出 C 末端 22 个氨基酸残基。网站预测 CI-9 的分子质量和 pI 分别为 7 170.98 Da 和 4.61。家蚕 CI-9 分子质量低，稳定性高，保守活性位点和 Kunitz 结构域显示其属于 Kunitz 型 CI。CI-9 和其他 Kunitz 型 CIs 之间的序列和 pI 的差异表明它是一种新型的胰凝乳蛋白酶抑制剂。

1 农业部蚕桑学重点实验室，西南大学，重庆
2 日本九州大学遗传资源研究所，九州大学，福冈

Characterization and localization of the vacuolar-type ATPase in the midgut cells of silkworm (*Bombyx mori*)

Yang HJ[1] Chen HQ[1] Chen KP[1*] Yao Q[1]
Zhao GL[1] Wu C[1] Lü P[1] Wang L[2]

Abstract: The vacuolar ATPase (V-ATPase) is a multifunctional enzyme that consists of several subunits. Subunit B is a part of the catalytic domain of the enzyme. The result of the RT-PCR suggested that the V-ATPase B subunit is a ubiquitous gene. 24 h after the larvae were infected with the *Bombyx mori* nucleopolyhedrovirus (*Bm*NPV), the expression level of the V-ATPase B subunit in the midgut of the resistant strain NB was about 3 times higher than in the susceptible strain 306, and then the expression level of the V-ATPase B subunit decreased rapidly to a very low level. This indicated that the virus may cause a lot of changes of physiological conditions in the midgut. Localization of the V-ATPase B subunit was attempted in midgut cells of *Bombyx mori* by immunohistochemistry. The immunohistochemical localization with the antibody against the B subunit revealed a positive staining in goblet cell apical membranes of *Bombyx mori* midgut cells as well as in the midgut of *Manduca sexta*. This sequence has been registered in GenBank under the accession number EU727173.

Published On: Zeitschrift Fur Naturforschung Section C-A Journal of Biosciences, 2009, 64 (11-12),899-905.

1 Institute of Life Sciences, Jiangsu University, Zhenjiang 212013, China.
2 Beijing Entry-Exit Inspection and Quarantine Bureau, Beijing 100026, China.
*Corresponding author E-mail: kpchen@ujs.edu.cn.

家蚕中肠细胞V型ATPase的表征和定位分析

杨华军[1] 陈慧卿[1] 陈克平[1*] 姚 勤[1]
赵国力[1] 吴 超[1] 吕 鹏[1] 汪 琳[2]

摘要: V 型 ATPase (V-ATPase)是由几个亚基组成的多功能酶。亚基B是酶催化结构域的一部分。RT-PCR 的结果表明 V-ATP 酶的B亚基是普遍表达的基因。幼虫感染家蚕核型多角体病毒(*Bm*NPV)24 h后,抗性株NB中肠中V-ATP酶B亚基的表达水平比敏感菌株306高出约3倍,然后V-ATP酶B亚基的表达迅速下降到非常低的水平。这表明病毒可能导致中肠许多生理条件的变化。B亚基抗体的免疫组化定位显示,与烟草天蛾中一致,V-ATP酶B亚基定位于家蚕中肠的杯状细胞顶端膜上。该序列已在GenBank注册,登录号为EU727173。

1 生命科学研究院,江苏大学,镇江
2 北京出入境检验检疫局,北京

Molecular cloning and characterization of *Bombyx mori* sterol carrier protein x/sterol carrier protein 2 (*SCPx/SCP2*) gene

Gong J Hou Y Zha XF Lu C Zhu Y Xia QY*

Abstract: Cholesterol transport is a very important process in insect. We have isolated the *Bombyx mori* sterol carrier protein x (*BmSCPx*) cDNA and sterol carrier protein 2 (*BmSCP2*) cDNA: a 1.7 kb clone encoding *SCPx*, a 3-ketoacyl CoA thiolase, and 0.6 kb clone presumably encoding *SCP2*, which is thought to be an intracellular lipid transfer protein. Interestingly, the identical gene *SCPx/SCP2* encodes the two types of transcripts by alternative splicing mechanism in *Bombyx mori*. The *SCPx* mRNA spans two exons in genome, and conceptual translation of the *SCPx* cDNA encodes a protein of 536 amino acids, which contains a thiolase domain and a SCP2 domain. Whereas the *SCP2* mRNA partly lakes the first exon, and the SCP2 is a 146 amino acids containing a SCP2 domain only. Both *Bm*SCPx and *Bm*SCP2 have a putative peroxisomal targeting signal in the C-terminal region. *Bm*SCPx shares 94 and 72% similarity to *Spodoptera littoralis* SCPx and human SCPx, respectively. RT-PCR analysis reveals that transcripts of *BmSCP2* were detected in all tissues analyzed. *BmSCPx* transcription expressed only in midgut and Malpighian tubules. However, the *BmSCPx* and *BmSCP2* express strong in midgut during the last instar larvae. The tissue-specific expression pattern of *BmSCPx* and *BmSCP2* is consistent with a role for these proteins in cholesterol metabolism. The results suggest that SCPx/SCP2 may play a key role in sterol absorption and intracellular carrier in silkworm.

Published On: DNA Sequence, 2006, 17(5), 326-333.

Key Sericultural Laboratory of Agricultural Ministry, Southwest University, Chongqing 400716, China.

*Corresponding author E-mail: xiaqy@swu.edu.cn.

家蚕甾醇转运蛋白*SCPx/SCP2*基因的克隆与分析

龚 竞 侯 勇 查幸福 鲁 成 朱 勇 夏庆友*

摘要:胆固醇转运是昆虫体内重要的生物学过程,本文克隆了家蚕体内的重要胆固醇转运蛋白SCPx/SCP2。编码该蛋白的基因是一个融合基因,在cDNA中一个1.7 kb片段编码的是一个3-酮脂酶辅酶A硫解酶蛋白SCPx,0.6 kb片段编码的是一个细胞内载脂蛋白SCP2。有趣的是,这两个基因是同一个基因通过选择性剪切而获得的。*SCPx*的mRNA跨越基因组上两个外显子,翻译成的蛋白质由536个氨基酸残基组成,包含一个硫解酶结构域和一个SCP2结构域。*SCP2*的mRNA较*SCPx*缺少第一个外显子,由146个氨基酸残基组成,其中包含SCP2结构域。无论SCPx还是SCP2,其C端均有一个过氧化物定位信号。家蚕的SCPx蛋白与棉铃虫和人的该同源蛋白比对,相似性分别为94%和72%。RT-PCR分析表明,家蚕*SCP2*在各个组织中均有表达,而*SCPx*只在家蚕中肠和马氏管中表达。并且无论是*SCPx*还是*SCP2*,都是在家蚕幼虫的末龄期表达量最高。这一组织时期特异的表达模式与该基因在胆固醇代谢中的作用是一致的。因此,这些结果暗示该蛋白质在家蚕胆固醇的吸收和细胞内转运过程中发挥了重要作用。

农业部蚕桑学重点实验室,西南大学,重庆

Expression and functional analysis of the cellular retinoic acid binding protein from silkworm pupae (*Bombyx mori*)

Wang XJ[1] Chen J[1] Lü ZB[1] Nie ZM[1] Wang D[1]
Shen HD[1] Wang XD[1] Wu XF[1,2] Zhang YZ [1*]

Abstract: Cellular retinoic acid binding protein (CRABP) is a member of intracellular lipid-binding protein (iLBP), and closely associated with retinoic acid (RA) activity. We have cloned the *CRABP* gene from silkworm pupae and studied the interaction between *Bombyx mori* CRABP (*Bm*CRABP) and all-trans retinoic acid (atRA). The MTT assay data indicated that when *Bm*CRABP is overexpressed in *Bm5* cells, the cells dramatically resisted to atRA-induced growth inhibition. Conversely, the cells were sensitive to atRA treatment upon knocking down the *BmCRABP* expression. Subcellular localization revealed that *Bm*CRABP is a cytoplasm protein, even when treated with atRA, the CRABP still remained in the cytoplasm. These data demonstrated that the function of *Bm*CRABP have an effect on the physiological function of atRA.

Published On: Journal of Cellular Biochemistry, 2007, 102(4), 970-979.

1 Institute of Biochemistry, College of Life Sciences, Zhejiang Sci-Tech University, Hangzhou 310018, China.
2 Institute of Biochemistry and Cell Biology, Chinese Academy of Sciences, Shanghai 200031, China.
*Corresponding author E-mail: yaozhou@chinagene.com.

蚕蛹细胞维甲酸结合蛋白的表达和功能分析

王学健[1] 陈 健[1] 吕正兵[1] 聂作明[1] 王 丹[1]
沈红丹[1] 王雪冬[1] 吴祥甫[1,2] 张耀洲[1*]

摘要：细胞维甲酸结合蛋白(CRABP)是细胞内酯结合蛋白(iLBP)的成员之一，与维甲酸(RA)活性密切相关。我们从蚕蛹中克隆了*CRABP*基因，并研究了家蚕CRABP(*Bm*CRABP)与全反式维甲酸(atRA)的相互作用。MTT测定数据表明，当*BmCRABP*在*Bm5*细胞中过表达时，细胞显著抵抗atRA诱导的生长抑制。相反，当下调表达*BmCRABP*表达时，细胞对atRA处理敏感。亚细胞定位显示*Bm*CRABP是细胞质蛋白，即使用atRA处理，CRABP仍然存在于细胞质中。这些数据表明*Bm*CRABP影响atRA的生理功能。

1 生命科学学院，浙江理工大学，杭州
2 生物化学与细胞生物学研究所，中国科学院，上海

Comparisons of contact chemoreception and food acceptance by larvae of polyphagous *Helicoverpa armigera* and oligophagous *Bombyx mori*

Zhang HJ[1,2] Cécile P. Faucher[2] Alisha Anderson[2] Amalia Z. Berna[2]
Stephen Trowell[2] Chen QM[1] Xia QY[1*] Sylwester Chyb[2*]

Abstract: We compared food choice and the initial response to deterrent treated diet between fifth instars of *Helicoverpa armigera*, a polyphagous generalist pest, and *Bombyx mori*, an oligophagous specialist beneficial. *Bombyx mori* was more behaviorally sensitive to salicin than to caffeine. The relative sensitivities were reversed for *H. armigera*, which was tolerant to the highest levels of salicin found in natural sources but sensitive to caffeine. A single gustatory receptor neuron(GRN) in the medial styloconic sensillum of *B. mori* was highly sensitive to salicin and caffeine. The styloconic sensilla of *H.armigera* did not respond consistently to either of the bitter compounds. Phagostimulants were also tested. *Myo*-inositol and sucrose were detected specifically by two GRNs located in *B. mori* lateral styloconic sensillum, whereas, in *H. armig*era,sucrose was sensed by a GRN in the lateral sensillum, and *myo*-inositol by a GRN in the medial sensillum. *Myo*-inositol responsiveness in both species occurred at or below 10^{-3} mmoL/L,which is far below the naturally occurring concentration of 1 mmoL/L in plants. Larval responses to specific plant secondary compounds appear to have complex determinants that may include host range, metabolic capacity, and gustatory repertoire.

Published On: Journal of Chemical Ecology, 2013, 39(8), 1070-1080.

1 State Key Laboratory of Silkworm Genome Biology, Southwest University, Chongqing 400716, China.
2 CSIRO Ecosystem Sciences, Canberra, ACT 2601, Australia.
*Corresponding author E-mail: xiaqy@swu.edu.cn; sylwester.chyb@ymail.com.

杂食性棉铃虫和寡食性家蚕幼虫接触化学感应和食物接受的比较

张辉洁[1,2] Cécile P.Faucher[2] Alisha Anderson[2] Amalia Z.Berna[2]
Stephen Trowell[2] 陈全梅[1] 夏庆友[1*] Sylwester Chyb[2*]

摘要：我们比较了五龄杂食性棉铃虫和寡食性家蚕的食物选择和对含有取食忌避物的食物的最初反应。在行为表现上家蚕对水杨苷比对咖啡因更为敏感，而棉铃虫刚好与家蚕相反，它能够对自然界中最高浓度的水杨苷耐受，却对咖啡因敏感。家蚕的中部栓锥感受器中有一个味觉受体神经元(GRN)对水杨苷和咖啡因高度敏感，而棉铃虫的栓锥感受器对这两种苦味的化合物都没有连续性反应。诱食剂也被用于测试，家蚕体内位于侧面栓锥感受器的两个味觉受体神经元负责特异性探测肌-肌醇和蔗糖，而在棉铃虫体内一个位于侧面感受器的味觉受体神经元用于探测蔗糖，而位于中部感受器的另一个味觉受体神经元负责探测肌醇。当浓度在 10^{-3} mmoL/L 及其以下时，家蚕和棉铃虫均能对肌醇产生反应，这远低于自然界中肌醇在植物中的浓度 1 mmoL/L，幼虫对于植物中特定的二级化合物的反应具有复杂的决定因素，可能包括宿主范围、代谢能力和味觉能力。

1 家蚕基因组生物学国家重点实验室，西南大学，重庆
2 澳大利亚联邦科学与工业研究组织系统生态所，堪培拉，澳大利亚

Microbial communities in the larval midgut of laboratory and field populations of cotton bollworm (*Helicoverpa armigera*)

Xiang H[1] Wei GF[2] Jia SH[1] Huang JH[1]
Miao XX[1] Zhou ZH[1] Zhao LP[2] Huang YP[1*]

Abstract: We compared the bacterial communities in the larval midgut of field and laboratory populations of a polyphagous pest, the cotton bollworm (*Helicoverpa armigera*), using denaturing gradient gel electrophoresis (DGGE) of amplified 16S rDNA sequences and 16S library sequence analysis. DGGE profiles and 16S rDNA library sequence analysis indicated similar patterns of midgut microbial community structure and diversity: specific bacterial types existed in both populations, and a more diverse microbial community was observed in caterpillars obtained from the field. The laboratory population harbored a rather simple gut microflora consisting mostly of phylotypes belonging to *Enterococcus* (84%). For the field population, phylotypes belonging to *Enterococcus* (28%) and *Lactococcus* (11%), as well as *Flavobacterium* (10%), *Acinetobacter* (19%), and *Stenotrophomonas* (10%) were dominant members. These results provided the first comprehensive description of the microbial diversity of the midgut of the important pest cotton bollworm and suggested that the environment and food supply might influence the diversity of the gut bacterial community.

Published On: Canadian Journal of Microbiology, 2006, 52(11), 1085-1192.

1 Shanghai Institute of Plant Physiology and Ecology, Shanghai Institute Biological Sciences, The Chinese Academy of Sciences, Shanghai 200032, China.
2 Department of Biotechnology, Shanghai Jiao Tong University, Shanghai 200240, China.
*Corresponding author E-mail: yongping@sippe.ac.cn.

棉铃虫实验室和田间种群幼虫中肠微生物群落研究

相　辉[1]　魏桂芳[2]　贾世海[1]　黄健华[1]
苗雪霞[1]　周志华[1]　赵立平[2]　黄勇平[1*]

摘要：我们使用变性梯度凝胶电泳（DGGE）对扩增的16S rDNA序列和16S文库序列进行了分析，比较了一种广食性害虫——棉铃虫（*Helicoverpa armigera*）的田间和实验室种群的幼虫中肠细菌群落。DGGE图谱和16S rDNA文库序列分析表明，这两个群体的中肠微生物群落结构和多样性具有相似的模式：两种群体中均存在特异性细菌类型，并且从田间获得的棉铃虫幼虫中观察到更多样化的微生物群落。实验室种群中含有一个相对简单的肠道微生物群落，主要由肠球菌属（84%）构成。对于田间种群，肠球菌属（28%）和乳杆菌属（11%）以及黄杆菌属（10%），不动杆菌属（19%）和嗜麦芽寡养单胞菌属（10%）为其主要构成。这些结果首次全面描述了棉铃虫的中肠微生物多样性，并提示环境和食物供应可能影响肠道细菌群落的多样性。

1　上海植物生理生态研究所，上海生命科学研究院，中国科学院，上海
2　生物科学与技术系，上海交通大学，上海

A novel method for isolation of membrane proteins: a baculovirus surface display system

Zhang YZ[1*] Lü ZB[1] Chen J[1] Chen Q[1] Quan YP[1] Kong LY[1] Zhang HH[1]
Li S[1] Zheng QL[1] Chen JQ[1] Nie ZM[1] Wang J[1] Jin YF[2] Wu XF[1]

Abstract: We have developed a novel baculovirus surface display (BVSD) system for the isolation of membrane proteins. We expressed a reporter gene that encoded hemagglutinin gene fused in frame with the signal peptide and transmembrane domain of the baculovirus gp64 protein, which is displayed on the surface of *Bm*NPV virions. The expression of this fusion protein on the virion envelope allowed us to develop two methods for isolating membrane proteins. In the first method, we isolated proteins directly from the envelope of budding *Bm*NPV virions. In the second method, we isolated proteins from cellular membranes that had disintegrated due to viral egress. We isolated 6 756 proteins. Of these, 1 883 have sequence similarities to membrane proteins and 1 550 proteins are homologous to known membrane proteins. This study indicates that membrane proteins can be effectively isolated using our BVSD system. Using an analogous method, membrane proteins can be isolated from other eukaryotic organisms, including human beings, by employing a host cell-specific budding virus.

Published On: Proteomics, 2008, 8(20), 4178-4185.

1 The Key Laboratory of Silkworm Bioreactor and Biomedicine, Zhejiang Province; Institute of Biochemistry, College of Life Sciences, Zhejiang Sci-Tech University, Hangzhou 310018, China.

2 College of Life Sciences, Zhejiang University, Hangzhou 310018, China.

*Corresponding author E-mail: yaozhou@chinagene.com.

一种分离膜蛋白的新方法：杆状病毒表面展示系统

张耀洲[1*] 吕正兵[1] 陈 健[1] 陈 琴[1] 全滟平[1] 孔令印[1] 张海花[1]
李 司[1] 郑青亮[1] 陈剑清[1] 聂作明[1] 王 江[1] 金勇丰[2] 吴祥甫[1]

摘要：我们研发了一种用于分离膜蛋白的新型杆状病毒表面展示(BVSD)系统。将编码血凝素基因的报告基因与杆状病毒gp64蛋白的信号肽和跨膜结构域融合表达，并成功显示在*Bm*NPV病毒粒子囊膜表面。这种将融合蛋白表达在病毒粒子囊膜表面的技术使我们能够开发出两种分离膜蛋白的方法。一种方法是，直接从出芽的*Bm*NPV病毒粒子的包膜中分离蛋白质；第二种方法是，从由于病毒流出而破裂的宿主细胞膜中分离出蛋白质。我们共分离出6 756个蛋白质。其中，1 883个与膜蛋白具有序列相似性，1 550个与已知的膜蛋白具有同源性。本研究结果表明，使用BVSD系统可以有效分离膜蛋白。通过类似的方法可以使用宿主细胞特异性出芽囊膜病毒从其他真核生物(包括人)中分离相应的膜蛋白。

1 家蚕生物反应器与生物医药重点实验室，杭州；生物化学研究所，生命科学学院，浙江理工大学，杭州
2 生命科学学院，浙江大学，杭州

Analysis of similarity/dissimilarity of DNA sequences based on a class of 2D graphical representation

Yao YH[1*] Dai Q[2] Nan XY[1] He PA[3]

Nie ZM[1] Zhou SP[3] Zhang YZ[1]

Abstract: On the basis of a class of 2D graphical representations of DNA sequences, sensitivity analysis has been performed, showing the high-capability of the proposed representations to take into account small modifications of the DNA sequences. And sensitivity analysis also indicates that the absolute differences of the leading eigenvalues of the L/L matrices associated with DNA increase with the increase of the number of the base mutations. Besides, we conclude that the similarity analysis method based on the correlation angles can better eliminate the effects of the lengths of DNA sequences if compared with the method using the Euclidean distances. As application, the examination of similarities / dissimilarities among the coding sequences of the first exon of *β-globin* gene of different species has been performed by our method, and the reasonable results verify the validity of our method.

Published On: Journal of Computational Chemistry, 2008, 29(10), 1632-1639.

1 College of Life Sciences, Zhejiang Sci-Tech University, Hangzhou 310018, China.

2 Deparment of Applied Mathematics, Dalian University of Technology, Dalian 116024, China.

3 College of Science, Zhejiang Sci-Tech University, Hangzhou 310018, China.

*Corresponding author E-mail: yaoyuhua2288@163.com.

基于一类二维图形表示的 DNA序列相似性/相异性分析

姚玉华[1*] 代 琦[2] 南旭莹[1] 贺平安[3]
聂作明[1] 周颂平[3] 张耀洲[1]

摘要：我们在一类DNA序列2D图展示的基础上进行了敏感性分析，经过实验证明，所提出的表示法能够考虑到DNA序列的微小变化。敏感性分析还表明，与DNA相关的L/L矩阵的首特征值的绝对差值随着碱基突变数目的增加而增加。此外，我们得出结论，如果与使用欧氏距离的方法相比，基于相关角的相似性分析方法可以更好地消除DNA序列长度带来的影响。对不同物种的β球蛋白基因第一外显子的编码序列之间的相似/相异性进行了测试，合理的结果验证了该方法的有效性。

1 生命科学学院，浙江理工大学，杭州
2 应用数学系，大连理工大学，大连
3 理学院，浙江理工大学，杭州

Analysis of similarity/dissimilarity of protein sequences

Yao YH[1*] Dai Q[2] Li C[3]
He PA[1] Nan XY[1] Zhang YZ[1]

Abstract: On the basis of a selected pair of physicochemical properties of amino acids, we introduce a dynamic 2D graphical representation of protein sequences. Then, we introduce and compare two numerical characterizations of protein graphs as descriptors to analyze and compare the nine ND5 proteins. The approach is simple, convenient, and fast.

Published On: Proteins, 2008, 73(4), 864-871.

1 College of Life Sciences, Zhejiang Sci-Tech University, Hangzhou 310018, China.
2 Department of Applied Mathematics, Dalian University of Technology, Dalian 116024, China.
3 Department of Mathematics, Bohai University, Jinzhou 121000, China.
*Corresponding author E-mail: yaoyuhua2288@163.com.

蛋白质序列相似性/相异性分析

姚玉华[1*]　代　琦[2]　李　春[3]
贺平安[1]　南旭莹[1]　张耀洲[1]

摘要：针对选定的一对氨基酸的理化性质，我们提出了一种蛋白质序列动态的2D图形表示。然后，我们引入和比较了两个数值特征作为描述符来分析9种NDS蛋白。该方法简单、方便且快速。

1　生命科学学院，浙江理工大学，杭州
2　应用数学系，大连理工大学，大连
3　数学系，渤海大学，锦州

Effect of extracts of mulberry leaves processed differently on the activity of α-glucosidase

Wang LL Zhou ZY*

Abstract: Mulberry, a deciduous perennial woody plant, belongs to the genus Morus of the family Moraceae. It is planted widely in China for rearing silkworms and its leaf is also used as herbal medicine by Chinese herbalists. Being the only plant leaf discovered so far containing 1-deoxynojirimycin(1-DNJ), a potent inhibitor of α-glucosidase, mulberry leaf is effective in lowering blood glucose, and thus can be used in the prevention and treatment of diabetes. Furthermore, mulberry leaf is also effective in preventing and treating some other diseases, as recorded in some Chinese famous ancient medical books. In this study, extracts of mulberry leaves processed differently including fresh, freeze-dried and natural dried mulberry leaves as well as mulberry tea, were used to study their efficiency in the inhibition of α-glucosidase in vitro. Results showed that (i) both fresh and freeze-dried mulberry leaves inhibited α-glucosidase significantly and were more effective than those natural dried; and (ii) mulberry tea showed the highest inhibition efficiency due to the addition of other medicinal plants during its processing. With mulberry leaf, whose high efficiency in the inhibition of α-glucosidase was shown in this study, there is a good prospect for the development of natural, nontoxic and side effect free plant foods to be used in lowering blood glucose.

Published On: Journal of Food Agriculture & Environment, 2008, 6(3-4), 86-89.

College of Life Sciences, Chongqing Normal University, Chongqing 400047, China.
*Corresponding author E-mail: zyzhou@cqnu.edu.cn.

不同加工方法的桑叶提取物对α-糖苷酶活性的影响

王林玲　周泽扬*

摘要：桑叶是一种多年生的落叶木本植物，属于桑科中的桑属植物。桑树在中国被广泛种植，桑叶用来养蚕也可作为中医入药。桑叶也是迄今为止唯一发现的含有1-脱氧野尻霉素（1-DNJ）的植物叶片，1-DNJ是一种强力的α-葡萄糖苷酶抑制剂。桑叶对降低血糖有效，因此可用于预防和治疗糖尿病。此外，根据中国古代著名医书的记载，桑叶对于预防和治疗一些其他疾病也是有效的。在本研究中，用经过不同处理的桑叶提取物，例如，新鲜的桑叶、冷冻干燥的桑叶和天然干燥的桑叶以及桑茶，来研究其在体外抑制α-葡萄糖苷酶的效率。结果显示：(i)新鲜的和冷冻干燥的桑叶比天然干燥的桑叶有着更显著的抑制α-葡萄糖苷酶的效果；(ii)桑茶由于在其加工过程中加入其他药用植物，显示出最高的抑制效率。本研究表明，桑叶具有高效抑制α-葡萄糖苷酶的作用，可以用于生产天然、无毒、无副作用的降血糖植物性食品并具有很好的发展前景。

生命科学学院，重庆师范大学，重庆

Bactrocerin-1: a novel inducible antimicrobial peptide from pupae of oriental fruit fly *Bactrocera dorsalis* Hendel

Dang XL[1,2] Tian JH[3] Yang WY[4] Wang WX[2] Jun Ishibashi[5] Ai Asaoka[5]

Yi HY[4] Li YF[2] Cao Y[4] Minoru Yamakawa[5,6] Wen SY[2*]

Abstract: A novel antimicrobial peptide, Bactrocerin-1, was purified and characterized from an immunized dipteran insect, Bactrocera dorsalis. Bactrocerin-1 has 20 amino acid residues with a mass of 2 325.95 Da. The amino acid sequence of Bactrocerin-1 showed very high similarity to the active fragment [46V-65S-NH(2)] of Coleoptericin A. The composition of amino acid residues revealed that Bactrocerin-1 is a hydrophobic, positively charged, and Lys / Ile / Gly-rich peptide. Minimal growth inhibition concentration (MIC) measurements for synthesized Bactrocerin-1 showed a very broad spectrum of anti-microbial activity against Gram-positive bacteria, Gram-negative bacteria, and fungi. Bactrocerin-1 did not show hemolytic activity toward mouse red blood cells even at a concentration of 50 μmol/L. Analysis of the Helical-wheel projection and the CD spectrum suggested that Bactrocerin-1 contains the amphipathic alpha-helix.

Published On: Archives of Insect Biochemistry and Physiology, 2009, 71(3), 117-129.

1 Zhejiang Institute of Subtropical Crops, Wenzhou, China.

2 Department of Entomology, South China Agricultural University, Guangzhou 51000, China.

3 Department of Material Science and Engineering, Jinan University, Guangzhou 51000, China.

4 College of Animal Science, South China Agricultural University, Guangzhou 51000, China.

5 Innate Immunity Research Unit, National Institute of Agrobiological Sciences, Ibaraki, Japan.

6 Graduate School of Life and Environmental Science, University of Tsukuba, Ibaraki, Japan.

*Corresponding author E-mail: shywen@scau.edu.cn.

一种来自桔小实蝇蛹的新诱导型抗菌肽

党向利[1,2]　田金环[3]　杨婉莹[4]　王文献[2]　Jun Ishibashi[5]　Ai Asaoka[5]
易辉玉[4]　李怡峰[2]　曹　阳[4]　Minoru Yamakawa[5,6]　温硕洋[2*]

摘要：本研究从免疫处理过的对双翅目昆虫桔小实蝇中纯化和表征了一种新型抗菌肽 Bactrocerin-1。Bactrocerin-1 包含 20 个氨基酸残基，分子质量为 2 325.95 Da。Bactrocerin-1 的氨基酸序列与鞘磷脂蛋白 A 的活性片段［46V-65S-NH（2）］显示出非常高的相似性。氨基酸残基的组成表明，Bactrocerin-1 是疏水性带正电荷的 Lys / Ile /Gly-rich。合成的 Bactrocerin-1 的最低生长抑制浓度（MIC）测量结果显示，其对革兰氏阳性菌、革兰氏阴性菌和真菌有非常广谱的抗微生物活性。即使浓度为 50 μmol/L，Bactrocerin-1 对小鼠红细胞也没有溶血活性。螺旋轮投影和 CD 光谱的分析表明，Bactrocerin-1 含有两亲性α-螺旋结构。

1　浙江省亚热带作物研究所，温州，中国
2　昆虫学系，华南农业大学，广州
3　材料科学与工程系，暨南大学，广州
4　动物科学学院，华南农业大学，广州
5　先天免疫研究组，国立农业生物科学研究所，茨城，日本
6　生命与环境科学研究所，筑波大学，茨城，日本

Molecular and functional characterization of a c-type lysozyme from the Asian corn borer, *Ostrinia furnacalis*

Wang WX[1] Wang YP[2,7] Deng XJ[3] Dang XL[4] Tian JH[5] Yi HY[1] Li YF[1]
He XF[1] Cao Y[3] Xia QY[6] Lai R[2] Wen SY[1*]

Abstract: Some lepidopteran lysozymes have been reported to display activity against Gram-positive and Gram-negative bacteria, in contrast to most lysozymes that are active only against Gram-positive bacteria. OstrinLysC, a c-type lysozyme, was purified from the Asian corn borer, *Ostrinia furnacalis* Guenée (Lepidoptera: Pyralidae), and shows activity against Gram-positive and Gram-negative bacteria. The NH_2-terminal amino acid sequence was determined by Edman degradation and used in a homology cloning strategy. The gene coding for *OstrinLysC* contains three exons and two introns. The expression profile of the *OstrinlysC* gene was examined by quantitative real-time PCR. Following injection of the larvae with bacteria, the *OstrinlysC* gene is strongly up-regulated in immune tissues. Transcripts were also detected in gut tissue. After feeding the larvae with bacteria, *OstrinlysC* transcripts increased in immune tissues. A very low level of transcript abundance was also detected in gut tissue. These results suggested that the *OstrinlysC* gene is involved in immune responses. The three dimensional structure of *OstrinLysC* was predicted. Based on comparison of the 3-D structure of OstrinLysC with that of silkworm lysozyme and chicken lysozyme, we hypothesize that the positive charge-rich surface and the short loop-2, which is close to the cluster of hydrophobic residues, may play important roles in the interaction with the outer membrane of Gram-negative bacterial cell walls.

Published On: Journal of Insect Science, 2009, 9.

1 Department of Entomology, College of Natural Resource and Environment, South China Agricultural University, Guangzhou 510642, China.
2 Kunming Institute of Zoology, Chinese Academy of Sciences. Kunming 650223, China.
3 Department of Sericulture Science, College of Animal Science, South China Agricultural University, Guangzhou 510642, China.
4 Zhejiang Institute of Subtropical Crops, Wenzhou 325005, China.
5 Department of Material and Engineering, Jinan University, Guangzhou 510632, China.
6 Key Sericultural Laboratory of Agriculture Ministry, College of Biotechnology, Southwest University, Chongqing 400716, China.
7 Graduate School of Chinese Academy Sciences, Beijing 100091, China.
*Corresponding author E-mail: shywen@scau.edu.cn.

一种亚洲玉米螟(*Ostrinia furnacalis*)溶菌酶的分子特征及其功能表征

王文献[1] 王一平[2,7] 邓小娟[3] 党向利[4] 田金环[5] 易辉玉[1] 李怡峰[1]
何晓芳[1] 曹 阳[3] 夏庆友[6] 赖 韧[2] 温硕洋[1*]

摘要:与大多数仅对革兰氏阳性细菌有活性的溶菌酶不同,一些鳞翅目溶菌酶对革兰氏阳性菌和革兰氏阴性菌都有活性。OstrinLysC是一种从鳞翅目昆虫亚洲玉米螟中纯化得到的c型溶菌酶,并显示出对革兰氏阳性菌和革兰氏阴性菌的抑制活性。通过Edman降解法测定NH_2末端的氨基酸序列,并用于同源重组策略。编码*OstrinLysC*的基因包含3个外显子和2个内含子。通过实时定量PCR分析了*OstrinlysC*基因的表达谱。在用细菌注射幼虫后,*OstrinlysC*基因在免疫组织中被显著上调,中肠也检测到该基因的表达。用细菌添食幼虫后,*OstrinlysC*转录本在免疫组织中上调,中肠组织中检测到非常低水平的转录本丰度。这些结果表明*OstrinlysC*基因参与了免疫应答。基于预测的OstrinLysC的3D结构,与蚕溶菌酶和鸡溶菌酶的3D结构比较,我们推测在与革兰氏阴性细菌细胞壁的外膜相互作用中,正电荷丰富的表面和靠近疏水残基簇的loop-2结构可能起重要作用。

1 昆虫学系,自然资源与环境学院,华南农业大学,广州
2 昆明动物研究所,中国科学院,昆明
3 蚕桑科学系,动物科学学院,华南农业大学,广州
4 浙江省亚热带作物研究所,温州
5 材料与工程系,暨南大学,广州
6 农业部蚕桑学重点实验室,蚕桑与生物技术学院,西南大学,重庆
7 中国科学院研究生院,北京

附图

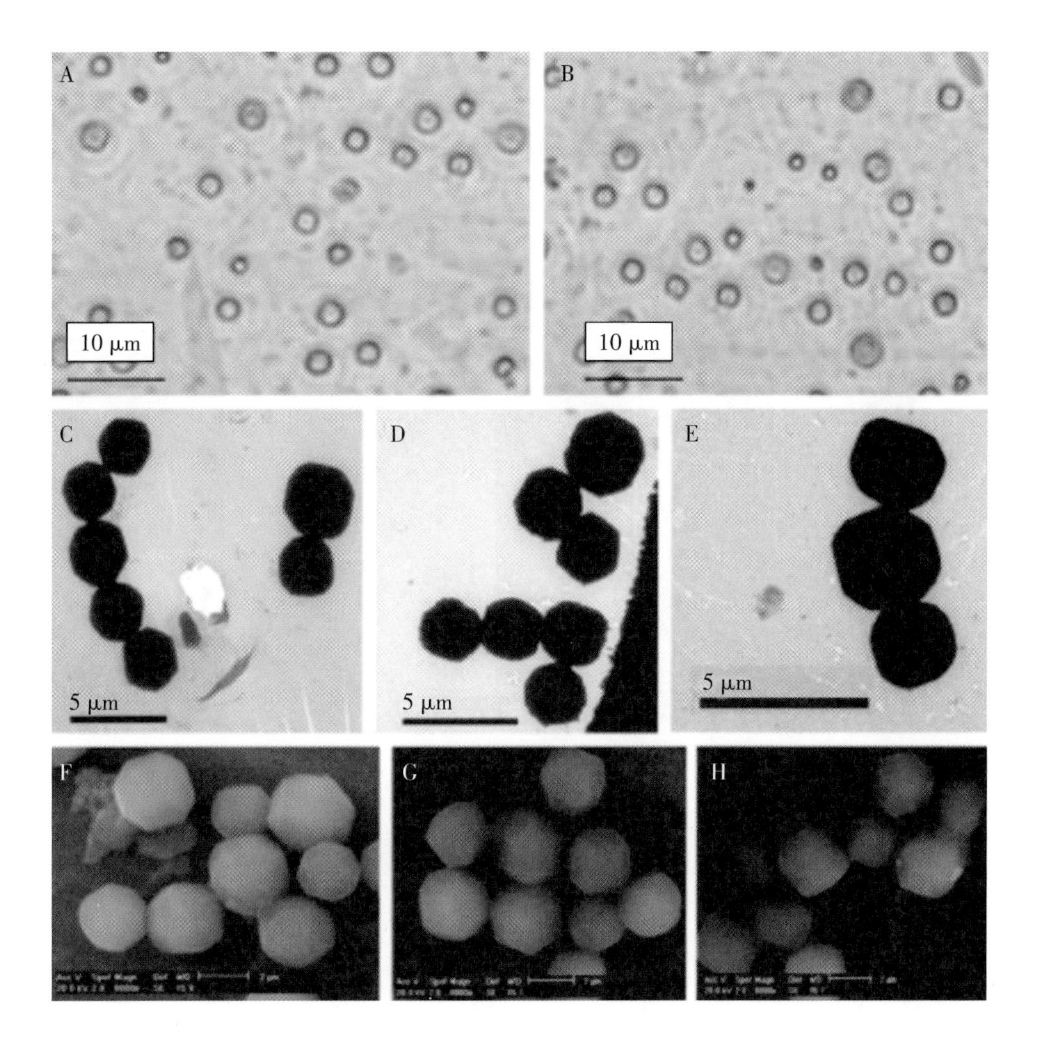

Figure 76. Photos of polyhedra after infection of dual expression bacmid baculoviruses. The larval hemolymph was checked for the presence of polyhedra under microscope and electronic microscope at 96 h post infection. a polyhedra under microscope. The recombinant bacmid baculoviruses (rBacmid/*Bm*NPV/SOD) DNA was transfected into *BmN* cells using FuGENE 6, and the recombinant viral was injected into the silkworm larvae by subcutaneous injection; b polyhedra in larvae hemolymph under microscope. The larvae was fed orally with hemolymph which was collected from the first batch of infected larvae; c polyhedra in hemolymph of wild NPV virus infected larvae under transmission microscope; d polyhedra under transmission microscope. The larvae was injected with recombinant bacmid baculoviruses as a; e polyhedra under transmission microscope. The hemolymph was collected as b; f polyhedra in hemolymph of wild NPV virus infected larvae under scanning electronic microscope; g polyhedra under scanning electronic microscope. The larvae was injected with recombinant bacmid baculoviruses as a; h polyhedral under scanning electronic microscope. The hemolymph was collected as b.（选自：*Improvement of recombinant baculovirus infection efficiency to express manganese superoxide dismutase in silkworm larvae through dual promoters of Pph and Pp10*，参见本书第472页。）

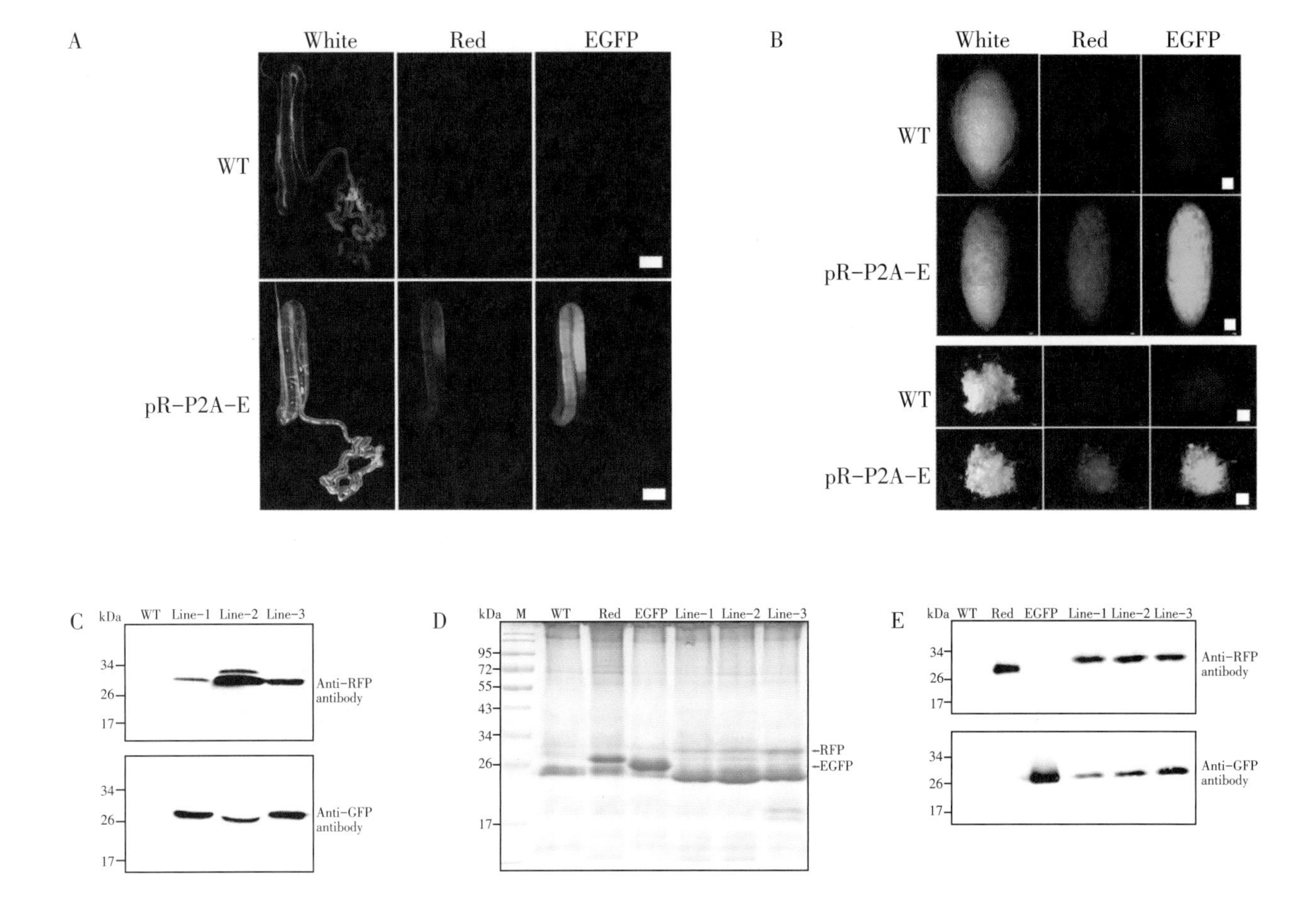

Figure 77. Expression of the sp*Ds*Red-P2A-spEGFP fusion gene in the transgenic silkworm. (A, B) Fluorescent images for the middle silk grand and cocoons of the transgenic *spDsRed-P2A-spEGFP* silkworm. WT and pR-P2A-E were the wild type and positive transgenic silkworm. (C) Protein analysis of samples from the middle silk grand with anti-RFP, anti-GFP antibodies. (D, E) Protein analysis for the samples from the cocoons by the coomassie brilliant blue (CCB) staining and western blot with anti-RFP, anti-GFP antibodies. Scale bar, 2 mm.（选自：*2A self-cleaving peptide-based multi-gene expression system in the silkworm Bombyx mori*，参见本书第502页。）

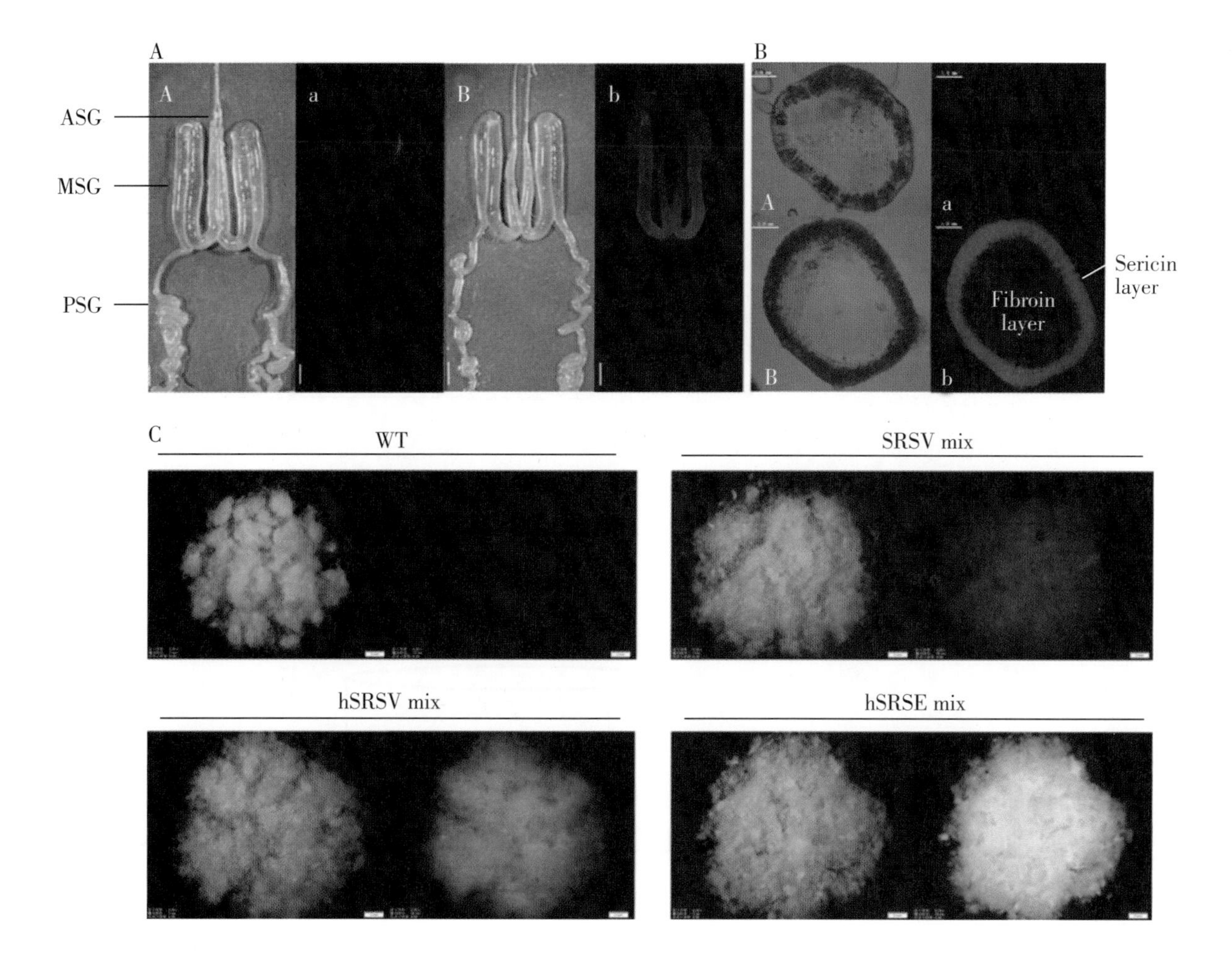

Figure 78. Detection of expression of *Ds*Red in transgenic silkworms. a Silk glands from wild-type silkworm and hSRSE transgenic silkworm on day 6 of the fifth instar were dissected and illuminated under white light (a, b) or *Ds*Red excitation wavelength light (a, b). The scale bar represents 0.5 cm. b The sectioned middle silk glands from wild-type silkworms and hSRSE transgenic silkworms on day 6 of the fifth instar were illuminated under white light (a, b) or *Ds*Red excitation wavelength light (a, b), Scale bar represents 1 mm. c Comparison of *Ds*Red fluorescence in cocoons of different transgenic silkworms. Randomly selected cocoons from WT and three lines of each SRSV, hSRSV, and hSRSE silkworms were cut into powder and photographed under white light or *Ds*Red excitation wavelength light. Scale bar represents 2 mm.（选自：*An optimized sericin-1 expression system for mass-producing recombinant proteins in the middle silk glands of transgenic silkworms*，参见本书第512页。）

Figure 79. Preparation of thin slices of cocoon shell for cell culture. Transgenic hFGF1 cocoons (A) were torn into small thin layers (B) and washed five times using double distilled water. The small pieces of cocoons were air dried and punched into 6 mm thick slices (C) The scale bar represents 1 cm. (D) Diagrammatic sketch for the culture of NIH/3T3 directly by raw silk thin slice.（选自：*Advanced silk material spun by a transgenic silkworm promotes cell proliferation for biomedical application*，参见本书第516页。）

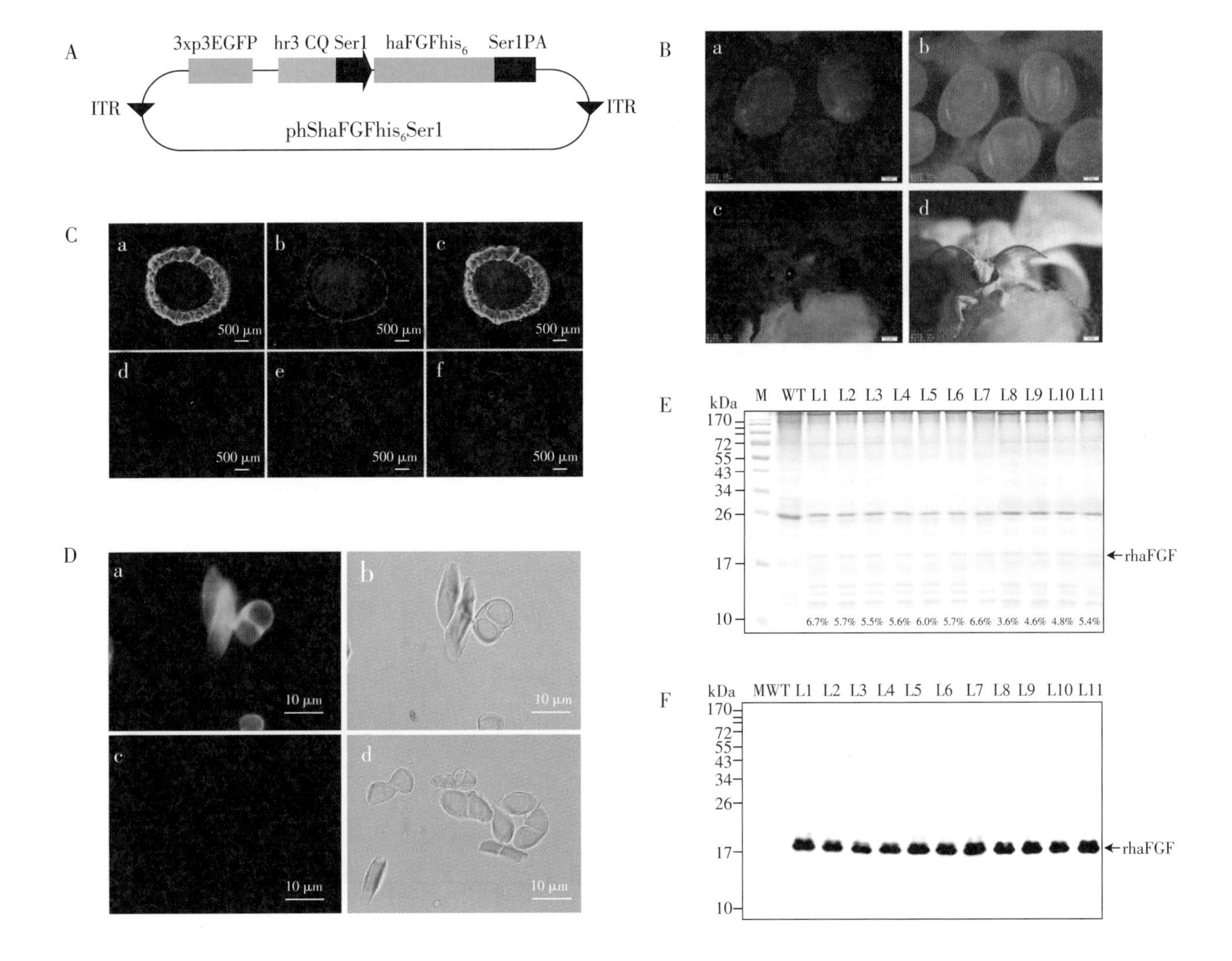

Figure 80. Generation of *ha*FGF transgenic silkworm and the expression analysis of recombinant proteins. (A) Structure map of the transgenic vector. (B) Fluorescent images of the transgenic silkworm in egg (a, b) and moth (c, d) stages, the scale bar represents 2 mm. (C) Immunohistochemical analysis of MSG cross sections of transgenic silkworm (a-c) and non-transgenic silkworm (d-f). The green fluorescence represents the immunoblot signals of *ha*FGF proteins; the DAPI stained by blue fluorescence represents the cell nucleus. Scale bar represents 500 μm. (D) Immunohistochemical analysis of raw silk cross sections of transgenic silkworm (a, b) and non-transgenic silkworm (c, d). Scale bar represents 10 μm. (E) SDS-PAGE analysis of the cocoon proteins from ten different transgenic lines; the percentages represent the *ha*FGF content of each line in the total cocoon extracts. (F) Western blot analysis of the *ha*FGF in cocoon extracts from ten different transgenic lines.（选自：*Large-scale production of bioactive recombinant human acidic fibroblast growth factor in transgenic silkworm cocoons*，参见本书第524页。）

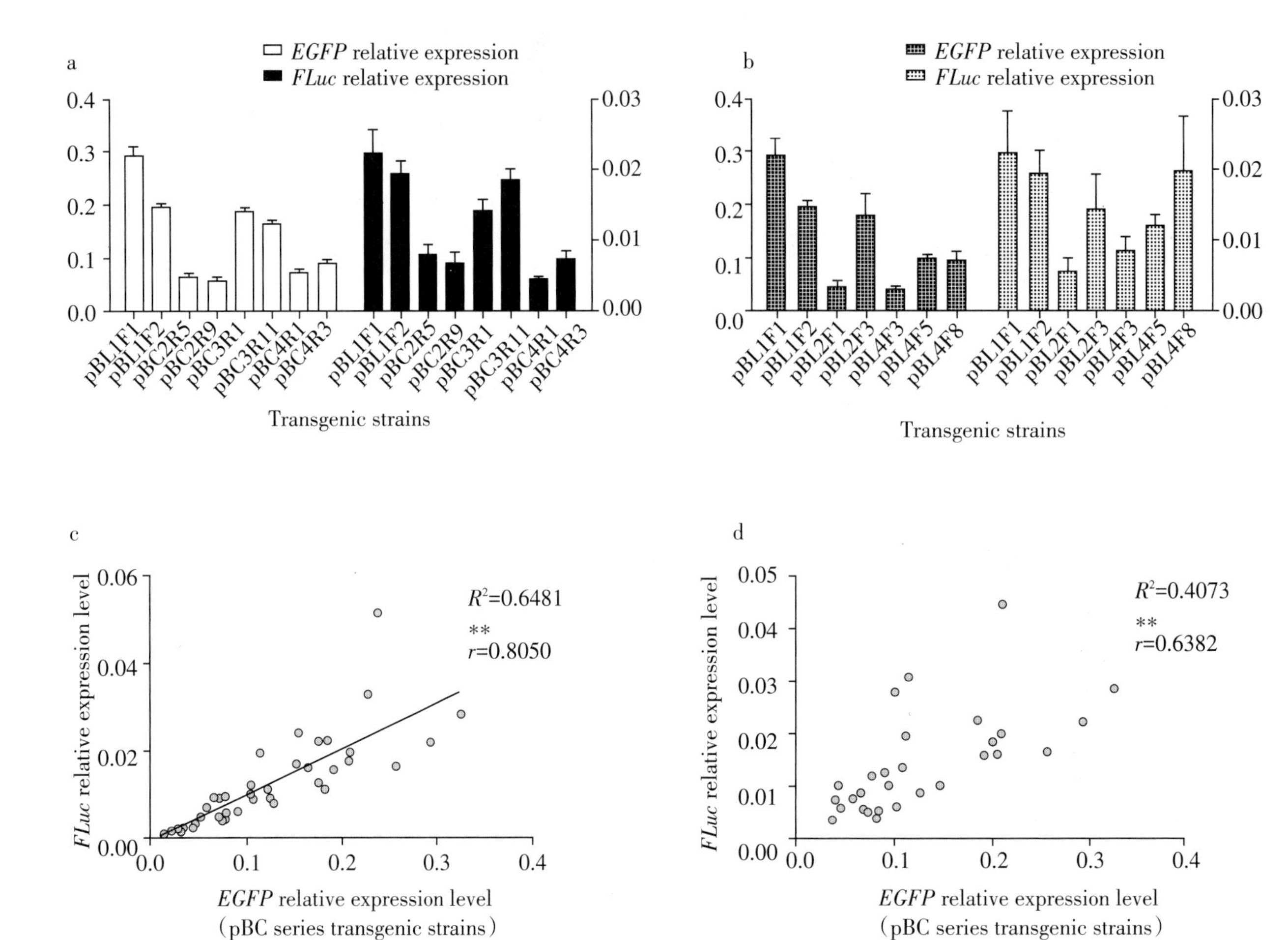

Figure 81. Correlation analysis between the marker (*EGFP*) and target (*FLuc*) genes at the transcriptional level. (a, c) Transcription pattern comparison of *EGFP* and *FLuc* in the pBC (including pBL1F) series transgenic silkworms. A significant correlation was revealed between the *EGFP* and *FLuc* expression levels with the two expression frames assembled in close proximity. (b, d) However, the correlation between *EGFP* and *FLuc* in the pBL series transgenic silkworms was relatively lower; the *Ser 1* promoters in front of *FLuc* were different, which affected the transcriptional level of *FLuc*. The qRT-PCR data are shown as the mean ± SD of three separate experiments performed in triplicate. $^{**}P<0.01$.（选自：*Analysis of the sericin1 promoter and assisted detection of exogenous gene expression efficiency in the silkworm Bombyx mori L.*，参见本书第530页。）

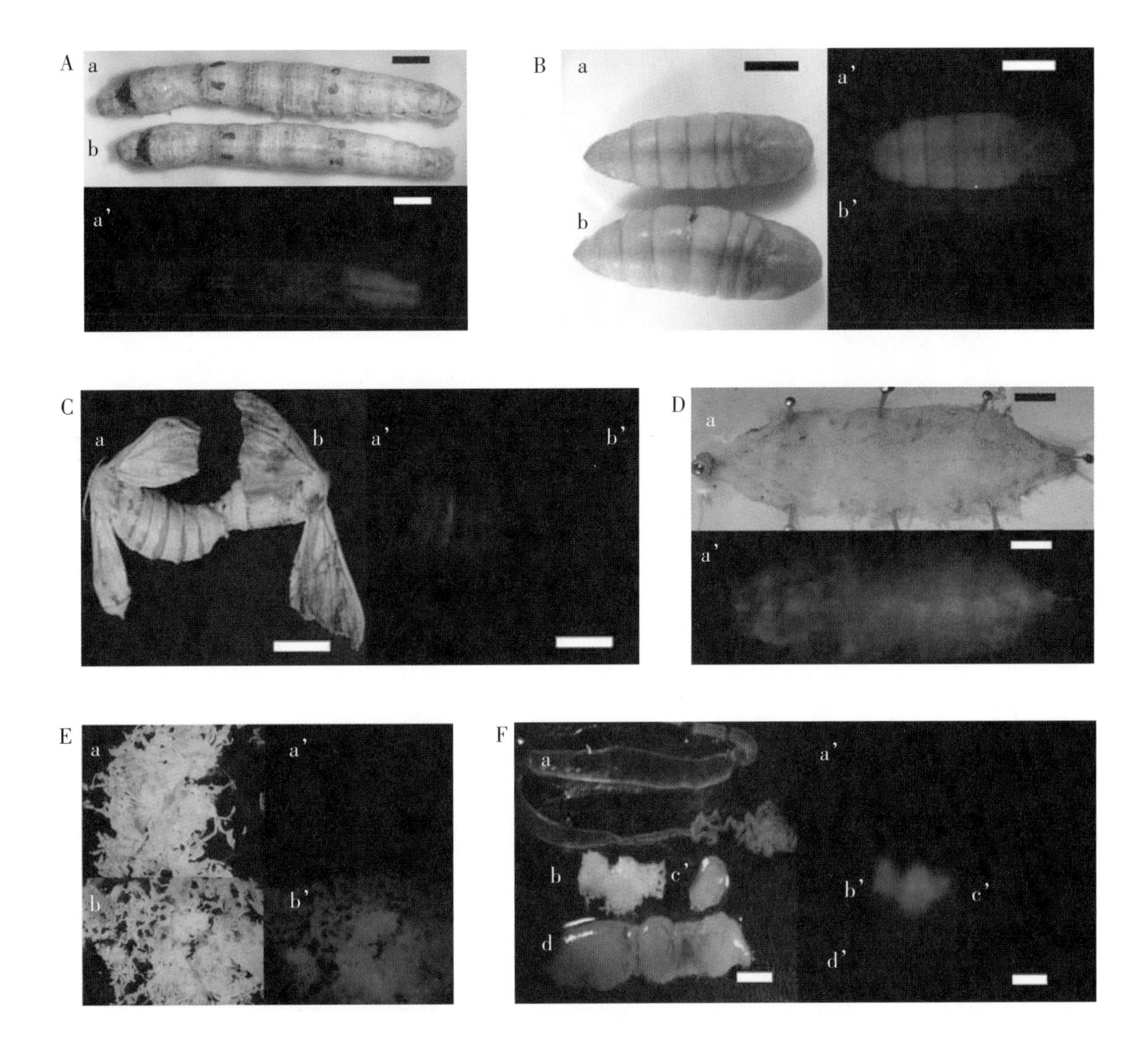

Figure 82. Fluorescent microscopic detection of *Ds*Red expression in transgenic silkworm. a Fluorescent detection of *Ds*Red in the transgenic day-4 fifth instar larvae. Wild type (a, a') and transgenic silkworm (b, b') were viewed under white light and 565 nm ultraviolet light (optimal for detecting RFP fluorescence). b Fluorescent detection of *Ds*Red in the transgenic pupae. Transgenic (a, a') and wild type pupae (b, b') were viewed under white light and 565 nm ultraviolet light. c Fluorescent detection of *Ds*Red in the transgenic adults. Transgenic (a, a') and wild type adults (b, b') were viewed under white light and 565 nm ultraviolet light. d Fluorescent detection of *Ds*Red in dissected transgenic day-7 fifth instar larvae. Transgenic larvae was viewed under white light (a) and 565 nm ultraviolet light (a'). e *Ds*Red fluorescent detection the isolated fat body from transgenic and wild-type day-7 fifth instar larvae. Fat body of wild type (a') and transgenic larvae (b, b') were dissected in 1×PBS and viewed under white light or 565 nm ultraviolet light. f Tissue-specific fl uorescent detection of *Ds*Red in the transgenic day-7 fifth instar larvae. Various tissues including the silk glands (a'), fat body (b, b'), gonads (c, c') and midgut (d, d') from transgenic larvae were viewed under white light and 565 nm ultraviolet light. The scale bar represents 0.5 cm.（选自：*The promoter of Bmlp3 gene can direct fat body-specific expression in the transgenic silkworm, Bombyx mori*，参见本书第534页。）

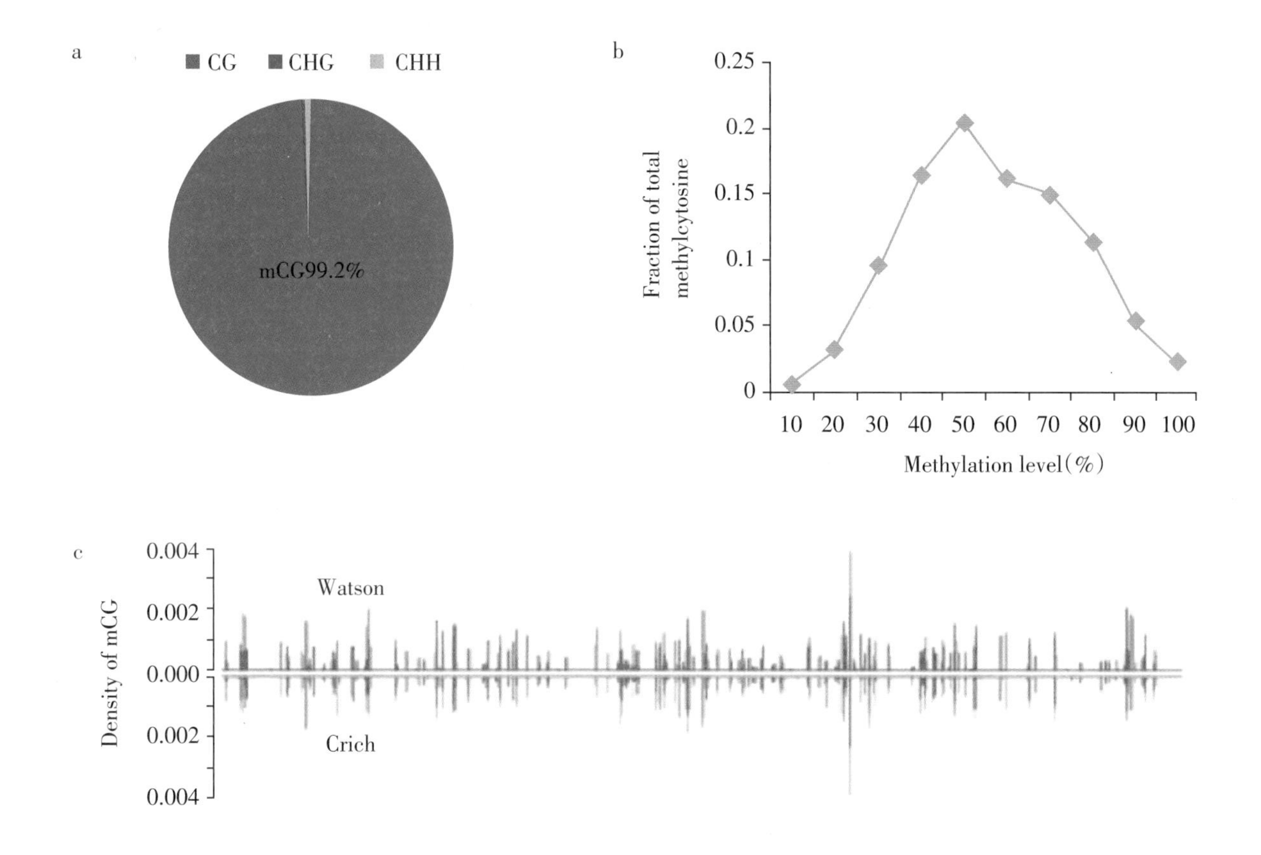

Figure 83. DNA methylation patterns and chromosomal distribution in *Bombyx mori*. (a) Fraction of mCs identified in each sequence context for the strain *Dazao*, indicating rather low and non-CG methylation, which are likely to be false positives. (b) Distribution of mCs (y axis) across methylation levels (x axis). Methylation level was determined by dividing the number of reads covering each mC by the total reads covering that cytosine. (c) Density of mCs identified on the two DNA strands (Watson and Crick) throughout chromosome 1 (out of 28). Density was calculated in 25-kb bins. The value refers to the number of mCs per base pair, as shown on the y axis.（选自：*Single base-resolution methylome of the silkworm reveals a sparse epigenomic map*，参见本书第548页。）

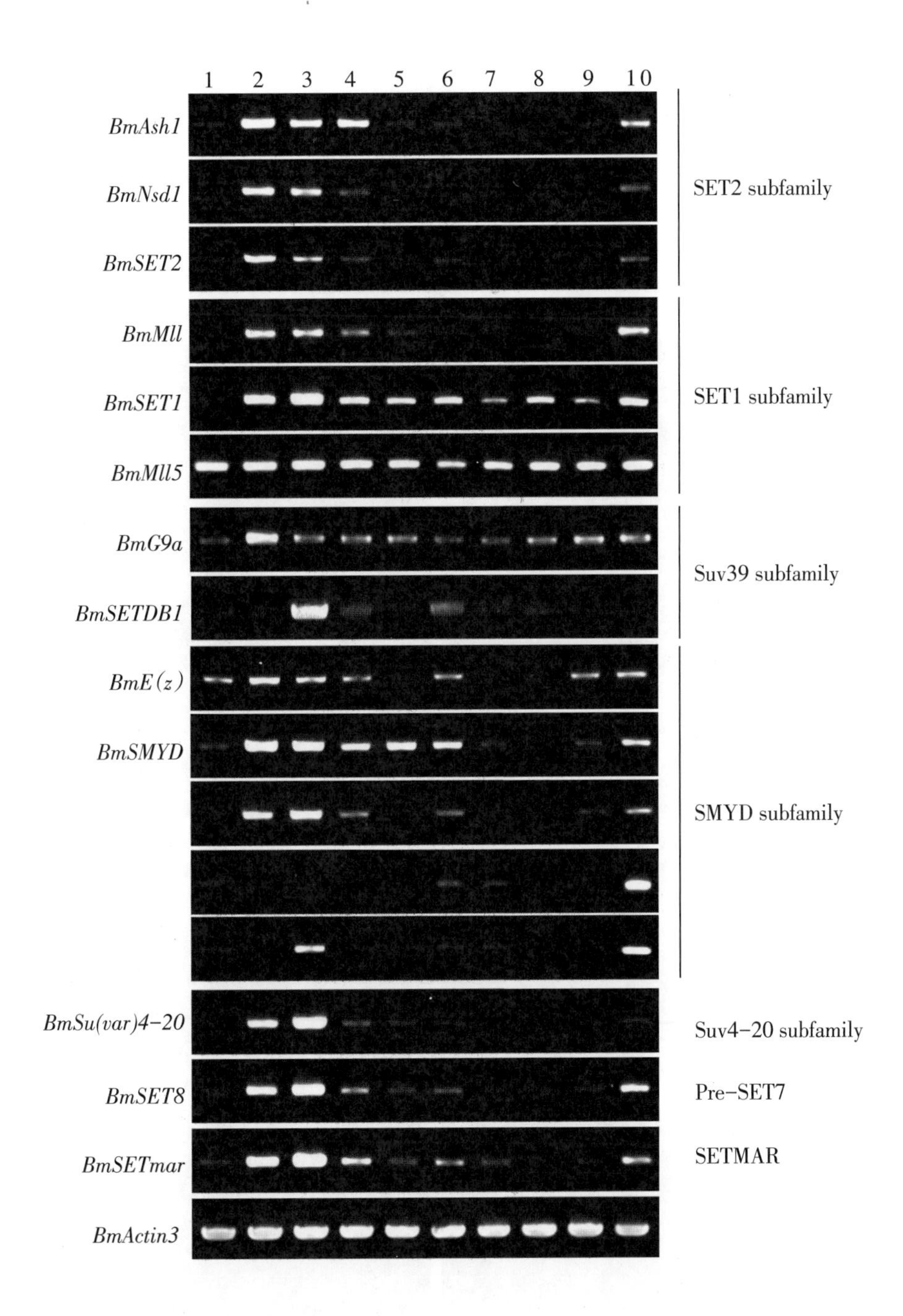

Figure 84. Developmental expression profile of silkworm SET-domain genes. 1: whole silkworm; 2: ovary; 3: testis; 4: blood; 5: midgut; 6: fat body; 7: body wall; 8: Malpighian tube; 9: silk gland; 10: head.（选自：*Identification and analysis of the SET-domain family in silkworm, Bombyx mori*，参见本书第560页。）

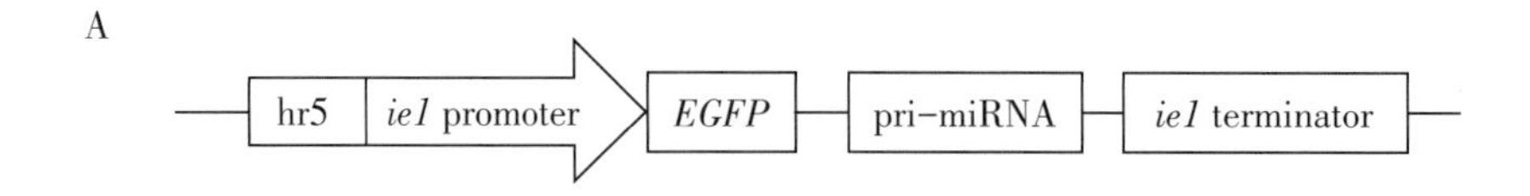

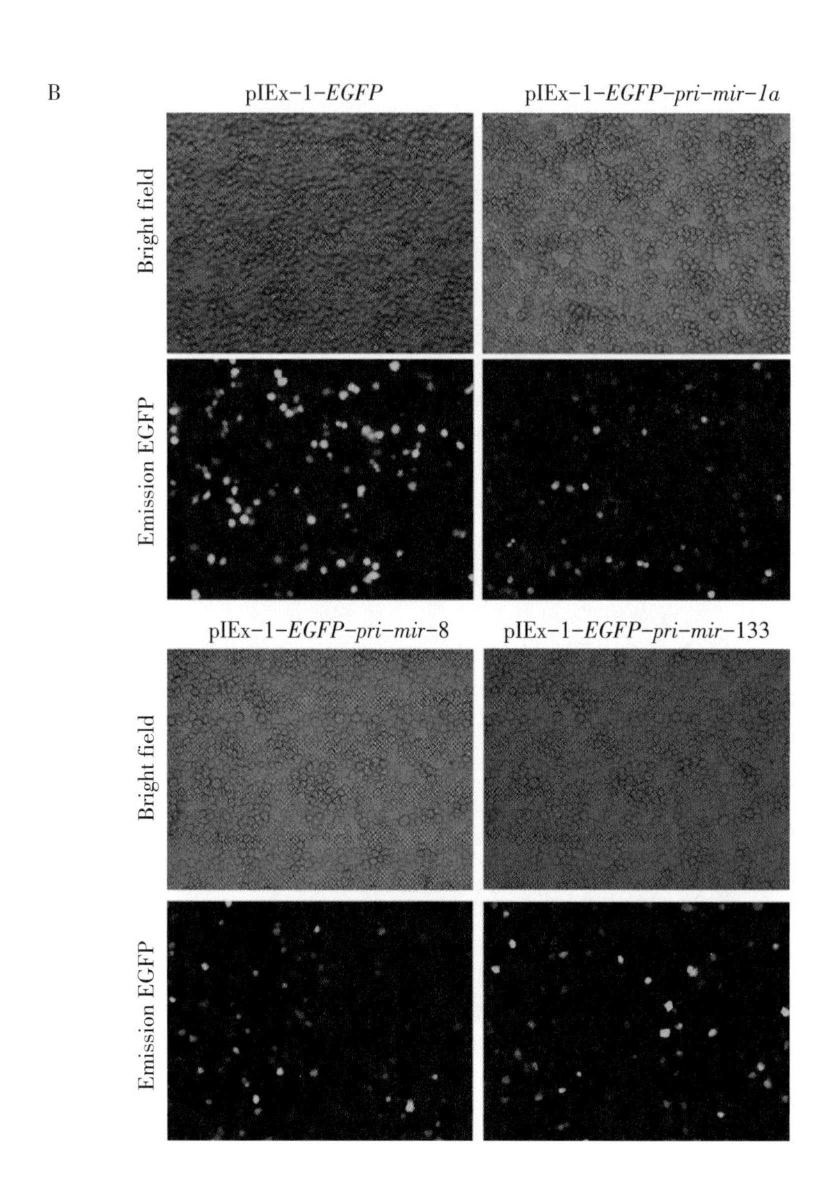

Figure 85. The representation of miRNAs overexpression in *BmN* cells. (a) Schematic diagram of pIEx-1-*EGFP-pri-mir-1a/8/133*. The miRNAs were driven by a baculovirus *ie1* promoter enhanced by hr5 enhancer. (b) The *BmN* cells transfected with pIEx-1-*EGFP* and pIEx-1-*EGFP-pri-mir-1a/8/133* vectors. The emission EGFP showed that the transfection efficiency is high and the EGFP protein was expressed in a high level.（选自：*Expression analysis of miRNAs in BmN cells*，参见本书第578页。）

A

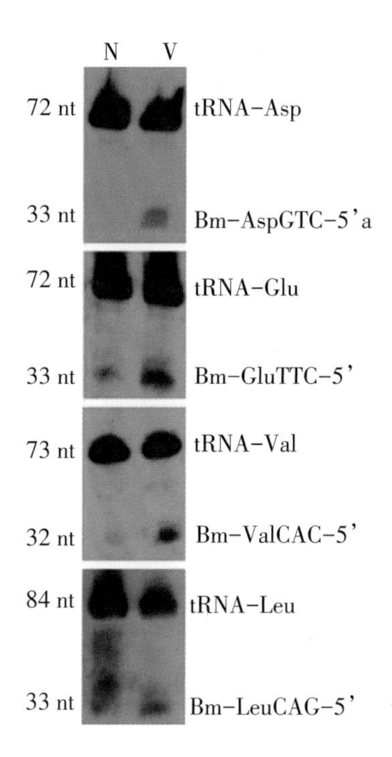
N
V
72 nt
tRNA-Asp
33 nt
Bm-AspGTC-5'a
72 nt
tRNA-Glu
33 nt
Bm-GluTTC-5'
73 nt
tRNA-Val
32 nt
Bm-ValCAC-5'
84 nt
tRNA-Leu
33 nt
Bm-LeuCAG-5'

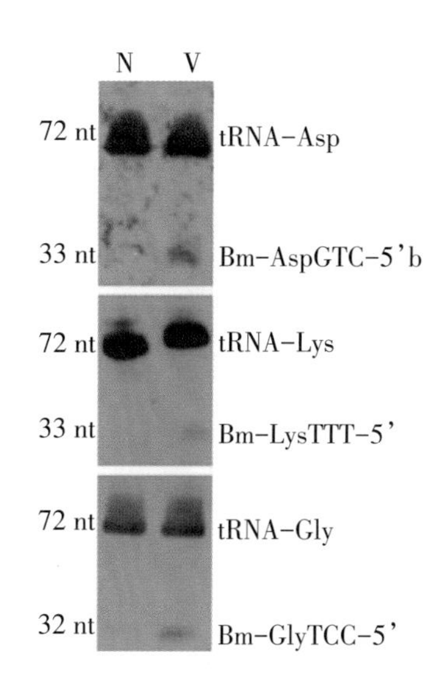
N
V
72 nt
tRNA-Asp
33 nt
Bm-AspGTC-5'b
72 nt
tRNA-Lys
33 nt
Bm-LysTTT-5'
72 nt
tRNA-Gly
32 nt
Bm-GlyTCC-5'

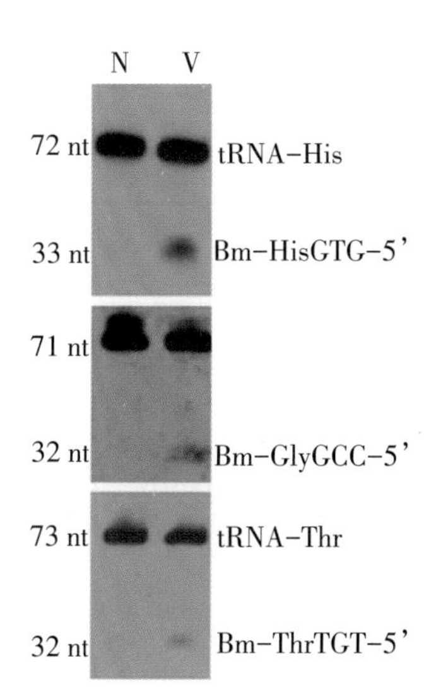
N
V
72 nt
tRNA-His
33 nt
Bm-HisGTG-5'
71 nt
32 nt
Bm-GlyGCC-5'
73 nt
tRNA-Thr
32 nt
Bm-ThrTGT-5'

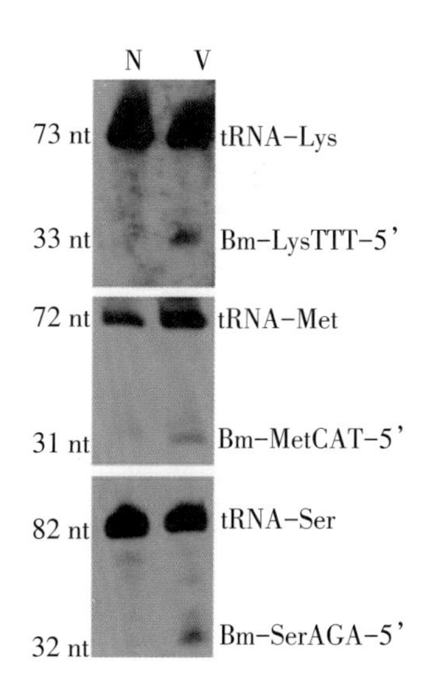
N
V
73 nt
tRNA-Lys
33 nt
Bm-LysTTT-5'
72 nt
tRNA-Met
31 nt
Bm-MetCAT-5'
82 nt
tRNA-Ser
32 nt
Bm-SerAGA-5'

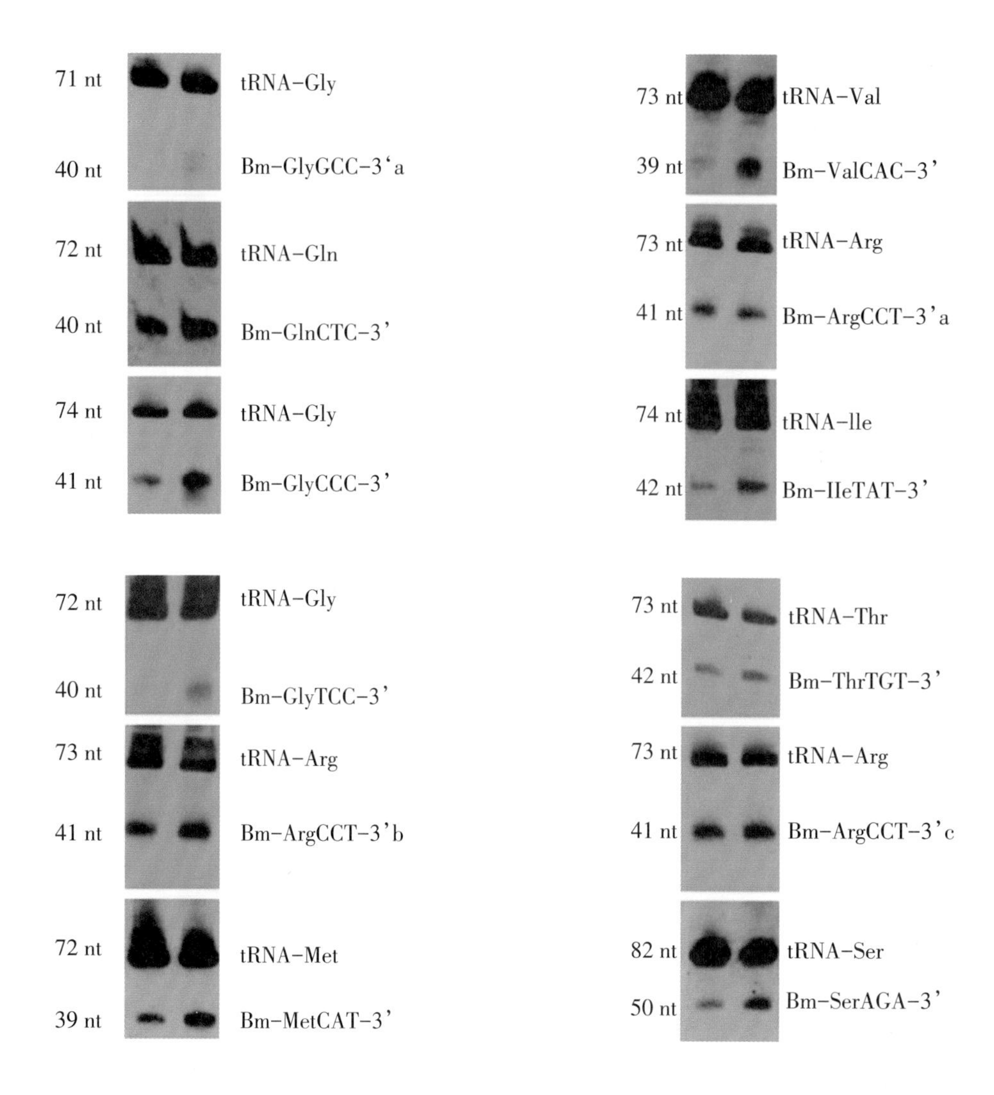

Figure 86. Identification of tRFs by Northern blot. (A) 5’ tRFs were identified by Northern blot with a length of ~33 bp. The majority of the identified 5’ tRFs were generated only in *Bm*NPV-infected cells, implying a *Bm*NPV-related role for these small RNAs. (B) 3’ tRFs were identified by Northern blot with a length of ~40 bp. However, the majority of the identified 3’ tRFs were generated in both the normal cells and in the *Bm*NPV-infected cells. The host tRNAs were also identified along with the tRFs. N: normal *BmN* cells; V: *Bm*NPV-infected *BmN* cells.（选自：*RIP-seq of BmAgo2-associated small RNAs reveal various types of small non-coding RNAs in the silkworm, Bombyx mori*，参见本书第588页。）

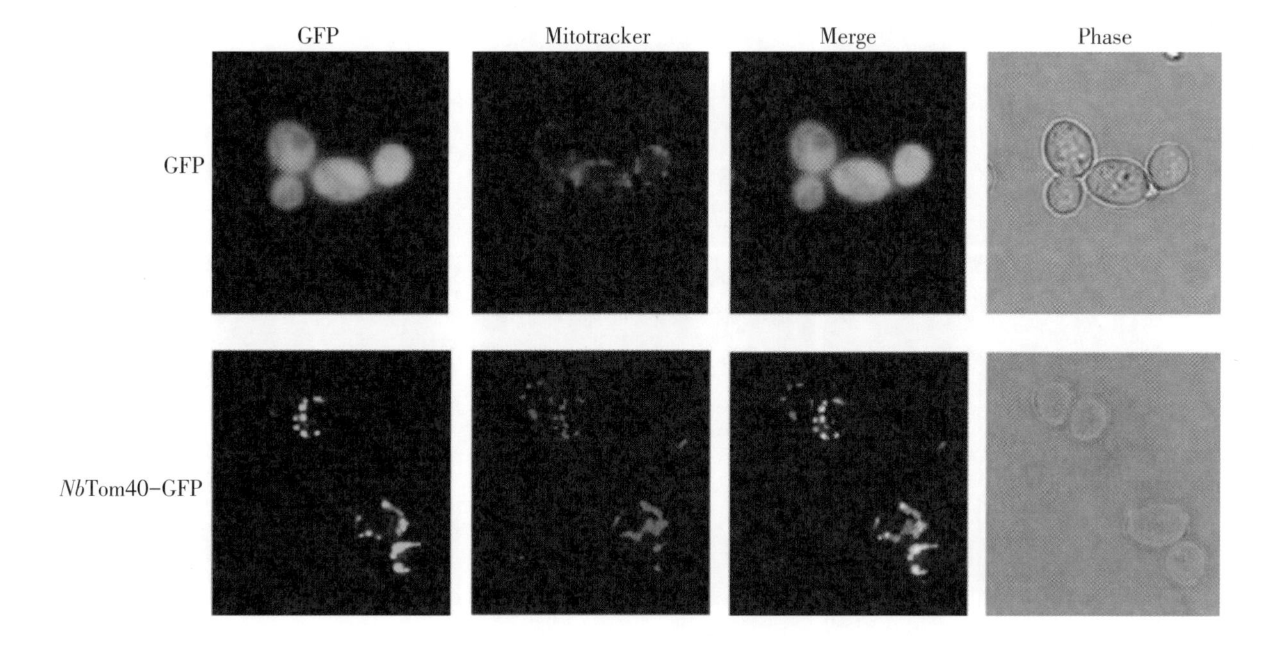

Figure 87. A GFP fusion protein of the translocase of outer mitochondrial membrane 40 kDa subunit homologue from *N. bombycis* (*Nb*Tom40) is targeted to yeast mitochondria. Yeast cells (pUG35-NbTom40-GFP or pUG35-GFP) were incubated in SC-Met-Ura medium at 30 °C overnight. GFP fluorescence was examined with a fluorescence microscope. Mitochondria were stained with the fluorescent dye MitoTracker Red. The green colour is used as the digital pseudocolour for the fluorescence emitted by GFP or *Nb*Tom40-GFP, red is the digital pseudocolour for fluorescence emitted by MitoTracker Red, and yellow results from their overlap, indicating the co-localization of *Nb*Tom40-GFP and MitoTracker Red in the mitochondria of transfected yeast cells. Yeast cells in the top row are expressing GFP as a control, demonstrating its cytosolic localization. The bottom row illustrates the co-localization of *Nb*Tom40-GFP and MitoTracker Red in the mitochondria of transfected yeast cells.（选自：*The protein import pore tom40 in the microsporidian Nosema bombycis*，参见本书第602页。）

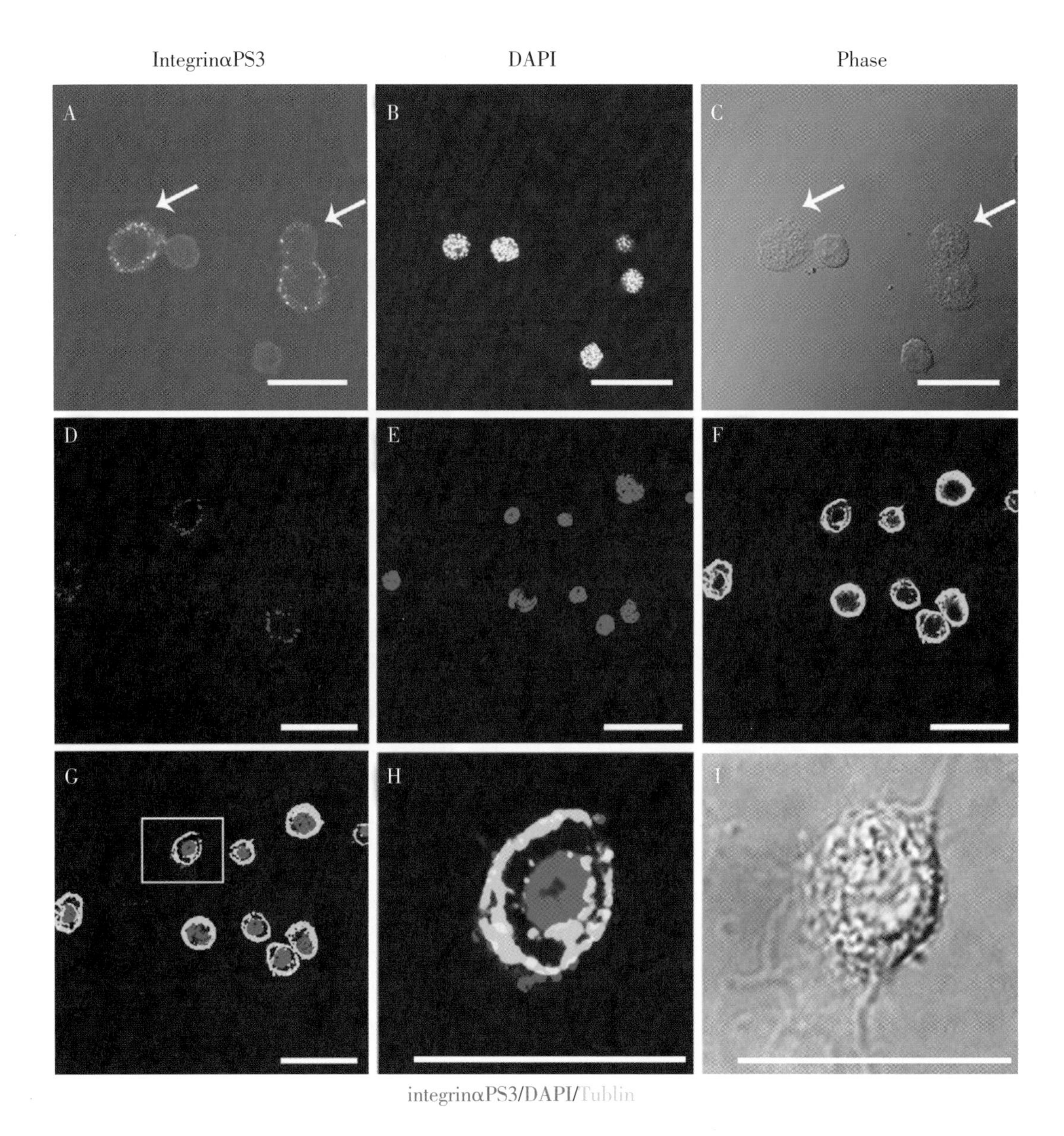

Figure 88. Localisation of *Bm*integrinαPS3 in the circulating haemocytes of larval silkworms. A double immunofluorescence analysis using *Bm*integrinαPS3 and tubulin antibodies showed that *Bm*integrinαPS3 is located on the cell membrane of granulocytes. Tubulin staining is shown in green, and DAPI counterstaining is shown in blue. Scale bar = 20 μm.(选自:*A novel granulocyte-specific α integrin is essential for cellular immunity in the silkworm Bombyx mori*,参见本书第606页。)

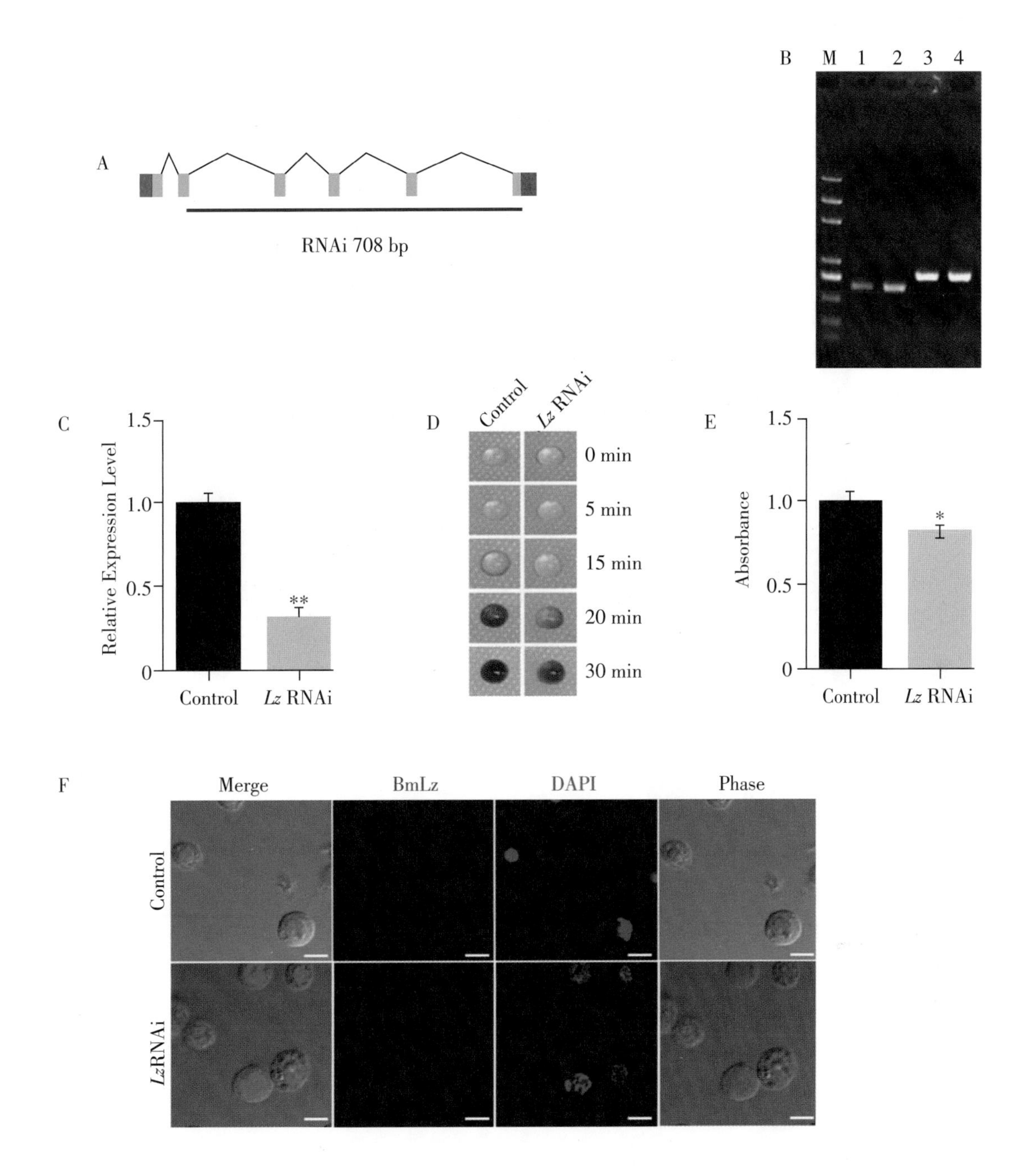

Figure 89. Down-regulation of *BmLz* by RNA interference in day 1 of 5th instar silkworm larva haemolymph. (A) The sequence of *BmLz* dsRNA. (B) The synthesis of dsRNA (1, 2 band were *EGFP* dsRNA; 3, 4 band were *BmLz* dsRNA). (C) qRT-PCR detected the expression level change of *BmLz* in hemocytes after RNAi. The data were analysed using Student's *t* test, $**P<0.01$. Data are shown as the mean ± SE from three independent experiments. (D) Survey of haemolymph melanism rate after *BmLz* was knockdown. (E) PO activity change after RNAi. (F) Immunofluorescence analysis using *BmLz* antibody in the circulating haemocytes after RNAi.（选自：*A novel Lozenge gene in silkworm, Bombyx mori regulates the melanization response of hemolymph*，参见本书第608页。）

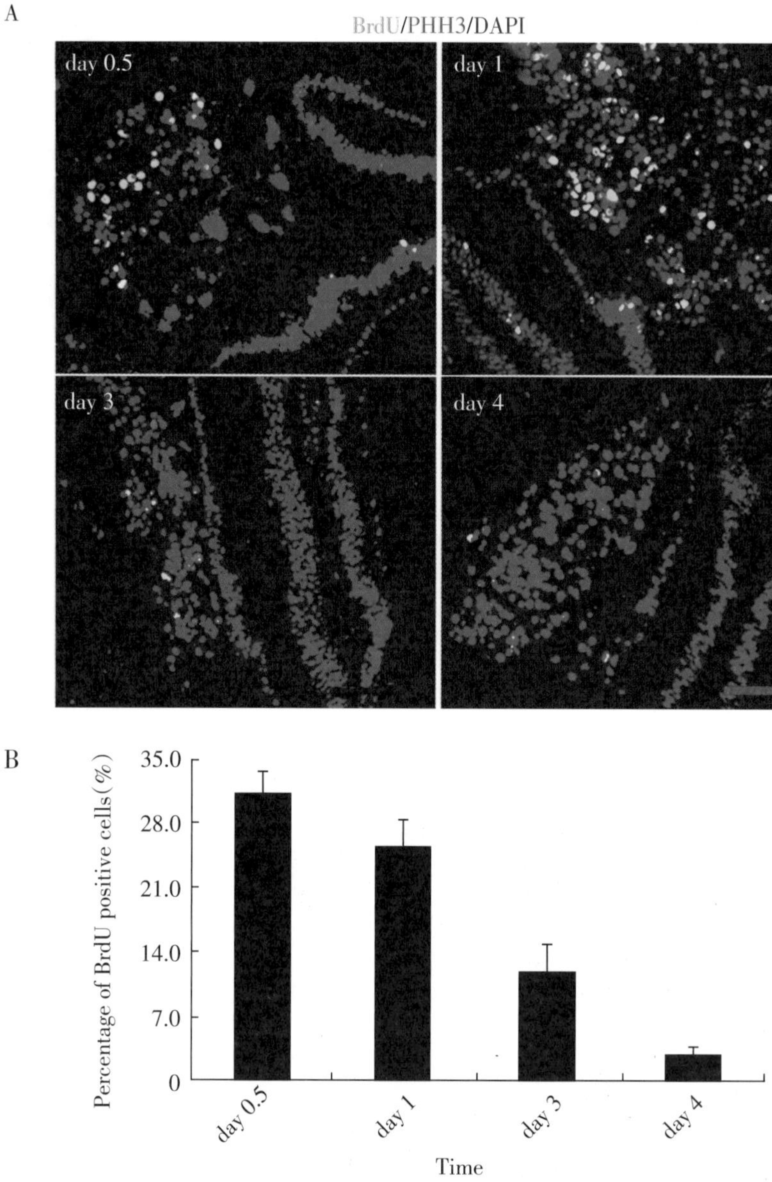

Figure 90. Identification and quantification of proliferating cells in HPOs. (A) Immunofluorescence staining for the proliferating cells in larval HPOs. The HPOs were collected at the indicated different time points after BrdU injection into larvae at L5D1, then sectioned and stained with the anti-BrdU (green) and anti-PHH3 (red) antibodies. Scale bar = 100 μm. (B) Quantification of the BrdU-positive cells in HPOs. The number of BrdU-positive cells dramatically decreased from day 0.5 to day 4. For each time point, four to six images were examined. Data are presented as means ± SD. (For interpretation of the references to colour in this figure legend, the reader is referred to the web version of this article.)（选自：*Characterization of hemocytes proliferation in larval silkworm, Bombyx mori*，参见本书第612页。）

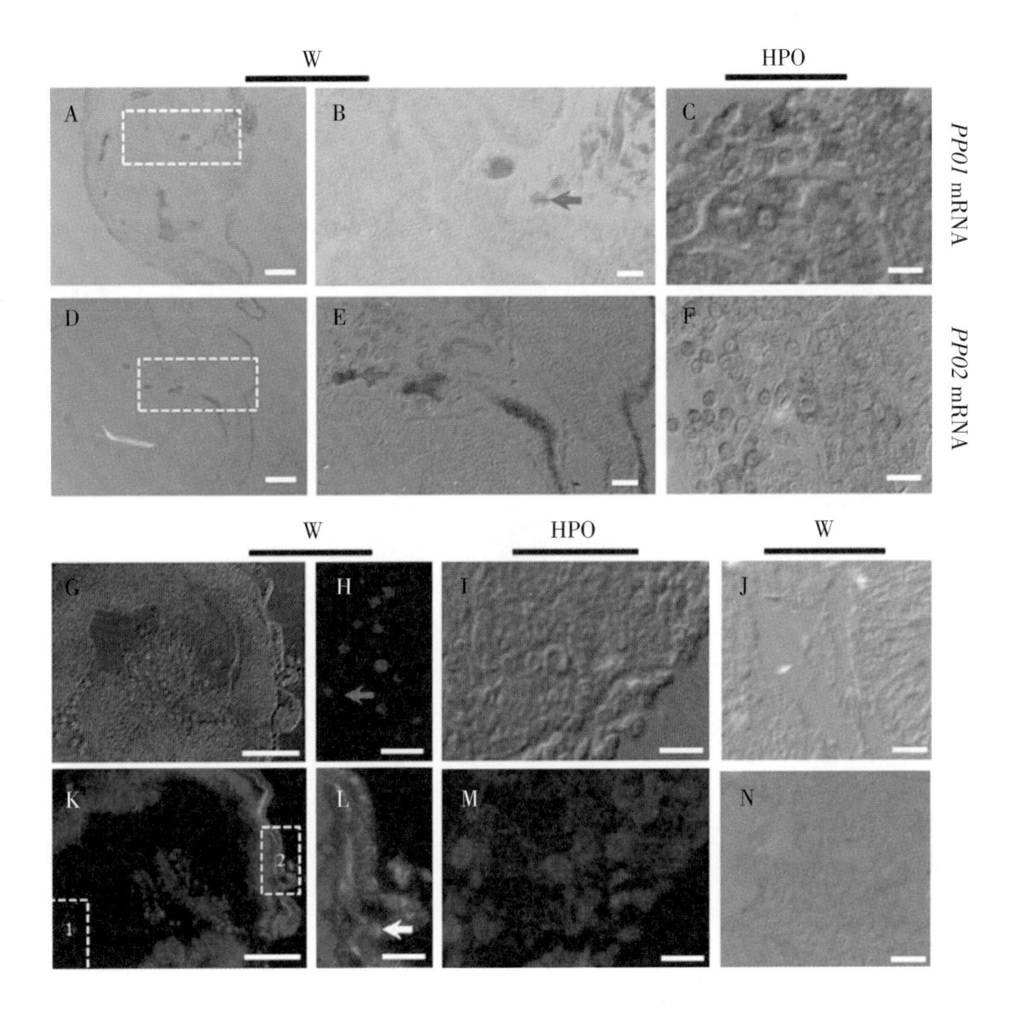

Figure 91. Location of PPO protein and *PPO* mRNA in wing discs. Wing discs (W) and hematopoietic organs (HPO) of larvae on V-3 were fixed and sectioned for PPO immuno-staining and *PPO* mRNA detection in the same condition. (A-F) Detection of *PPO1* (A-C) and *PPO2* (D-F) mRNA in the wing discs (A, B, D, E) and hematopoietic organs (C, F). In the hematopoietic organs, there are many cells showing *PPO1* and *PPO2* mRNA signals (C, F). Wing discs imaged at low magnification were shown (A, D). The framed areas in (A) and (D) were shown in (B) and (E) respectively at high magnification. The arrow-indicated several cells accumulated together in the cavity of wing discs had *PPO* mRNA signals (B, E). (J and N) Negative control using sense mRNA probes. (G-I, K-M) Detection of PPO proteins in cells of wing disc (G, H, K, L) and hematopoietic organs (I, M). Cells with red fluorescence in wing discs and hematopoietic organs have PPO proteins. The enlarged picture in (H) was from the dot-lined frame 1 in (K). A few cells in the cavity of wing discs (H) were found to have PPO. Among cells on the brisk of the wing discs, the PPO signal was also strong (L), which is shown in the dot-lined frame 2 in (K). DAPI (blue fluorescence) was used for nuclei counterstaining. The arrows point to some cells containing PPO proteins. Bar: (A, D, G, K) 80 μm; (B, E) 50 μm; All others: 20 μm.（选自：*Existence of prophenoloxidase in wing discs: a source of plasma prophenoloxidase in the silkworm, Bombyx mori*，参见本书第620页。）

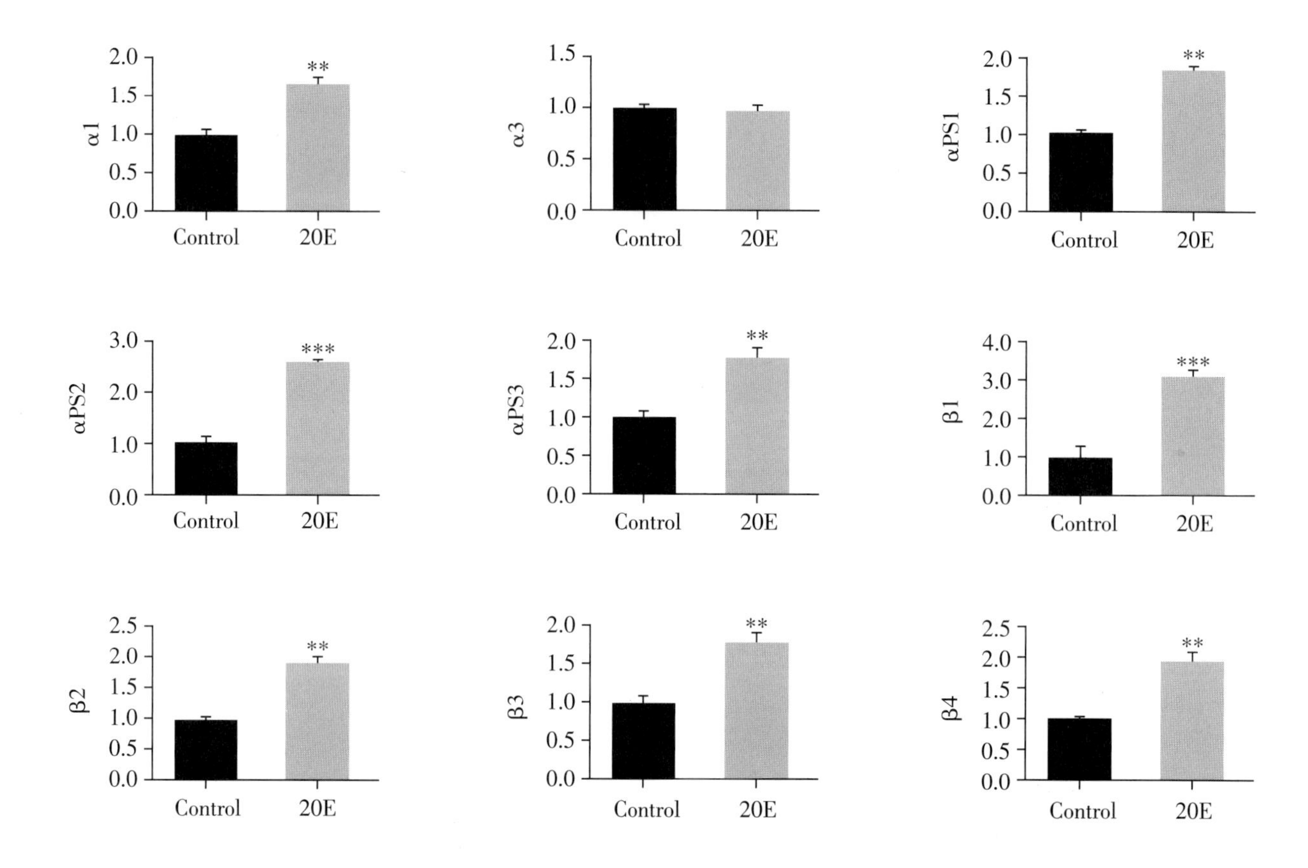

Figure 92. Relative transcript levels of integrins in response to 20-E treated 18 h *in vivo*. The corresponding amount of alcohol was used as a control. The differences between the experimental and the control groups were analyzed by the Student's *t* test, $**P<0.01$, $***P<0.001$. All experiments were repeated at least three times.（选自：*Characterization and identification of the integrin family in silkworm, Bombyx mori*，参见本书第622页。）

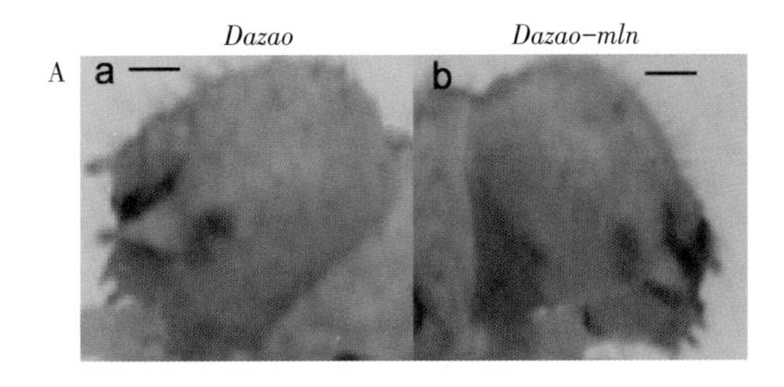

A
Dazao
Dazao-mln
a
b

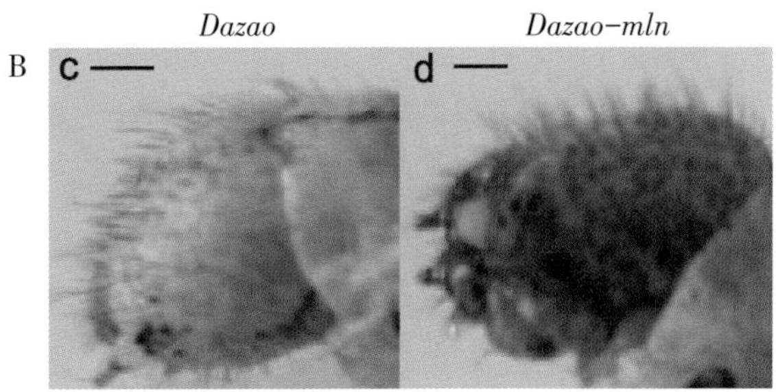

B
Dazao
Dazao-mln
c
d

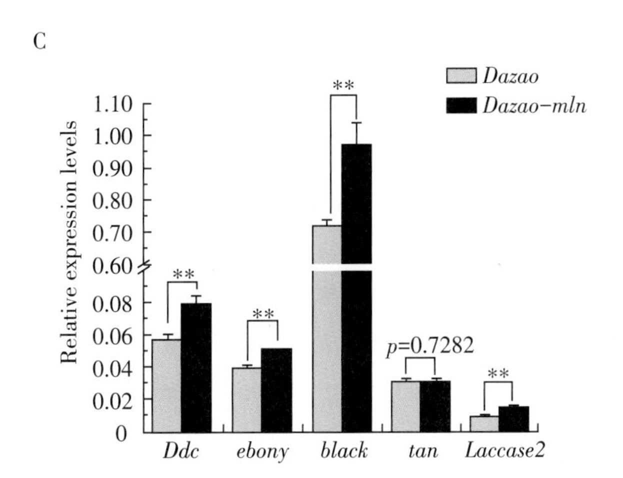

C
Dazao
Dazao-mln
Relative expression levels
p=0.7282
Ddc
ebony
black
tan
Laccase2

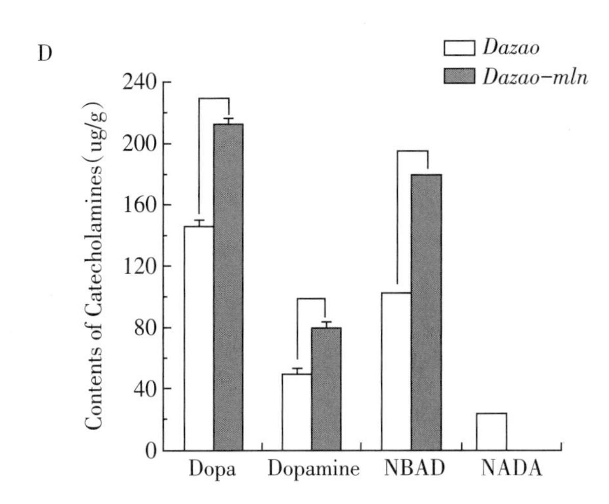

D
Dazao
Dazao-mln
Contents of Catecholamines(ug/g)
Dopa
Dopamine
NBAD
NADA

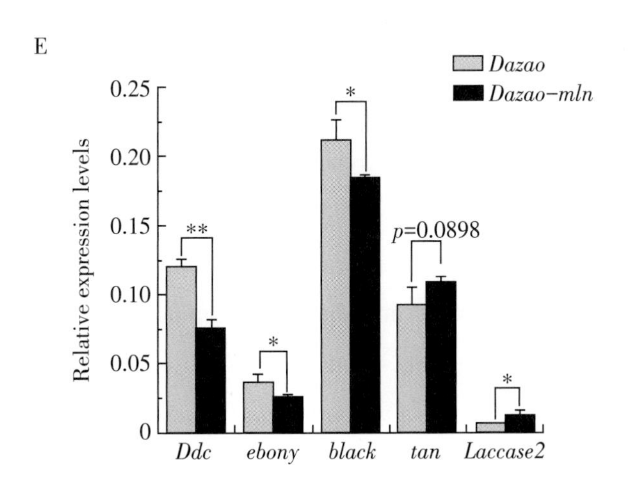

E
Dazao
Dazao-mln
Relative expression levels
p=0.0898
Ddc
ebony
black
tan
Laccase2

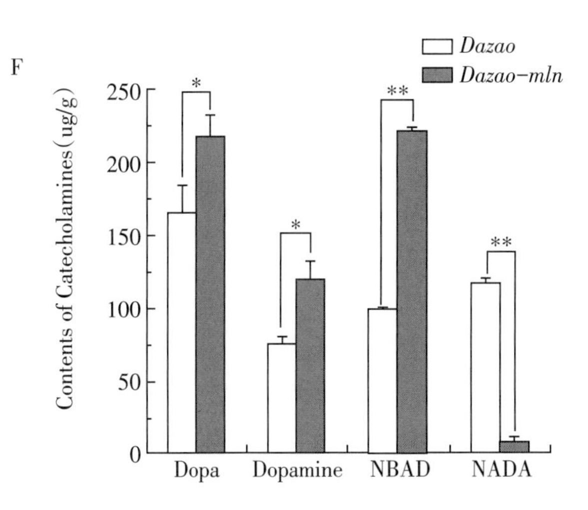

F
Dazao
Dazao-mln
Contents of Catecholamines(ug/g)
Dopa
Dopamine
NBAD
NADA

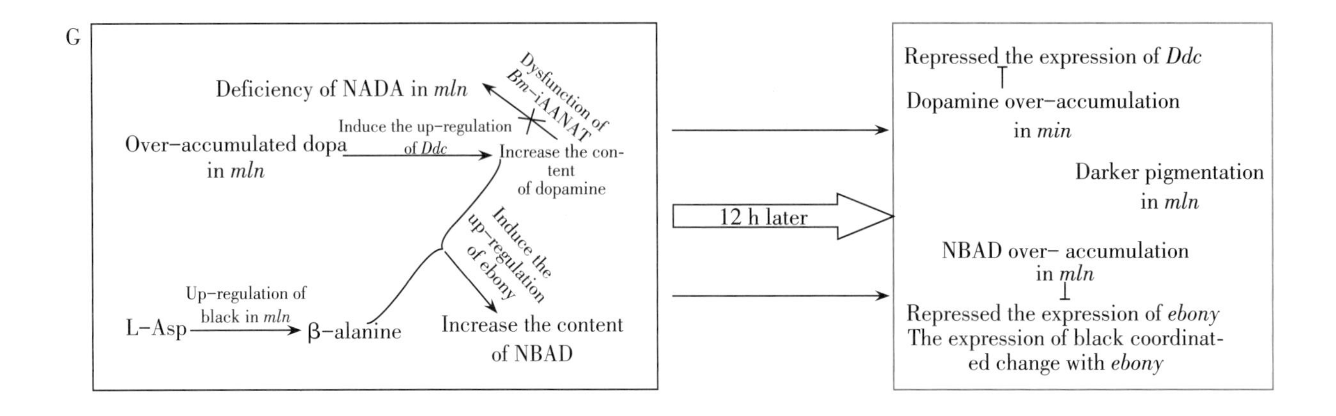

Figure 93. Pigmentation of head and catecholamine metabolism in fifth instar larvae after molt. (A). Color patterns of the heads from both wild-type and *mln* mutant strains immediately after the fourth molt. a and b indicate wild-type and mutant, respectively. Scale bar: 1 mm. (B). Color pattern of the heads from 12 h of fifth instars of wild-type and *mln* mutant. c and d indicate the wild-type and mutant, respectively. Scale bar: 1 mm. (C). Differences in expression profiles of melanin metabolism genes in the fifth instars just after molt in both wild-type and mutant. (Student's *t*-test. n=3. ** P<0.01. Data are presented as mean ±S.D.). (D). Differences in catecholamines between fifth instars just after molt of both wild-type and mutant. (Student's *t*-test. * P<0.05, ** P<0.01. Data are presented as mean ±S.D. of three separate experiments.). (E). Differences in the expression patterns of melanin genes between 12 h of fifth instars of both wild-type and mutant. (Student's *t*-test. n=3. * P<0.05, ** P<0.01). Data are presented as mean ±S.D.). (F). Differences in catecholamine content at 12 h of fifth instars between wild-type and mutant strains. (Student's *t*-test. * P<0.05, ** P<0.01. Data are presented as mean ±S.D. of three separate experiments.). (G). Schematic diagram of the melanism process of the heads of mln fifth instar larva. Blue and red boxes indicate fifth instar immediately after molt and 12 h, respectively. The blunt symbol indicates inhibitory action, while the fork signifies dysfunction of the *Bm-iAANAT* gene.（选自：*Effects of altered catecholamine metabolism on pigmentation and physical properties of sclerotized regions in the silkworm melanism mutant*, 参见本书第640页。）

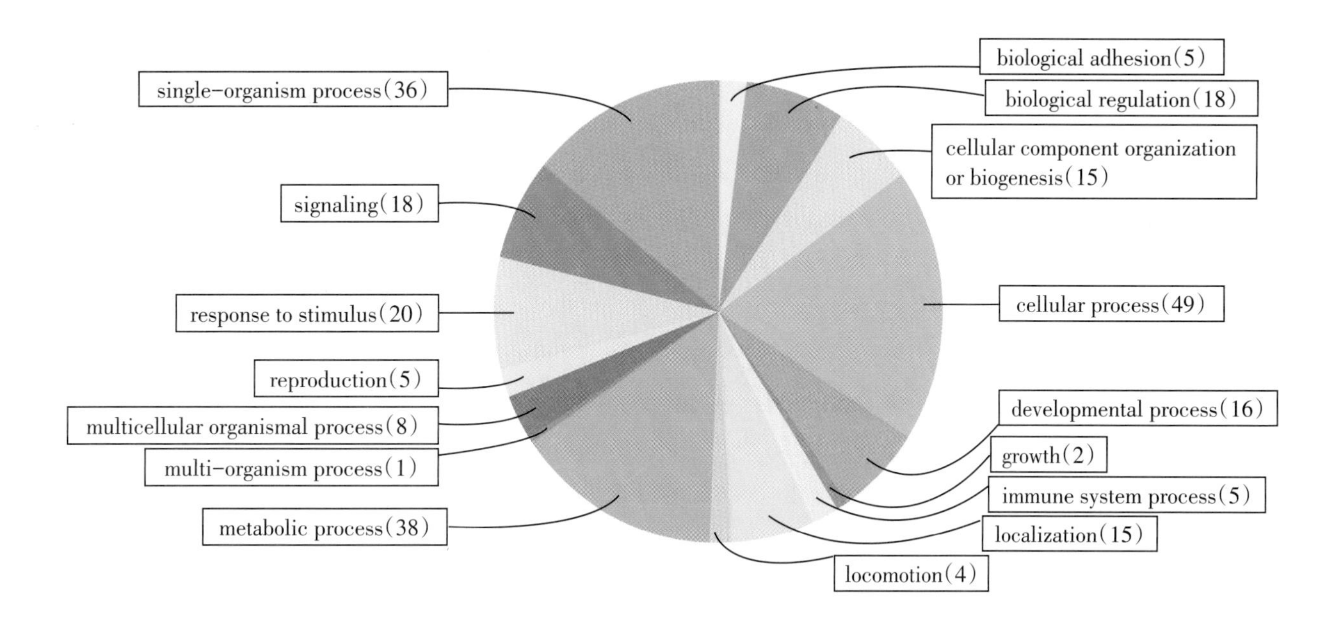

Figure 94. The category of biological process for the unique DEGs in the *cot* strain after hyperthermia induction. DEGs (Table S4) were broadly categorized according to their biological process. Numbers represent the actual number of DEGs.（选自：*Transcriptome analysis of neonatal larvae after hyperthermia-induced seizures in the contractile silkworm, Bombyx mori*, 参见本书第644页。）

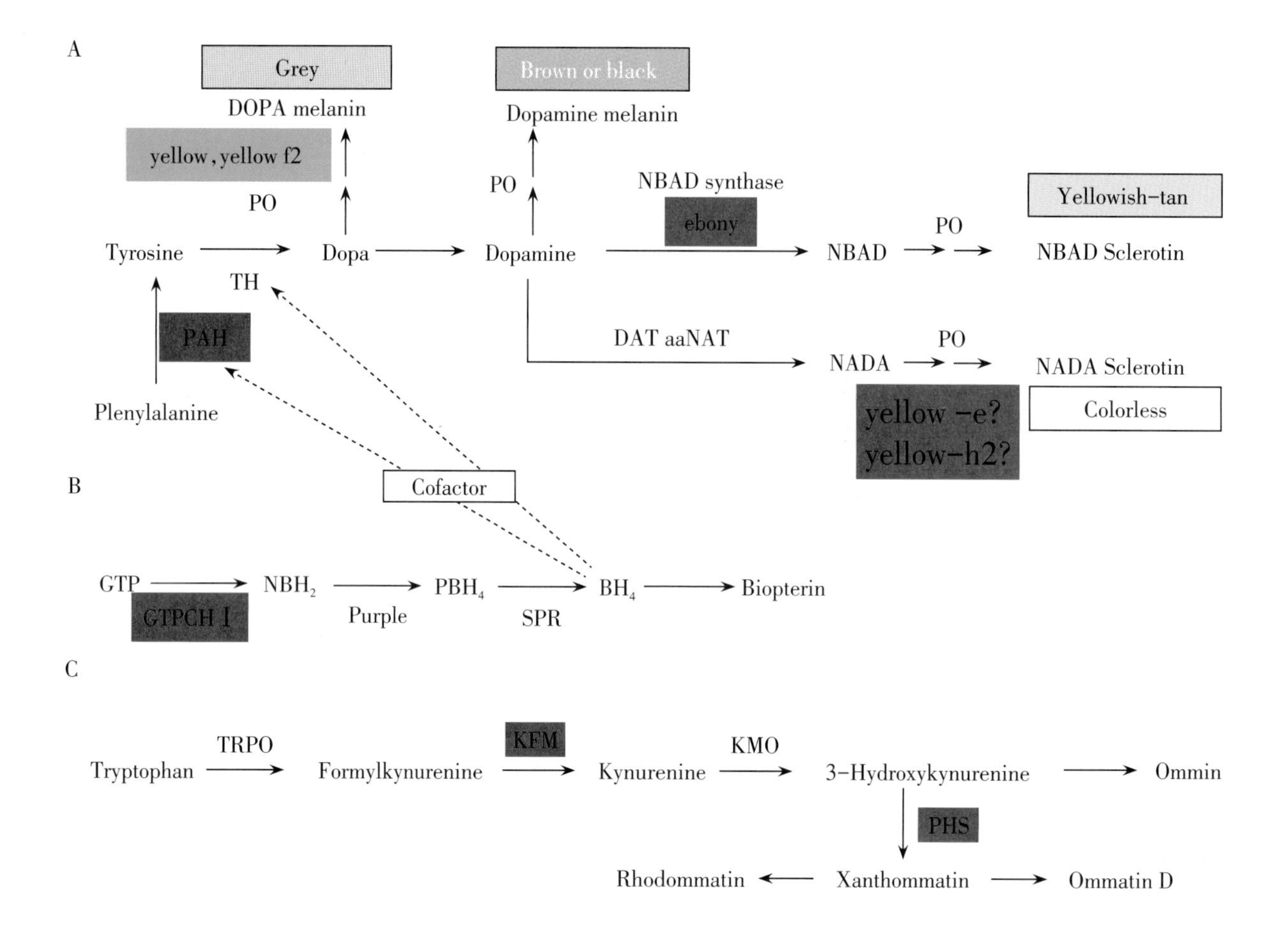

Figure 95. Differentially expressed genes in pigment pathways. (A) Simplified and synthetic melanin; (B) pteridine; (C) ommochrome pathways. Red, upregulated genes; green, downregulated genes in quail mutant. PAH, phenylalanine hydroxylase; TH, tyrosine hydroxylase; DAT, dopamine acetyltransferase; DOPA, dihydroxyphenylalanine; NADA, N-acetyl dopamine; NBAD, N-β-alanyl dopamine; PO, phenoloxidases; GTP, guanosine triphosphate; GTPCH I, GTP cyclohydrolase I; NBH2, dihydroneopterin triphosphate; BH4, tetrahydrobiopterin; TRPO, tryptophan pyrrolase; KFM, kynurenine formamidase; KMO, kynurenine 3-monooxygenase; PHS, phenoxazinone synthetase.［选自：*Transcriptome analysis of integument differentially expressed genes in the pigment mutant (quail) during molting of silkworm, Bombyx mori*，参见本书第646页。］

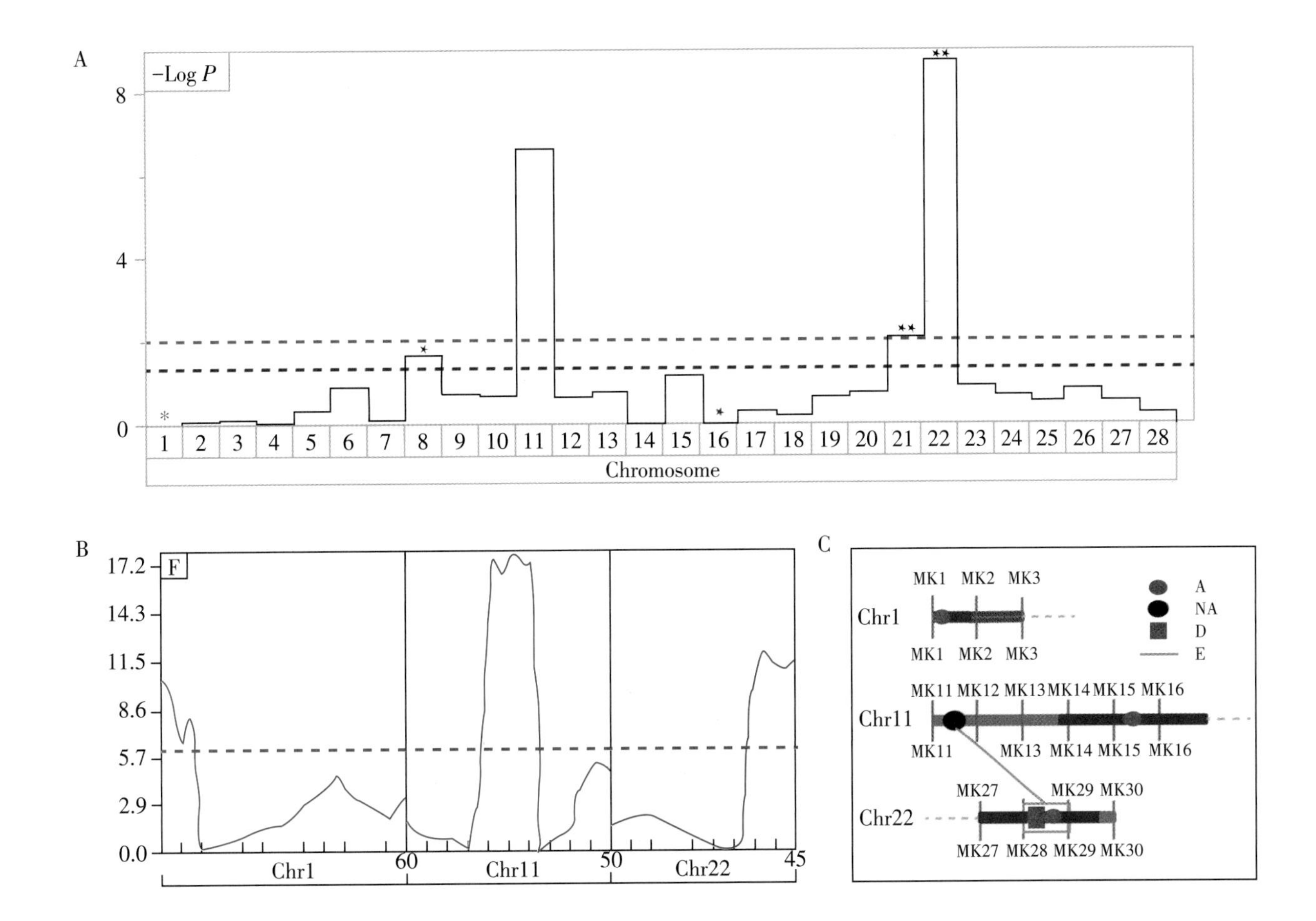

Figure 96. Results of the linkage and mapping analysis. (a) Significance values for each chromosome (Student' s *t*-test; *P<0.05; **P<0.01). The red star represents the sex chromosome. (b) *F*-value of the intervals on the linked chromosomes. (c) QTLNETWORK on the linked chromosomes; the red circle (A) represents the addictive effect; the red square (D) represents the dominance effect; the orange line (E) represents the epistasis effect; and the black circle (NA) is the non-addictive effect, which means that the locus has no single effect but can interact with other loci to affect the value of trait of interest (Yang *et al*. 2007, 2008).（选自：*A composite method for mapping quantitative trait loci without interference of female achiasmatic and gender effects in silkworm, Bombyx mori*，参见本书第666页。）

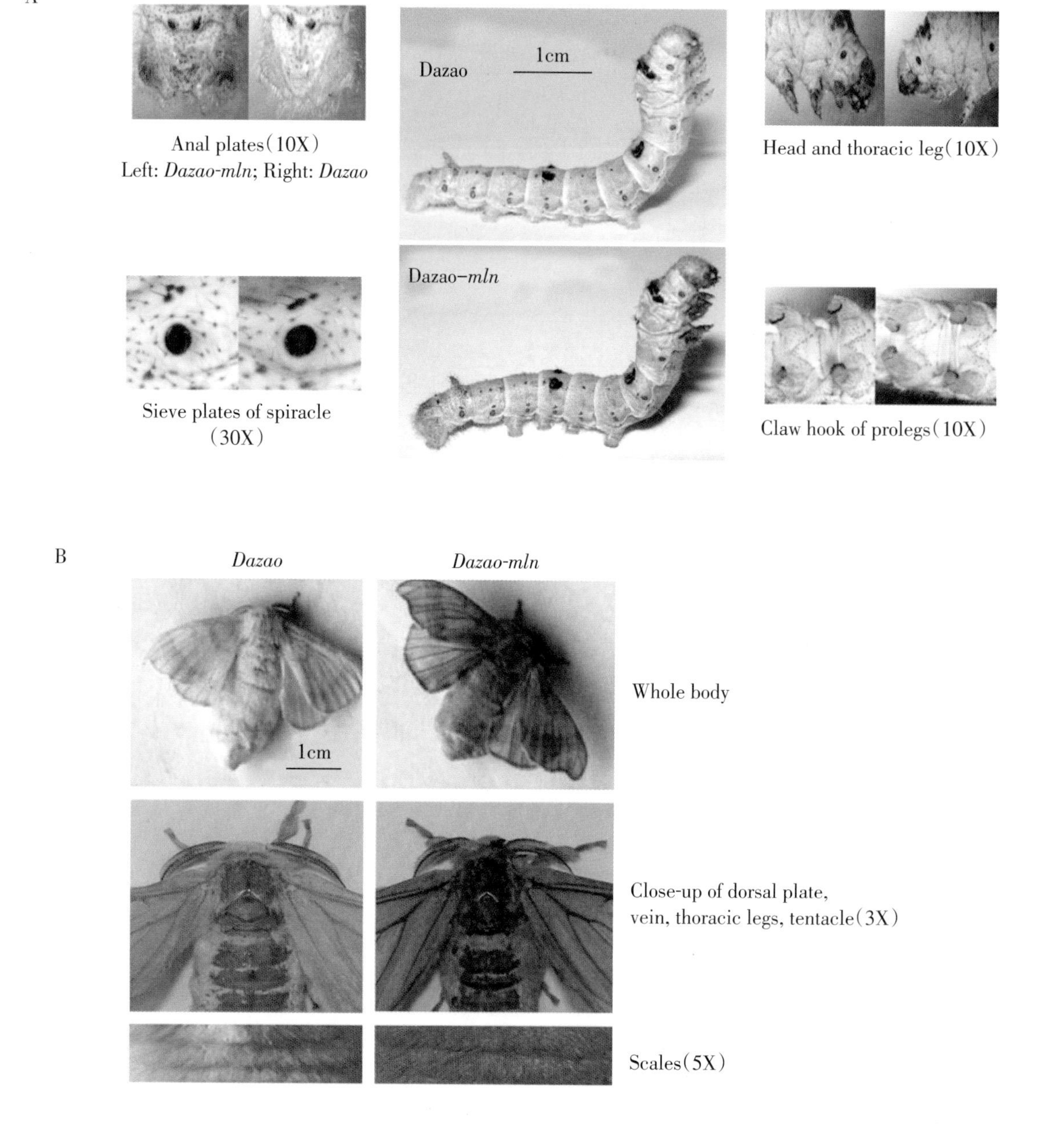

Figure 97. Phenotype of WT and mln mutant strain. A, WT (*Dazao*, $+^{mln}/+^{mln}$) and *mln* mutant strain (*Dazao-mln*, *mln*/*mln*) larvae are shown on day 1 of the 5*th* instar. In the *mln* mutant, the head, anal plate, thoracic legs, sieve plates of the spiracle, and claw hook of pro-legs of larvae are blacker or more buffy compared with the WT. B, body, scales, dorsal plate, vein, thoracic legs, and tentacle of the *mln* adult are also blacker than the WT. Scale bar, 1 cm. Magnification, indicated in the panels.（选自：*Mutations of an Arylalkylamine-N-acetyl transferase, Bm-iAANAT, are responsible for silkworm melanism mutant*，参见本书第668页。）

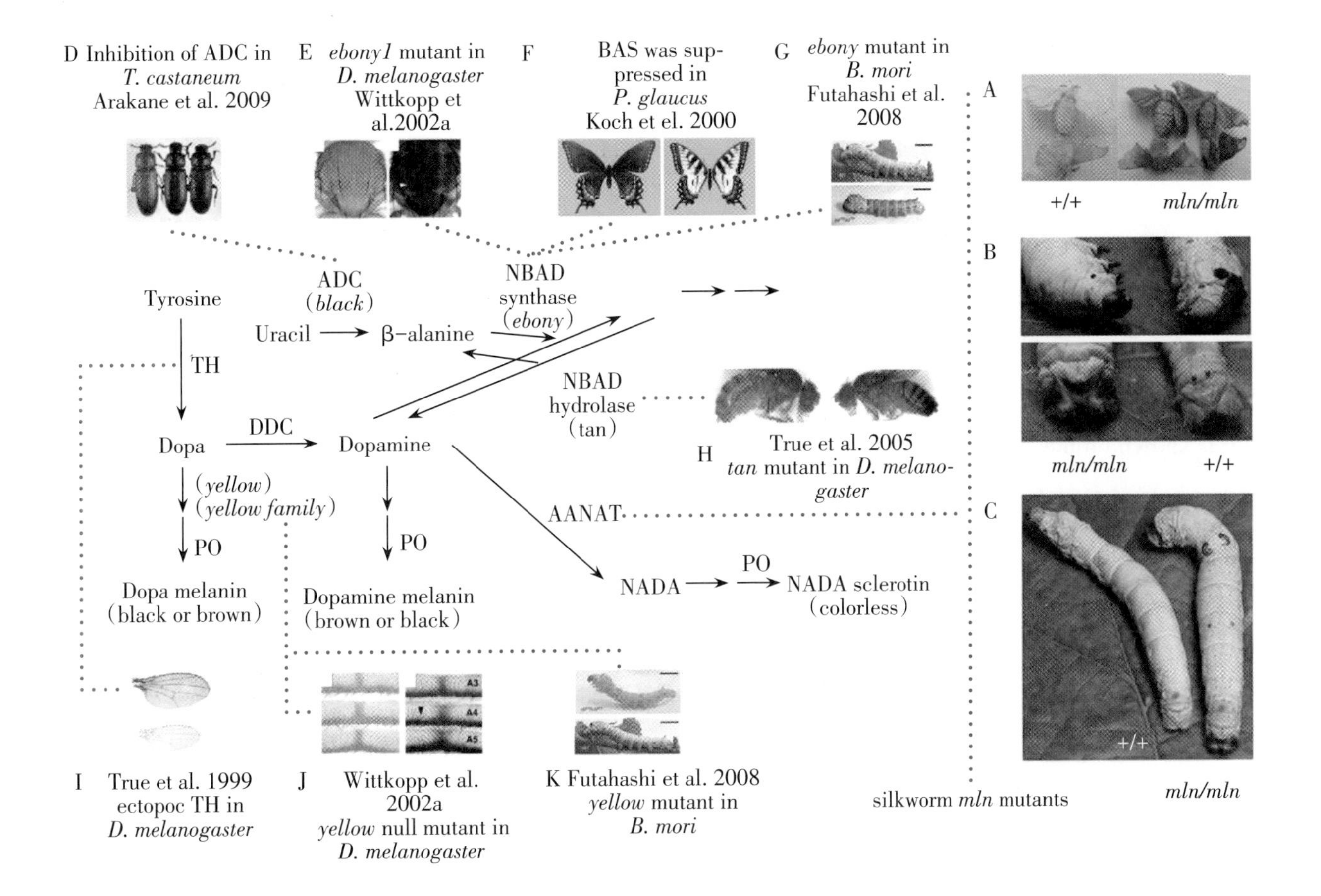

Figure 98. Insect pigmentation pathway and melanin synthesis enzyme-induced phenotypes. The proposed insect pigmentation pathway [largely according to published literature (True, 2003; Wittkopp *et al.*, 2003a; Arakane *et al.*, 2009)] is illustrated in the center (see Discussion). (A) Adult wild-type (+/+) and mln mutant (*mln/mln*) *Bombyx mori*. (B) Regional pigmentation phenotypes of *mln* and wild-type larvae. In *mln*, the head and forelegs are intensely black and the tail spot is brown, instead of colorless as in the wild type. (C) The *mln* locus has been introduced into a line for silk production. The left larva is *Jingsong*, a traditional high-yielding strain in China. (D-K) According to published reports, several representative phenotypes of insects are presented as follows: (D) the middle specimen indicates *Tribolium castaneum* injected with dsRNA for *ADC*, and the right-hand specimen indicates the *T. castaneum* black mutant [reproduced with permission (Arakane *et al.*, 2009)]; (E) on the right, the thorax phenotype of the *D. melanogaster* *ebony*[1] mutant [reproduced with permission (Wittkopp *et al.*, 2002a)]; (F) on the left, NBAD synthase (BAS) activity is suppressed, resulting in a melanic female *Papilio glaucus* phenotype [reproduced with permission (Koch *et al.*, 2000)]; (G) the phenotype of the *B. mori* so mutant is shown below and is determined by the deletion of the silkworm *ebony*

gene [reproduced with permission (Futahashi *et al.*, 2008)]; (H) the phenotype of the *D. melanogaster tan*^{5} mutant is shown on the right [reproduced with permission (True *et al.*, 2005)]; (I) ectopic melanin pattern typical of *hsp70-GAL4/+*; *UAS-TH/+* (ectopic tyrosine hydroxylase) flies is shown above [reproduced with permission (True *et al.*, 1999)]; (J) on the left is the thorax of a *D. melanogaster yellow* mutant [reproduced with permission (Wittkopp *et al.*, 2002a)]; (K) the top specimen indicates the *B. mori ch*^{k12} mutant caused by a *yellow* mutation [reproduced with permission (Futahashi et al., 2008)]. The insects in D-K not specifically mentioned above are wild type. Our findings provide the first model of an arylalkylamine N-acetyltransferase (AANAT) mutation contributing to an ectopic color pattern. Red dotted lines connect pairs of enzymes and their corresponding phenotype. ADC, amino acid decarboxylases; DDC, DOPA decarboxylase; LM, long mutant; NADA, N-acetyl dopamine; NBAD, N-β-alanyl dopamine; PO, phenoloxidase.（选自：*Disruption of an N-acetyltransferase gene in the silkworm reveals a novel role in pigmentation*，参见本书第670页。）

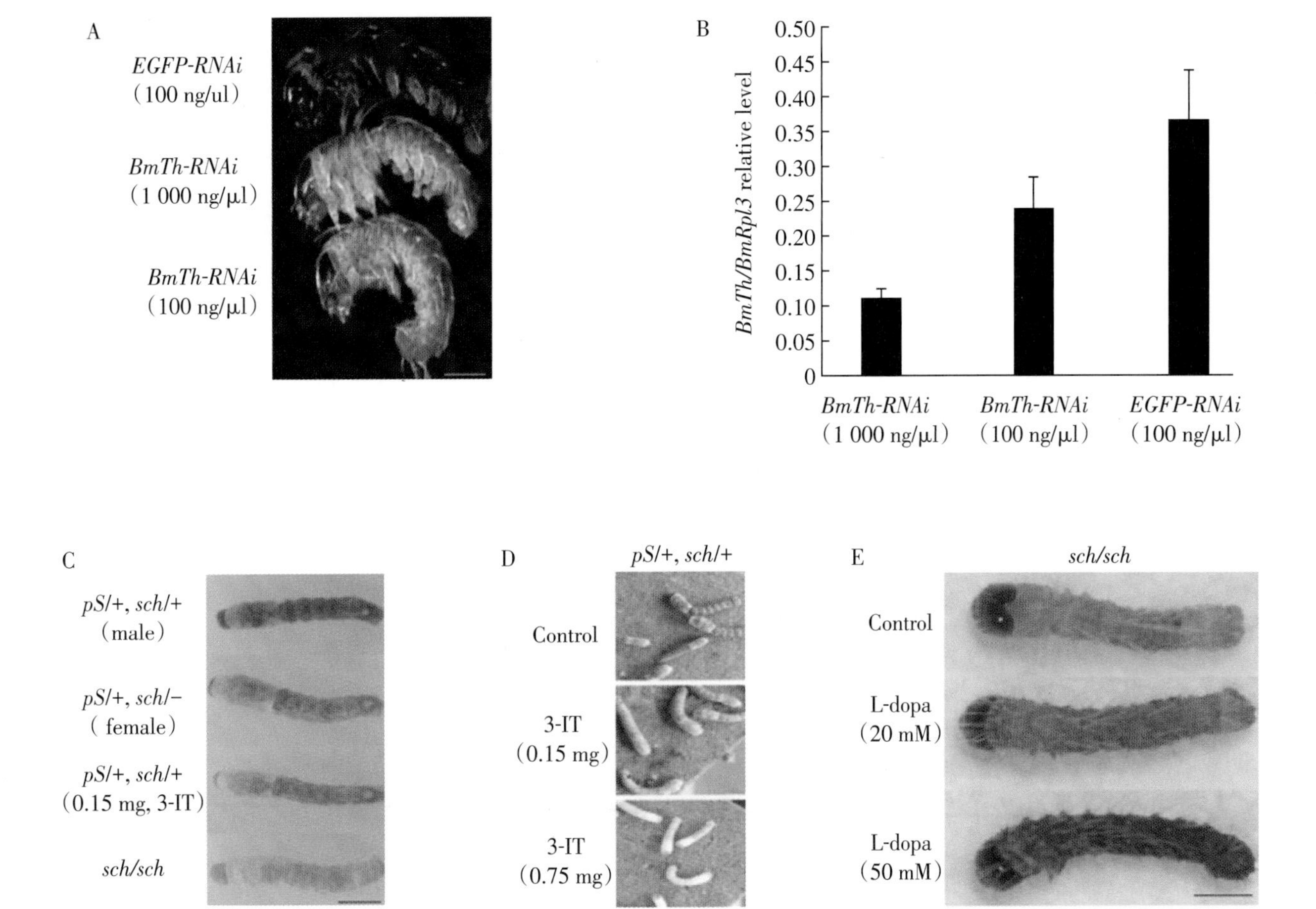

Figure 99. Functional analysis of the *BmTh* gene. (A) RNAi effects on coloration in neonate larvae. Double-stranded RNA for *EGFP* (0.3 ng per individual, Top) or *BmTh* (3 ng per individual, Middle, and 0.3 ng per individual, Bottom) were injected into eggs within 4 h after laying. (Scale bar, 0.5 mm.) (B) Dose-dependent RNAi-induced suppression of targeted gene transcripts. *EGFP*-RNAi is shown as control. (C) Effects of the TH inhibitor, 3-iodo-tyrosine (3-IT), on the cuticular pattern of second instar day 0 larvae. (Scale bar, 1.5 mm.) (D) Dosage-dependence of treatment with 3-IT on the black stripe of the larval cuticle on second instar day 0 larvae. (E) Rescue of the *sch* mutant phenotype from chocolate to black color by feeding L-dopa to *sch* neonate larvae. (Scale bar 0.5 mm.)(选自：*Repression of tyrosine hydroxylase is responsible for the sex-linked chocolate mutation of the silkworm, Bombyx mori*，参见本书第672页。)

A

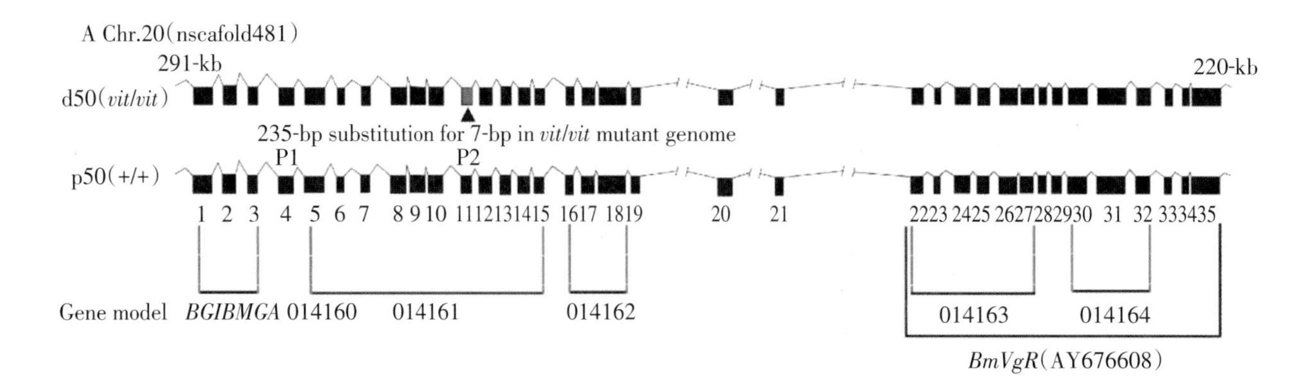

B

S LBD1 EGF1 LBD2 EGF2 O T C

PaVgR NH2 COOH
BgVgR NH2 COOH
RmVgR NH2 COOH
PhVgR NH2 COOH
NlVgR NH2 COOH
DmYPR NH2 COOH
AaVgR NH2 COOH
AgVgR NH2 COOH
ApVgR NH2 COOH
AsVgR NH2 COOH
BmVgR NH2 COOH
SlVgR NH2 COOH
AmVgR NH2 COOH
AfVgR NH2 COOH
BiVgR NH2 COOH
MrVgR NH2 COOH
SiVgR NH2 COOH
HsVgR NH2 COOH
NvVgR NH2 COOH
TcVgR NH2 COOH

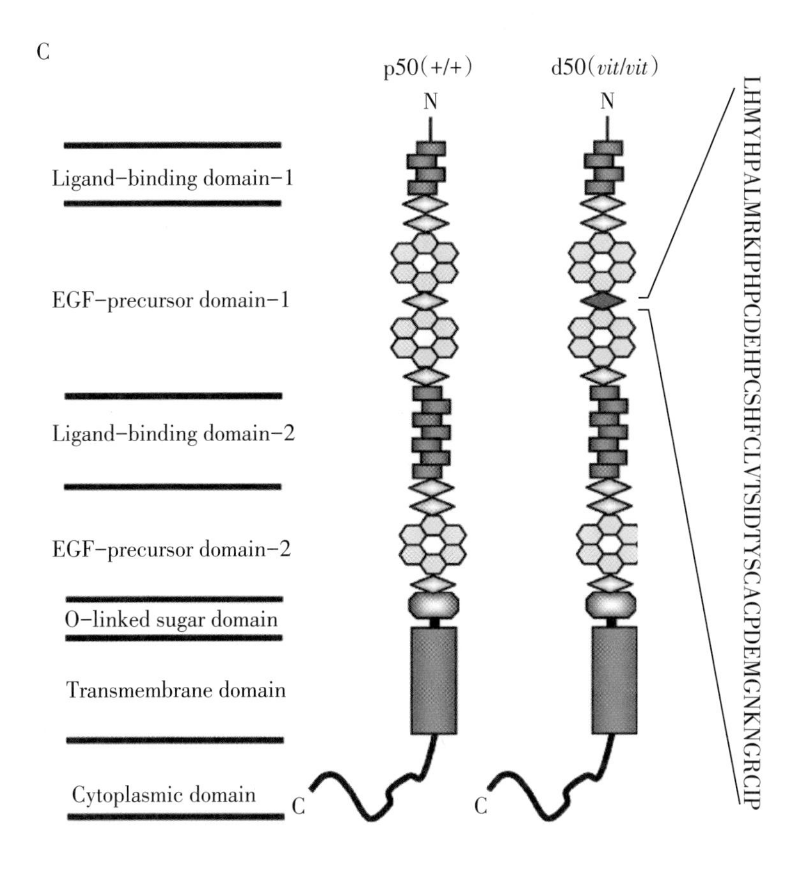

Figure 100. Sequence and structural analysis of *BmVgR* of WT and the *vit* mutant strains. A, cloning of complete *BmVgR* cDNA and the gene structure alignment between d50 (*vit*/*vit*) and p50 (WT) strains. Total RNA was extracted from ovaries of d50- and p50-strain moths on the day of eclosion. Degenerate primers were designed on the basis of the coding sequence (CDS) sequence of *BmVgR* (*BGBMGA014160-4*), AY676608, and genomic sequence (nscafold481) downloaded from the silkworm genome database, SilkDB v2.0. P, probe (numbers are order number). B, schematic alignment of modular domains from *Bm*VgR and other insects VgR / YPR. *P. americana* VgR (*Pa*VgR, BAC02725), *B. germanica* VgR (*Bg*VgR,CAJ19121), *Rhyparobia maderae* VgR (*Rm*VgR, BAE93218), *Pediculus humanus* corporis VgR (*Ph*-VgR, EEB10383), *Nilaparvata lugens* VgR (*Nl*VgR, ADE34166), *D. melanogaster* YPR (*Dm*YPR, AAB60217), *Aedes aegypti* VgR (*Aa*VgR, AAK15810), *Anopheles gambiae* VgR (AgVgR, EAA06264), *A. pernyi* VgR (*Ap*VgR, AEJ88360), *A. selene* VgR (*As*VgR, AFV32171), *B. mori* VgR (*Bm*VgR, ADK94452), *S. litura* VgR (*Sl*VgR, ADK94033), *A. mellifera* VgR (*Am*VgR, XP_001121707), *Apis florae* VgR (*Af*VgR, XP_003690500), *Bombus impatiens* VgR (*Bi*VgR, XP_003489577), *Megachile rotundata* VgR (*Mr*VgR, XP_003704660), *S. invicta* VgR (*Si*VgR, AAP92450), *Harpegnathos saltator* VgR (*Hs*VgR, EFN84770), *Nasonia vitripennis* VgR (*Nv*VgR, XP_001602954), and *Tribolium castaneum* VgR (*Tc*VgR, XP_968903). *S*, signal peptide; *LBD*, class A (complement-type) Cys-rich repeats; *EGF*, class B Cys-rich (EGF-type) repeats; *O*, O-linked sugar domain; *T*, transmembrane domain; *C*, cytoplasmic tail. The functional domains of *Bm*VgR are similar tothe VgR/YPR of other insect species but differ in having four class A repeats in LBD1 and seven class A repeats in LBD2, which might be unique to Lepidoptera. C, predicted structural organization of *Bm-VgR* with reference to Tufail *et al* (5). *Bm*VgR has five modular domains, similar to other LDLR family members. It harbors two of the LBD and EGF domains and has four class A repeats in LBD1 and seven class A repeats in LBD2. A deletion of 50 amino in the *vit* mutant compared with WT (shown in *red*), which belong to the third class B region of the EGF1 domain of *Bm*VgR.（选自：*Vitellogenin receptor mutation leads to the oogenesis mutant phenotype "scanty vitellin" of the silkworm, Bombyx mori*，参见本书第674页。）

A

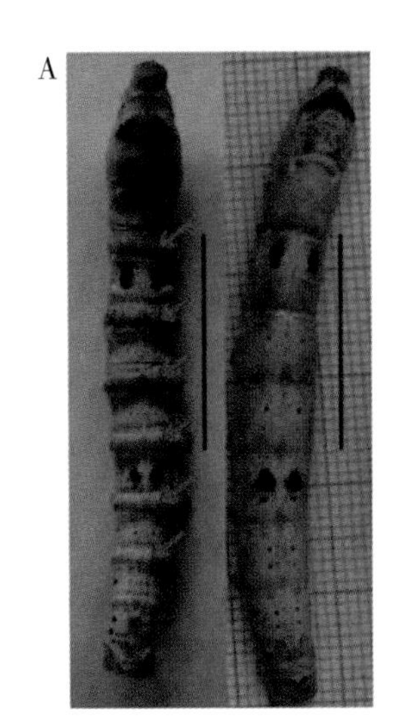

B

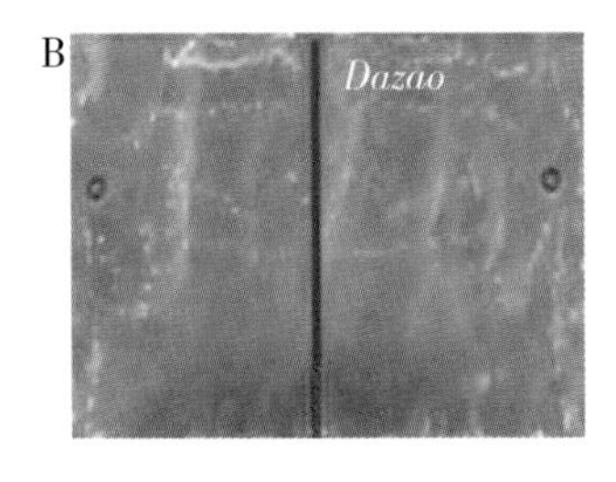

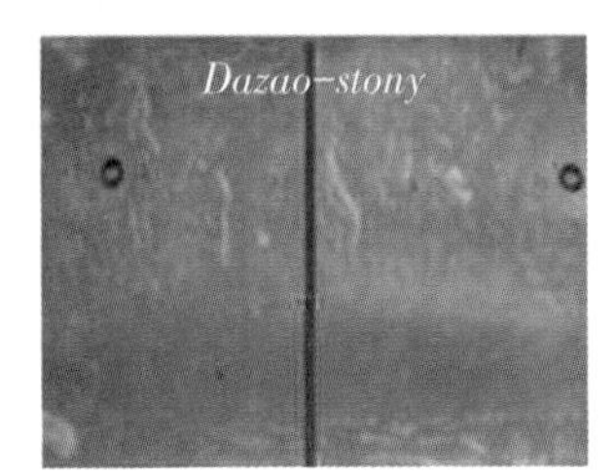

C

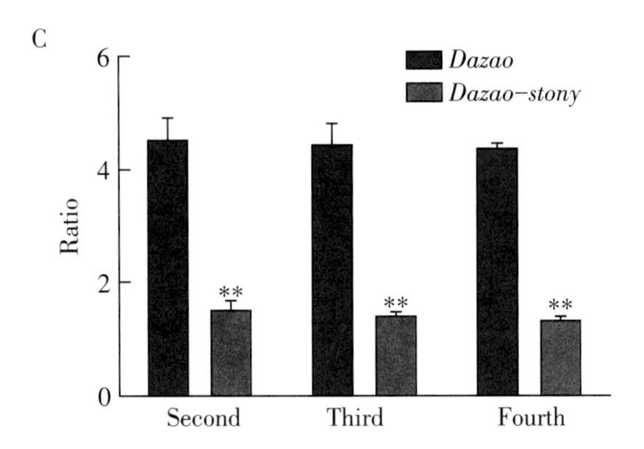

D

E

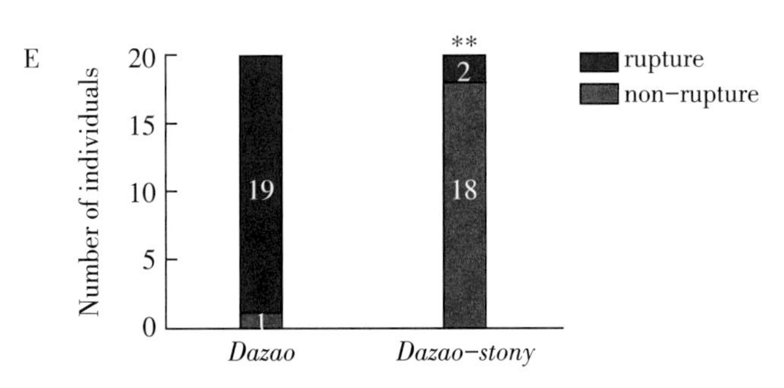

F

N=65

T > 30s

n=5 n=58**

P< 0.0001

n=55 n=7

90° 60°

T< 1s

n=5

Time / second

35 30 25 20 15 10 5 0

Dazao Dazao-stony

G

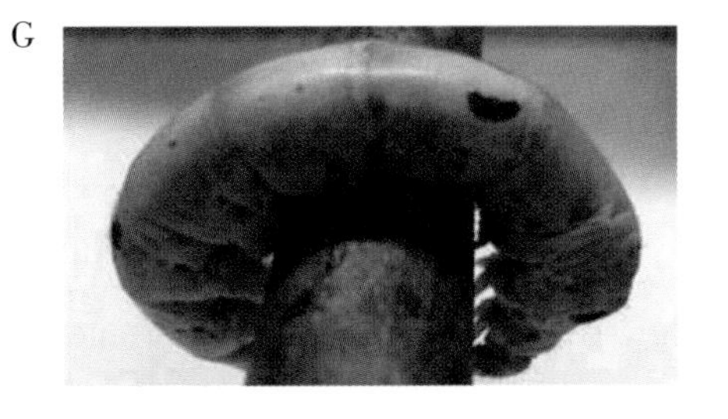

H

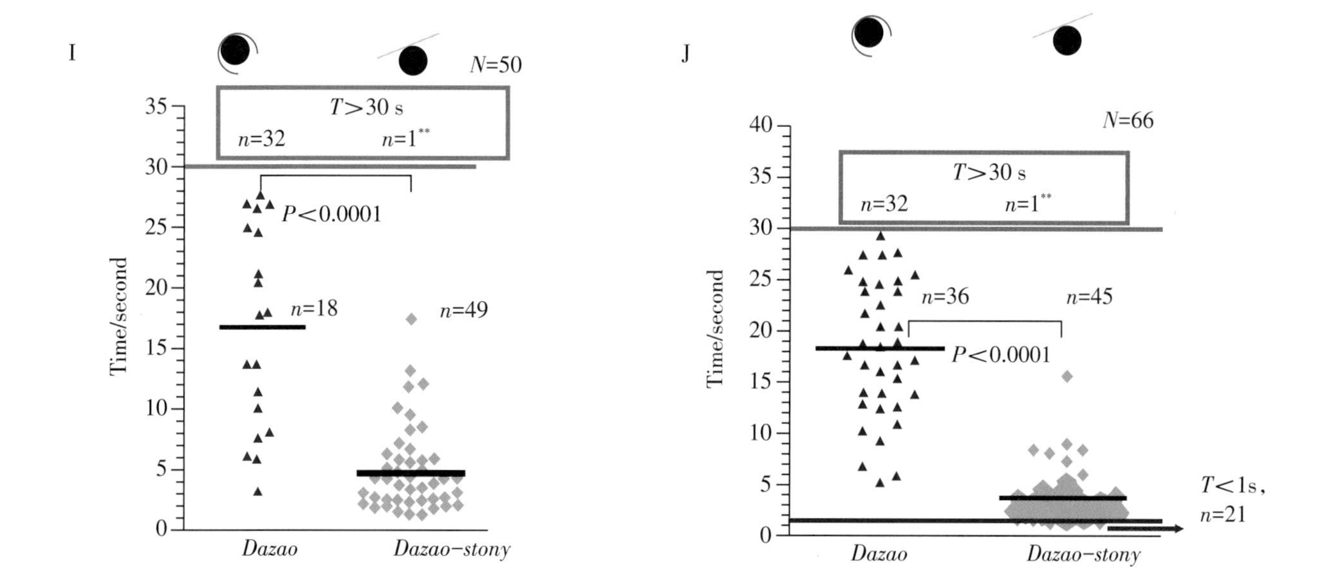

Figure 101. Characterization of *Dazao-stony* mutant. (A) Phenotype of the dorsal side of *Dazao* (right) and *Dazao-stony* (left). Red arrows indicate bulges. Scale bar, 2 cm. (B) Dorsal cuticle anatomy characteristics of *Dazao* and *Dazao-stony.* The blue and red lines represent the length spanned by internode and intersegmental fold, respectively. (C) Ratios between the length of internode and intersegmental fold in the second, third, and fourth abdomen segments of *Dazao* and *Dazao-stony*, respectively (n = 5). (D) Ratio of the midgut content mass and larval volume between *Dazao* and *Dazao-stony* (n = 6). Data represent mean values ±SD. Student's t-test; *, P<0.05; **, P<0.01 (C and D). (E) Statistics of breached Dazao and Dazao-stony larvae in dropping experiment. (χ^2-test; **, P<0.01). (F) Analysis of the larval abdominal crooking test. Both red and blue lines represent threshold time. The numbers in the red box indicate individuals that can-not crook to abdomen in 30 sec. (χ^2-test; **, P<0.01). The counterclockwise and clockwise arrowheads represent the crooking angles in *Dazao* and *Dazao-stony*, respectively. Both black lines represent the average time of abdomen bending between *Dazao* and *Dazao-stony* within the threshold time. (Mann-Whitney U-test, P < 0.0001). (G and H) Phenotypes of grasping mallet of *Dazao* and *Dazao-stony*, respectively. (I and J) Analysis of grasping tests on the large and small diameter mallets of *Dazao* and *Dazao-stony*, respectively. Both the red and blue lines signify threshold time. The numbers in the red box represent individuals that grasp for >30 sec. (χ^2-test; **, P<0.01). Both black lines represent the aver-age time of grasping between *Dazao* and *Dazao-stony* within the threshold time (Mann-Whitney U-test, P<0.0001). The black solid circle represents the cross-section of the mallet. Blue and yellow curves represent the ventral bending of *Dazao* and *Dazao-stony*, respectively.（选自：*Mutation of a cuticular protein, BmorCPR2, alters larval body shape and adaptability in silkworm, Bombyx mori*，参见本书第678页。）

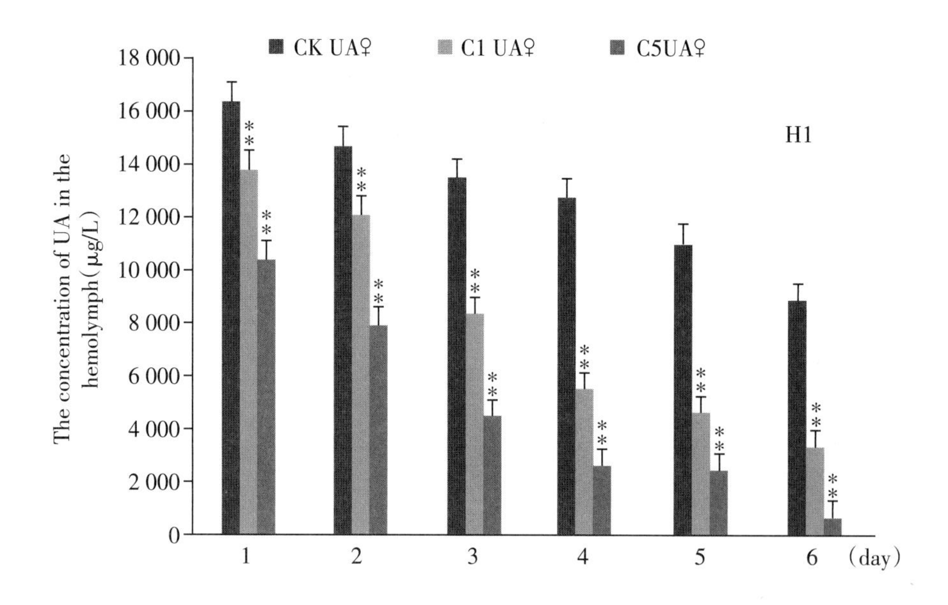
CK UA♀
C1 UA♀
C5UA♀
H1
The concentration of UA in the hemolymph(μg/L)
18 000
16 000
14 000
12 000
10 000
8 000
6 000
4 000
2 000
0
1
2
3
4
5
6
(day)

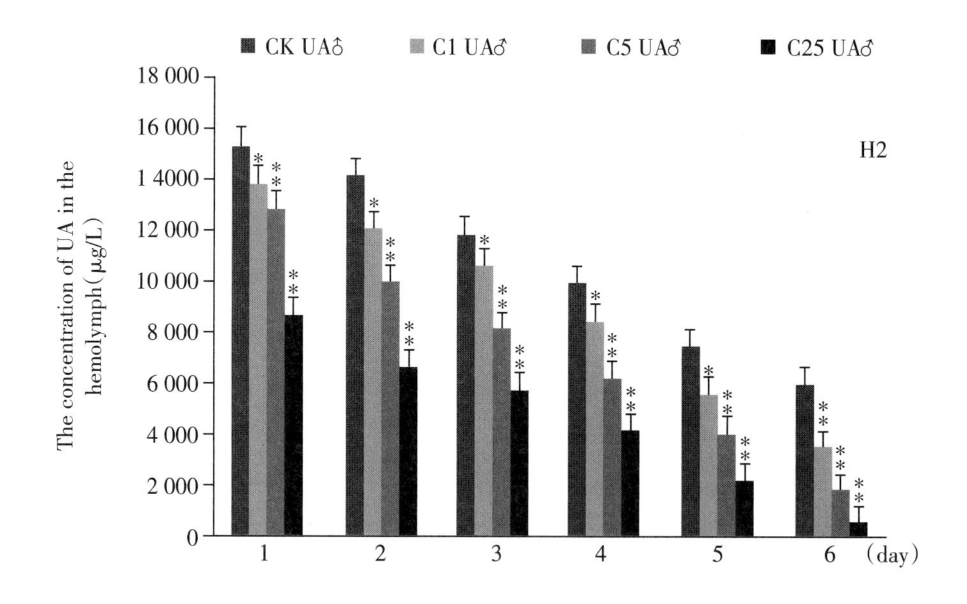
CK UA♂
C1 UA♂
C5 UA♂
C25 UA♂
H2
The concentration of UA in the hemolymph(μg/L)
18 000
16 000
1 4000
12 000
10 000
8 000
6 000
4 000
2 000
0
1
2
3
4
5
6
(day)

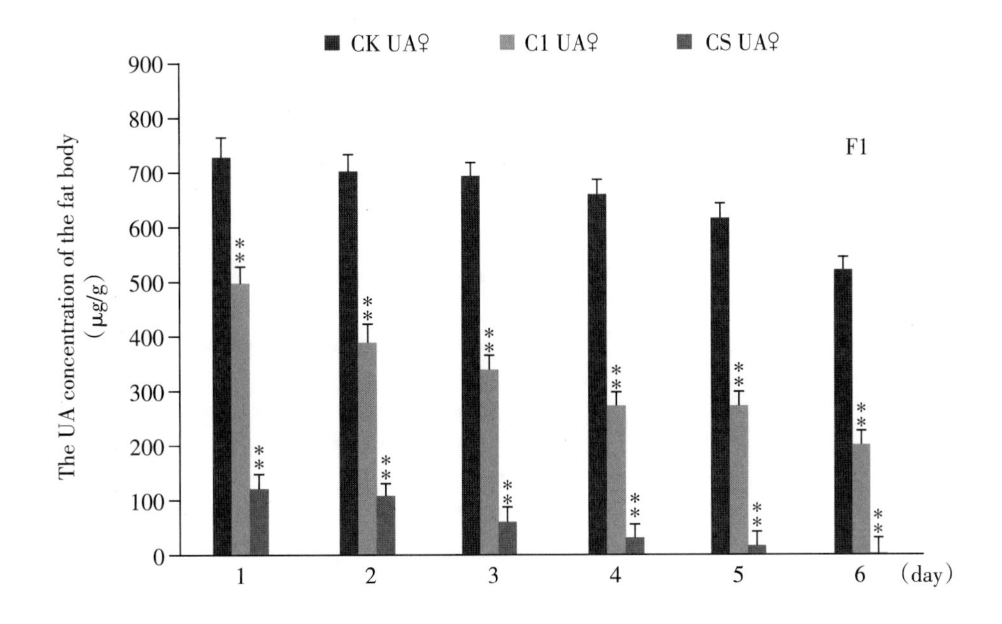

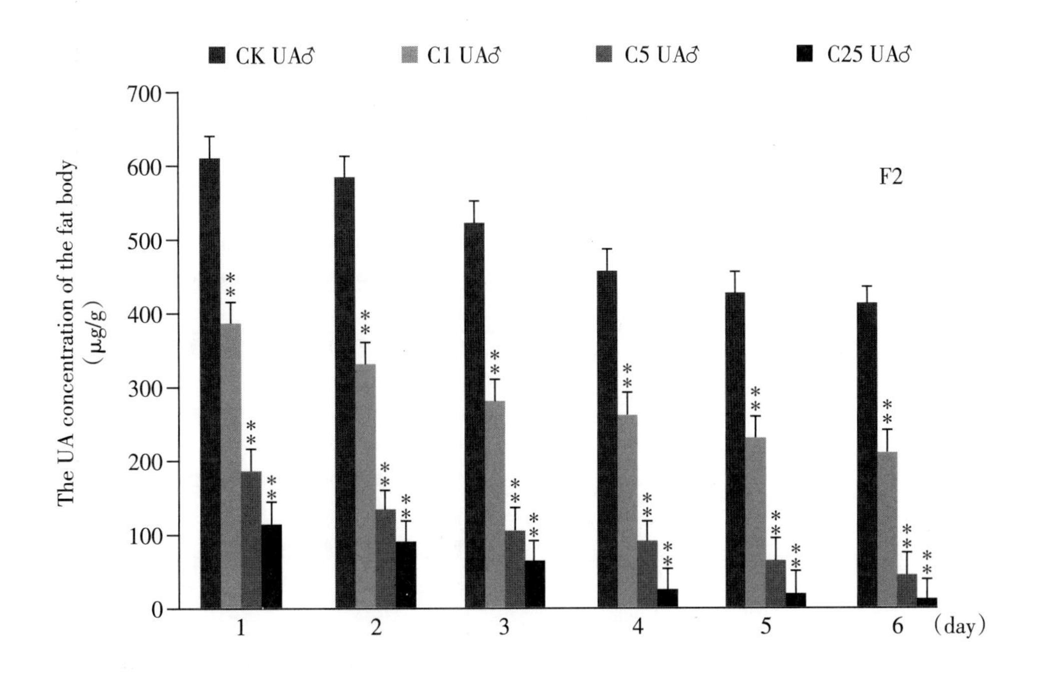

Figure 102. Effects of treatment of allopurinol on the content of uric acid in the hemolymph and fat body of 5th instar *Bombyx mori*. H1 and H2 were for the content of uric acid in the hemolymph of the female and male, respectively. F1 and F2 were for the content of uric acid in the fat body of the female and male, respectively. CK: the control group, C1, C5, and C25 denoted the concentration of allopurinol were 1, 5, and 25 mg/mL, respectively (silkworm variety: *Dazao*, $n = 5$). High quality figures are available online.（选自：*Silkworms can be used as an animal model to screen and evaluate gouty therapeutic drugs*，参见本书第682页。）

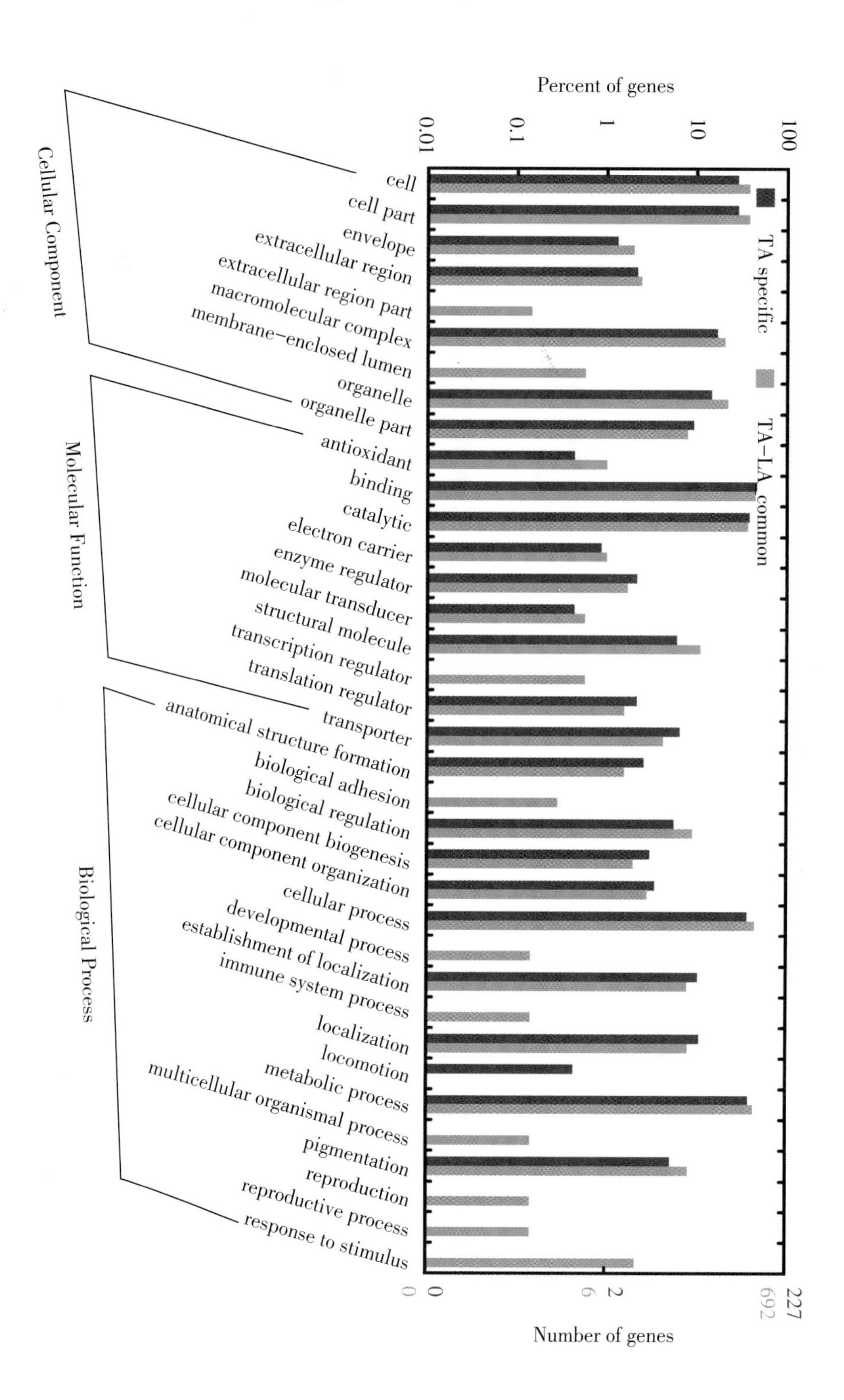

Figure 103. GO classifications of the identified proteins from the *B. mori* embryos at TA. The TA-specific and TA-LA common proteins were classified into cellular component, molecular function and biological process categories by WEGO according to the GO terms.(选自:*Shotgun proteomic analysis on the embryos of silkworm Bombyx mori at the end of organogenesis*,参见本书第694页。)

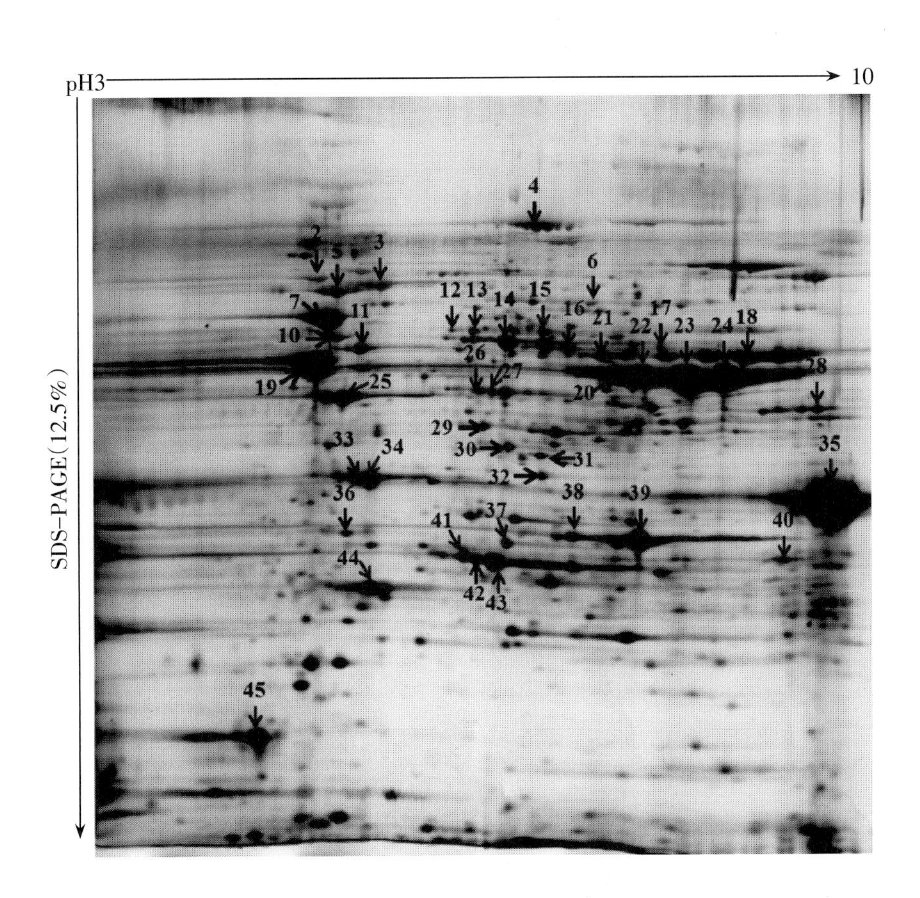

Figure 104. 2-DE analysis of silkworm MT proteins. The protein samples were added to the pH 3 to 10 NL IPG strips, followed by SDS-PAGE on a 12.5% polyacrylamide gel. The gels were stained using silver staining. Spots analyzed by MALDI-TOF MS were designated by numbers.（选自：*Proteomic-based insight into malpighian tubules of silkworm Bombyx mori*, 参见本书第712页。）

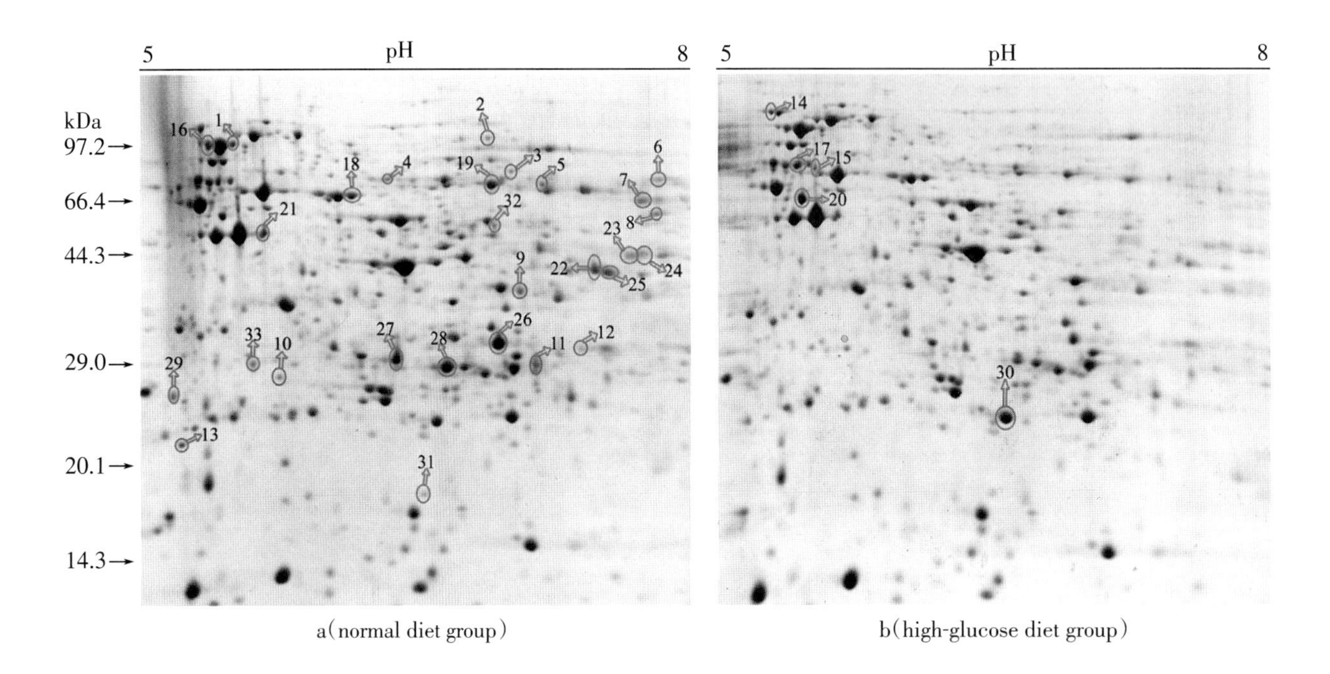

Figure 105. The 2-DE of silkworm midgut proteins. The protein extraction and 2-DE were performed as described in the Materials and Methods section. The differentially expressed proteins are indicated by arrows and numerically label. (a) Normal diet group. (b) High-glucose diet group. (选自：*Comparative proteomic analysis reveals the suppressive effects of dietary high glucose on the midgut growth of silkworm*, 参见本书第714页。)

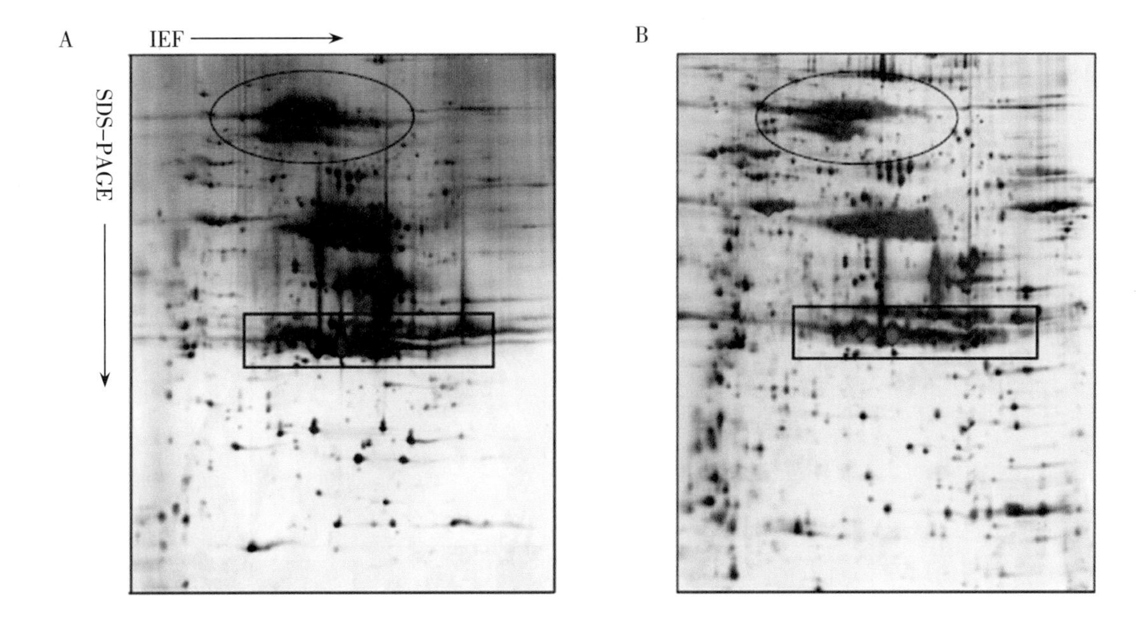

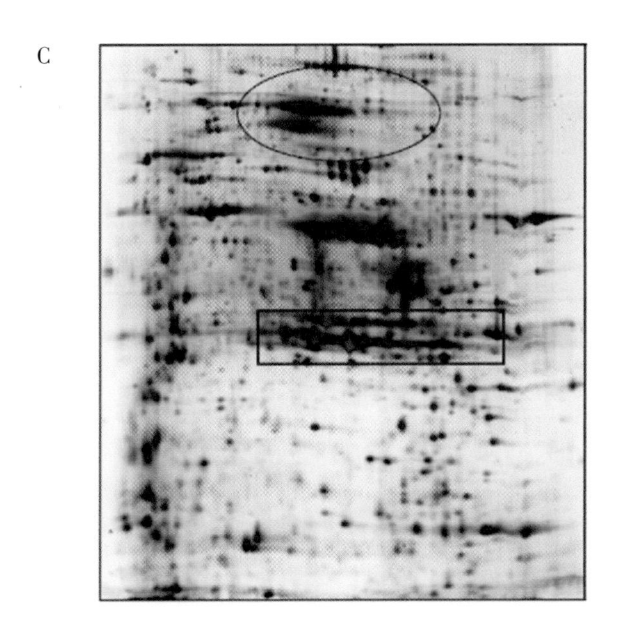

Figure 106. Comparison of 2-DE patterns of silkworm eggs prepared by sample preparation method I (A); and sample preparation method II with Tris-soluble fraction in the extract mixture at various proportions: 20% (B) and 40% (C). The same amount of proteins (220 μg) was loaded on each gel. IPG Strip (18 cm, pH 3-10 NL), SDS-PAGE vertical gel (12.5%), silver stain.（选自：*Development of an effective sample preparation approach for proteomic analysis of silkworm eggs using two-dimensional gel electrophoresis and mass spectrometry*，参见本书第716页。）

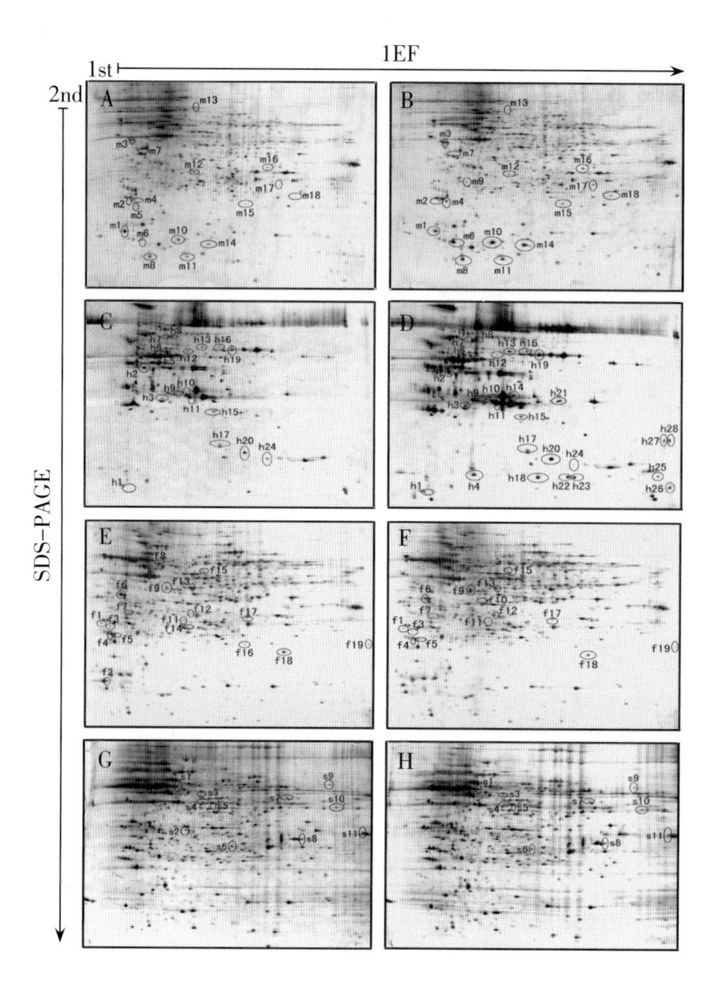

Figure 107. 2-DE pattern of the proteins in various silkworm tissues. The proteins of midgut (A and B), hemolymph (C and D), fat body (E and F) and posterior silk gland (G and H) of silkworms reared on fresh mulberry leaves (A, C, E, and G) and artificial diet (B, D, F, and H). Sixty micrograms of protein was applied to the IPG strip (24 cm, pH 3-10, L) and 12.5% SDS-PAGE was carried out for separation in the second dimension. The differentially expressed protein spots were shown in gels with a small letter and an Arabic numeral, the small letter ‘m’, ‘h’, ‘f’ and ‘s’ represent midgut, hemolymph, fat body and posterior silk gland, respectively. The circle or ellipse indicates the successfully identified protein spots. [选自：*Comparative proteomic analysis between the domesticated silkworm (Bombyx mori) reared on fresh mulberry leaves and on artificial diet*，参见本书第720页。]

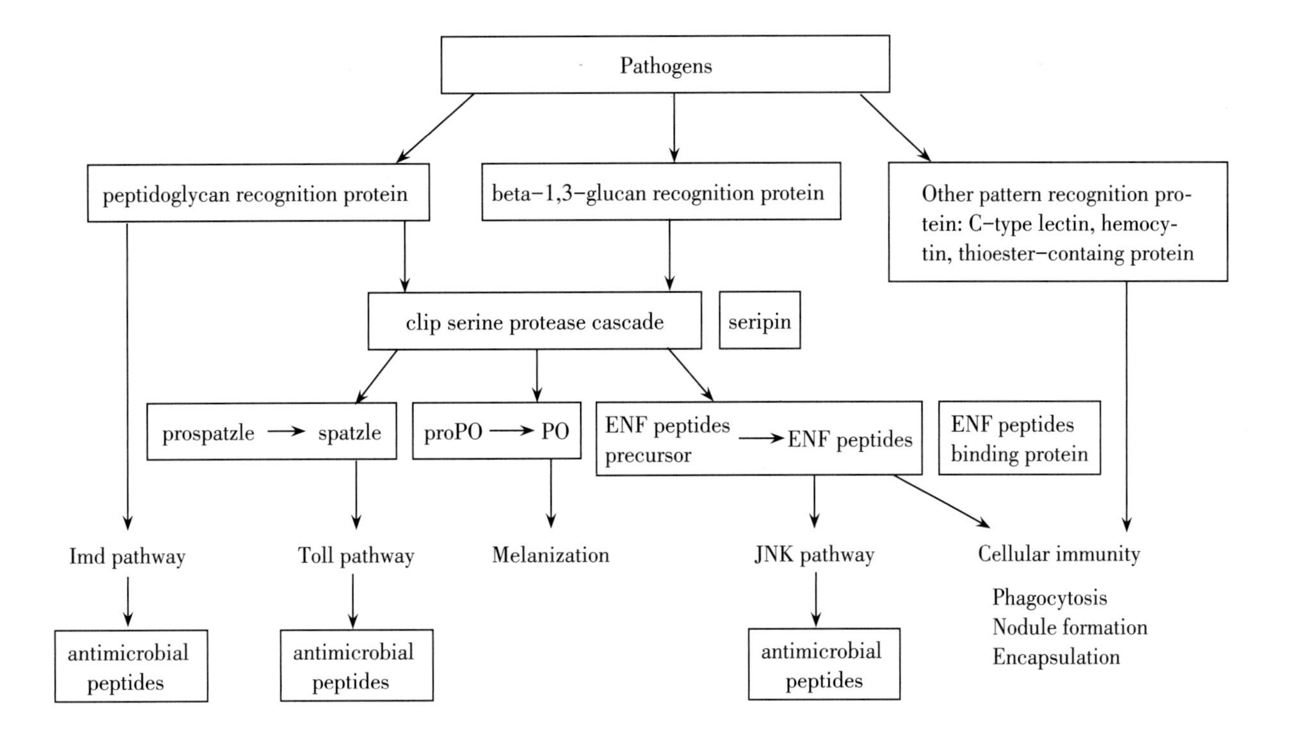

Figure 108. Pathways of immune responses in *B. mori* hemolymph. This figure presents both humoral immunity pathways (Imd pathway, Toll pathway, Melanization, JNK pathway) and cellular immunity pathways in hemolymph. PO represents phenoloxidase. Serpins act as regulators of Clip serine proteases, and ENF peptides-binding, and cellular immunity proteins act as regulators of ENF peptides.（选自：*Proteomics of larval hemolymph in Bombyx mori reveals various nutrient-storage and immunity-related proteins*，参见本书第 726 页。）

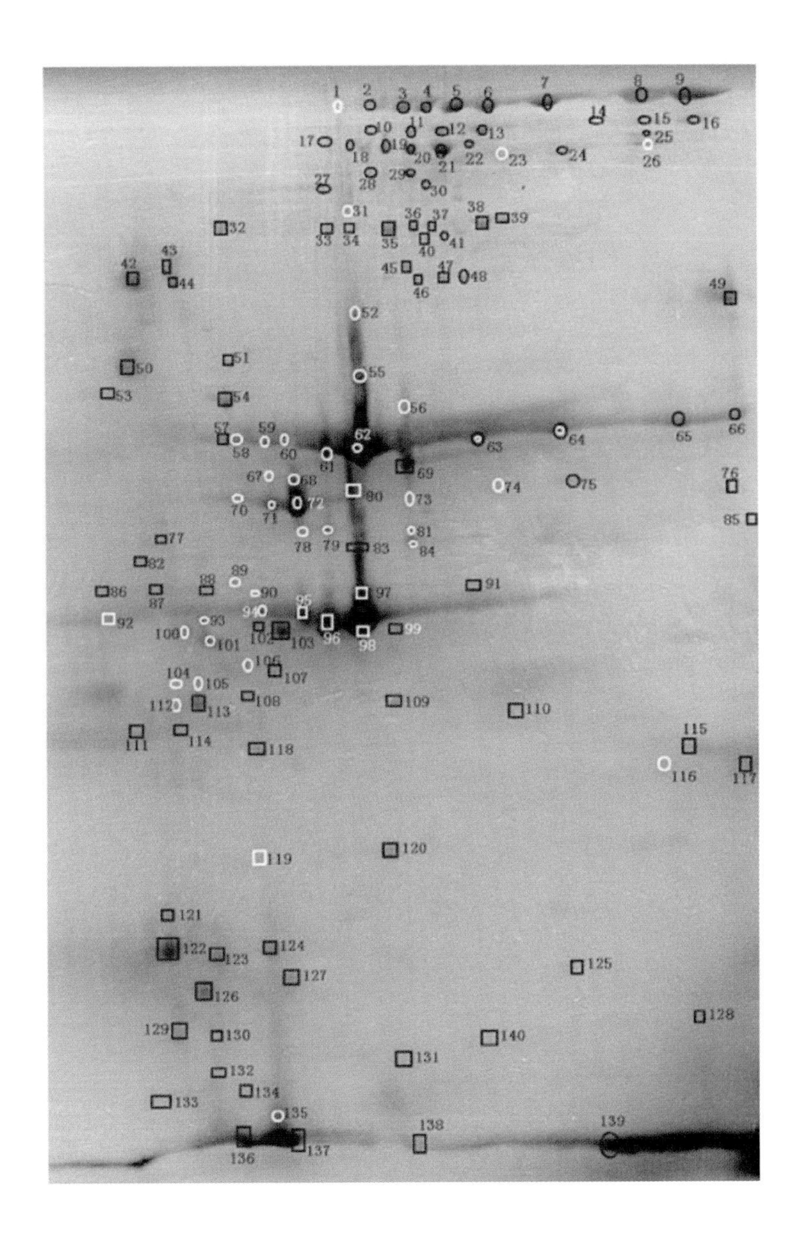

Figure 109. ODV proteins separated on a 2-DE and stained with Coomassie blue. The enclosed spots were cut for mass spectral identification. See tables 1 and 2 for identification of these bands. Protein spots 8, 46, 47, 54, 69, 77, 82, 83, 86, 91, 99, 109, 110, 115, 117, 122, 123, 124, 126, 129, 130, 136 and 138 did not match any proteins in NCBI databases. Black circles = GP41/P40; white circles = VP39; white frame = polyhedron, and black frame = other proteins.（选自：*Determination of protein composition and host-derived proteins of Bombyx mori nucleopolyhedrovirus by 2-dimensional electrophoresis and mass spectrometry*，参见本书第728页。）

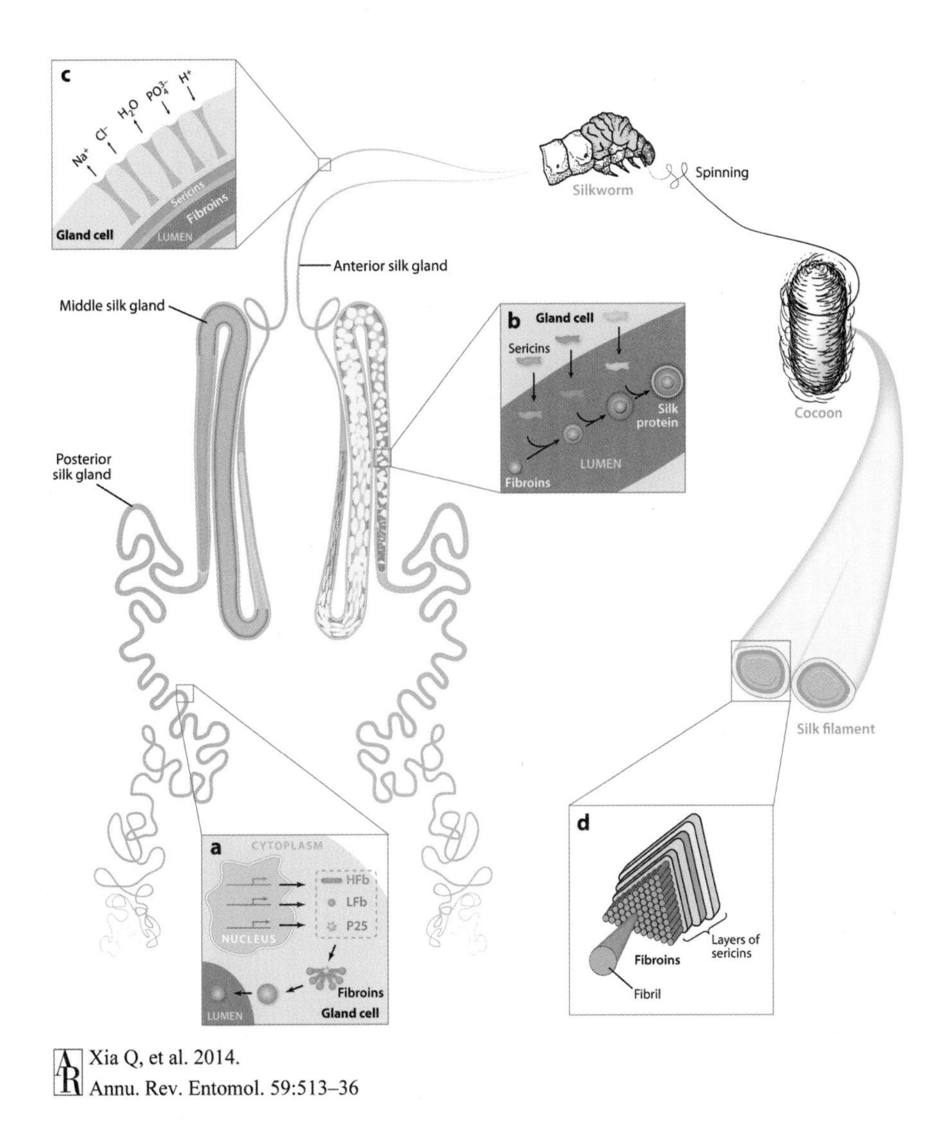

Figure 110. A representative schematic of silk protein synthesis and secretion in the silk glands of *Bombyx mori*. (a) The fibroins (heavy chain fibroin, light chain fibroin, and P25 protein) are synthesized, transformed from a random coil to β-sheets, and assembled as a complex composed of disulfide-linked heavy (H-fibroin, Hfb) and light (L-fibroin, Lfb) peptides and a fibrohexamerin (P25) protein at a molar ratio of 6∶6∶1 in the posterior silk gland (PSG). (b) The fibroin complexes are transported to the middle silk gland (MSG), where the sericin proteins are synthesized, are secreted into the lumen, and then coat the surface of the fibroins to form the silk protein complexes. (c) The silk protein complexes are transported and further assembled into silk fibers under the proper conditions of pH and metal ion concentrations along the anterior silk gland. (d) The silk fiber, which consists of fibroins and sericins, is finally spun by mature fifth instar larvae to form a cocoon. The silk gland at the left illustrates the regions where different sericins (*green, blue, and yellow*) are synthesized and secreted into the lumen, where they coat the surface of fibroins to form silk fibers (*pink*). The silk gland at the right illustrates the process of integration and formation of the silk fiber from the silk proteins along the MSG.（选自：*Advances in silkworm studies accelerated by the genome sequencing of Bombyx mori*，参见本书第744页。）

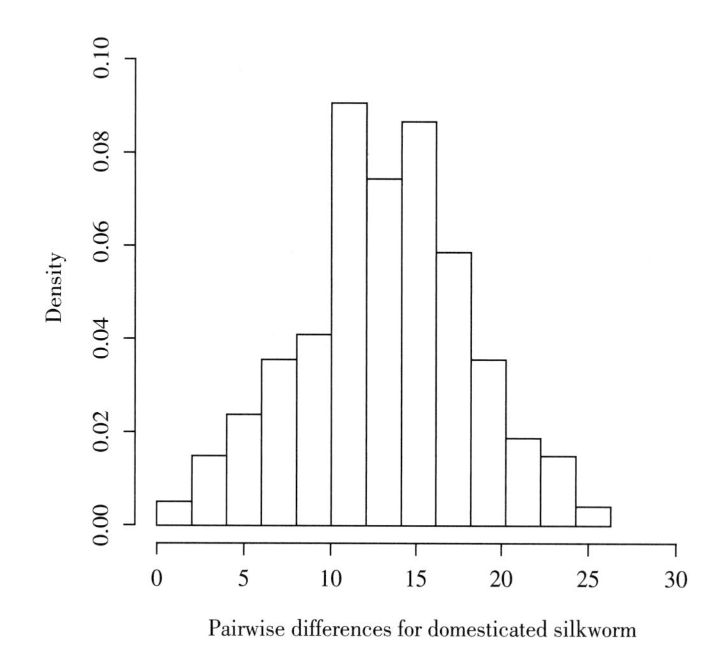

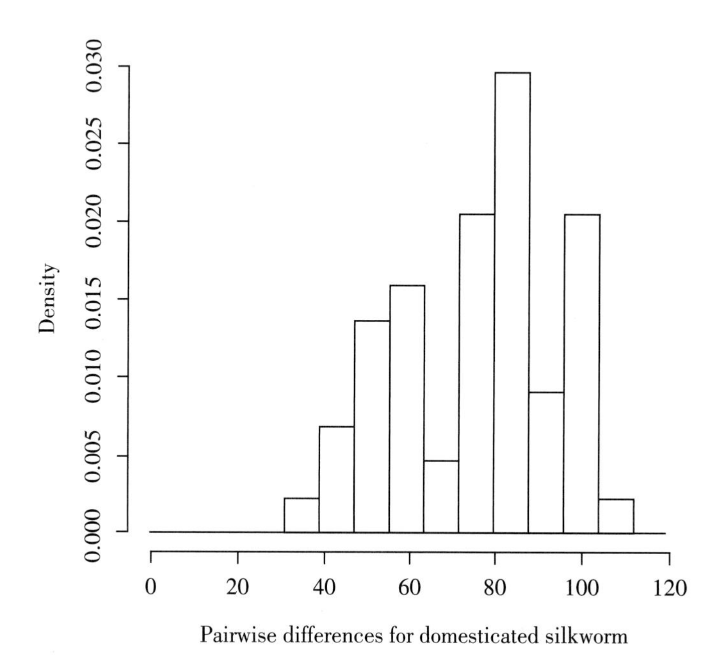

Figure 111. Pairwise difference distribution for domesticated silkworm genome sequences (top) and Chinese wild silkworm genome sequences (bottom). (选自：*Genetic diversity, molecular phylogeny and selection evidence of the silkworm mitochondria implicated by complete resequencing of 41 genomes*，参见本书第746页。)

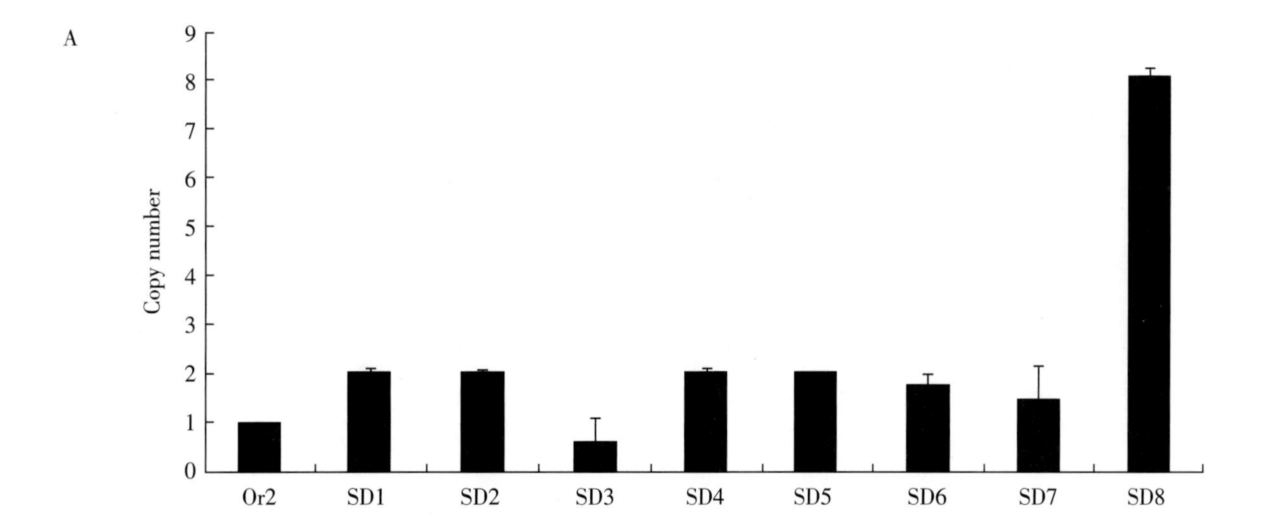

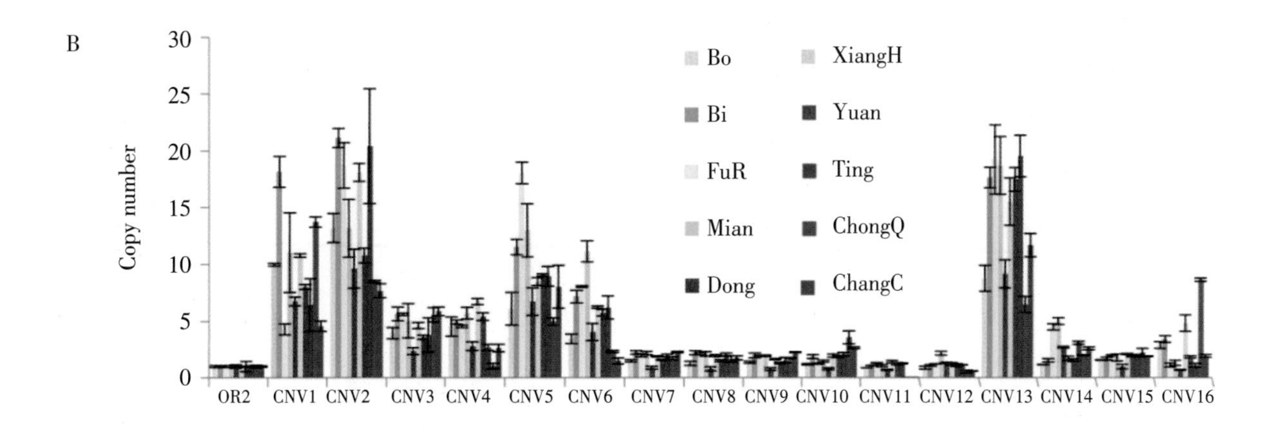

Figure 112. CNV validation using q-PCR. Aa Validation of CNV regions in *DZ* genome. Each bar represents one CNV region. *Bb* CNV investigation among ten different silkworm strains. The x-axis represents *OR2* gene and 16 selected CNV regions. *Each color* represents one silkworm strain. The domesticated strains include *Bo, Bi, FuR, Mian, Dong, XiangH, Yuan* and *Ting*. The wild strains include *ChongQ* and *ChangC*. The y-axis represents relative copy numbers (Color figure online).（选自：*Genome-wide patterns of genetic variation among silkworms*，参见本书第748页。）

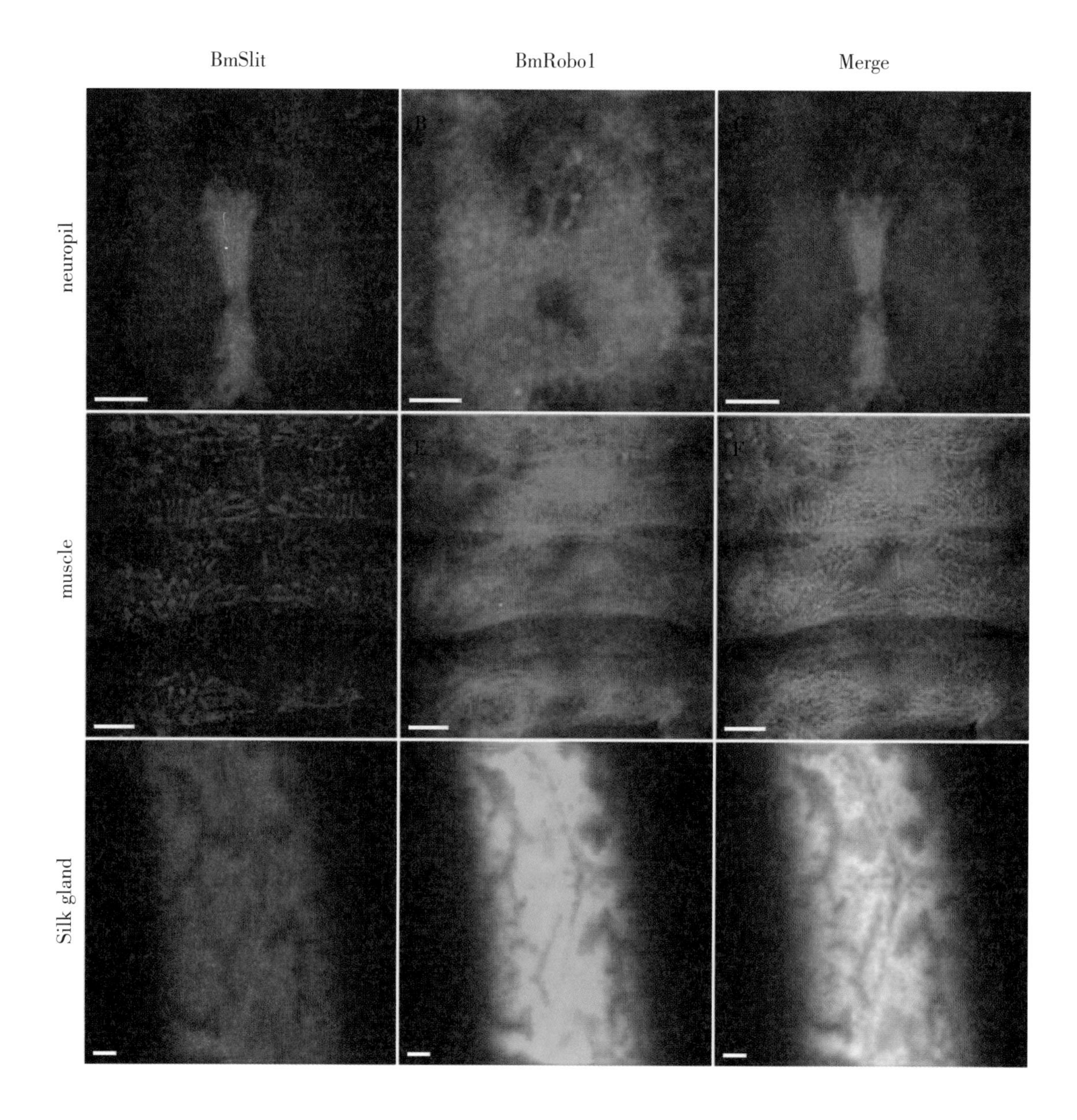

Figure 113. Colocalization of *Bm*Slit and *Bm*Robo1. (A-C) *Bm*Slit (A) and *Bm*Robo1 (B) were colocalized in the neuropil. (D-F) *Bm*Slit (D) and *Bm*Robo1 (E) were colocalized in the muscle. (G-I) *Bm*Slit (G) and *Bm*Robo1 (H) were colocalized in the silk gland. (C, F and I) Merged images of (A) and (B), (D) and (E), (G) and (H), respectively. Embryos of stage 22 and 23 and the silk gland of day 3 fifth instar larvae were stained by anti-*Bm*Slit and anti-*Bm*Robo1 antibodies. Scale bars represent 100 μm (A-C, G-I) and 400 μm (D-F).（选自：*Evolutionarily conserved repulsive guidance role of Slit in the silkworm Bombyx mori*，参见本书第754页。）

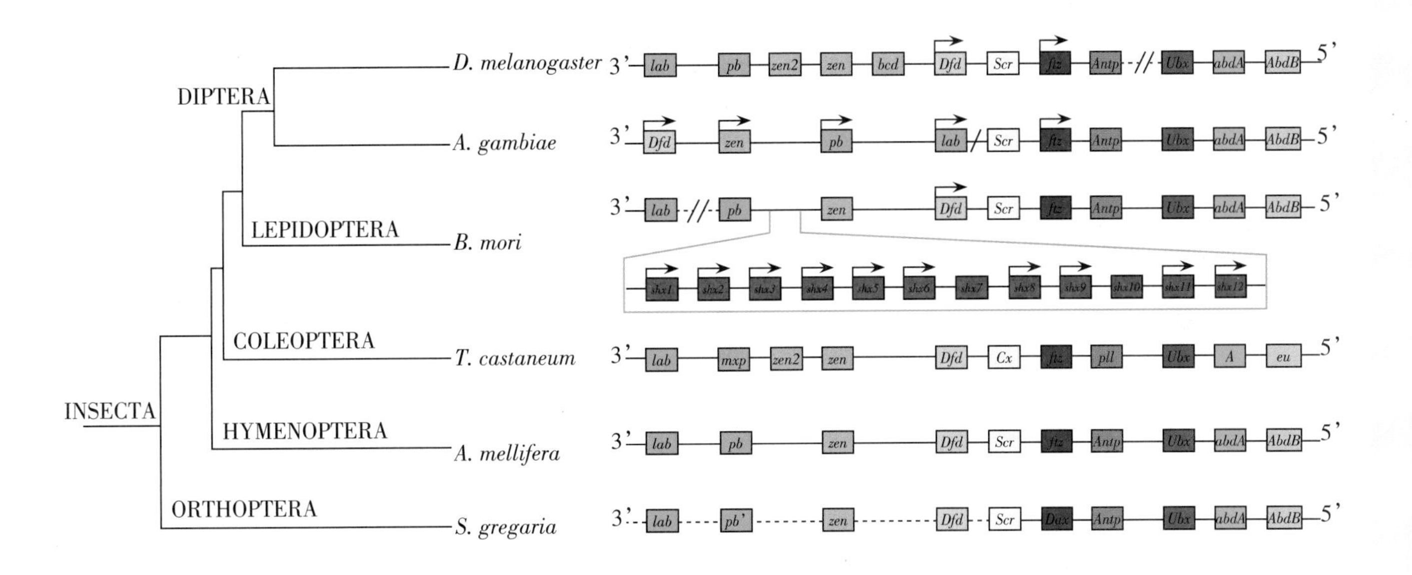

Figure 114. Alignment of homeodomain sequences from genes of silkworm *Hox* cluster. Amino acids identical with Ubx homeodomain sequence were shown as black points, while those specially conserved in the *Bmshx* genes are shown in a yellow background. The numbering scheme at the top of the figure refers to the amino acid positions within the 60-residue canonical homeodomain. The positions of the three alpha helices found in the X-ray structure of engrailed (Kissinger *et al.*, 1990) were shown above the figure.（选自：*A genomewide survey of homeobox genes and identification of novel structure of the Hox cluster in the silkworm, Bombyx mori*，参见本书第796页。）

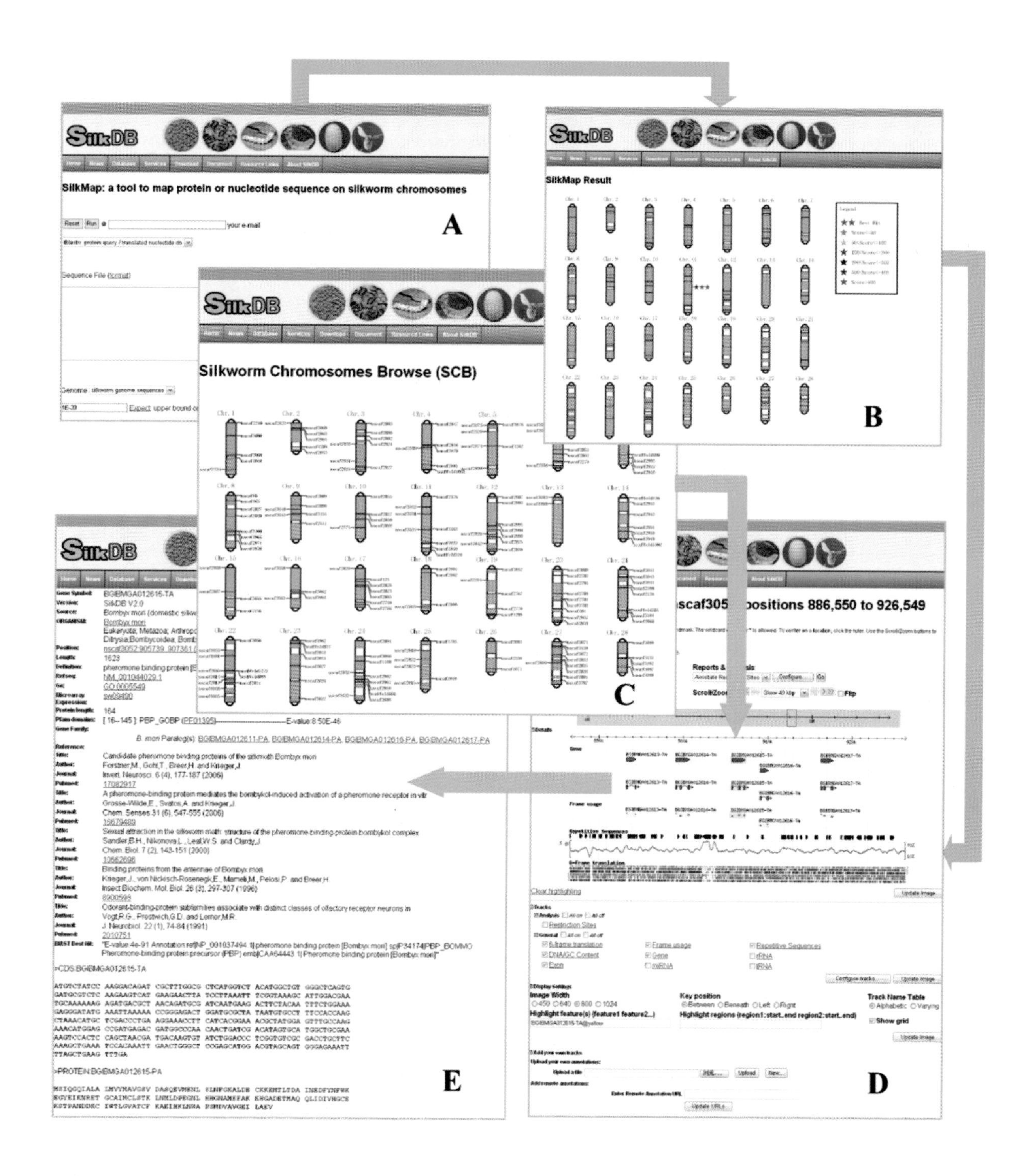

Figure 115. Example of searching and browsing of SilkDB. (A) Snapshot of the tool of SilkMap. This tool enables users to anchor protein or nucleotide sequence on silkworm chromosomes. (B) An example of search result for SilkMap. The pentagrams indicate the position on chromosome for query sequence, and provide the hyperlink to corresponding GBrowse view. (C) Snapshot of the tool of SCB. This tool enables users to click on any part of chromosome to access the page of detailed view by GBrowse. (D) Visualization of genome annotation by GBrowse. (E) An example of Gene Page. [选自：*SilkDB v2.0: a platform for silkworm (Bombyx mori) genome biology*，参见本书第798页。]

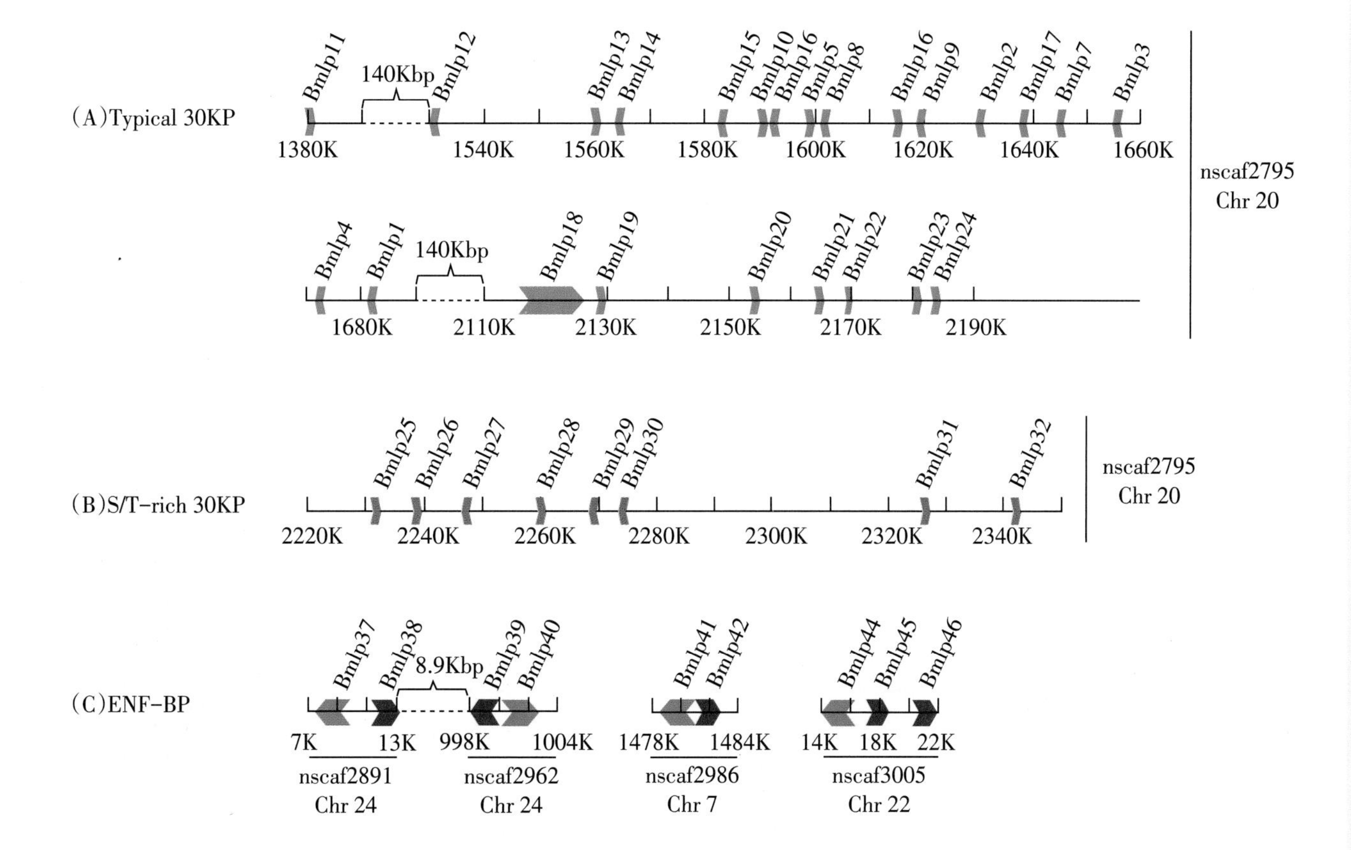

Figure 116. Genome locations of silkworm low molecular weight lipoprotein genes. Most of typical *30KP* and *S/T-rich 30KP* genes distribute on nscaf2795 of chromosome 20 (A, B). The typical *30KP* genes are shown in green and *S/T-rich 30KP* genes in purple. C *ENF-BP* genes on chromosome 24, 7, and 22. The genes of *ENF-BP* cluster I are in blue; the members of cluster II are in red. The direction of gene transcription is indicated by the arrow.（选自：*Identification of novel members reveals the structural and functional divergence of lepidopteran-specific Lipoprotein_11 family*，参见本书第806页。）

Figure 117. Phenotypes observed after injection of silkworm *βFTZ-F1* dsRNA. A dosage of 25 μg dsRNA for silkworm *βFTZ-F1* was injected in the silkworm individuals at just beginning wandering (W0). The silkworm individuals treated with 0.85% NaCl at W0 were used as controls. (a) *BmβFTZ-F1* dsRNA-treated silkworm individuals (right) can normally develop to the completion of spinning like controls (left) at W48 (48 h after wandering, just completing spinning). (b-d) The silkworm individuals after *BmβFTZ-F1* dsRNA injection failed to complete larva-pupa transition (right) at W72 (12 h after pupation of the controls), W96, or W120.（选自：*Nuclear receptors in Bombyx mori*: *Insights into genomic structure and developmental expression*，参见本书第810页。）

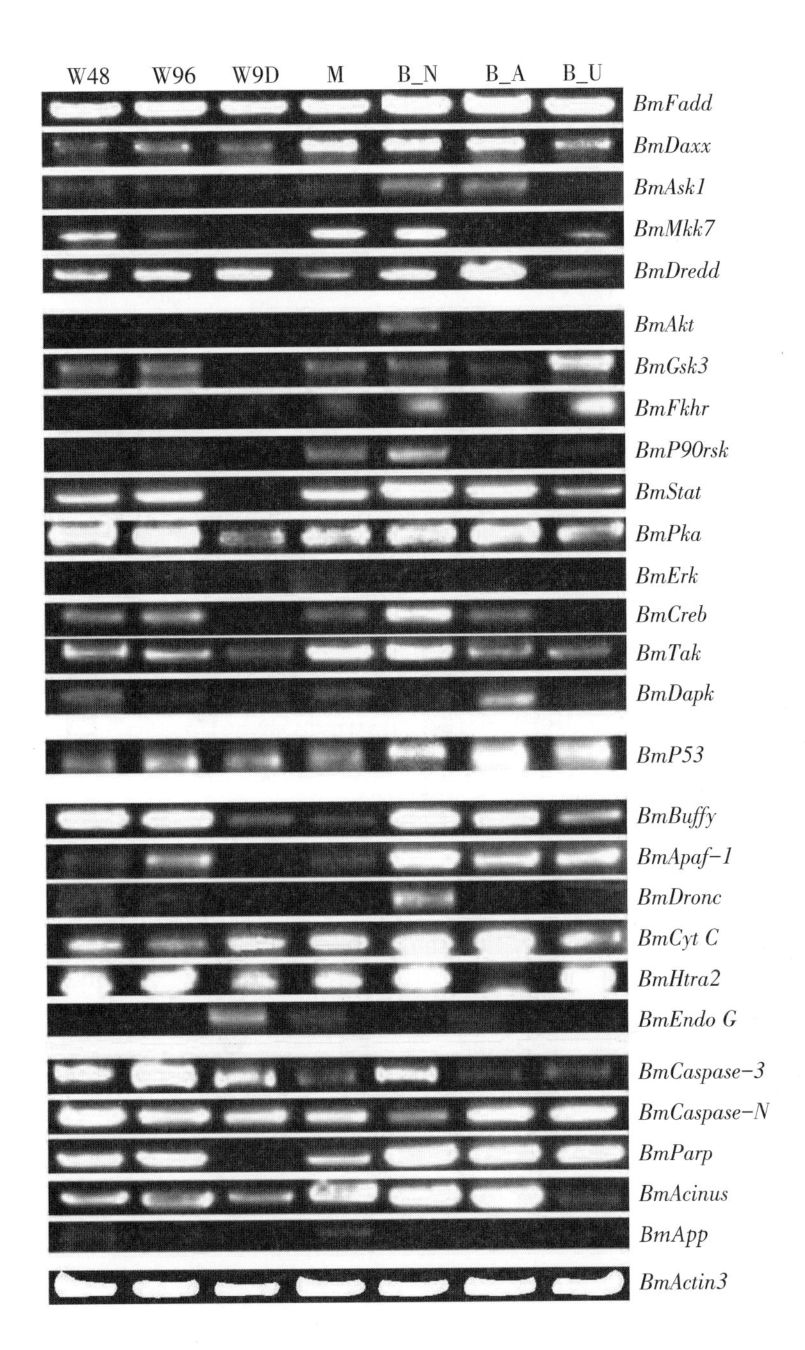

Figure 118. RT-PCR analysis of the expression of the putative silkworm apoptosis-related genes during different growth stages and in *BmE* cells. Total RNA was isolated as described in the Methods section and analyzed by RT-PCR. W48: about 48 h after wandering; W96: about 96 h after wandering; W9D: about 9th day after wandering; M: silk moth; B_N: normal *BmE* cells; B_A: *BmE* cells exposed to actinomycin D; B_U: *BmE* cells exposed to UV irradiation.（选自：*The genomic underpinnings of apoptosis in the silkworm, Bombyx mori*，参见本书第812页。）

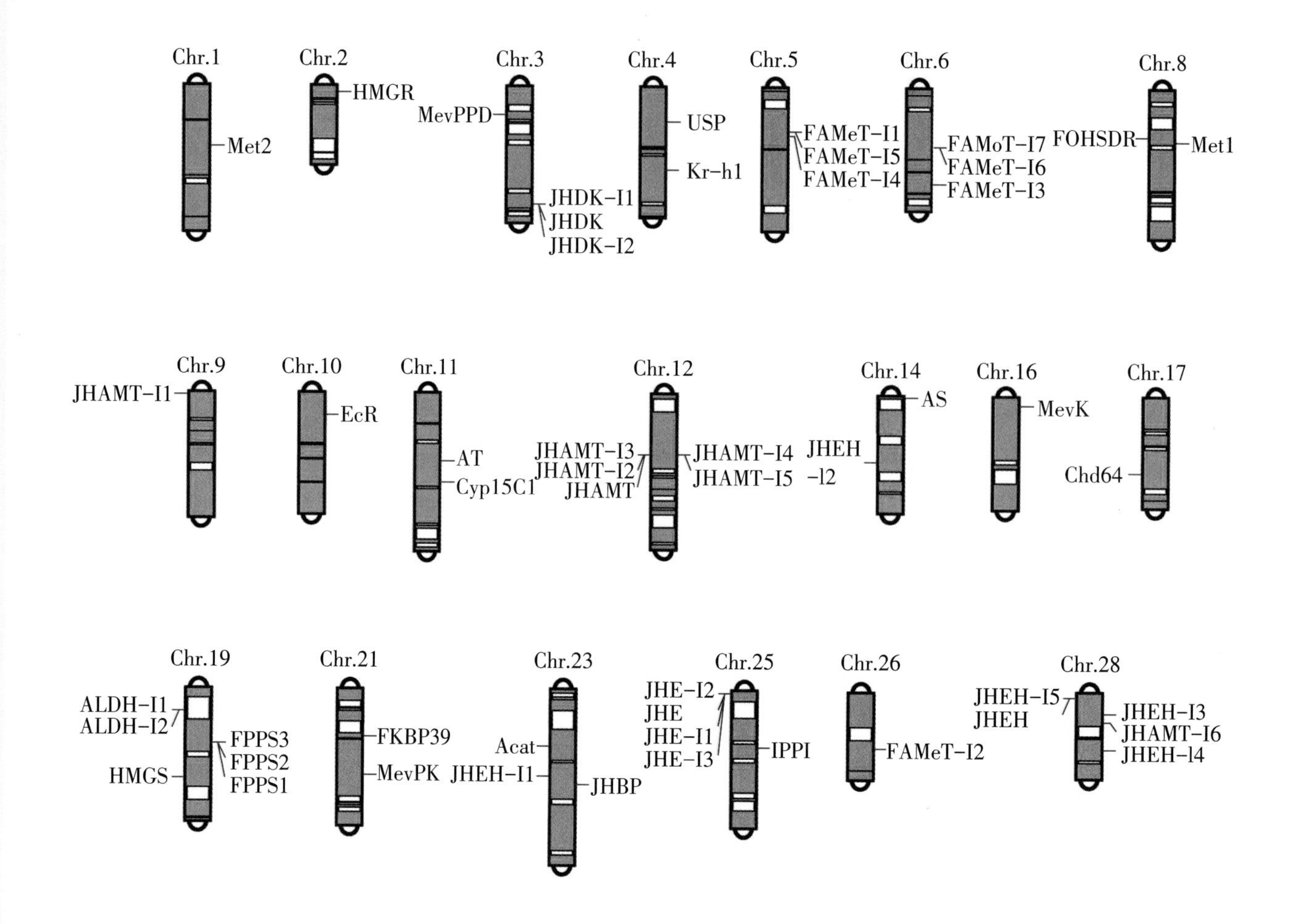

Figure 119. Chromosomal distribution of JH-related genes in *B. mori*. Based on the assembly of the whole-genome sequence and single-nucleotide polymorphism (SNP) markers linkage map for *B. mori*, a total of 52 JH-related genes were mapped on the different chromosomes of *B. mori*. Several different copies of each of the JH-related genes with multiple copies are clustered in tandem on the chromosomes.（选自：*Genome-wide comparison of genes involved in the biosynthesis, metabolism, and signaling of juvenile hormone between silkworm and other insects*，参见本书第816页。）

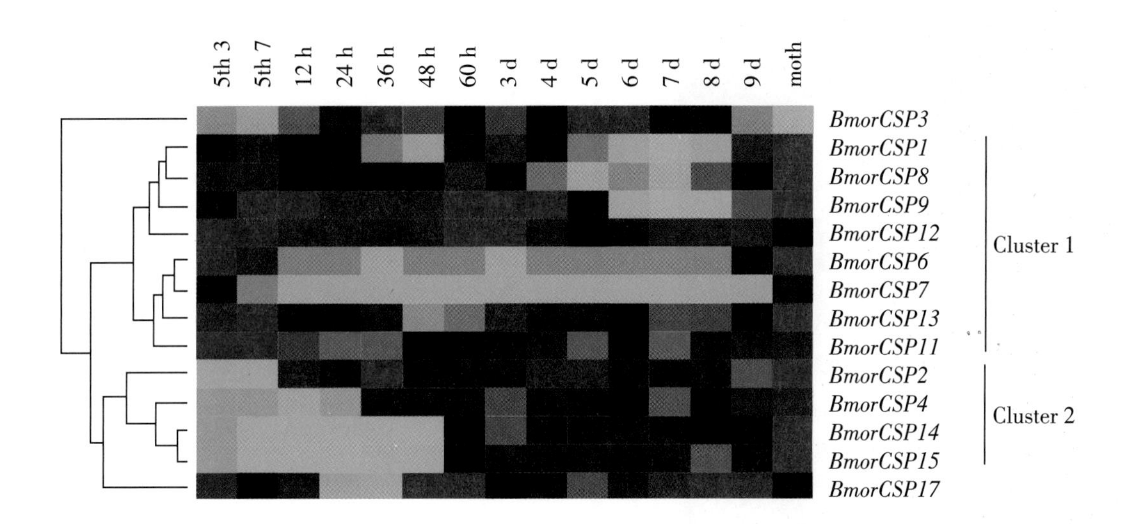

Figure 120. Expression levels for silkworm *CSP* genes at different developmental stages. The expression levels are multiples of the median for that gene.（选自：*Identification and expression pattern of the chemosensory protein gene family in the silkworm, Bombyx mori*，参见本书第824页。）

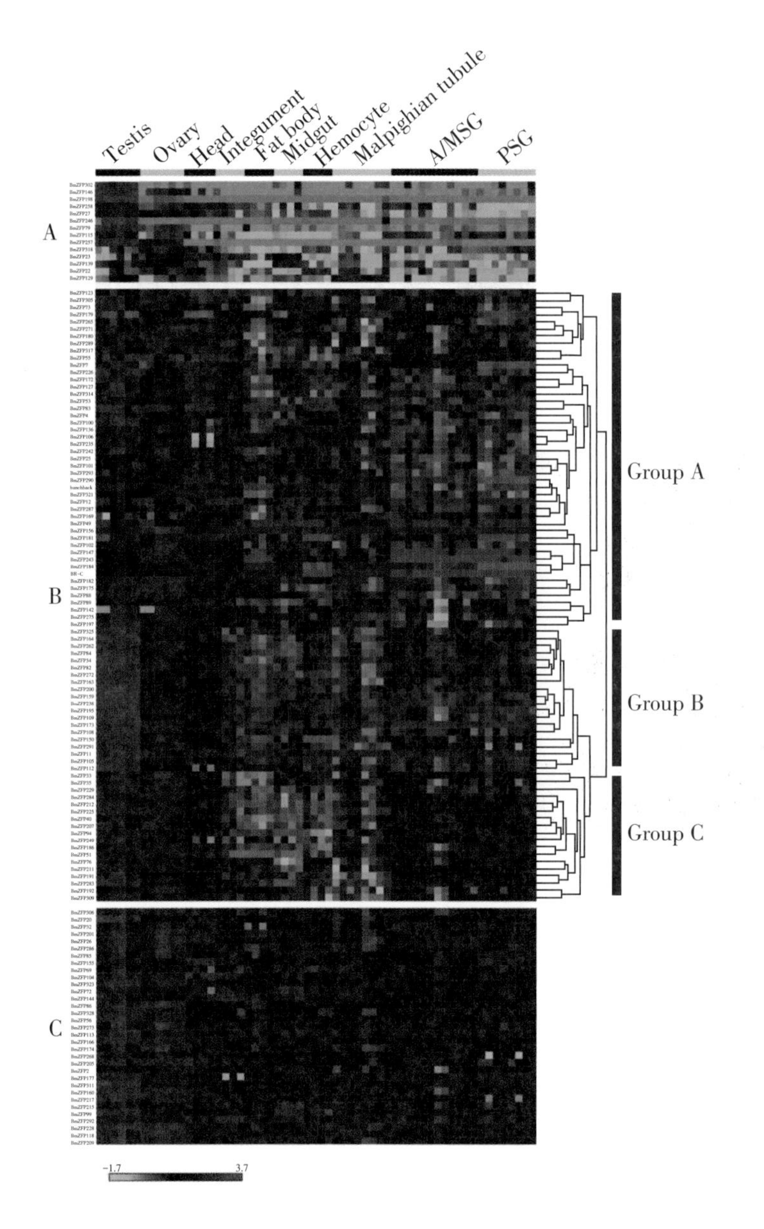

Figure 121. Analysis of expression patterns for *C2H2 ZFP* genes of silkworm. Hierarchical clustering with the average linkage method was performed using Cluster software (http://genome www. stanford. edu/clustering/), and the data was visualized by matrix2png (http://bioinformatics. ubc. ca/matrix2png/). Gene expression levels are represented by red and green boxes (denoting higher and lower expression levels, respectively). Vertical columns of boxes represent different genes, and horizontal rows represent different samples. (A) The *C2H2 ZFP* genes that showed tissue-specific expression. First, One-way ANOVA tests were performed to identify the differentially expressed genes across all investigated tissues ($P < 0.001$). Then, the expression of genes that were significantly different than other tissues with significance level as $P < 0.01$ and with change ratio > two-fold were considered as tissue-specific genes. (B) The *C2H2 ZFP* genes that showed differential expression, i. e., the expression was restricted to some samples. Based on the expression, we could class these genes into three groups: A, B and C. Group A is a miscellaneous, and genes' expressions were restricted to some tissues. Genes in groups B and C were detected to express in testis and ovary, whereas group C has weak expression in A/MSG and PSG. (C) The *C2H2 ZFP* genes that showed expression in all our investigated tissues. (选自：*Species-specific expansion of C2H2 zinc-finger genes and their expression profiles in silkworm, Bombyx mori*，参见本书第828页。)

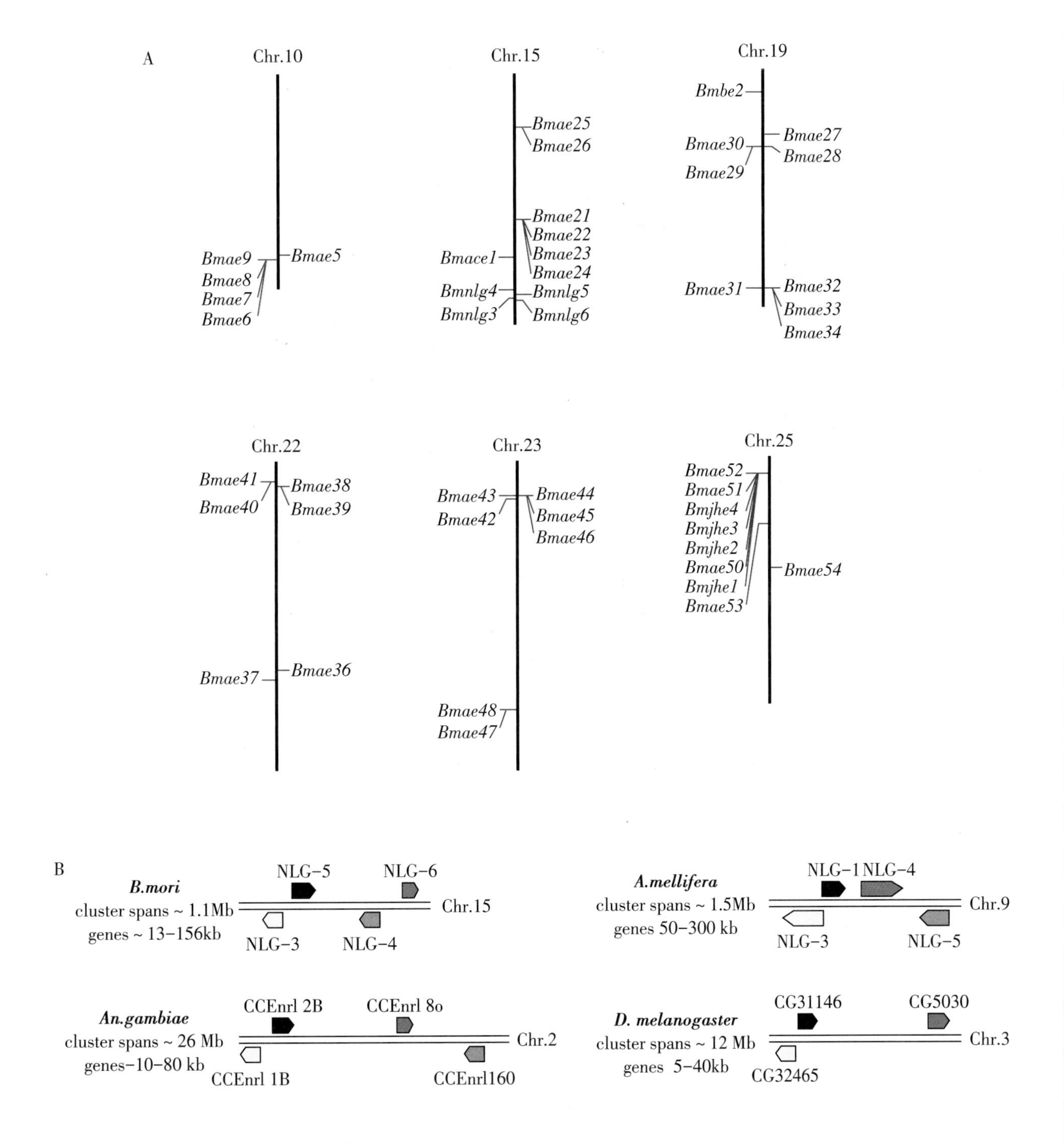

Figure 122. The cluster organization of *BmCOEs* and microsynteny of neuroligins. (A) The cluster organization of *BmCOEs* in the silkworm genome. Only those genes involved in clusters of five or more COE genes on the same chromosomes are shown. (B) microsynteny of neuroligins among *B. mori*, *D. melanogaster*, *An. gambiae*, and *Ap. Mellifera*. The arrows represent gene localization and transcriptional orientation.（选自：*Annotation and expression of carboxylesterases in the silkworm, Bombyx mori*，参见本书第840页。）

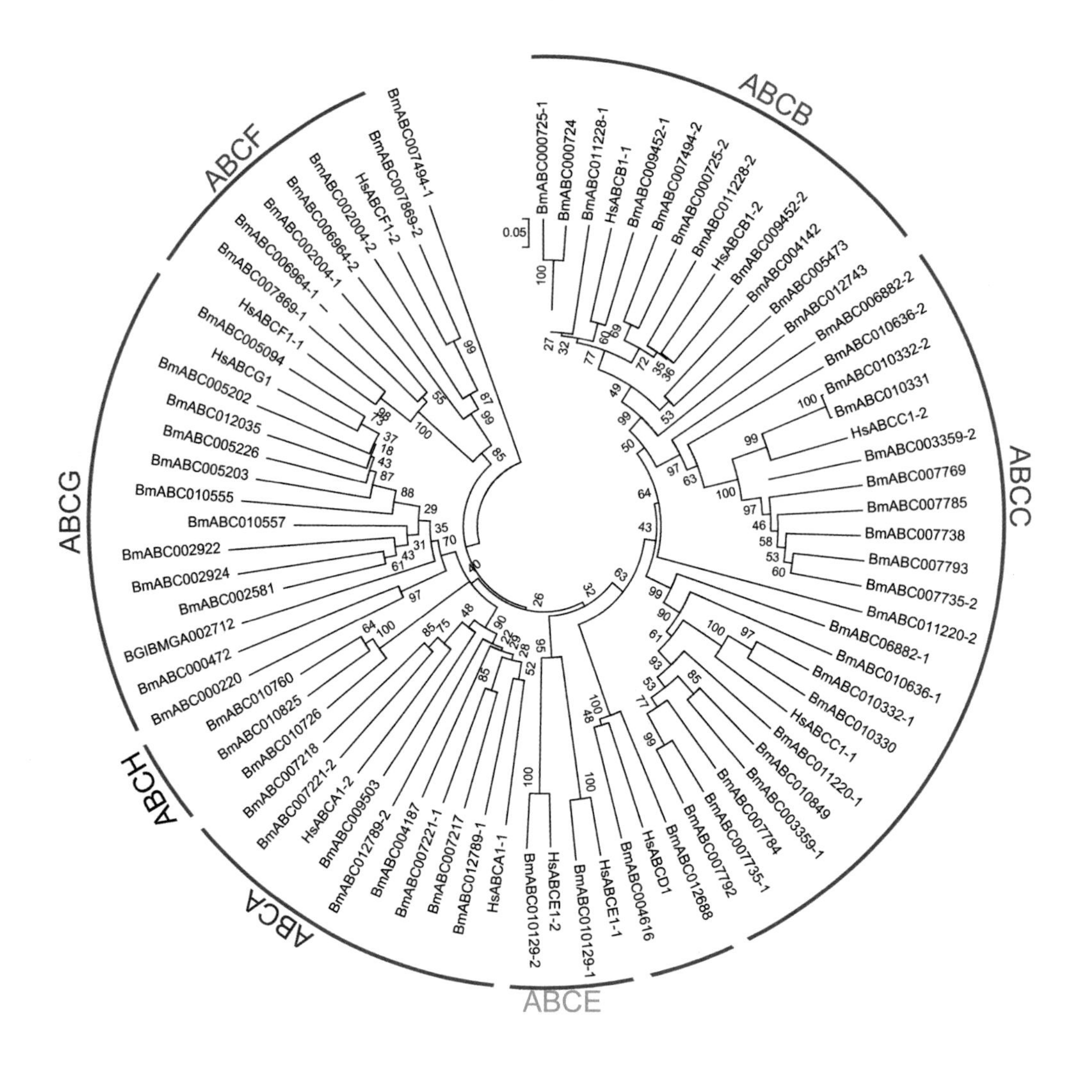

Figure 123. Phylogenetic tree of the silkworm ABC transporters. The phylogenetic tree was constructed using a neighbor-joining technique to analyze the amino acid sequences of the nucleotide binding domain (NBD). Analysis was performed with the program package MEGA4.0. The number at the branch point of the node represents the value resulting from 1 000 replications and gaps were deleted with pairwise deletion method.（选自：*Genome-wide identification and characterization of ATP-binding cassette transporters in the silkworm, Bombyx mori*，参见本书第842页。）

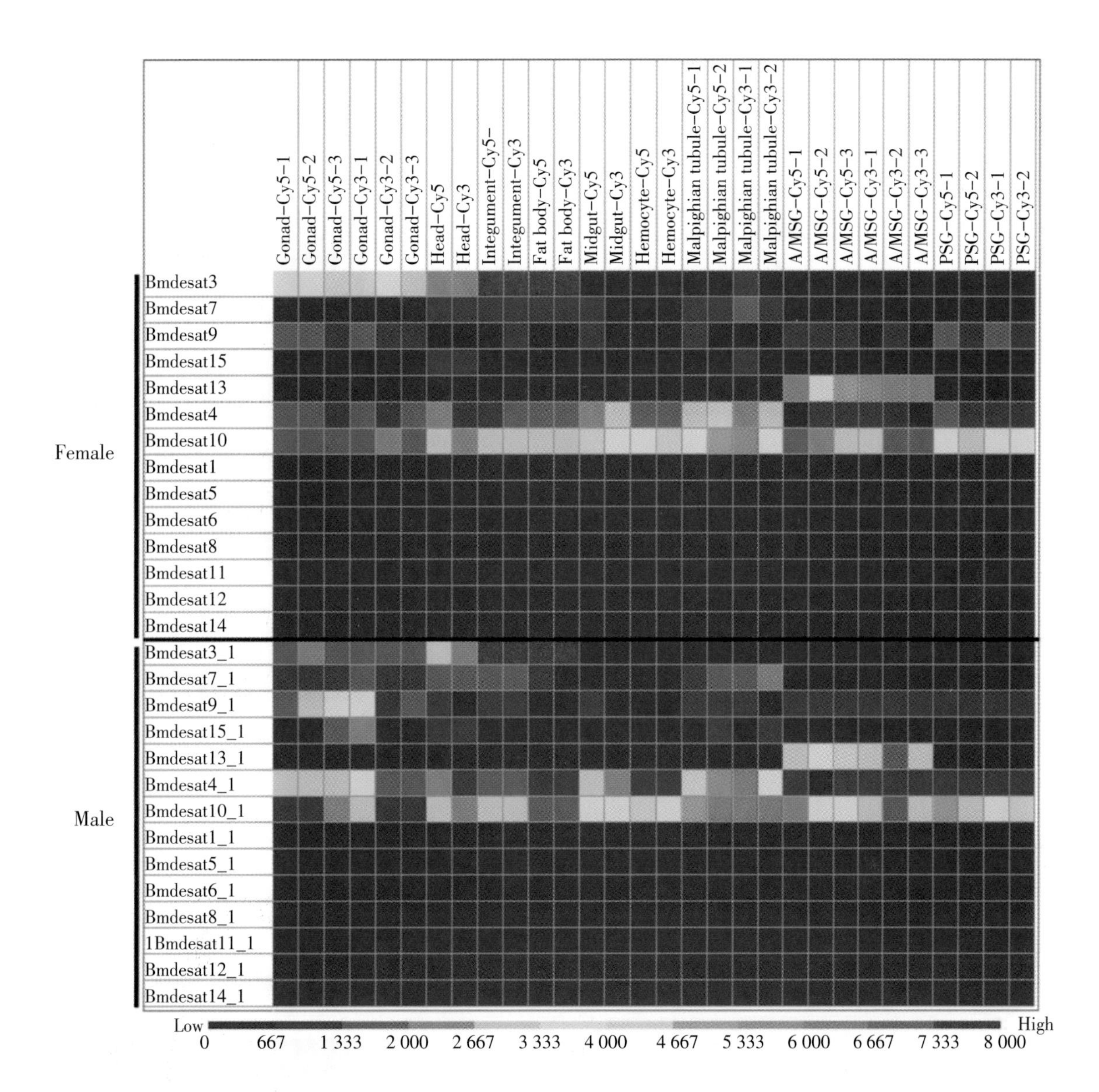

Figure 124. Expression profiles of desaturase genes in multiple larval tissues on day 3 of fifth instar of silkworm by microarray data. The expression levels are illustrated by red (higher expression) and blue (lower expression) boxes. A/MSG = anterior/median silk gland; PSG = posterior silk gland. F = female and M = male.（选自：*Genome-wide identification and expression profiling of the fatty acid desaturase gene family in the silkworm, Bombyx mori*, 参见本书第846页。）

(a) *Cellular immune response*

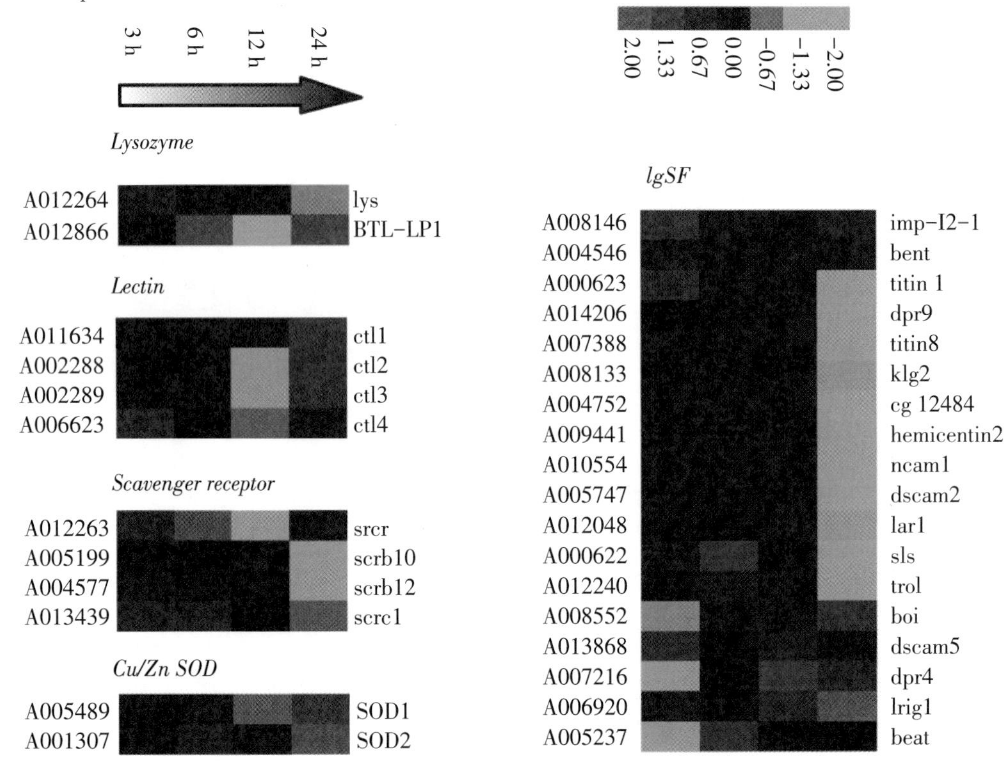

(b) *Senne protease cascade melanization pathway*

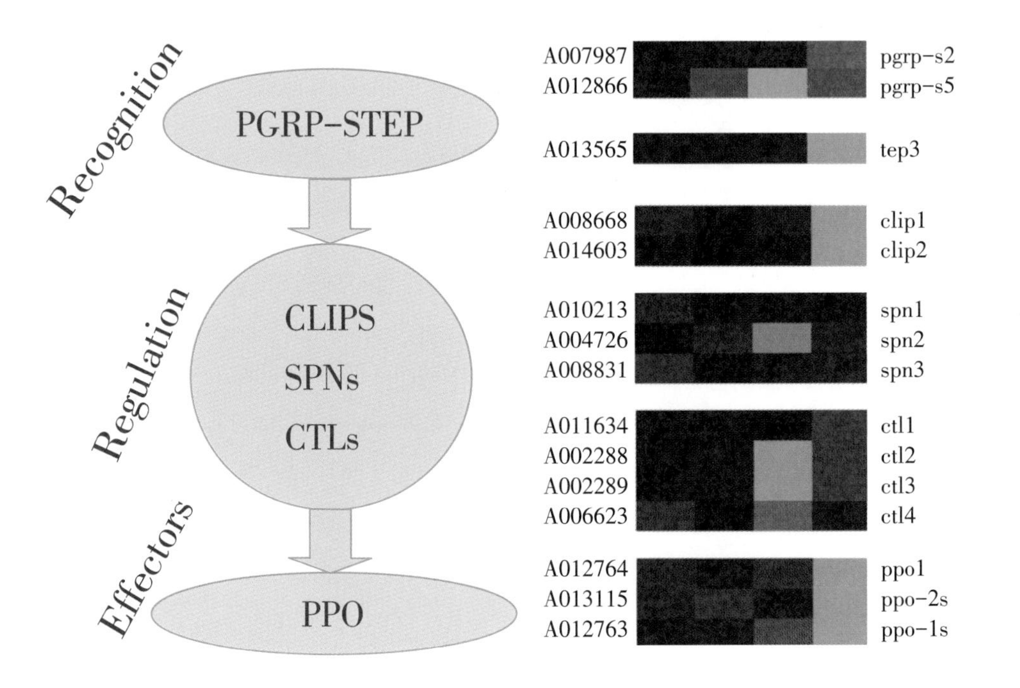

(c)*Hemolymph color darker*

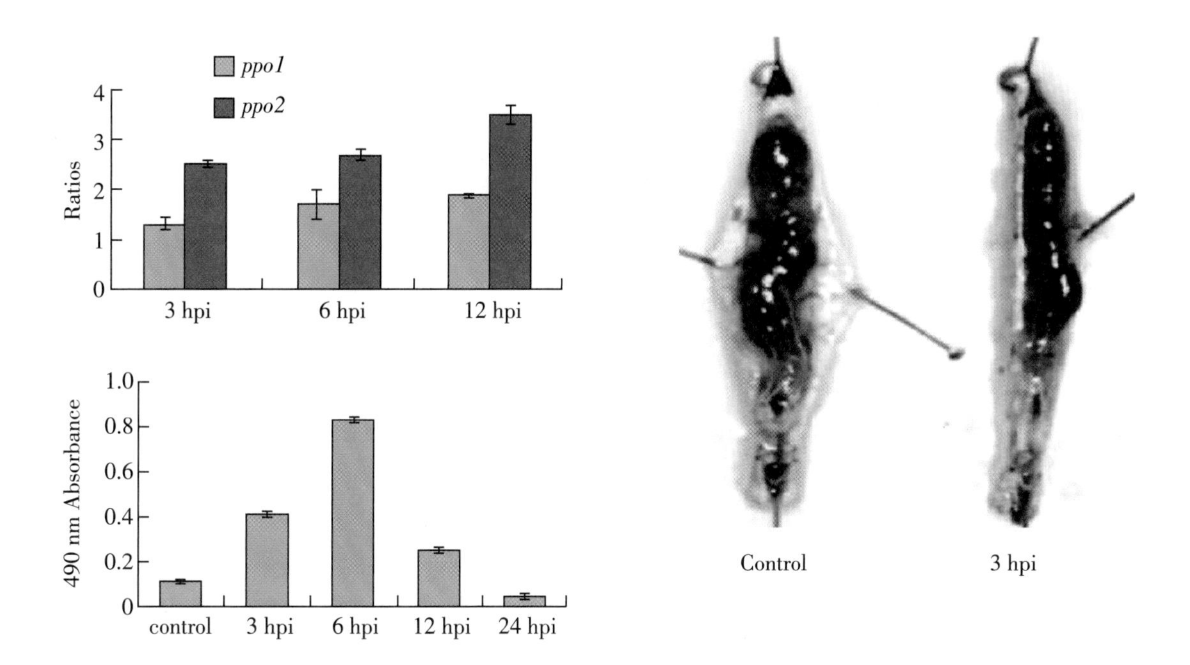

Figure 125. *Bb* induced silkworm cellular immune response and serine protease cascade melanization pathway. (a) Cluster of cellular immune response families. (b) Cluster diagram of the induced synthesis of melanin by serine protease cascade melanization pathway related genes. For a detailed view of the cluster ratios, see Table S4. (c) Real time PCR analysis of *ppo1* and *ppo2* ratios of *Bb* infected whole larvae comparing to non-induced control and 490 nm absorbance of hemolymph during the infection, and the picture of the hemolymph melanization 3 h after *Bb* oral infection.(选自:*A genome-wide survey for host response of silkworm, Bombyx mori during pathogen Bacillus bombyseptieus infection*,参见本书第860页。)

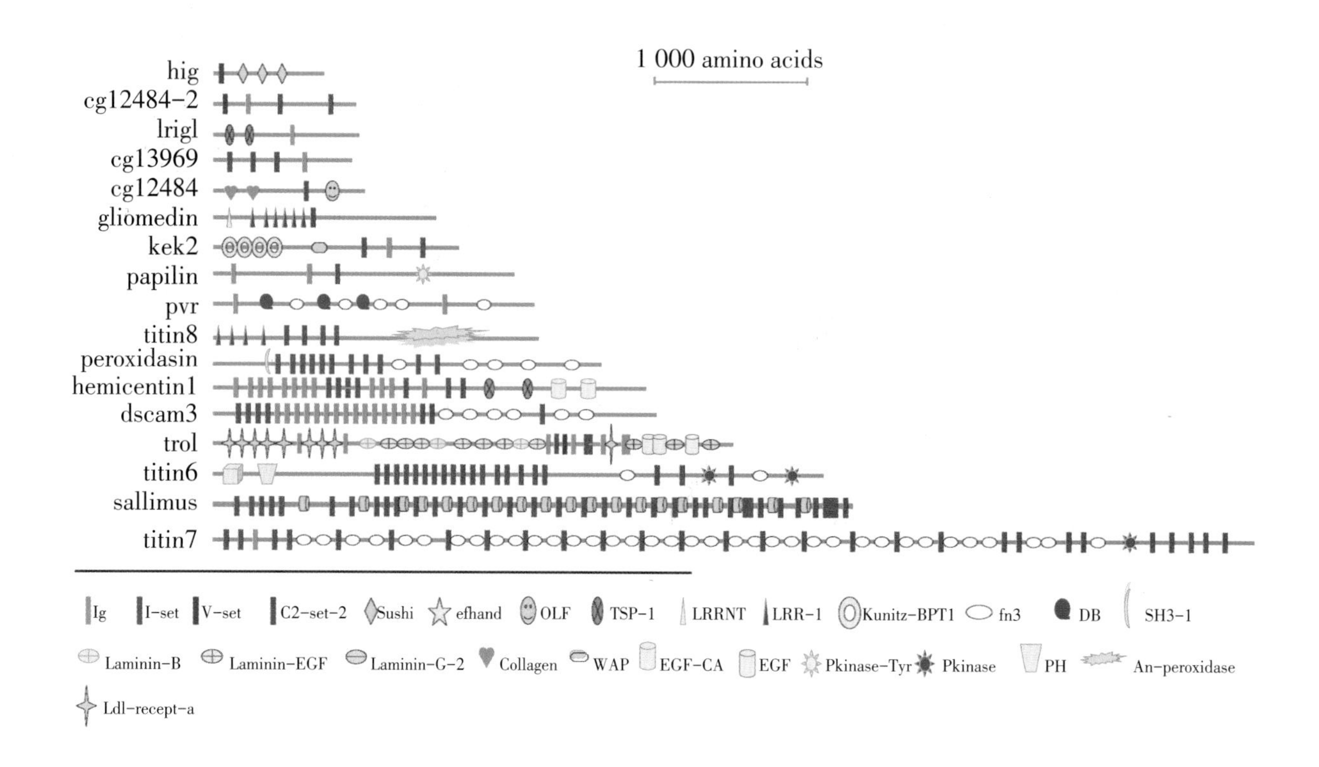

Figure 126. The typical domain architectures of silkworm immunoglobulin superfamily (IgSF) members. For each protein, the green horizontal line represents the sequence length, each symbol represents each domain.（选自：*Immunoglobulin superfamily is conserved but evolved rapidly and is active in the silkworm, Bombyx mori*，参见本书第862页。）

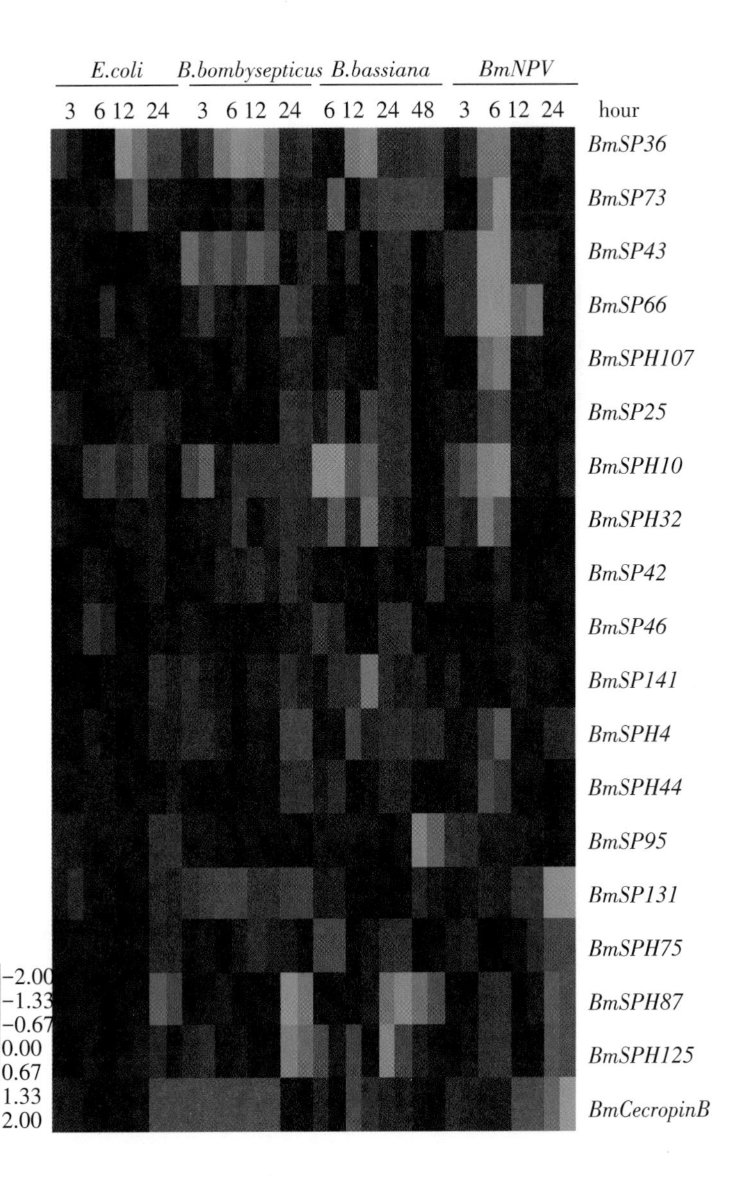

Figure 127. Hierarchical cluster analysis of *SP* and *SPH* genes after microorganism infection. Red represents that gene expression is up-regulated; black represents that gene expression is not changed; green represents that gene expression is download-regulated; gray represents data missing. Hierarchical clustering was performed using Cluster version 3.0 and visualized by TreeView.（选自：*Genome-wide identification and expression analysis of serine proteases and homologs in the silkworm Bombyx mori*，参见本书第864页。）

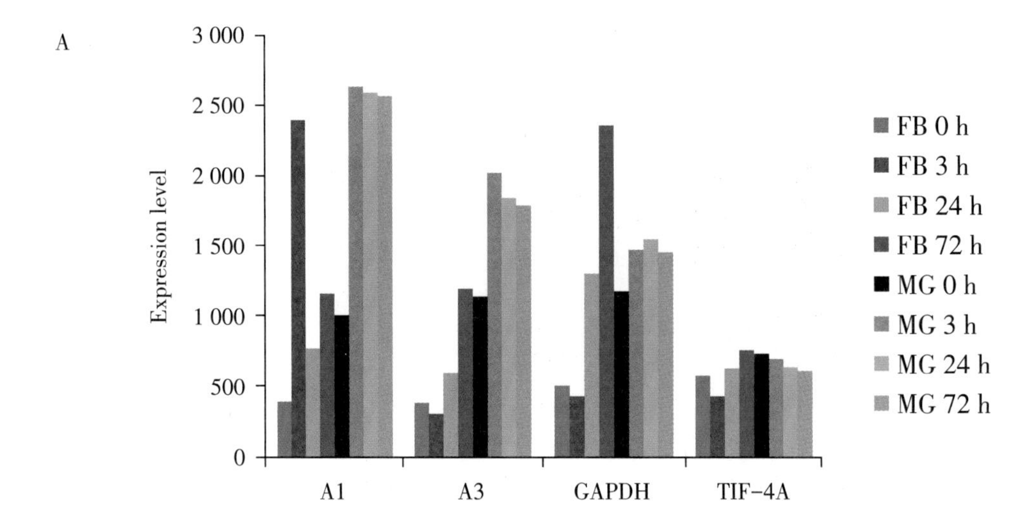

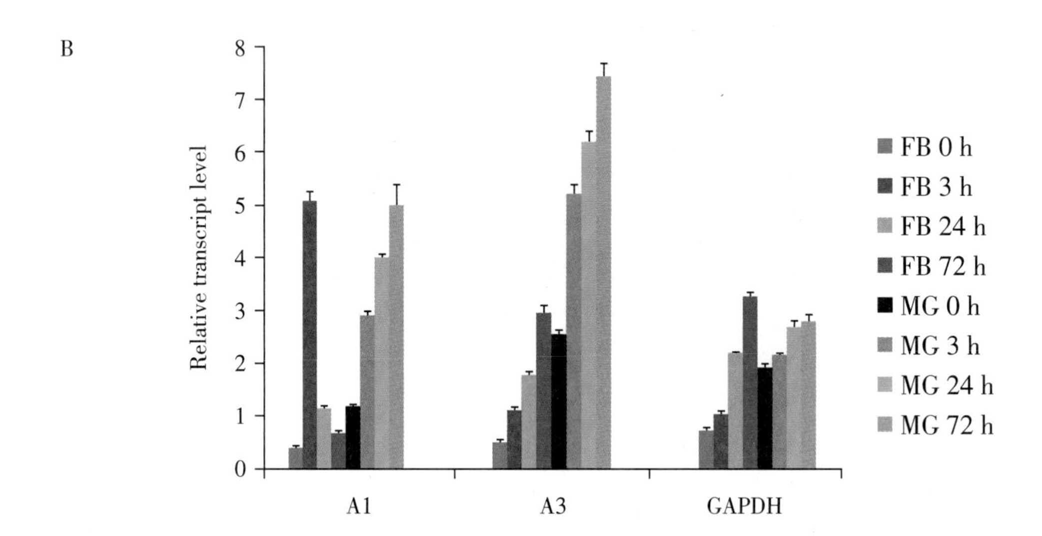

Figure 128. The expression patterns of candidate RGs after *Bm*CPV challenge. The MG and FB of silkworms were extracted at 0, 3, 24, and 72 h after *Bm*CPV infection. a The expression levels of candidate RGs in RNA-seq data. The data were from our recent study (Jiang *et al.*, unpublished data). b Detection of the expression levels of candidate RGs by qPCR. TIF-4A was used for qPCR data normalization. Bars indicate standard deviations.（选自：*Selection of reference genes for analysis of stress-responsive genes after challenge with viruses and temperature changes in the silkworm Bombyx mori*，参见本书第880页。）

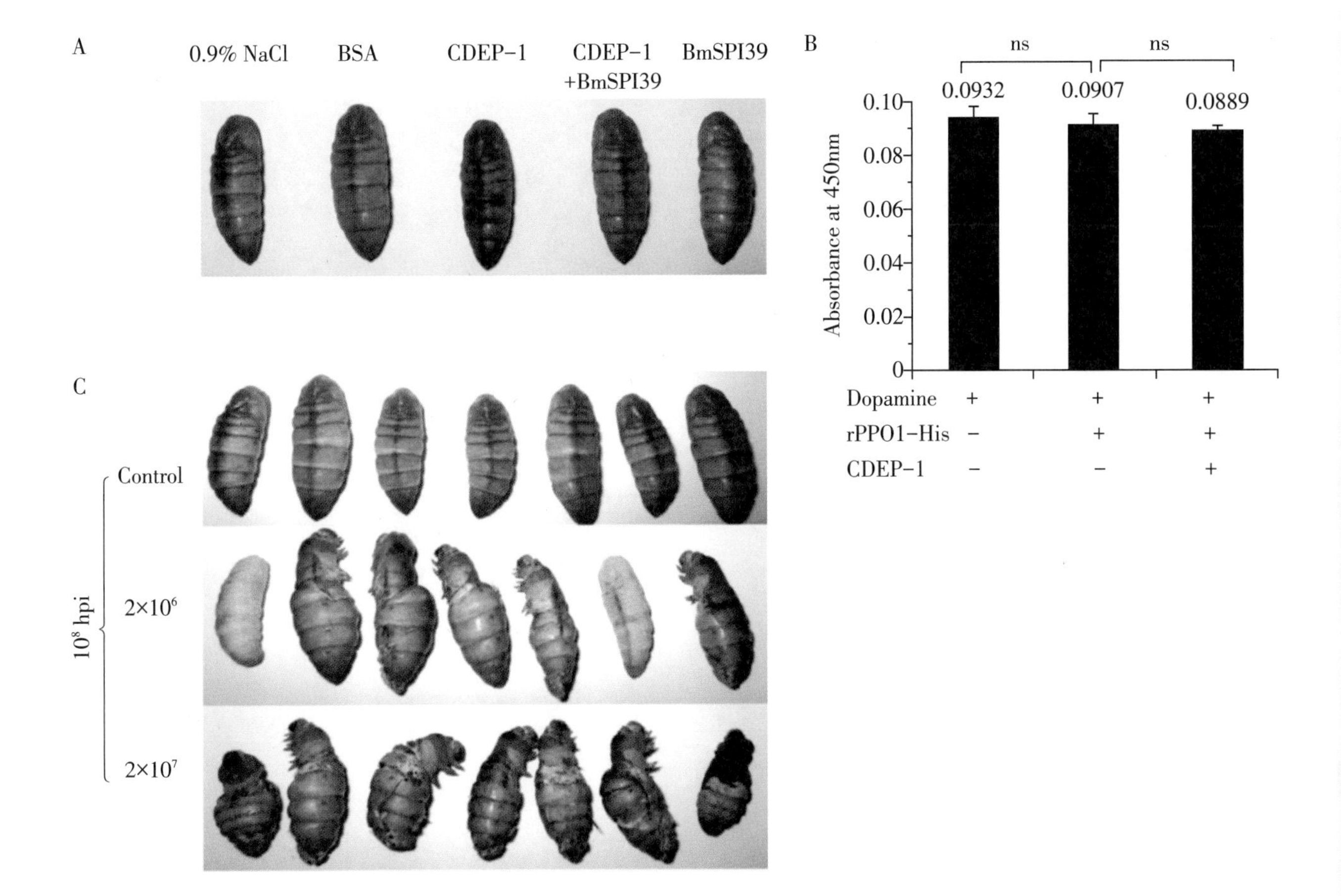

Figure 129. Phenol oxidase activation studies. (A) Inhibition of CDEP-l-induced melanization by BmSPI39. At least eight insects were used for each treatment and each treatment was repeated independently. One insect in each treatment group was selected randomly as representative of that group and shown here. The variation observed in any two independent experiments was only in the amount of melanization of the cuticle. (B) Activation of rPPO1 in vitro. rPPO1-His, *Drosophila* rPPO1-His (His-tag at the C-terminus). (C) Effect of *B. bassiana* infection on silkworms. Conidial suspensions were prepared at concentrations of 2×10^6 and 2×10^7 conidia/mL with ddH_2O. ddH_2O was used as the blank control. "hpi", hours post-infection.（选自：*TIL-type protease inhibitors may be used as targeted resistance factors to enhance silkworm defenses against invasive fungi*，参见本书第884页。）

A

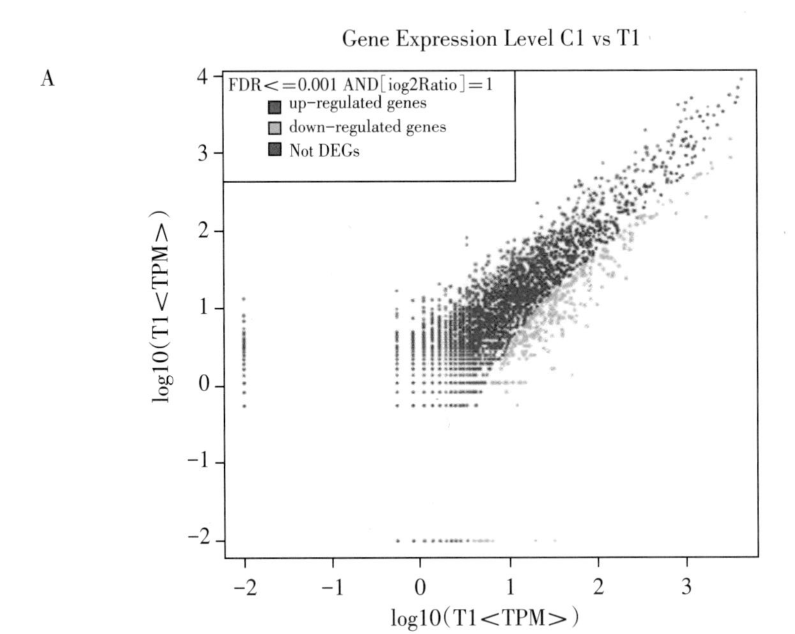

B

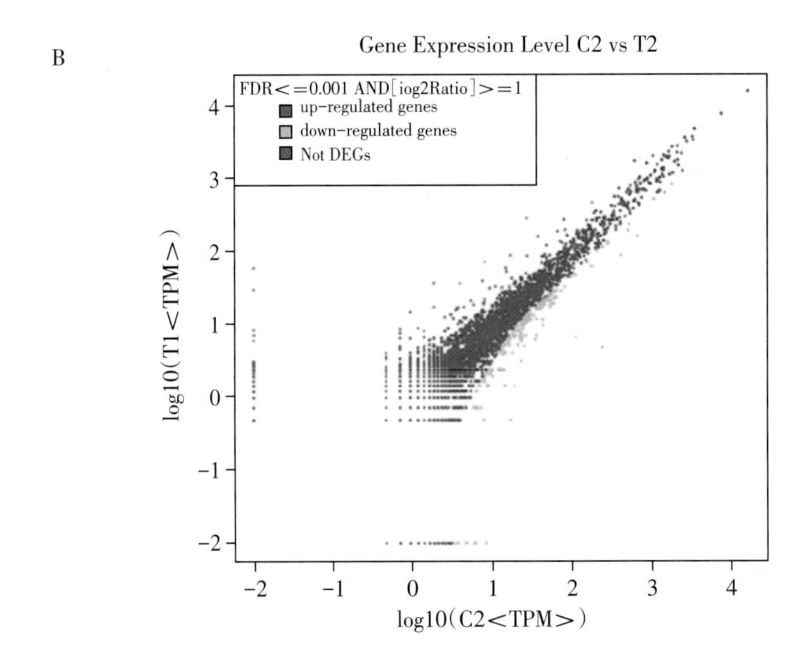

Figure 130. Differentially expressed genes. The red part represents those genes up-regulated in *Bm*CPV-infected group compared to control group. The green part shows those genes down-regulated in *Bm*CPV-infected group. The blue part shows those genes without expression difference between these two samples. (C1): Control *4008*. (T1): *Bm*CPV-infected *4008*. (C2): Control *p50*. (T2): *Bm*CPV-infected *p50*.（选自：*Cytoplasmic polyhedrosis virus-induced differential gene expression in two silkworm strains of different susceptibility*，参见本书第890页。）

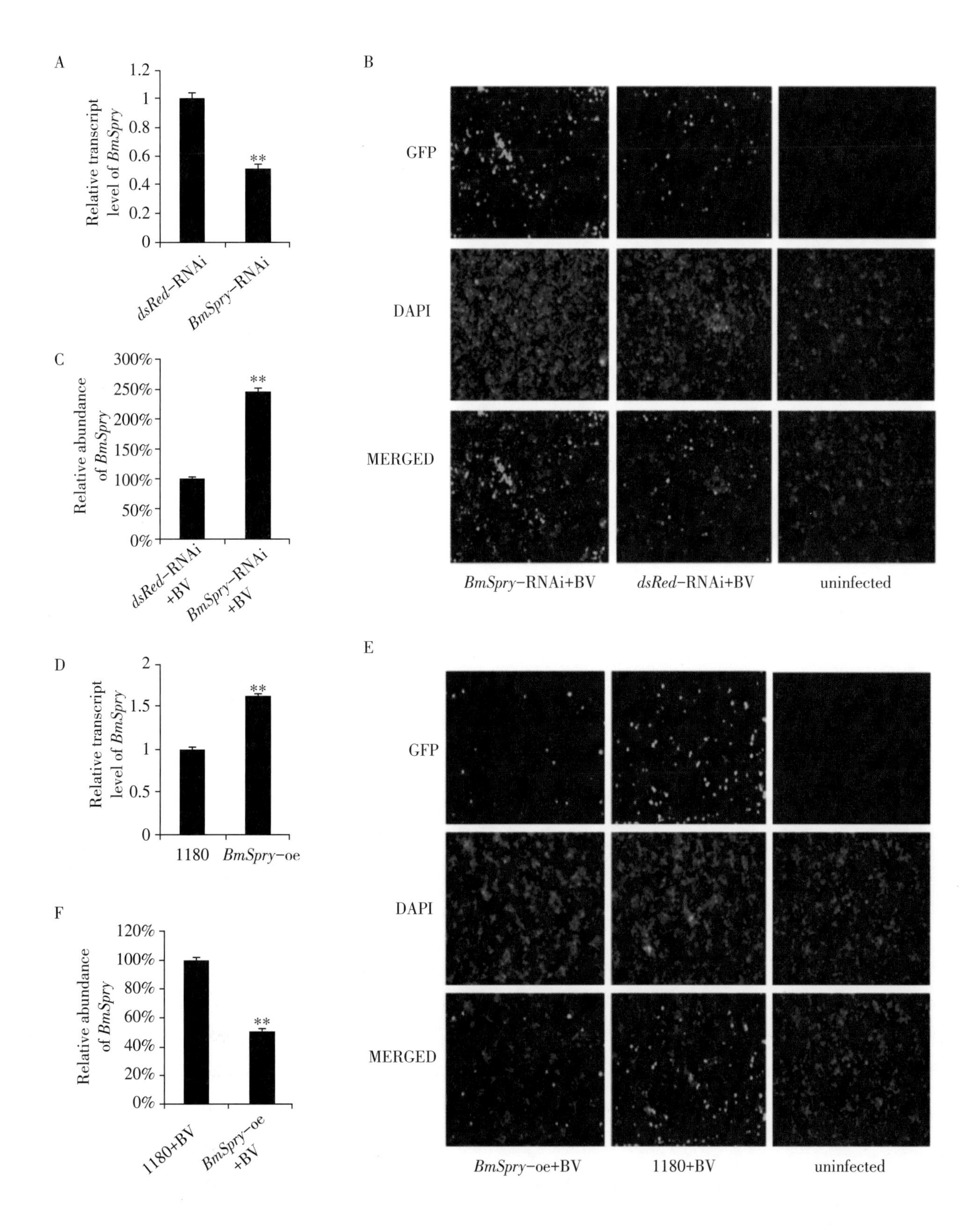
A
1.2
1
0.8
0.6
0.4
0.2
0
Relative transcript level of BmSpry
**
dsRed-RNAi
BmSpry-RNAi
B
GFP
DAPI
MERGED
BmSpry-RNAi+BV
dsRed-RNAi+BV
uninfected
C
300%
250%
200%
150%
100%
50%
0%
Relative abundance of BmSpry
**
dsRed-RNAi +BV
BmSpry-RNAi +BV
D
2
1.5
1
0.5
0
Relative transcript level of BmSpry
**
1180
BmSpry-oe
E
GFP
DAPI
MERGED
BmSpry-oe+BV
1180+BV
uninfected
F
120%
100%
80%
60%
40%
20%
0%
Relative abundance of BmSpry
**
1180+BV
BmSpry-oe +BV

Figure 131. *BmSpry* inhibited *Bm*NPV replication in cultured cells. (A) *BmE* cells pretreated with dsRNA as indicated, the dsRNA of ds*Red* was used as a negative control. At 3 days post transfection, total RNA was extracted and qPCR was used to analyze the *BmSpry* expression level. (B) *BmE* cells treated with dsRNA against the indicated genes were infected with BmNPV-GFP at MOI of 1 for 3 days and processed for immunofluorescence. (C) *BmE* cells treated with the indicated dsRNA were infected with *Bm*NPV-GFP at MOI of 1 and infection total genomes were extracted for qPCR at 3 days post. (D) *BmE* cells were used for transient transfection and the empty vector 1180 was used as a negative control. *BmSpry*-oe was an overexpression vector of *BmSpry*. At 3 days post-transfection, total RNA was extracted for qPCR. (E) *BmE* cells were subjected to transient transfection with *BmSpry* expression vectors and the empty 1180 vector, as indicated. At 2 days post-transfection, the cells were infected with *Bm*NPV-GFP at MOI.（选自：*Identification of a new sprouty protein responsible for the inhibition of the Bombyx mori nucleopolyhedrovirus reproduction*，参见本书第904页。）

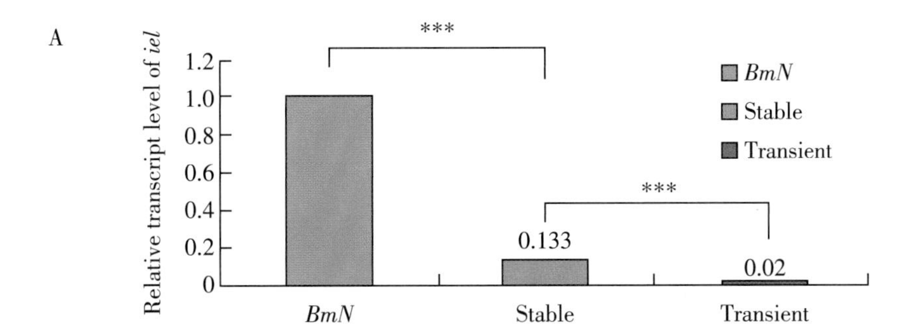
A
***
***
Relative transcript level of iel
1.2
1.0
0.8
0.6
0.4
0.2
0
BmN
Stable
Transient
0.133
0.02
BmN
Stable
Transient

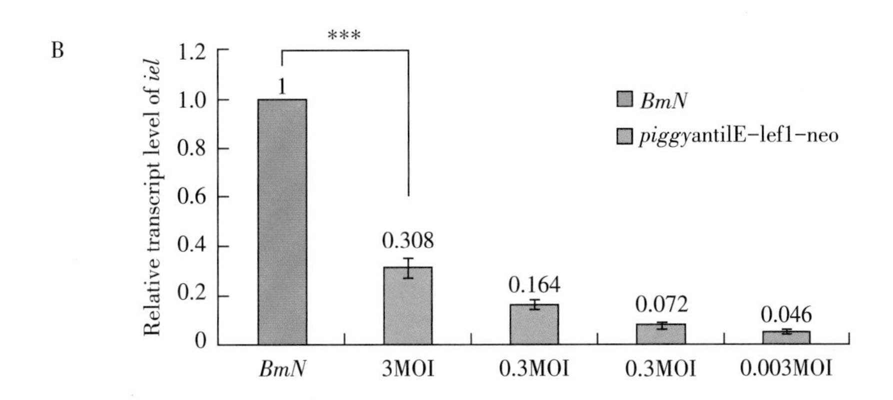
B
***
Relative transcript level of iel
1.2
1.0
0.8
0.6
0.4
0.2
0
1
0.308
0.164
0.072
0.046
BmN
piggyantilE-lef1-neo
BmN
3MOI
0.3MOI
0.3MOI
0.003MOI

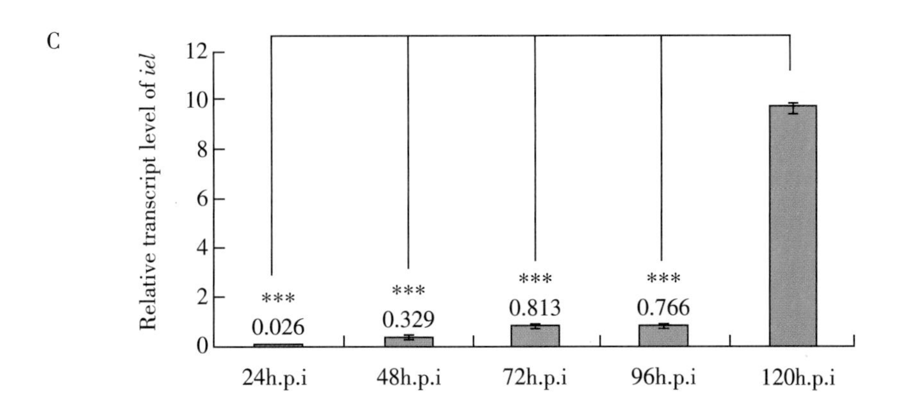
C
Relative transcript level of iel
12
10
8
6
4
2
0
***
***
***
***
0.026
0.329
0.813
0.766
24h.p.i
48h.p.i
72h.p.i
96h.p.i
120h.p.i

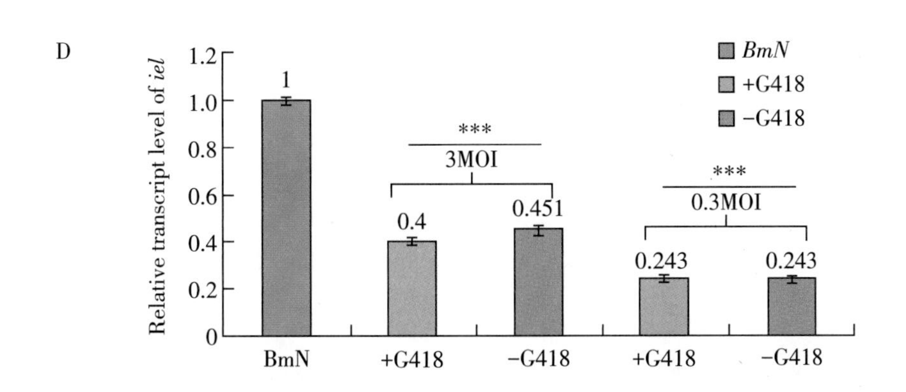
D
Relative transcript level of iel
1.2
1.0
0.8
0.6
0.4
0.2
0
BmN
+G418
−G418
1
***
3MOI
***
0.3MOI
0.4
0.451
0.243
0.243
BmN
+G418
−G418
+G418
−G418

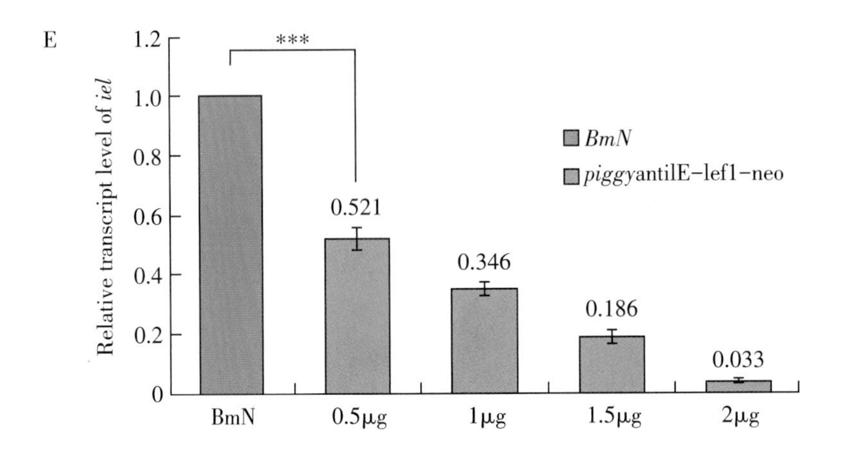

Figure 132. The resistance of transfected cells to *Bm*NPV. (a) The abrogation of *Bm*NPV proliferation between stable transformed cells and transiently transfected cells. *BmN* cells (1×10^5) transiently transfected with *piggy*antiIE-lef1-neo plasmid were infected with *Bm*NPV (MOI 0.03) at 48 h after transfection, and then incubated at 26 ℃ for 48 h and an equal amount of stable transformed cell line was treated in the same way. An equal amount of *BmN* cell line was used as a control. (b) The resistance of 1×10^5 *BmN* cells transiently transfected with transgenic vector *piggy*antiIE-lef1-neo to *Bm*NPV at various titers. *BmN* cells (1×10^5) transiently transfected with *piggy*antiIE-lef1-neo plasmid were infected with *BmNPV* at various degrees of virus titer. The methods and qPCR analysis were as described for Figure 2. (c) The time course of *BmNPV* infection. *BmN* cells (1×10^5) transiently transfected with transgenic vector *piggy*antiIE-lef1-neo were infected with *Bm*NPV (MOI 0.03). The relative abundance of *ie-1* transcript levels was obtained as mentioned in Figure 2. The most efficient in suppressing virus proliferation was found at 24 h.p.i. and the *ie-1* expression levels did not vary much until 120 h.p.i., when it reached 9.71-fold greater compared with control *BmN* cells as the recovery of *Bm*NPV replication leading to the lysis of the infected cells. (d) The effect of G418 in a pressure filter on the neomycin resistance gene. Stable transformed cells transfected with plasmid *piggy*antiIE-lef1-neo were screened by G418 over 3 months, and then G418 was removed for another 3 months of growth. The cell line was designated as -G418; stable transformed cells containing plasmid *piggy*antiIE-lef1-neo were screened continuously by G418, the cell line was designated as +G418. The 1×10^5 +G418 and 1×10^5-G418 cell lines were infected with *Bm*NPV MOI 3 and 0.3 for 48 h, respectively. Thereafter, total RNA was extracted and cDNA was synthesized to estimate the relative expression level of the *ie-1* gene. (e) The inhibitory effect depends on the load of transgenic vector used. *BmN* cells (1×10^5) transiently transfected with *piggy*antiIE-lef1-neo plasmid were infected with *Bm*NPV (MOI 3). The amount of plasmid used in transfection is shown in the figure. qPCR analysis was carried out as described for Figure 2. ****P*<0.01.（选自：*Resistance of transgenic silkworm to BmNPV could be improved by silencing ie-1 and lef-1 genes*，参见本书第914页。）

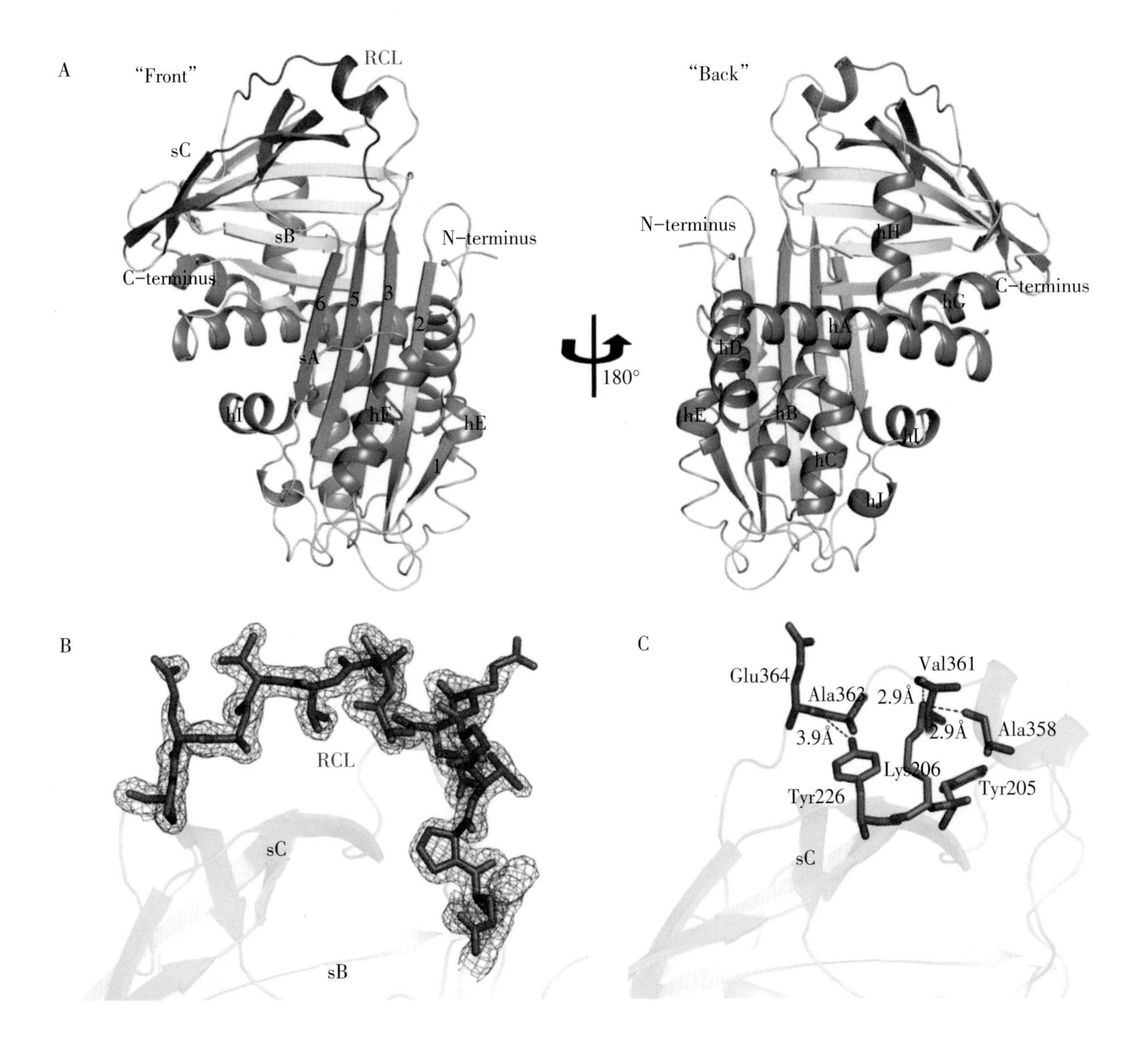

Figure 133. Overall structure of serpin18. (A) Schematic representation of serpin18 fold. Blue, helical region; red, RCL; and the three β-sheets are distinguished by different colour (sA, orange; sB, yellow; sC, magenta). Secondary elements are labeled, and the specific numbering of strands 1-3, 5, and 6 of sA is shown. (B) The residues in RCL are emphasized in stick presentation with the corresponding electron density at a sigma level of 1.0 in the *2Fo-Fc* map. (C) Residues involved the interactions between the center of RCL and sC are shown with sticks, and the hydrogen bonds are black dashed lines. All figures were prepared using PyMOL.(选自：*Structural insights into the unique inhibitory mechanism of the silkworm protease inhibitor serpin18*，参考本书第930页。)

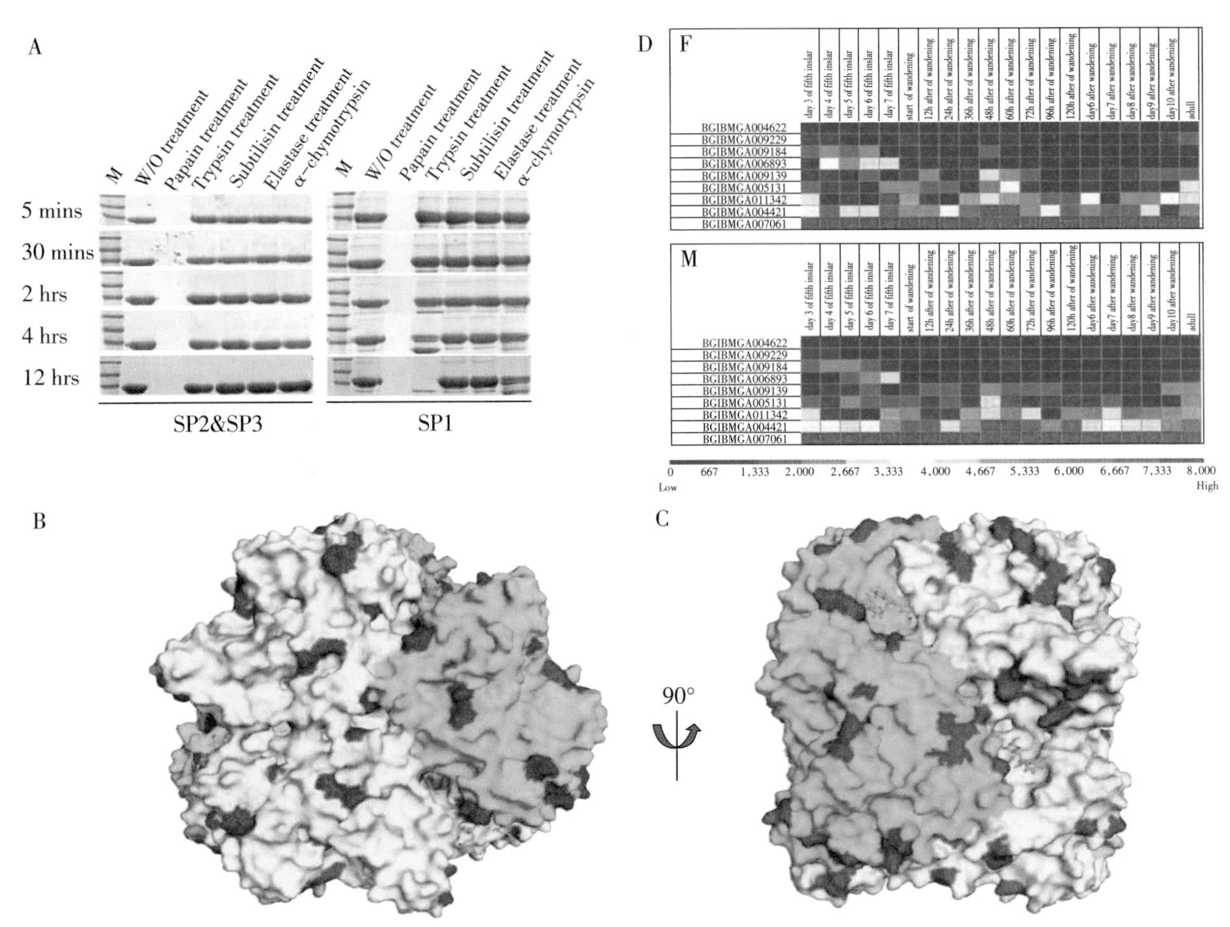
A
M
W/O treatment
Papain treatment
Trypsin treatment
Subtilisin treatment
Elastase treatment
α-chymotrypsin
5 mins
30 mins
2 hrs
4 hrs
12 hrs
SP2&SP3
SP1
D
F
M
BGIBMGA004622
BGIBMGA009229
BGIBMGA009184
BGIBMGA006893
BGIBMGA009139
BGIBMGA005131
BGIBMGA011342
BGIBMGA004421
BGIBMGA007061
0 667 1,333 2,000 2,667 3,333 4,000 4,667 5,333 6,000 6,667 7,333 8,000
Low
High
B
C
90°
Red spot:Papain cleavage sites

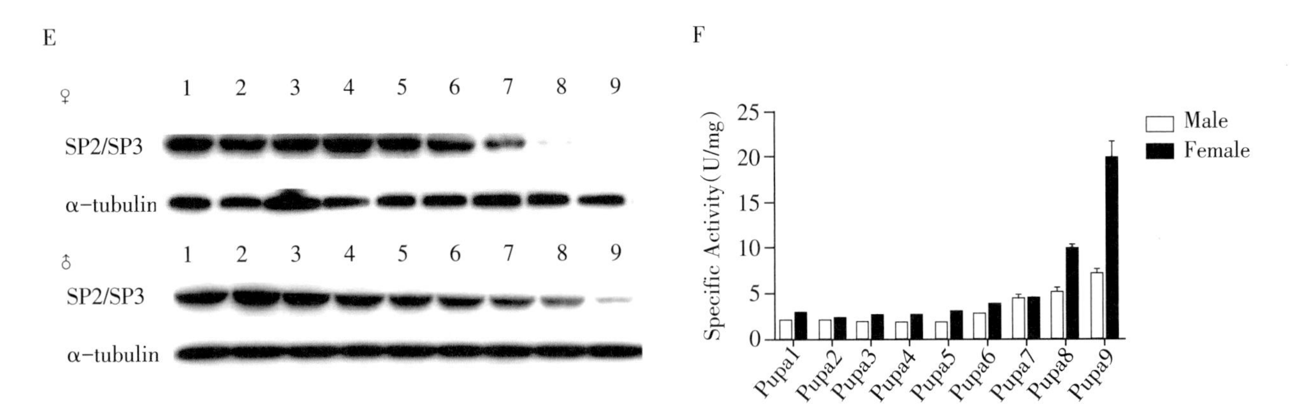
E
♀
1 2 3 4 5 6 7 8 9
SP2/SP3
α-tubulin
♂
F
Specific Activity (U/mg)
Male
Female
Pupa1
Pupa2
Pupa3
Pupa4
Pupa5
Pupa6
Pupa7
Pupa8
Pupa9

Figure 134. *Bombyx mori* SP2/SP3 complex is the substrate for papain-like protease. A: Limited proteolytic assay on purified *Bombyx mori* SP2/SP3 complex (left) and SP1 (right) by incubation with Papain, Trypsin, Subtlisin, Elastase, and α-chymotrypsin at different time scale. B: Surface view of *Bombyx mori* SP2/SP3 heterohexamer with Papain cleavage sites mapping at the surface. The SP2 molecules are colored in cyan, whereas the SP3 molecules are colored in yellow. The bound oligosaccharide chain is shown in stick mode and colored in green. C: Surface view of *Bombyx mori* SP2/SP3 heterohexamer with a 90° rotation of (B) along *Y* axis. The same color code as that in (B). D: Expression profile of papain-like proteases in *Bombyx mori* during different developmental stages. The columns represents 20 different sample time points: day 3, 4, 5, 6, 7 of the fifth instar, start of wandering, 14 different times after wandering: 12 h, 24 h, 36 h, 48 h, 60 h, 3 days, 4 days, 5 days, 6 days, 7 days, 8 days, 9 days, 10 days, and adult. "M" represents the sample collected from male silkworm, whereas "F" represents the sample collected from female silkworm. E: Expression profile of *Bombyx mori* arylphorins at different pupa developmental stages. Arylphorin proteins extracted from pupa fat body (20 μg) at each stage were separated by SDS-PAGE and transferred to membrane, followed by hybridized using SP2/SP3 polyclonal antibodies. 1-9 represent the samples extracted from 1st to 9th day of pupa, respectively. F: Protease activity of *Bombyx mori* fat at different pupa developmental stages. Protein samples extracted from the fat body of pupa were used to measure cathespin B and cathespin L proteolytic activity against Z-Phe-Arg-MCA. One unit of enzyme activity represents the fluorescence released from 5 nmol substrate over per 10 min. （选自：*Crystal structure of Bombyx mori arylphorins reveals a 3:3 heterohexamer with multiple papain cleavage sites*，参见本书第932页。）

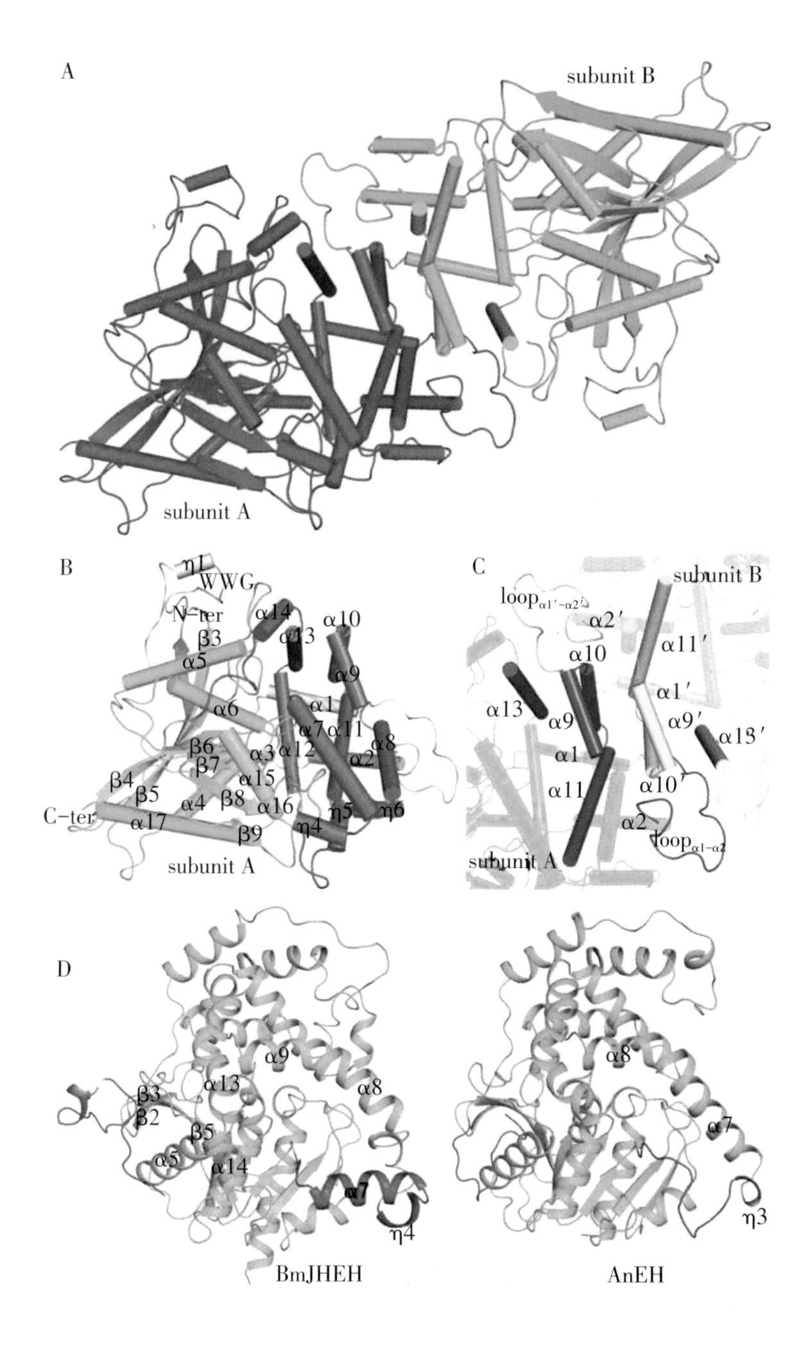

Figure 135. Overall structure of *Bm*JHEH. (A) The dimeric structure. The subunits A and B are colored red and green, respectively. (B) Three parts of subunit A are colored gray for the N-terminal region (Leu33-Tyr118), cyan for the α/β domain (Pro119-Leu255 and Gln392-Asn456) and pink for the lid domain (Ser256-Val391), respectively. The "WWG" motif is labeled in orange. The secondary structural elements are labeled sequentially. (C) Residues involved in dimeric interaction are colored blue in subunit A and yellow in subunit B, respectively. The other residues of two subunits were shown with transparency. (D) Structural comparison between *Bm*JHEH and *An*EH. The similar parts are colored cyan, whereas the different segments are colored red.（选自：*Crystal structure of juvenile hormone epoxide hydrolase from the silkworm Bombyx mori*，参见本书第934页。）

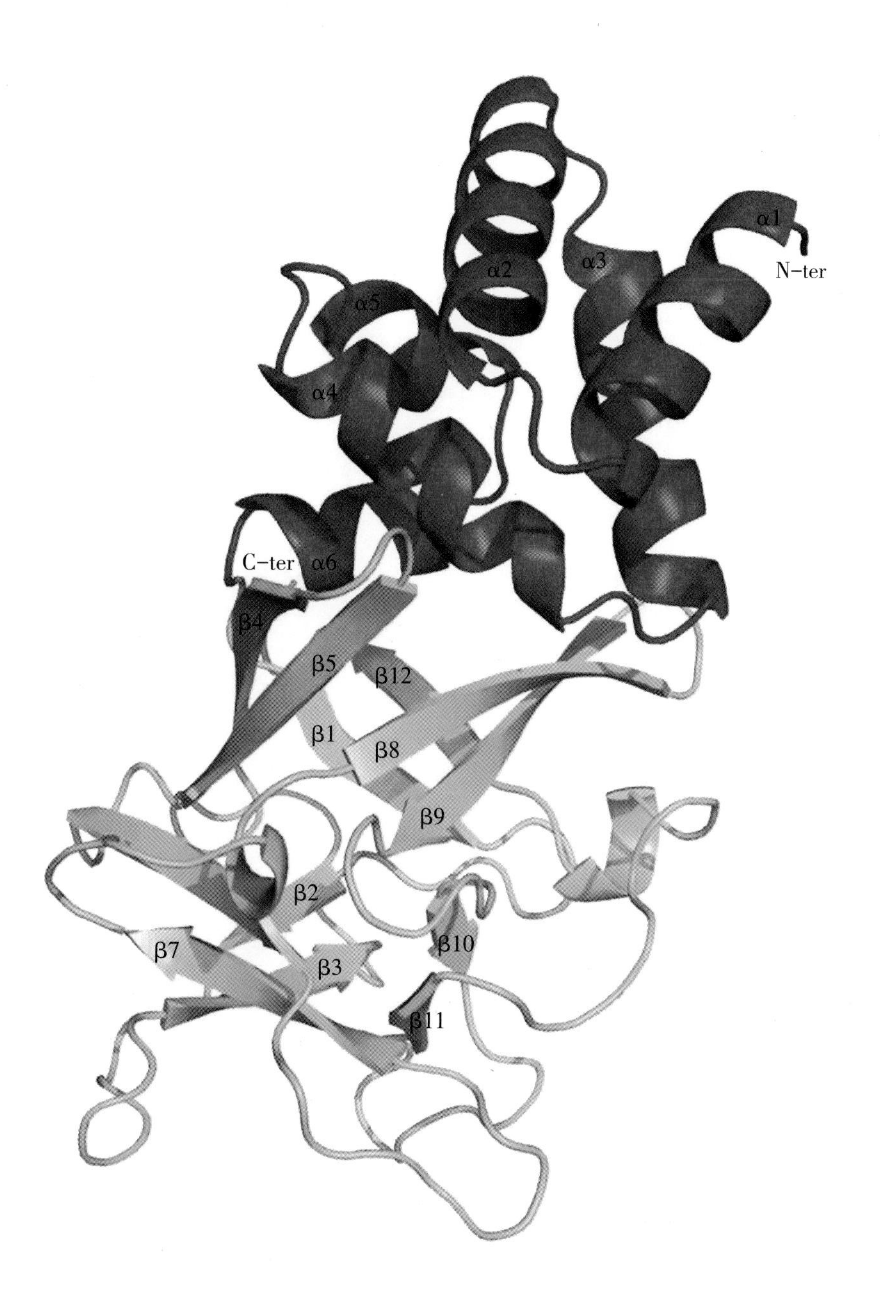

Figure 136. Overall structure of *Bm*lp7. The NTD (red) contains six helices, arranged as a right-handed superhelix. The CTD (green) is a typical β-trefoil barrel. The two domains are mainly linked via hydrophobic interactions.（选自：*Crystal structure of the 30K protein from the silkworm Bombyx mori reveals a new member of the beta-trefoil superfamily*, 参见本书第936页。）

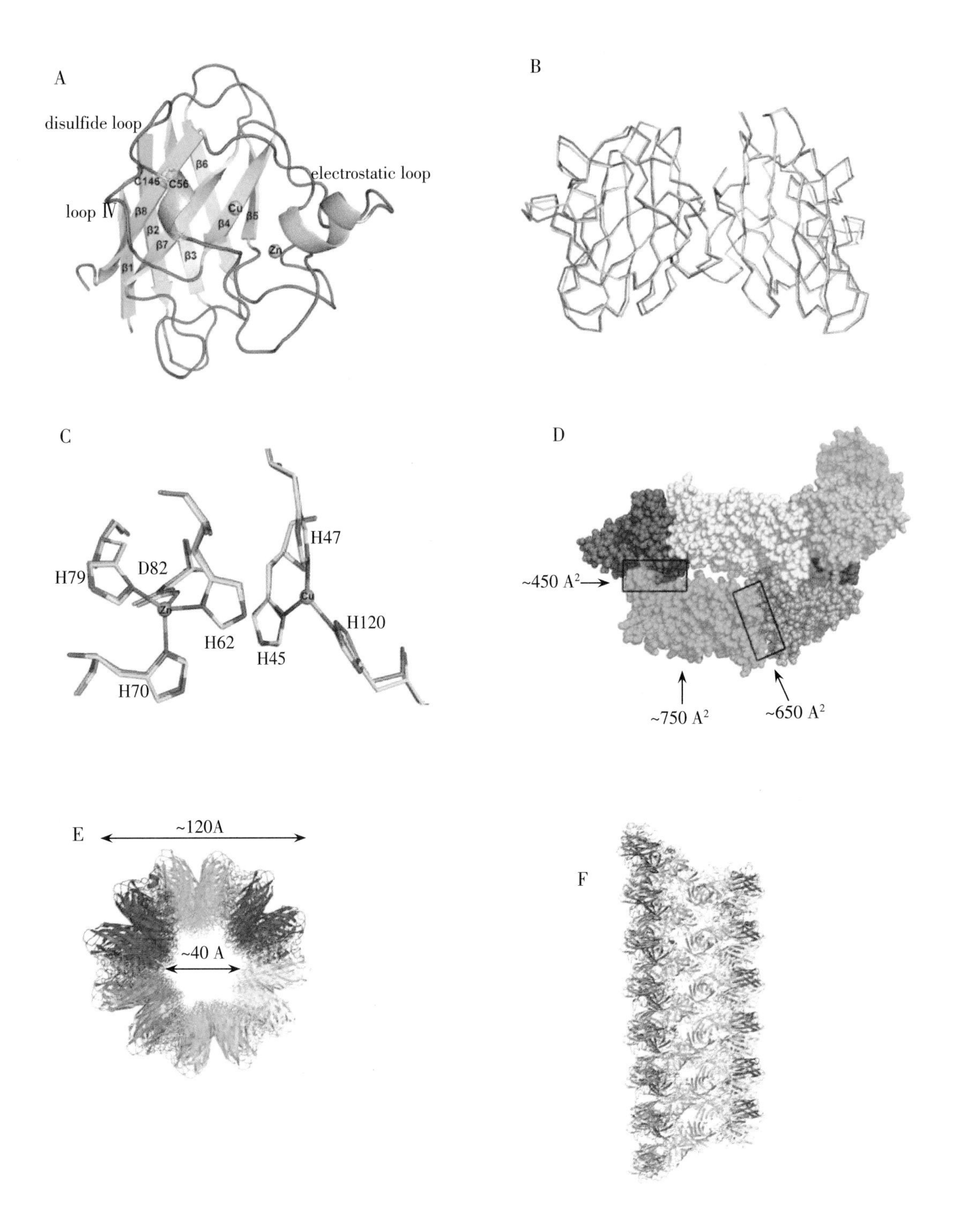

Figure 137. A: Cartoon representation of the holo *Bm*Sod1 monomer. The bound zinc and copper ions were shown in sphere. B: Superposition of the dimers of holo (cyan) and Cu-deficient *Bm*Sod1 (pink). C: Superposition of the zinc and copper binding sites of holo (cyan) and Cu-deficient *Bm*Sod1 (pink). D: Fibril assembly interfaces of *Bm*Sod1, the nonnative interfaces were boxed with black rectangles. E and F: Top view and front view of the water-filled nanotube of *Bm*Sod1.（选自: *Crystal structures of holo and Cu-deficient Cu/Zn-SOD from the silkworm Bombyx mori and the implications in amyotrophic lateral sclerosis*, 参见本书第938页。）

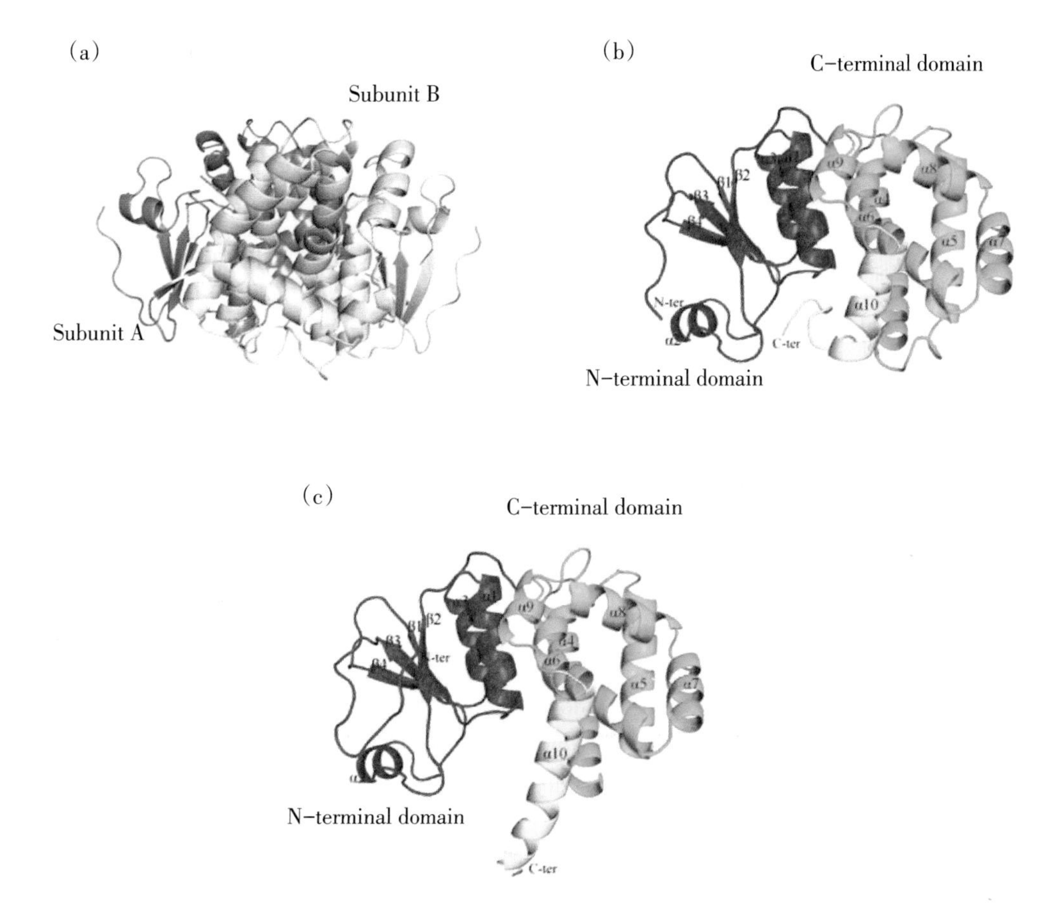

Figure 138. Overall structure of *bm*GSTO3-3. (a) Schematic representation of the *bm*GSTO3-3 homodimer. The *bm*GSTO3-3 monomers of (b) chain A and (c) chain C, with the two domains in cyan and red, respectively. The C-terminal segments of chain A and chain C have different conformations, as colored yellow.（选自：*Structure-guided activity restoration of the silkworm glutathione transferase Omega GSTO3-3*，参见本书第956页。）

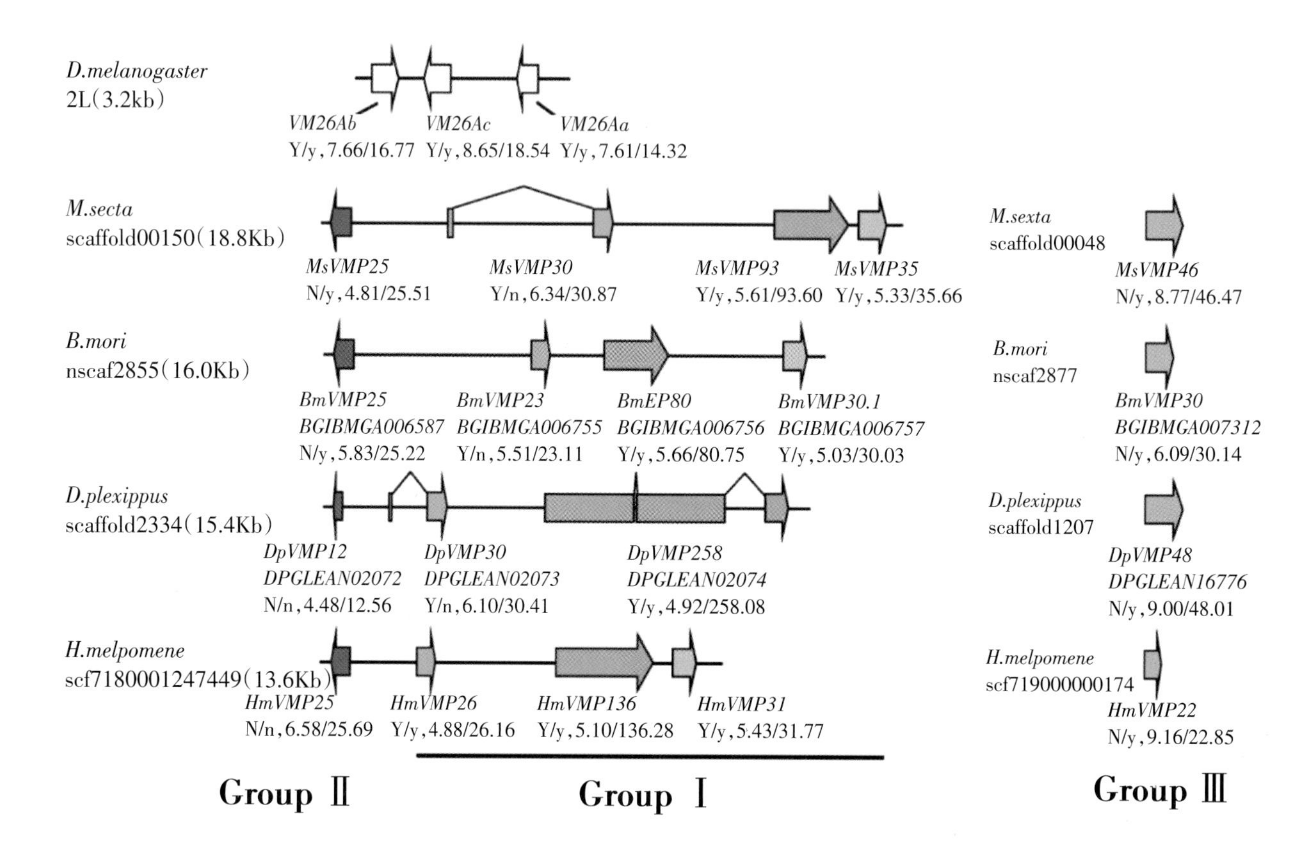

Figure 139. The VMPs coding regions in *D. melanogaster* and four Lepidopteran insects. The size and scaffold IDs for the VMPs coding regions are given. Each homologous gene among the four Lepidoptera is depicted in the same color and their NCBI GenBank accession numbers are listed in Table 1S. In "*BmEP80, BGIBMGA006756*, Y/y, 5.66/80.75", *BGIBMGA006756* is the ID number for *BmEP80* in SilkDB, "Y/y" means that the VM domain/signal peptide is present in *BmEP80*, and "5.66/80.75" means that the PI/MW of *BmEP80* is 5.66/80.75. The group numbers are indicated as I, II, III.（选自：*A syntenic coding region for vitelline membrane proteins in four lepidopteran insects*，参见本书第958页。）

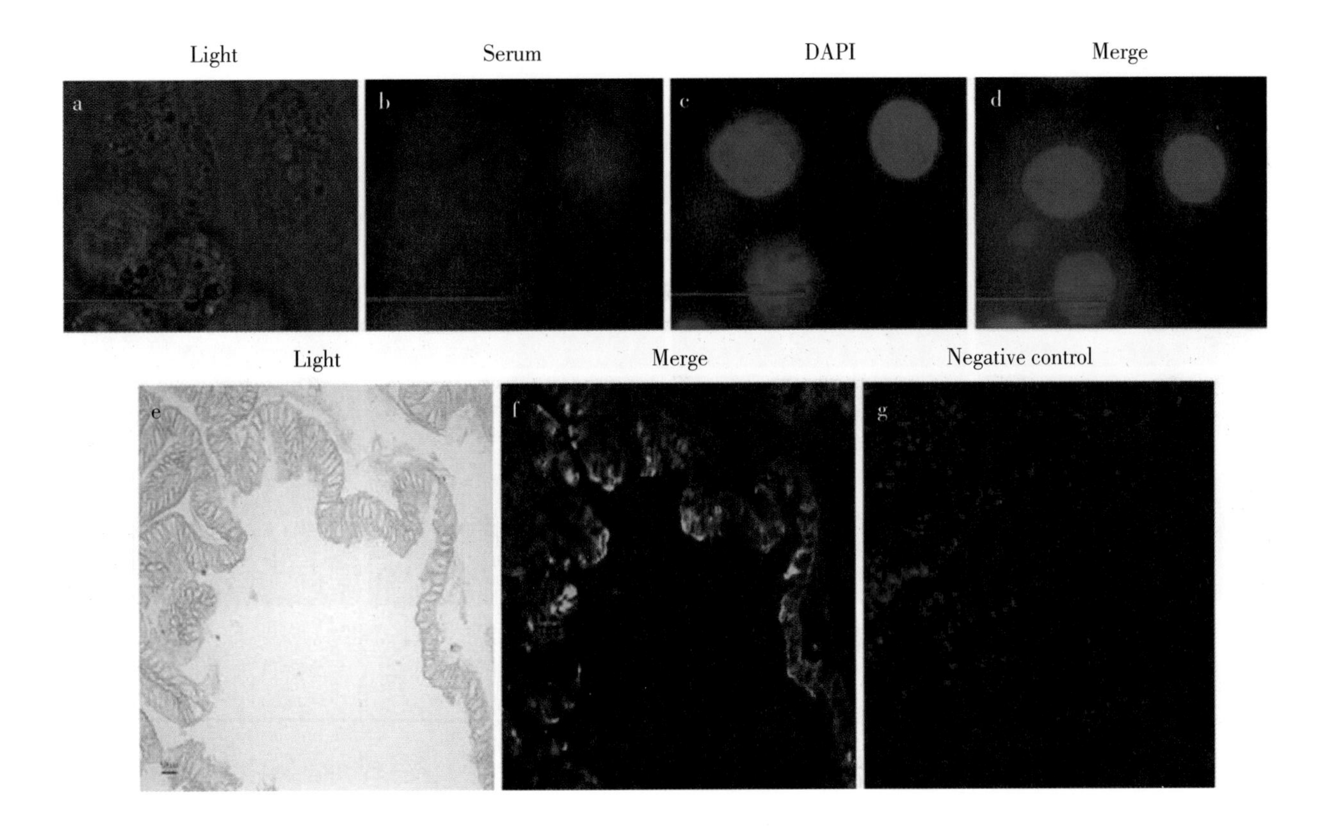

Figure 140. Subcellular localization of *Bm*Mtch in *BmN* cells and in the midgut by immunofluorescence. (a) Cells under normal light. (b) *Bm*Mtch subcellular localization as indicated by Rhodamine-labeled secondary antibody. *Red fluorescence* represents the positive protein. (c) DAPI staining. (d) Merged image. (e)Midgut section under normal light. f *Bm*Mtch localization in the midgut. *Green fluorescence* represents the positive protein. (g) Negative control. *Blue fluorescence* represents the nuclei. (Color figure online).(选自:*Molecular characterization of a peritrophic membrane protein from the silkworm, Bombyx mori*,参考本书第960页。)

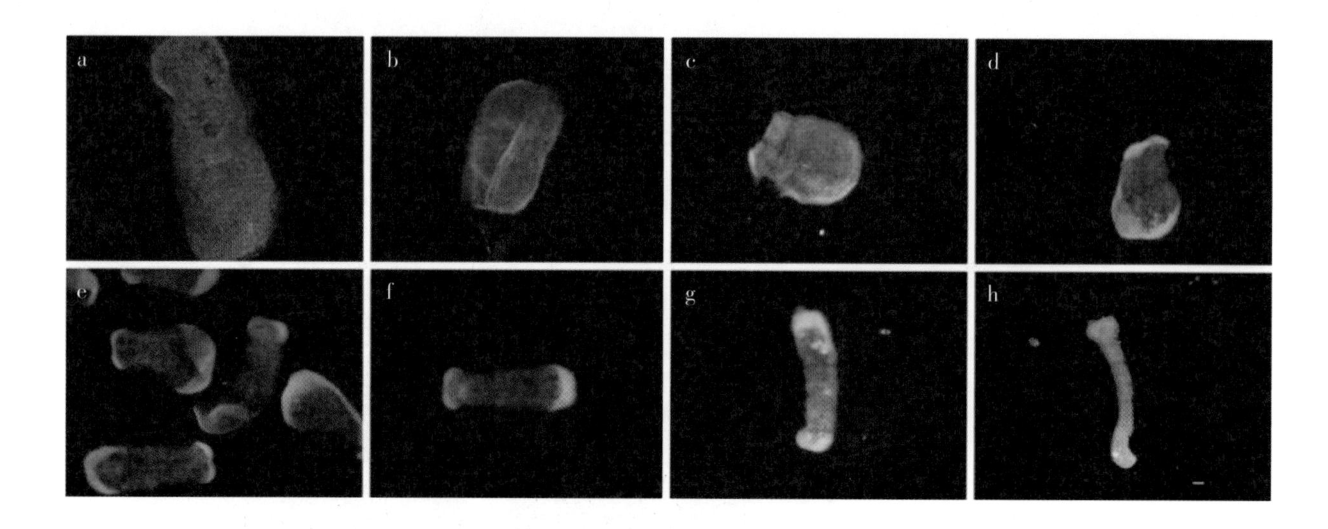

Figure 141. The early intact HNMW embryos. a-h The statuses respectively at 16 h, 20 h, 24 h, 28 h, 32 h, 36 h, 40 h and 48 h after egg-laying embryos. Scale bar represents 100 mm.（选自：*A novel method of silkworm embryo preparation for immunohistochemistry*，参考本书第962页。）

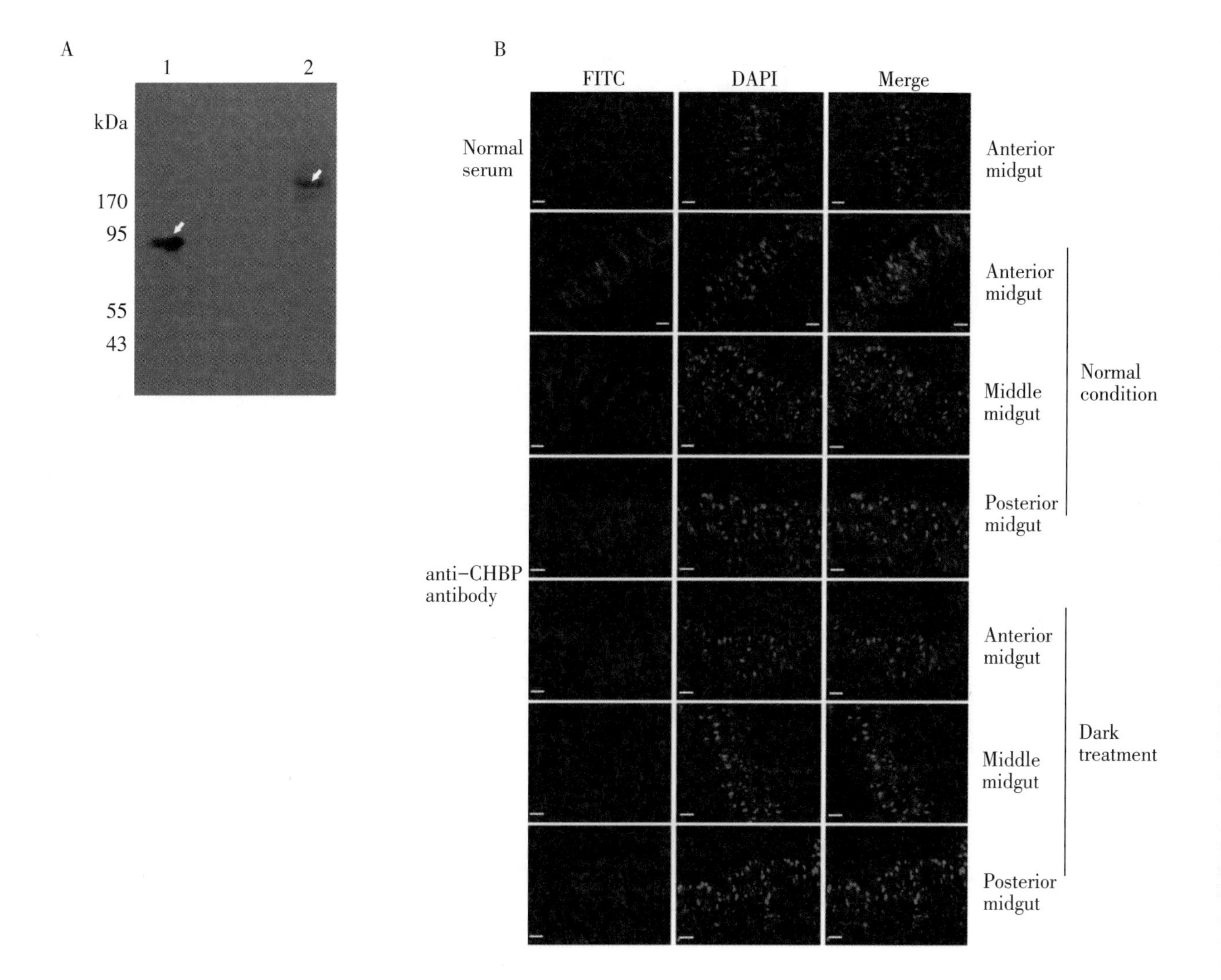

Figure 142. Detection of CHBP in the midgut of the silkworm. (A) Western blot analyses of the expression of CHBP. Lane 1: recombinant GST-CHBP-I, lane 2: CHBP extracted from the midgut. The bands are indicated by arrows. (B) Expression pattern of CHPB at different parts of the midgut by fluorescence immunohistochemistry. CHPB protein was detected in anterior, middle, and posterior midgut with antiCHPB antibody. The silkworms were reared under normal or dark conditions. For the control, the anterior midgut was detected with normal rabbit serum. Scale bar: 50 μm.（选自：*Expression analysis of chlorophyllid α binding protein, a secretory, red fluorescence protein in the midgut of silkworm, Bombyx mori*，参考本书第964页。）

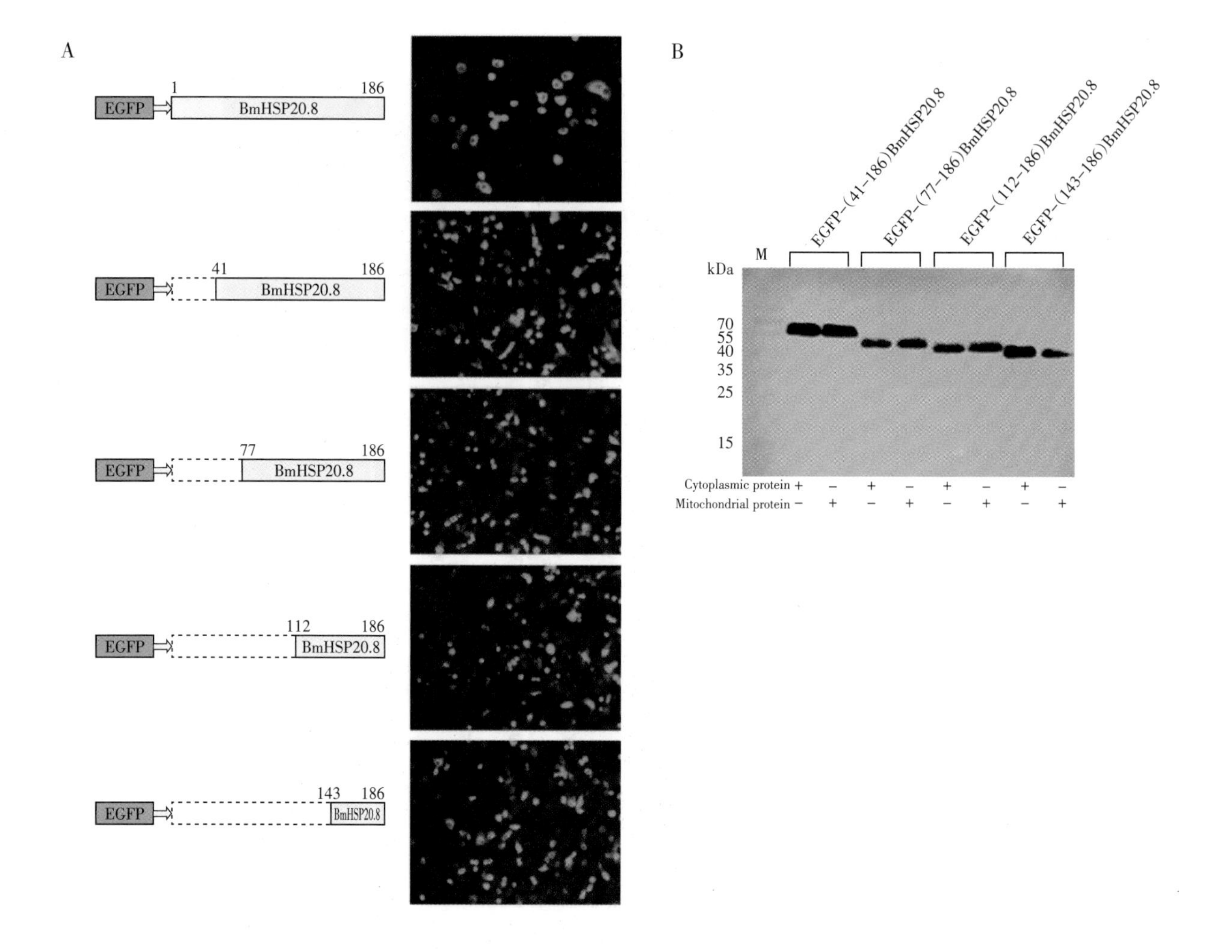

Figure 143. The determination of mitochondrial localization signal domain of BmHSP20.8. (A): The expression cassette of EGFP-fused truncated *Bm*HSP20.8 and their expression in *BmN* cells. The green fluorescence of EGFP-truncated *Bm*HSP20.8 proteins could be observed under the fluorescence microscope. (B): The mitochondrial localization of EGFP-truncated BmHSP20.8 proteins in *BmN* cells identified by *westem* blot. The EGFP-(41-186) *Bm*HSP20.8, EGFP-(77-186) *Bm*HSP20.8, EGFP-(112-186) *Bm*HSP20.8, and EGFP-(143-186) *Bm*HSP20.8 protein were all identified in the mitochondrial portion of *BmN* cells suggesting a localization signal domain in the range of amino acids from 143 to 186 in *Bm*HSP20.8.（选自：*BmHSP20.8 is localized in the mitochondria and has a molecular chaperone function in vitro*，参考本书第968页。）